OFFICIAL GAZETTE ✦ UNITED STATES PATENT OFFICE

September 6, 1955 Volume 698 Number 1

PATENTS

NOTICES

Trademark Rules

New Rules of Practice in Trademark Cases became effective on August 15, 1955. Copies are available without charge upon request addressed to the U. S. Patent Office. Distribution is generally limited to one copy per person, and in no event will more than five copies be furnished without charge on a single order. Quantities in excess of five copies may be purchased from the U. S. Patent Office at $.15 each. Copies of this edition will not be sold by the Superintendent of Documents.

Disclaimers

2,584,597.—*Rolf W. Landauer,* New York, N. Y. TRAVELING WAVE TUBE. Patent dated Feb. 5, 1952. Disclaimer filed Aug. 9, 1955, by the assignee, *Sylvania Electric Products, Inc.*

Hereby enters this disclaimer to claim 7 of said patent.

2,685,894.—*David B. Parlin,* Thompsonville, Conn. MANUFACTURE OF SINGLE AND MULTIFRAME JACQUARD WOVEN CARPETS. Patent dated Aug. 10, 1954. Disclaimer filed Aug. 8, 1955, by the inventor; the assignee, *Bigelow-Sanford Carpet Company, Inc.,* assenting and disclaiming.

Hereby enter this disclaimer to claims 1, 2, 4, 7, 13, and 15 of said patent.

Adjudicated Patents

(D. C. R. I.) Wentworth Patent No. 2,496,969 for spectacle sweat bar. *Held* valid and infringed. *Welsh Mfg. Co.* v. *Atlantic Optical Products,* 131 F. Supp. 462; — USPQ —.

(D. C. R. I.) Lindblom Patent No. 2,507,474 for brow bar support for goggles. Claims 1, 2, and 4 *Held* valid and infringed. *Id.*

Patents Available for Licensing or Sale

2,549,972. Releasing Tap Holder. Thomas R. Jones. Correspondence to John P. Smith, 855 First National Bank Building, Chicago 3, Ill.

2,645,079. Method of Operating Jet Propulsion Motors. Union Oil Company of California. Correspondence to: Ross J. Garofalo, Patent Counsel, Union Oil Company of California, P. O. Box 218, Brea, Calif. (For non-exclusive licensing.)

General Electric Company is prepared to grant non-exclusive licenses under the following two patents upon reasonable terms. Applications for license should be addressed to: General Electric Company, Patent Counsel, Measurements and Industrial Products Division, 920 Western Ave., West Lynn 3, Mass.

2,656,494. Blocking Layer Rectifier.

2,669,126. Electrical Control Arrangement.

New Applications Received During July 1955

Patents	6,091
Plants	8
Reissues	17
Designs	456
Total	6,572

Issue

Patents	634	—No.	2,716,748	to No.	2,717,381, incl.
Designs	38	—No.	175,488	to No.	175,525, incl.
Plants	2	—No.	1,417	to No.	1,418, incl.
Reissues	3	—No.	24,058	to No.	24,060, incl.
Total	677				

CONDITION OF PATENT APPLICATIONS AS OF JULY 29, 1955

Total number of pending applications (excluding Designs)	222,873
Total number of pending Design applications	7,182
Total number of applications awaiting action (excluding Designs)	139,581
Total number of Design applications awaiting action	2,809
Date of oldest new application	June 1, 1954
Date of oldest amended application	Aug. 24, 1953

ROSA, M. C., Executive Examiner

PATENT EXAMINING GROUPS, AND SUPERVISORY EXAMINERS	DIVISIONS
I. STONE, I. G., CHEMICAL AND RELATED ARTS	6, 31, 38, 43, 50, 56, 59, 63, 64.
II. STRACHAN, O. W., COMMUNICATIONS, RADIANT ENERGY AND ELECTRICAL ARTS	16, 23, 26, 37, 42, 48, 51, 54, 69, 70.
III. YUNG KWAI, B., MECHANICAL MANUFACTURING, MACHINE ELEMENTS AND DESIGNS	2, 12, 13, 14, 21, 24, 57, 58, 61, Designs.
IV. FREEHOF, H. B., MATERIAL HANDLING AND TREATING, OPTICS, RAILWAYS AND AMUSEMENT DEVICES.	7, 11, 17, 27, 34, 35, 39, 53, 62.
V. HULL, J. S., STATIC STRUCTURES AND INSTRUMENTS OF PRECISION	8, 20, 29, 33, 36, 40, 41, 52, 66.
VI. MURPHY, T. F., AGRICULTURE, TRANSPORTATION, PUMPS AND MOTORS	1, 4, 5, 9, 18, 22, 28, 45, 47.
VII. KAUFFMAN, H. E., HEATING AND COOLING, PLASTIC SHAPING AND COATING, SEPARATION AND MIXING, BODY TREATMENT AND CARE.	3, 15, 19, 25, 30, 32, 49, 55, 67.

DIVISIONS, EXAMINERS AND SUBJECTS OF INVENTION (Roman numerals in parentheses indicate Examining Group)	Oldest Application	
	New	Amended
1. (VI) GOLDBERG, A. J., Excavating; Planting; Plows; Harrows; Earth Rollers; Plant Husbandry; Scattering Unloaders; Sewage	11-1-54	4-2-54
2. (III) HERRMANN, D., Fishing, Trapping and Vermin Destroying; Presses; Tobacco; Textile Wringers	1-31-55	8-16-54
3. (VII) LE ROY, C. A., Metal Founding and Treatment; Metallurgy (Process and Apparatus); Alloys; Sintered Metal Stock; Miscellaneous Heating; Coating or Plastic Compositions (part), e. g., Inorganic, Mold and Mold Coating Compositions	11-1-54	11-25-53
4. (VI) FALLER, E. A., Hoists; Power Driven Conveyors; Handling Apparatus; Elevators; Feeding of Indefinite Lengths	11-2-54	11-24-53
5. (VI) ROBINSON, C. W., Harvesters; Potato Diggers; Stalk Pullers and Choppers; Stone Gatherers; Threshing; Knotters; Animal Husbandry; Bee Culture; Dairy; Butchering; Vegetable and Meat Cutters and Comminutors; Fences; Gates	12-8-54	4-12-54
6. (I) SURLE, H., Carbon Chemistry (part), e. g. Natural Resins, Proteins, Heterocyclic, Amides, Amines, General Organic Processes	8-20-54	3-2-54
7. (IV) GONSALVES, J. E., Optics, Photographic Apparatus	11-5-54	12-2-53
8. (V) LEWIS, R. O., Beds; Chairs and Seats; Cabinets; Tables; Miscellaneous Furniture	9-2-54	3-3-54
9. (VI) BRANSON, J. H., Pumps; Fans; Turbines	11-1-54	2-5-54
11. (IV) BENHAM, E. V., Boots, Shoes and Leggings; Shoe and Leather Manufacture; Button, Eyelet and Rivet Setting; Nailing, Stapling and Clip Clenching; Cutlery; Cleaning and Liquid Treatment of Solids	3-3-55	6-28-54
12. (III) SPINTMAN, S., Machine Elements; Engine Starters; Clutches; Interrelated Clutch and Motor Controls	10-14-54	10-28-53
13. (III) BEALL, T. E., Gear Cutting; Electric Lamp and Tube Manufacture; Needle and Pin Making; Metal Working (part), e. g. Special Work, Forging, Plastic Working, Drawing, Sawing, Milling, Planing, Turning	10-18-54	1-15-54
14. (III) MANIAN, J. C., Metal Working (part), e. g. Sheet Metal, Wire, Bending, Miscellaneous Processes, Assembly and Disassembly Apparatus; Wire Fabrics; Air Brakes	12-31-54	3-2-54
15. (VII) BRINDISI, M. V., Plastics; Plastic Block and Earthenware Apparatus; Glass	11-15-54	4-7-54
16. (II) LOVEWELL, N. N., Television; Telephony; Recorders	10-11-54	10-12-53
17. (IV) LEIGHEY, R. A., Paper Manufactures; Packaging; Typewriters; Printing; Type Casting and Setting; Sheet Material Association or Folding; Sheet or Web Feeding	11-24-54	4-12-54
18. (VI) KURZ, J. A., Power Plants; Fluid Transmissions; Servomotor Systems; Jet Motors; Combustion Turbines; Speed Responsive Devices, Brakes	10-1-54	10-28-53
19. (VII) PATRICK, P. L., Stoves and Furnaces; Boilers; Concentrating Evaporators; Fluid Fuel Burners	11-24-54	3-29-54
20. (V) BROWN, L. M., Miscellaneous Hardware; Closure Fasteners; Locks; Safes; Bank Protection; Bread, Pastry and Confection Making; Tents and Canopies; Umbrellas; Canes; Undertaking	12-15-54	4-15-54
21. (III) MADER, R. C., Textiles	11-16-54	3-22-54
22. (VI) MARLAND, M. L., Aeronautics; Boats; Buoys; Ships; Marine Propulsion; Propellers; Windmills; Fluid Diaphragms and Bellows; Boring and Drilling	1-7-55	2-25-54
23. (II) ANDRUS, L. M., Cash and Fare Registers; Calculators and Counters; Education	6-1-54	8-27-53
24. (III) DRACOPOULOS, P. T. (HICKEY, T. J., acting), Apparel; Apparel Apparatus; Sewing Machines; Textiles, Ironing or Smoothing	12-8-54	9-13-54
25. (VII) NEVIUS, R. D., Coating—Processes, Miscellaneous Products and Apparatus; Distillation; Wood Treating Apparatus	10-27-54	1-5-54
26. (II) YOUNG, R. R., Electricity—Generation, Motive Power, Transmission Systems, Voltage and Phase Control Systems, Furnaces, Batteries, Battery Charging and Discharging, Arc Lamps, Resistors and Rheostats, Prime Mover Dynamo Plants; Elevators (part), e. g. Miscellaneous Electric Control Mechanism	1-3-55	7-13-54
27. (IV) JAMES, S., Brushing, Scrubbing and General Cleaning; Brush, Broom and Mop Making	1-11-55	6-7-54
28. (VI) BRAUNER, R. H., Internal Combustion Engines; Expansible Chamber Motors; Fluid Servomotors; Spring, Weight and Animal Powered Motors; Cylinders; Pistons; Drive Shafts; Flexible Shaft Couplings; Chucks or Sockets; Chute, Skid, Guide and Way Conveyers; Fluid Current Conveyers; Pneumatic Dispatch; Store Service; Wheel Substitutes	11-29-54	6-7-54
29. (V) HABECKER, L. B., Tools; Woodworking; Button, Barrel and Wheel Making; Rubber Tire Removing Tools; Washing Machines; Baggage; Cloth, Leather and Rubber Receptacles; Package and Article Carriers	10-27-54	2-23-54
30. (VII) O'LEARY, R. A., Refrigeration; Heating Systems; Automatic Temperature and Humidity Regulation, Thermostats, Humidistats; Illuminating Burners; Fluid Sprinkling, Spraying and Diffusing	1-24-55	3-3-54

DIVISIONS, EXAMINERS AND SUBJECTS OF INVENTION (Roman numerals in parentheses indicate Examining Group)	Oldest Application	
	New	Amended
31. (I) HUTCHISON, E. W., Mineral Oils; Carbon Chemistry (part), e. g. Urea Adducts, Silicon Containing Carbon Compounds, Hydrogenation of Carbon Oxides, Partial Oxidation of Non-Aromatic Hydrocarbon Mixtures, Hydrocarbons, Halogenated Hydrocarbons	10-8-54	2-10-54
32. (VII) BERMAN, H., Gas and Liquid Contact Apparatus; Heat Exchange; Gas Separation; Agitation; Fluid Pressure Modulators; Self Proportioning Fluid Systems; Liquid Level Responsive Systems; Fire Extinguishers	12-21-54	6-7-54
33. (V) MUSHAKE, W. L., Bridges; Hydraulic and Earth Engineering; Building Structures; Roads and Pavements	10-14-54	1-29-54
34. (IV) SAPERSTEIN, S., Railways—Draft Appliances, Switches and Signals, Surface Track, Rolling Stock, Track Sanders; Electricity, Transmission to Vehicles; Dumping Vehicles; Vehicle Fenders; Hand and Hoist Line Implements	10-28-54	12-29-53
35. (IV) BROMLEY, E. D., Dispensing; Filling and Closing Receptacles; Toilet, Kitchen and Table Articles	1-3-55	2-24-54
36. (V) McFADYEN, A. D., Measuring and Testing	12-14-54	7-19-54
37. (II) LEVY, M. L., Electricity—Switches, Welding, Heating	8-11-54	3-22-54
38. (I) MARMELSTEIN, N., Carbon Chemistry (part), e. g. Lignins, Azo, Carbohydrate Derivatives; Carbocyclic or Acyclic Compounds (part), e. g. Anthrones, Triarylmethanes, Esters, Acids, Ketones, Aldehydes, Ethers, Phenols, Alcohols	10-5-54	1-3-54
39. (IV) WEIL, I., Fluid-Pressure Regulators; Valves; Fluid Handling (except Pressure Modulating Relays, Self-Proportioning Systems, Float Valves, Diaphragms and Bellows)	1-21-55	4-2-54
40. (V) DRUMMOND, E. J., Receptacles—Metallic, Paper, Wooden, Glass; Special Receptacles and Packages	1-3-55	5-3-54
41. (V) GURLEY, R. B., Coin Controlled Apparatus; Dispensing Cabinets; Coin Handling; Mail, Fare or Other Collection Boxes or Chutes; Buckles, Buttons and Clasps; Racks; Fire Escapes; Ladders; Scaffolds	7-7-54	10-12-53
42. (II) MARANS, H., Electric Signaling; Signals and Indicators; Telegraphy; Electrical Connectors	10-29-54	4-19-54
43. (I) ARNOLD, D., Medicines, Poisons, Cosmetics; Sugar and Starch; Bleaching, Dyeing, Fluid Treatment of Textiles, Skins, and Leathers; Preserving, Sterilizing and Disinfecting (except Wood Treatment Apparatus)	11-1-54	11-4-53
45. (VI) MANIAN, J. A., Wheels, Tires and Axles; Railway Wheels and Axles; Lubrication; Bearings and Guides; Belt and Sprocket Gearing; Spring Devices; Animal Draft Appliances	12-17-54	7-27-54
47. (VI) KANOF, W. J., Mining, Quarrying, and Ice Harvesting; Motor Vehicles; Land Vehicles	12-16-54	6-11-54
48. (II) BERNSTEIN, S., Electricity—Conversion Systems, Protective Systems; Measuring and Testing (except Meters); Spark Plugs and Ignition Systems, Switchboards, Relays, Magnets, Inductors, Transformers, Condensers, Transistors, Barrier Layer Rectifiers	7-8-54	8-24-53
49. (VII) BENDETT, B., Drying and Gas or Vapor Contact with Solids; Ventilation; Wells; Earth Boring	11-10-54	5-27-54
50. (I) BENGEL, W. G., Carbon Chemistry (part), e. g. Synthetic Resins, Natural or Synthetic Rubber	11-30-54	4-14-54
51. (II) YAFFEE, S., Radio Transmitters, Receivers and Tuners; Oscillators; Modulators; Piezoelectric Devices; Music	11-1-54	4-2-54
52. (V) NEFF, P. R., Supports; Joint Packing; Valved Pipe Joints or Couplings; Rod Joints or Couplings; Tool Handle Fastenings; Pipes and Tubular Conduits; Shaft Packing	12-6-54	4-2-54
53. (IV) REYNOLDS, E. R., Label Pasting and Paper Hanging; Card, Picture and Sign Exhibiting; Books and Book Making; Manifolding; Printed Matter; Stationery; Paper Files and Binders; Flexible or Portable Closures or Partitions; Doors, Windows, Awnings and Shutters; Harness; Whip Apparatus	12-1-54	1-4-54
54. (II) NILSON, R. G., Electric Lamps; Electronic Tubes; Miscellaneous Discharge Devices; Lamp, Cathode Ray and Gas Discharge Device Circuits; Ray Energy (e. g. X-Ray, Ultraviolet, Radioactive) Applications	10-19-54	1-19-54
55. (VII) KLINE, J. R., Surgery; Dentistry; Artificial Body Members; Separating and Sorting Solids; Centrifugal Bowl Separators; Comminutors	7-6-54	12-31-53
56. (I) KEELY, J. E., (SPECK, J. R., acting), Electrical and Wave Energy Chemistry; Liquid Separation or Purification	12-10-54	4-15-54
57. (III) MILLER, A. B., Cutting and Punching; Bolt, Nut, Rivet, Nail, Screw, Chain and Horseshoe Making; Driven and Screw Fastenings; Nut and Bolt Locks; Jewelry; Pipe Joints or Couplings	3-7-55	6-3-54
58. (III) DOWELL, E. F., Rolls and Rollers; Making Metal Tools and Implements; Stone Working; Abrading Processes and Apparatus; Food Apparatus; Closure Operators; Baths, Closets, Sinks and Spittoons	11-16-54	1-4-54
59. (I) HENKIN, B., Inorganic Chemistry; Fertilizers; Gas, Heating and Illuminating	12-27-54	2-23-54
61. (III) MORSE (Miss), E. L., Winding and Reeling; Pushing and Pulling; Horology; Time Controlling Apparatus; Railway Mail Delivery	1-3-55	7-1-54
62. (IV) SHAPIRO, A., Games; Toys; Amusements and Exercising Devices; Mechanical Guns and Projectors; Illumination	12-10-54	4-23-54
63. (I) WINKELSTEIN, A. H., Foods and Beverages; Carbon Chemistry (part), e. g. Fats and Metal Containing Carbocyclic or Acyclic Carbon Compounds; Abrading Compositions; Coating or Plastic Compositions (part), e. g. Pigments, Fillers, Driers, and Organic Compositions	9-15-54	12-24-53
64. (I) GORECKI, G. A., Fuels; Miscellaneous Compositions	11-23-54	12-23-53
66. (V) LISANN, I., Geometric Instruments; Automatic Weighing Scales; Acoustics	10-26-54	3-8-54
67. (VII) KRAFFT, C. F., Laminated Fabrics; Photographic Processes and Products; Ornamentation; Paper Making	11-8-54	5-27-54
69. (II) GALVIN, D. J., Wave Guides; Amplifiers; Electric Meters; Sound Recording; Conductors; Insulators	10-18-54	11-19-53
70. (II) BREWRINK, J. L., Explosive Weapons, Ammunition, Charges and Composition; Explosive Charge Manufacturing; Jet Motor Processes; Torpedoes; Radar; Sonar; Automatic Pilots; Antennas; Actinide Series (e. g. Fissionable) Compounds; Irradiation Chemistry; Mass Spectrometers	8-20-54	9-8-53
DESIGNS: A—BREHM, G. L., Industrial Arts	1-7-55	1-17-55
DESIGNS: B—GRAY, M. A., Household, Personal and Fine Arts	1-19-55	11-23-54

The following divisions have been abolished: 10, 44, 46, 60, 65 and 68

EXPIRATION OF PATENTS

The patents within the range of numbers indicated below expire during September 1955, except those which may have been extended under the provisions of the Veterans Patent Extension Act (64 Stat. 316 as amended by 66 Stat. 321) and those which may have expired earlier due to shortened terms under the provisions of Public Law 690. A list of Veterans' patents which have been extended appears in the *Annual Index of Patents—1953*.

Patents ---------- Numbers 2,128,889 to 2,131,721, inclusive
Plant Patents ---------- Numbers 290 to 292, inclusive

DECISIONS IN PATENT CASES

U. S. Court of Customs and Patent Appeals

In re Fairbanks

No. 6103. Decided May 25, 1955

[222 F.2d 725; 106 USPQ 94]

Patentability—Particular Subject Matter—Method of Photographing a Picture Play.

Claims to method of photographing a picture play *Held* properly rejected as unpatentable over the cited prior art.

Appeal from the Patent Office. Serial No. 125,997.

AFFIRMED.

Conder C. Henry for Fairbanks.

E. L. Reynolds (*H. S. Miller* of counsel) for the Commissioner of Patents.

Before Garrett, *Chief Judge*, and O'Connell, Johnson, Worley, and Cole, *Associate Judges*.

Garrett, *Chief Judge*, delivered the opinion of the court.

This is an appeal from the decision of the Board of Appeals of the United States Patent Office affirming that of the Primary Examiner finally rejecting all of the claims, Nos. 9, 10, 11, 12, and 15, of appellant's application for patent, Serial No. 125,997, filed November 7, 1949, for "Motion Picture Photographing."

Claims 9 and 15 were cited by the Board of Appeals as illustrative, and read as follows:

9. The method of continuously photographing action occurring in a number of different sets in which long, intermediate, and close-up shots are to be photographed comprising positioning a plurality of cameras on the floor of said sets in accordance with a pre-arranged starting schedule, providing extendible cables above said sets for energizing said cameras from a position not to interfere with the movement of said cameras on said floor as the action in said sets moves from set to set, and maneuvering at least one of said cameras over a pre-arranged path during the action in one of said sets to maintain the nature of any particular shot during the photographing of the action by said cameras, each of said cameras being simultaneously controlled by a director.

15. The method of photographing a picture play which includes a plurality of different sets in which the action to be photographed continues between sets comprising positioning cameras to obtain long, intermediate, and close-up shots of the portion of said sets to be photographed, and controlling the starting and maneuvering of said cameras according to a time and position schedule as the action progresses between said sets to maintain the nature of any particular shot being photographed by a particular camera, all of said cameras being simultaneously controlled by one director.

Claim 10 is similar to claim 9, but indicates that the cameras are not all running continuously, but that some are started and stopped during the sequence of action on different sets. Claim 11 is directed to action occurring on one set where "simultaneously operating" cameras photograph different aspects of the action on the set, the cameras being described as "synchronously operated." Claim 12 is similar to claim 11 but includes the limitation of "controlling the operation of each of said cameras, including the starting, stopping, and movement thereof, in accordance with a predetermined schedule, each of said cameras and its condition of operation or non-operation being simultaneously visible to a director."

All of the claims have been rejected as unpatentable over the cited art, and claim 15 has been additionally rejected on the ground that it incorporates an unpatentable "mental step."

The following references were relied upon: De Mille, "Motion Picture Directing", Transactions of the Society of Motion Picture Engineers, Vol. 12 (1928), pages 295 to 309; Willat, "Mechanical Problems of a Director", ibid., pages 285 to 294; Haskin et al., 2,293,207, Aug. 18, 1942; Del Riccio, 2,382,616, Aug. 14, 1945.

Appellant summarizes the references in his brief in the following language, which may be considered adequate for present purposes. (References to the printed record are omitted.)

Willat is relied upon by the Board as showing what the appellant freely admits, that one step in a method of making a motion picture is to mount a camera on a moveable platform and follow the actors around the set. Another step is to make shots from different distances. Otherwise, Willat merely discusses some of the ordinary problems encountered by a motion picture director in making any photoplay.

* * * * * * *

As correctly stated by the Board, "The De Mille article deals mainly with the problems of producing big feature films and teaches the use of a plurality of camera, the simultaneous taking of short and long shots and the moving of a camera to take shots at different distances from the scene to be photographed." However, it should be noted that De Mille, who is associated with Paramount, is concerned only with the general problems of any movie director in making a motion picture by separately and noncontinuously taking shots of the various scenes at different times. This article, it seems, is the best of the references and * * * accurately describes the usual practice of making motion pictures.

* * * * * * *

Haskin et al. shows a mount provided with wheels for moving picture apparatus. The apparatus is automatically movable over a predetermined path.

* * * * * * *

Del Riccio discloses a method of photographing a horse race by placing in spaced relation, a plurality of motion picture cameras around a race track, each camera being operated by a cameraman who uses his own discretion in taking a section of the track, using one type of shots, and subsequently assembling the pictures so taken to make a continuous film.

Appellant's method is concerned with the making of motion pictures, particularly, it appears from the briefs, with the making of motion pictures for use on television. It is conceded that many of the steps recited in the claims, such as the use of a plurality of cameras, the taking of long or close-up shots, and the maneuvering of the cameras, are old, appellant stating that such steps are included in the method "merely for the purpose of making it complete." The features of claimed novelty in appellant's method appear to be that appellant continues to photograph action as it goes from one set to an adjoining one, and that in so doing certain of the cameras are started and stopped as desired by the director.

On first impression it would seem that the size or number of sets to be covered at one taking would be determined by the director simply as a matter of preference. If a script should contain a sequence in which the action moves from one room to another, it would seem simply a matter of judgment whether to photograph the action in the two rooms separately as two scenes, or whether to construct one large set containing the two rooms, and to photograph the action in the two rooms as one scene. Likewise, if only one close-up shot was desired, it would seem obvious that money could be saved by operating the close-up camera only long enough to take that shot. However, appellant urges the existence of invention so persistently, that it is necessary to scrutinize his arguments for possible justification.

The gist of appellant's argument supporting patentability is that the method must be patentable since it had never been used before despite "the stimulus of a driving need."

With regard to following action from set to set without interruption, the Primary Examiner stated:

> Applicant stresses rapidity and efficiency in making a moving picture without interrupting the photographing of the various sequences constituting the final story. Due to human requirements such as rest and food and due to maintenance requirements of mechanical equipment, it is to be assumed that the final story is very limited as to time and as to scope of scenery. Under such conditions it is believed that a good director would obviously do what applicant purports to be a patentable method.

The Board of Appeals approved this holding, stating:

> Photographing two different sets with different cameras, whether concurrently or sequentially, involves, in our opinion, only a matter of choice, apparently depending on the number of cameras and camera men and the amount of floor space that are available to the director. Such matters are deemed within the judgment of a skilled director.

We are in complete accord with the holding of the lower tribunals on this point. The size of a set, or the number of sub-divisions of a set to be used, seems to be merely a matter for the judgment of a director, to be exercised according to the requirements of the plot, the number of cameras and the amount of space available, and the number of retakes which will be expected to be necessary to get a satisfactory result. Willat indicates in his article that scenes may be frequently broken up in a different manner than was originally planned, according to practical considerations. As far as we can see, all appellant has done is to make one scene out of what would normally be treated as two scenes. As indicated by the Board, this is simply a matter of choice.

The second feature of appellant's method for which patentability is claimed is the idea of starting one or more cameras after the other cameras had commenced photographing a scene, so as to take only a particular short camera shot as might be desired.

Appellant devotes a good deal of space in his brief to discussing the prior art methods of making motion pictures. It appears that when a sound motion picture is to be taken, it is necessary, in order to get a proper synchronization of sound and action, at the very beginning of the scene to focus all cameras upon a pair of sticks which are struck together near a sound microphone, thereby giving a point of reference common to all cameras and to the sound track to facilitate matching. One result of this procedure, of course, is that one camera cannot be started or stopped intermittently during the scene without a loss of synchronization. The pertinency of this discussion (which has little foundation in the specification) appears to be that appellant has invented and patented a method of synchronizing sound and film which permits cameras to be cut in or out as desired during a scene. Appellant urges that we consider as evidence of invention the fact that "to practice his invention effectively it was necessary for him to make another invention."

Aside from the fact that the referred to patent has not been incorporated into the record, we could not accept it as having any bearing at all upon the instant case, even if it were properly in the record. If the allowance of some claims by the Patent Office in the same application cannot be considered on appeal from the rejection of claims, a fortiori claims allowed on a different application cannot be considered.

Furthermore, we must note that the appealed claims are not limited to the taking of motion pictures with accompanying sound, and therefore the patentability of the claims must be determined irrespective of difficulties introduced by the use of sound tracks. The De Mille article, which is primarily concerned with the production of silent movies, does indeed state that all cameras operate throughout a scene. As one reason for this, De Mille says that it can result in finding several good moments in one camera view, other than the one that had been planned. De Mille warns, however, that the director must have good judgment in this, or the waste of film can ruin the organization, for "cameramen love to turn the handle." We are in full agree-
commercial success is nearly always the result of a
ment with the following holding of the Examiner:

> * * * whether the cameras are operated simultaneously or at chosen intervals becomes a matter of judgment on the part of the director depending on how much film may be used as expressed [by] De Mille.

Appellant has alluded to the great success of his method in producing motion pictures for use on television, and argues that this is evidence of invention. It is well established that commercial success is important only when the question of invention is in doubt. The reason for this rule is not that the courts belittle the importance of commercial success, but because large number of co-acting factors, not all of which have a bearing on the question of invention. In this case we are unable to conclude that appellant's alleged success is the result of invention, rather than the result of the peculiar requirements of the television market, appellant's patent, or appellant's personal skill in direction.

The various requirements of the claims, other than those discussed above, do not seem to us to be sufficient upon which to base a finding of patentability, and we do not deem it necessary to discuss them in detail.

For the above reasons we are of the opinion that all of appellant's claims were properly rejected as lacking in invention over the prior art. In this view of the case it is unnecessary to consider the rejection of claim 15 on the ground that it incorporates a mental step.

[1] The decision of the Board of Appeals is affirmed.

AFFIRMED.

PATENT SUITS

Notices under 35 U. S. C. 290; Patent Act of 1952

1,510,441, 1,683,072, E. H. Hebern, Electric coding machine; 1,861,857, same, Cryptographic machine; 2,269,341, same, Message transmission device; 2,373,890, same, Cipher machine; 2,267,196, W. N. Fanning, Remote control system, filed May 19, 1953, Ct. Cls., Doc. 213/53, *Ellie L. Hebern, Executrix of the Estate of Edward H. Hebern, deceased, et al.* v. *The United States.* Petition dismissed as to patents 1,510,441 and 1,683,072 June 7, 1955.

1,600,588, 1,600,589, 2,008,210, O. F. Hipkins, Traction device, filed Nov. 1, 1950, Ct. Cls., Cong. 17866, *Otho F. Hipkins et al.* v. *The United States.* Payment recommended to Congress Apr. 7, 1953.

1,600,589. (See 1,600,588.)

1,627,184. (See 2,029,778.)

1,627,185. (See 2,029,778.)

1,653,361, H. E. Krammer, Shock absorber, filed Sept. 1, 1949, Ct. Cls., Doc. 49297, *Henry E. Krammer* v. *The United States.* Petition dismissed June 2, 1953.

1,660,857, J. L. Breese, Jr., Fluid-fuel-burning apparatus; 1,702,929, same, Combustion; 2,182,465, J. L. Breese, Burner for liquid fuel; 2,226,216, same, Method of making burner pots; 2,396,818, same, Generator type heater; 2,393,231, same, Spherical liquid fuel burner; 2,397,529, same, Burner for army type stoves; 2,393,232, same, Stove structure; 2,410,478, same, Generator type burner; 2,073,270, B. Valjean, Combustion apparatus; 2,393,248, Hayter and Perry, Horizontal pot type burner; 2,386,556, M. D. Huston, same; 2,348,721, Breese and Hayter, Horizontal hydroxylating burner; 2,422,653, same, Method of burning liquid hydrocarbon, filed June 12, 1951, Ct. Cls., Doc. 50191, *Breese Burners, Inc.* v. *The United States.*

1,672,163. (See 2,029,778.)

1,683,072. (See 1,510,441.)

1,690,881, G. Thilo, Circuit for amplifying direct or alternating currents by vacuum tubes; 1,720,351, Rottgardt and Kühn, Method and means for controlling discharge gaps; 1,828,094, H. Andrewes, Electrical frequency-changing apparatus of the thermionic type; 1,862,394, E. Asch, Continuous current amplifier; 1,862,393, same, Thermionic amplifying circuits; 1,898,046, Geffcken and Richter, Electric relay device for indicating weak currents; 2,120,916, R. E. Bitner, Light frequency converter; 2,164,402, G. Guanella, Electrical circuit; 2,241,615, J. Plebanski, Electric phase control system; 2,258,436, M. von Ardenne, Phototube for translation of images, filed July 2, 1951, Ct. Cls., Doc. 50216, *Radio Patents Corp. et al.* v. *The United States.*

1,698,934. (See 1,698,935.)

1,698,935, P. C. Chesterfield, High speed alloy; 1,698,934, same, Alloy and method of making the same, filed July 7, 1949, Ct. Cls., Doc. 49230, *Percy C. Chesterfield* v. *The United States.*

1,702,929. (See 1,660,857.)

1,713,589, E. V. Bereslavsky, Low-compression fuel, filed June 25, 1948, Ct. Cls., Doc. 48,722, *Euphime V. Bereslavsky* v. *The United States.* Patent held valid Oct. 5, 1954.

1,715,283, W. A. Fletcher, Communicating circuits, filed May 28, 1952, Ct. Cls., Doc. 268/52, *William Arthur Fletcher* v. *The United States.*

1,720,351. (See 1,690,881.)

1,744,016, A. P. Steckel, Metal rolling; 1,779,195, same, Method and apparatus for rolling thin sheet like material, filed Mar. 28, 1949, Ct. Cls., Doc. 49085, *The Union National Bank of Youngstown, Ohio, Trustee* v. *The United States.* Same, filed Aug. 15, 1949, Ct. Cls., Doc. 49281, *The Union National Bank of Youngstown, Ohio, Trustee* v. *The United States.*

1,758,795, L. L. Irvin, Parachute-rip-cord construction; 1,842,611, 2,016,235, 2,016,236, 2,095,135, same, Parachute apparatus; 1,958,000, H. G. Hamer, deceased, by M. A. Hamer, administratrix, same; 1,899,656, Wigley and Austing, Parachute harness, filed Feb. 3, 1950, Ct. Cls., Doc. 49479, *The Irving Air Chute Co., Inc.* v. *The United States.*

1,776,228, R. A. Coffman, Starter; 1,946,309, same, Motor; 2,005,913, same, Motor and motive system; 2,164,700, same, Device for discharging explosive or combustible charges; 2,175,743, same, Priming system for internal combustion engines; 2,283,185, same, Diesel engine starter; 2,284,640, same, Breech mechanism for power generating units; 2,299,464, same, Power generating unit, filed July 1, 1949, Ct. Cls., Doc. 49223, *Roscoe A. Coffman* v. *The United States.* Judgment for plaintiff June 7, 1955.

1,779,195. (See 1,744,016.)

1,828,094. (See 1,690,881.)

1,842,611. (See 1,758,795.)

1,861,857. (See 1,510,441.)

1,862,393. (See 1,690,881.)

1,862,394. (See 1,690,881.)

1,898,046. (See 1,690,881.)

1,899,656. (See 1,758,795.)

1,946,309. (See 1,776,228.)

1,958,000. (See 1,758,795.)

2,005,913. (See 1,776,228.)

2,008,210. (See 1,600,588.)

2,016,235. (See 1,758,795.)

2,016,236. (See 1,758,795.)

2,029,778, 1,627,185, 1,672,163, H. E. Krammer, Aircraft; 1,627,184, same, Sectional flying boat, filed July 5, 1949, Ct. Cls., Doc. 49225, *Henry E. Krammer* v. *The United States.*

2,073,270. (See 1,660,857.)

2,095,135. (See 1,758,795.)

2,120,916. (See 1,690,881.)

2,164,402. (See 1,690,881.)

2,164,700. (See 1,776,228.)

2,175,743. (See 1,776,228.)

2,182,465. (See 1,660,857.)

2,226,216. (See 1,660,857.)

2,241,615. (See 1,690,881.)

2,258,436. (See 1,690,881.)

2,267,196. (See 1,510,441.)

2,269,341. (See 1,510,441.)

2,283,185. (See 1,776,228.)

2,284,640. (See 1,776,228.)

2,299,464. (See 1,776,228.)

2,348,721. (See 1,660,857.)

2,373,890. (See 1,510,441.)

2,386,556. (See 1,660,857.)

2,393,231. (See 1,660,857.)

2,393,232. (See 1,660,857.)

2,393,248. (See 1,660,857.)

2,396,818. (See 1,660,857.)

2,397,529. (See 1,660,857.)

2,410,478. (See 1,660,857.)

2,422,653. (See 1,660,857.)

REISSUES

SEPTEMBER 6, 1955

Matter enclosed in heavy brackets [ ] appears in the original patent but forms no part of this reissue specification; matter printed in italics indicates additions made by reissue.

24,058
AUTOMATIC TONG EXTRACTING AND RESETTING MECHANISM

Ernest E. Eckstein, Elgin, Oreg.
Original No. 2,656,212, dated October 20, 1953, Serial No. 302,707, August 5, 1952. Application for reissue January 4, 1954, Serial No. 402,198
11 Claims. (Cl. 294—110)

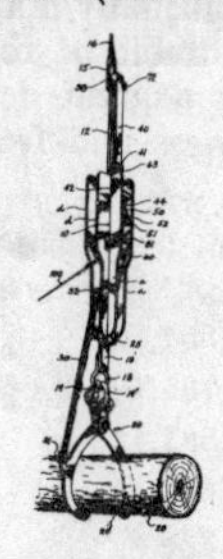

11. In combination, a load lifting cable, a pair of tongs, a load lifting connector interconnecting the cable and tongs, a flexible tong extracting connector secured at one end to one of the tongs and secured at its other end to the load lifting connector, said load lifting connector comprising a housing and bar means projecting through the housing, said bar means including first and second portions relatively movable within said housing, latch means carried by the housing in fixed position thereon, and means associated with the first portion of the bar means adapted to engage with the latch means to hold said first portion of the bar means within the housing, and thereby determine the effect of the lifting connector so as to produce load lifting or tong extracting, and latch disengagement means carried by the housing and adapted to be engaged by the second portion of said bar means as the lifting connector is engaged with a log to cause relative movement between said housing and said second portion of the bar means and engage said second portion of the bar means with the latch disengaging means to automatically release the means associated with said first portion of the bar means.

24,059
GATE HINGE

Thomas J. Pinion, Birmingham, and Jack D. Killough, Adamsville, Ala.; said Pinion assignor to said Killough
Original No. 2,702,399, dated February 22, 1955, Serial No. 327,946, December 26, 1952. Application for reissue April 11, 1955, Serial No. 500,716
3 Claims. (Cl. 16—153)

2. For use as a gate supporting and gravity closing hinge, a pair of sleeves disposed for mounting one above the other about a vertically disposed post, the upper sleeve being rotatable and slidable on the post and adapted for attachment to the gate and the lower sleeve being adapted for attachment to the post, an intermediate sleeve rotatable and slidable on the post between the said upper and lower sleeves, coacting interengaging cam surfaces on adjacent ends of said sleeves urging the gate toward closed position upon rotation of the upper and intermediate sleeves, and coacting means on said sleeves limiting relative rotation therebetween to the extent that said respective cam surfaces remain in said co-acting interen-

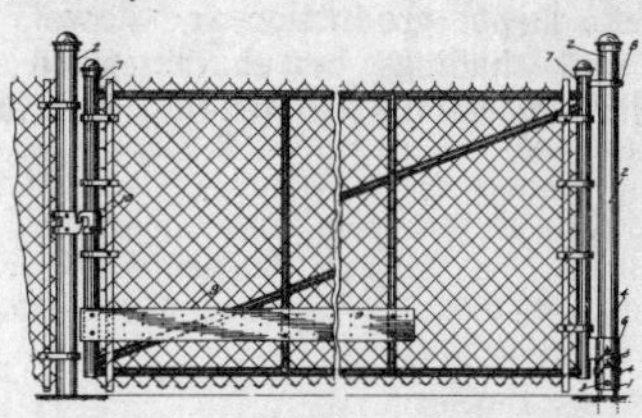

gaging relationship, said coacting means including a stop member on each of said sleeves normally circumferentially spaced from and engaging a stop member on an adjoining sleeve during relative rotation of said sleeves whereby rotation of said upper sleeve with the gate will turn said intermediate sleeve until its stop member engages the stop member on the lower sleeve.

24,060
APPARATUS FOR FORMING GLASS FIBERS

Robert G. Russell, Granville, Ohio, assignor to Owens-Corning Fiberglas Corporation, a corporation of Delaware
Original No. 2,634,553, dated April 14, 1953, Serial No. 65,139, December 14, 1948. Application for reissue April 1, 1955, Serial No. 498,790
17 Claims. (Cl. 49—17)

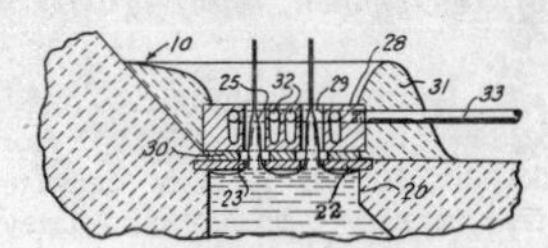

15. Apparatus for producing glass filaments comprising a container for molten glass, a row of projecting nipples in one wall of the container having orifices therein through which the glass flows in the form of streams and from which the streams in the form of cones are attenuated to fine filaments, a fluid cooled unit mounted adjacent the face of said wall of the container, said unit having a portion extending on one side of said row of nipples, said portion having at least one surface extending generally parallel to the direction of flow of said streams from said orifices in the region of said cones, said surface extending to at least the level of the tips of said cones, means for circulating a cooling fluid through said unit to cool said surface and the cones adjacent thereto, and means for attenuating the streams to fine filaments.

PLANT PATENTS

GRANTED SEPTEMBER 6, 1955

Owing to the fact that almost all of the illustrations of the plant patents are in colors, it is not practicable to print a cut of the drawing.

1,417
ROSE PLANT
Arthur Preston Howard, Sierra Madre, Calif., assignor to Howard & Smith, Montebello, Calif.
Application August 31, 1954, Serial No. 453,451
1 Claim. (Cl. 47—61)

The new and distinct variety of hybrid tea rose plant, essentially like its parent with its combination of strong growth, prolific flower production and flowers of golden yellow with pink shadings, but characterized as to novelty by its climbing habit as shown and described.

1,418
VIOLA PLANT
Edith Wintermute Pawla, Capitola, Calif.
Application September 7, 1954, Serial No. 454,648
1 Claim. (Cl. 47—60)

A new and distinct variety of viola plant substantially as herein described and illustrated, characterized by large fragrant blooms on unusually long stems, with a root structure which has a habit of rooting by thick lateral stolons, roots reaching a great depth making the plant highly resistant to dryness and frost.

PATENTS

GRANTED SEPTEMBER 6, 1955

GENERAL AND MECHANICAL

2,716,748
TAB FORMING, STAPLING AND PRINTING DEVICE

Gerald John Sutton, London, England, assignor to The National Marking Machine Company, Cincinnati, Ohio, a corporation of Ohio

Application August 23, 1951, Serial No. 243,321

10 Claims. (Cl. 1—2)

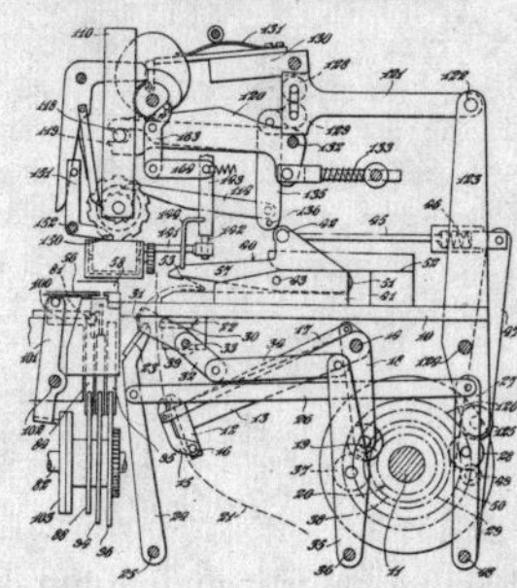

1. A machine for attaching a tab, label or the like applied to a fabric or fabric article which includes a pair of spaced needles each having a longitudinal groove therein, means to feed a U-shaped staple to said needles so that the limbs of said staple are located within the grooves of said needles, means to move said needles reciprocatingly through said fabric and applied tab, a reciprocable support for said staple movable towards said fabric simultaneously with the movement of said needles towards said fabric, whereby to cause said staple and said needles to move together without relative movement therebetween through said fabric and said tab, said reciprocable support holding the staple in engaged relationship with said fabric and said tab on the reverse movement of said needles and positively actuated means to close the limbs of said staple while said staple is held in said engaged relationship.

2,716,749
STAPLING MACHINE FOR SIMULTANEOUSLY INSERTING AT LEAST TWO STAPLES

Ewald Rudolf Timmerbeil, Arnhem, Netherlands

Application May 17, 1952, Serial No. 288,376

Claims priority, application Netherlands February 18, 1952

4 Claims. (Cl. 1—49)

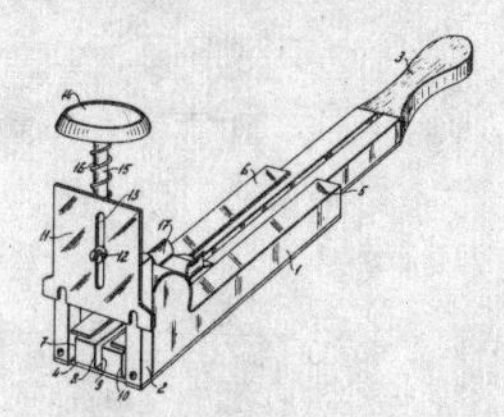

1. A stapling machine for stapling together a plurality of bodies; said stapling machine comprising means defining at least two parallel guides each forming a staple receiving space which, at least in part, adjoins the corresponding part of the staple receiving space formed by another of said guides to receive a strip of separable U-shaped staples with the staples in each guide having portions thereof abutting corresponding portions of the staples in the other of said guides, a staple driving mechanism at one end of said guide defining means and including a common driving member extending across all of said guide to simultaneously drive staples from the several strips and in contact with each other, and feeding mechanism operative to effect simultaneous feeding movement of the staple strips in said guides toward said staple driving mechanism.

2,716,750
TOOL FOR HOLDING A FASTENER WHILE STARTING IT INTO WORK

William B. Biblis, North Lawrence, Ohio

Application July 3, 1953, Serial No. 365,867

2 Claims. (Cl. 1—49.8)

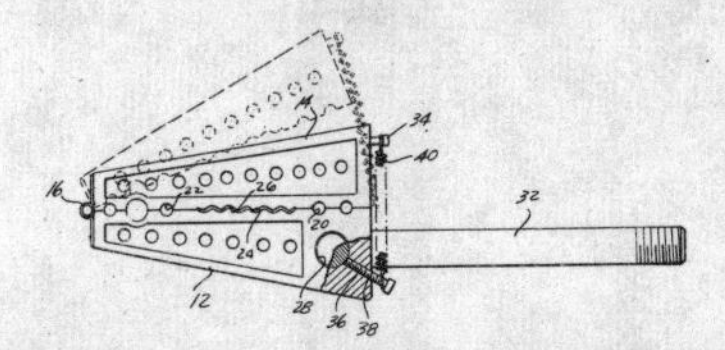

1. A tool for holding a fastener while starting it into work comprising a pair of jaws having mating fastener receiving recesses in adjacent edges, a pivot connecting the jaws together adjacent one end of the tool, and yielding means connecting the opposite ends of the jaws and urging said jaws toward one another and into clamping engagement with fasteners seated in the recesses, one of said jaws having an internally screw threaded socket extending thereinto, and a handle threadedly engaged with the threads in the socket and extending outwardly from the tool adjacent one end thereof.

2,716,751
MACHINE FOR BANDING MEAT CASINGS

Howard Kelem, Far Rockaway, N. Y.

Application September 17, 1954, Serial No. 456,714

8 Claims. (Cl. 1—215)

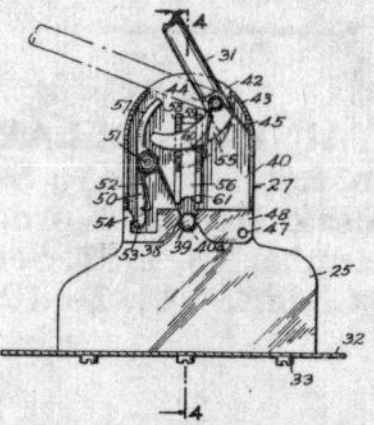

1. In a machine for banding a meat casing, a fixed anvil having a slot extending downwardly from the upper edge thereof for the reception of the crimped end part of the casing and of sufficient height to extend above the crimped part inserted into the slot, a locking tooth on the anvil, a fastener-stapling head pivoted to the anvil for movement into a position to expose the top of the slot, a spring-pressed pawl carried by the head in position to engage the tooth of the anvil and to the lock the head releasably to the anvil, means carried by the head to exert downward pressure on a fastener inserted into the slot above and independently of the casing to drive the fastener along the anvil and about the crimped cas-

ing, the fastener-driving means comprising a manually operable lever pivoted to the head, a rotatable member associated with the lever, a fastener-driver mounted in the head and engaged by the member for reciprocation thereby, and an operative connection between the member and the pawl, said member being movable with the driver on operation of the lever in one direction to control the removal of the pawl from the tooth and thereby to unlock the head from the anvil.

2,716,752
RINGING DEVICE FOR CATTLE
Elvin W. Robertson, Warrenton, Va.
Application October 21, 1954, Serial No. 463,775
7 Claims. (Cl. 1—260)

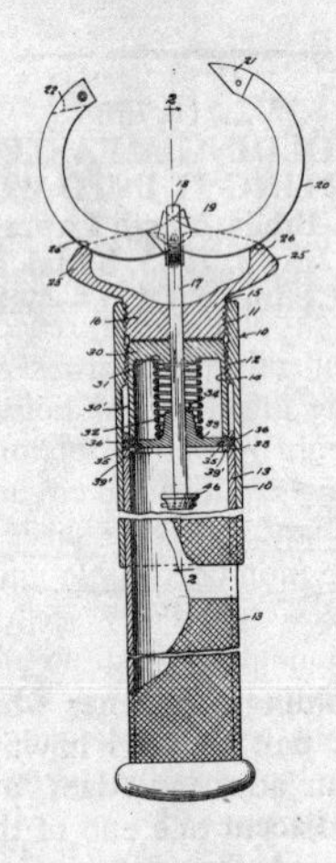

1. In a ringing device for animals, a tubular member having its interior provided with reversely disposed upper and lower threads, a closure plug engaging the upper of said threads and having an aperture therethrough, a second plug engaging said lower threads and having an aperture therethrough in alignment with said first mentioned aperture, a pair of fulcrum points extending outwardly of said plug wherein the sides of a separable pivoted ring are adapted to seat, a hooked rod extending through said aperture engaging said ring at the pivot point thereof whereby retractive movement of said rod will close said ring through the nostrils of an animal, a rotatable sleeve engaging said second plug and spring means within said second plug rotatable to a position to retract said rod.

2,716,753
SHAPE RETAINING COLLAPSIBLE CAP
Herman Gordon, Louisville, Ky.; Selma S. Gordon, executrix of said Herman Gordon, deceased
Application December 18, 1952, Serial No. 326,697
2 Claims. (Cl. 2—195)

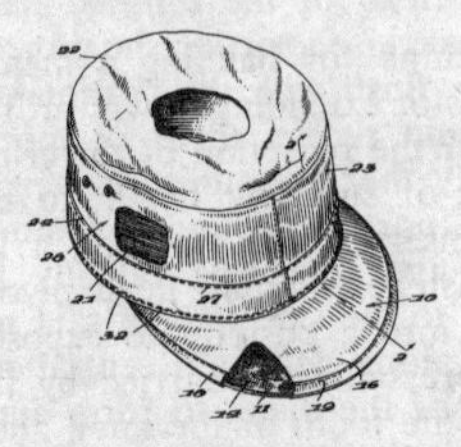

1. In a shape retaining flexible cap, hair cloth having hair picks therein extending in a vertical direction forming a band and a foundation for a crown of the cap, a fabric ring covering an inner face of the hair cloth, another fabric ring covering the outer face of the hair cloth, a fabric crown cover, stitching securing top edges of the fabric rings to the hair cloth and to each other and securing the perimeter of the fabric crown cover to the top edges of the fabric rings and the hair cloth, a crescent, shaped flexible sheet forming a foundation for a visor, textile cloth covering faces of said flexible sheet having concave inner edge portions, and stitching securing said concave edge portions to the lower edge portion of the hair cloth and to the lower edge portions of said fabric rings.

2,716,754
SKI PANTS
Harold S. Hirsch, Portland, Oreg.
Application August 11, 1952, Serial No. 303,731
3 Claims. (Cl. 2—227)

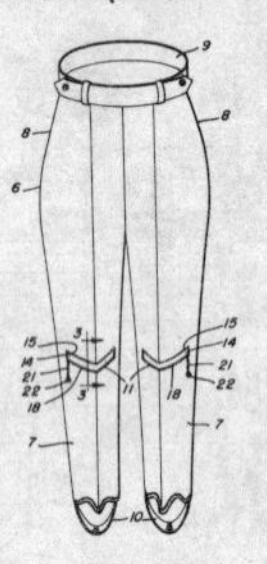

1. A pair of skiing pants including hip, knee and ankle portions and providing anchoring means at said hip and ankle portions whereby to effect a tautness of the frontal portion of said pants rendering it possible to bend the knee portion only slightly; a relief area at the frontal portion of each knee portion, said relief area having a transversely extending slit formed therein, a slit closure providing interengageable means carired adjacent the opposed slit edges, a flap overlying said closure means and stitched to said pants at one side only of said slit, said closure means and related flap being co-extensive in length with said slit, a pleat underlying related closure means and extending beyond the ends of said slit and stitching securing the ends of related closure means, said pleat, and said flap on opposite sides of said slit whereby to prevent elongation of said slit by ripping of the pants material.

2,716,755
SIMULATED SHOW HANDKERCHIEF DEVICE
James J. Foglio, New Rochelle, N. Y.
Application May 10, 1954, Serial No. 428,487
2 Claims. (Cl. 2—279)

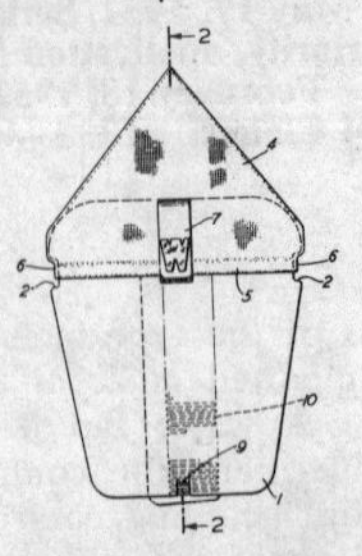

1. A simulated show handkerchief device for use in pockets of garments comprising a flat area of material of such size and dimensions as to fit wholly within a pocket, a hook means integral with the central portion of the upper edge of the area and projecting from one face thereof, and having an upwardly facing open spring groove adjacent the area and a downwardly facing open spring groove remote from the area, the downwardly facing open spring groove being adapted to receive the

upper edge of the front wall of the pocket, a fabric corner adapted to be displayed from the pocket and having a lower edge of substantially the width of the area and pocket, an elastic band attached to the sides of the lower edge of the fabric corner and extending therebetween, the bottom of the upwardly facing open spring groove being adapted to support the lower edge of the fabric corner and of one long side of the elastic band, and a pair of notches in the sides of the area near its upper end adapted to be positioned below the upper edge of the pocket on insertion of the area into the pocket, the upwardly facing groove being adapted removably to receive the lower edge region of the fabric corner with the elastic band about the face of the area opposite to that from which the hook means extends with the elastic band engaging the notches.

2,716,756
ADJUSTABLE BELT
Charles B. Gainsburgh, Merrick, N. Y.
Application August 15, 1952, Serial No. 304,560
1 Claim. (Cl. 2—321)

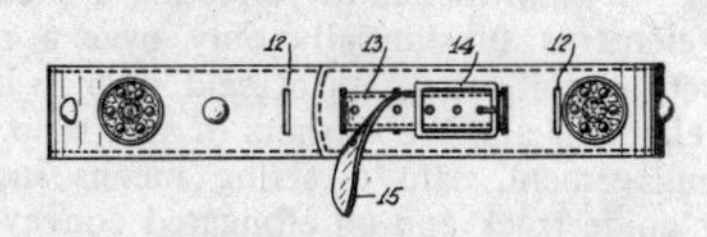

In a belt of two pieces having front and rear ends, the rear ends adapted to overlie to a varying degree, and each of the rear ends provided with a series of selectively registrable similar slots, means to secure the rear ends together in varying positions of adjustment, said means comprising a flexible strap having a buckle on one end and having its other end free, said free end being threaded through spaced selectively registered corresponding slots to hold said overlapped ends together, and the ends of said strap being detachably connected.

2,716,757
LAVABOS
Maurits S. Eriksson, Ulriksdal, Sweden
Application December 18, 1952, Serial No. 326,623
Claims priority, application Sweden December 21, 1951
3 Claims. (Cl. 4—166)

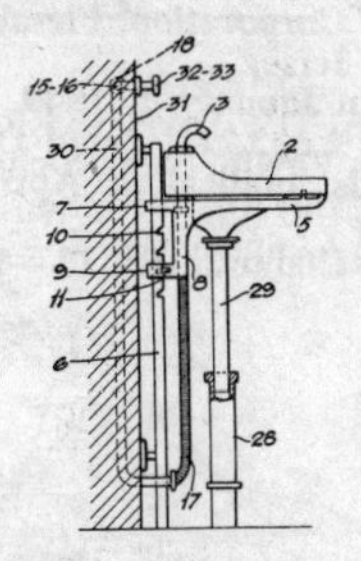

3. In a lavabo of the type including a vertically adjustable basin supported adjacent a wall, the improvements comprising a water tap fixedly carried by the basin and movable therewith, controllable water supply means for the tap including control cocks fixedly supported at a level above the highest adjusted position of the basin, a pipe having its outlet below the lowest vertically adjusted position of the basin and a flexible conduit means extending between the outlet of said pipe and the tap whereby the control of the water supplied to the basin is effected at a fixed level above the basin while the tap and the discharge of water therefrom into the basin remains at a fixed level relative to the basin regardless of the vertically adjusted position of the basin.

2,716,758
MARKER BUOY
Thomas H. Hajecate, Houston, Tex., assignor to The Light House, Inc., a corporation of Texas
Application February 20, 1953, Serial No. 337,991
3 Claims. (Cl. 9—8)

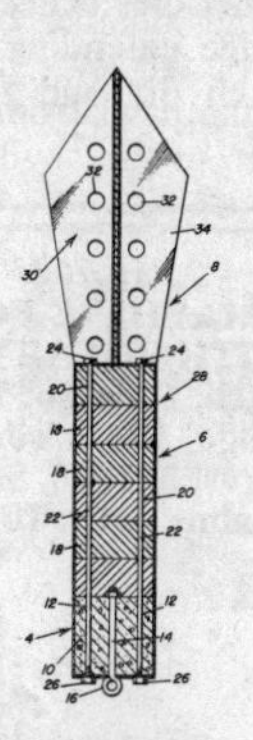

1. A marker buoy comprising, in combination, a buoyant body section having a ballast section at its lower end, and light and radar reflecting section on its upper end, the reflective properties of the reflectors in latter section being effective for both day and night requirements, said light reflecting section comprising a plurality of upstanding equidistant circumferentially spaced fins covered with light reflective sheet material, the latter being salt resisting, non-corrodible and visible in a circle ranging 360 degrees, said body section being composed of separable components superimposed atop one another, said components being separably bolted together and being interchangeable and replaceable, and a non-corrodible, non-electrolytic jacket completely encasing said ballast section and body section.

2,716,759
DIES FOR MAKING HEADED FASTENERS
Alfred G. Merlin and Joseph P. Mamere, Cleveland, Ohio, assignors to The National Screw & Manufacturing Company, Cleveland, Ohio, a corporation of Ohio
Application July 19, 1951, Serial No. 237,634
1 Claim. (Cl. 10—26)

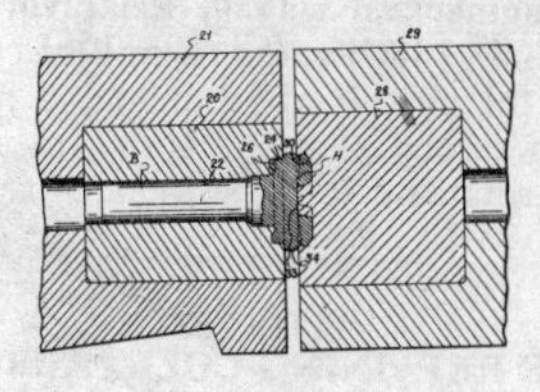

In apparatus for forming threaded fasteners of the protruding head type, a hammer die having a recess adapted to engage the outer end of the blank to shape the same, said recess being approximately one-third of the height of the head of the finished fastener, said recess having a central raised portion the projecting end of which is substantially flush with the portion of the hammer die surrounding said recess, three or more web-like ridges of substantially the same uniform height as that of said central raised portion arranged symmetrically with respect thereto and extending radially from said central raised portion to the side wall of said recess and adapted to be pressed along with said central raised portion into the outer end of the blank to form diagonal slots in the head of the blank the depth of which is approximately one-third of the height of the head of the finished fastener, and an anvil die having an aperture to receive the shank of the blank and a counterbore or recess to receive the head of the blank the inner or bottom wall of which recess has an annular ridge surrounding said aperture to form a recess in the underside of the head of the blank upon relative movement of said dies towards each other with a blank therein, the depth of the recesses in the anvil and hammer dies being substantially the same with said annular ridge extending above the bottom of the recess in the anvil die and terminating below the face of said anvil die.

2,716,760
METHOD AND MACHINE FOR ASSEMBLING WASHERS WITH ROTARY FASTENERS
Ougljesa Jules Poupitch, Chicago, Ill., assignor to Illinois Tool Works, Chicago, Ill., a corporation of Illinois
Application August 3, 1950, Serial No. 177,536
37 Claims. (Cl. 10—155)

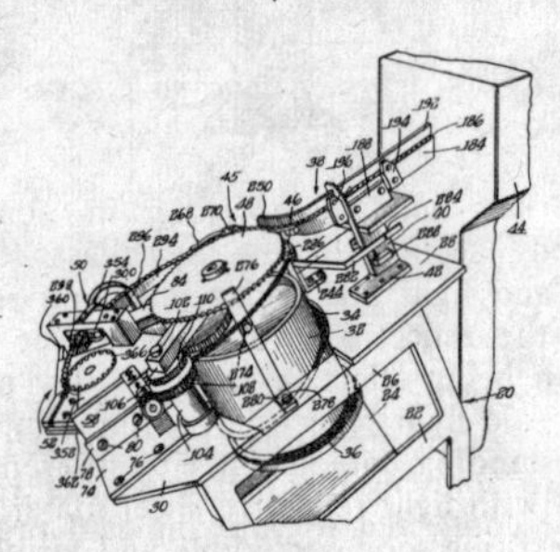

1. Apparatus for assembling fastening elements and washers, comprising means for receiving a continuous helically coiled strip of washers, means for continuously feeding said strip from the coil over a spiral path, and means for directing a succession of fastening elements over a curved path converging with the spiral path of said strip in timed relation to the movement of said strip to bring said fastening elements and the washers of said strip into telescoping relation.

2,716,761
METHOD AND MACHINE FOR ASSEMBLING WASHERS WITH ROTARY FASTENERS
Moritz H. Nielsen, Chicago, Ill., assignor to Illinois Tool Works, Chicago, Ill., a corporation of Illinois
Application August 3, 1950, Serial No. 177,538
27 Claims. (Cl. 10—155)

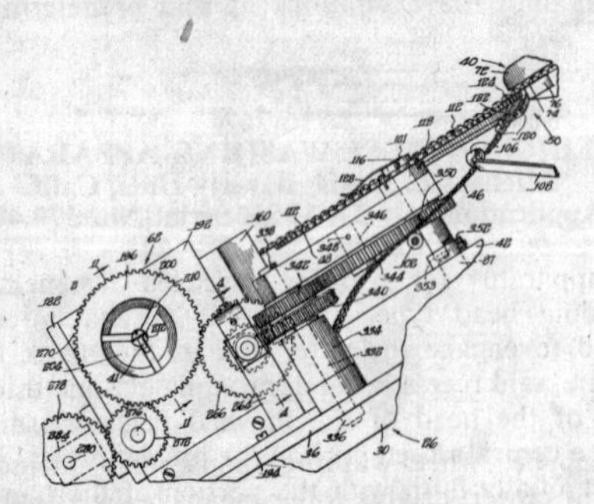

1. In apparatus for producing unit assemblies of washers and rotary fasteners, the combination comprising means for continuously feeding washers adjoined to one another in the form of a longitudinal strip at a uniform constant rate, means for relatively twisting the washers of an adjacent pair from the plane of said strip while said strip is advanced at said uniform constant rate to fracture the interconnection and sever the washers from said strip, and means for operating said feeding means and said twisting means in proper timed relation.

2,716,762
METHOD AND MACHINE FOR ASSEMBLING WASHERS WITH ROTARY FASTENERS
William Stern, Park Ridge, Ill., assignor to Illinois Tool Works, Chicago, Ill., a corporation of Illinois
Application August 3, 1950, Serial No. 177,503
26 Claims. (Cl. 10—155)

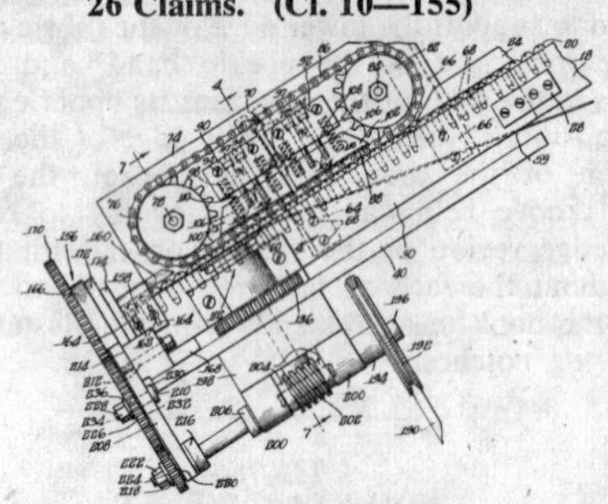

1. Apparatus for assembling fastening elements and washers, comprising means for continuously feeding a strip of washers over a predetermined path at a uniform continuous rate, and means for directing a succession of fastening elements substantially only over a rectilinear path converging with the path of said strip to bring said fastening elements and the washers of said strip into telescoping engagement, said directing means including a rectilinear guide track and an elongated conveyor means having a substantially straight reach parallel to and spaced from said guide track, said conveyor means including a plurality of elements spaced equally to the spacing between axes of the washers of said strip for engaging the fastening elements to maintain the fastening elements at right angles to the track and in proper spaced relation to the washers of said strip, said conveyor means being operated in timed relation with the movement of said strip, and said strip path and said track converging at a small acute angle so that the fastening elements initially telescope with the washers of said strip at a point spaced from the initial engagement of the conveyor means with the fastening elements.

2,716,763
DEVICES FOR USE IN MAKING SLIP-LASTED SHOES
Paul W. Senfleben, Beverly, Mass., assignor to United Shoe Machinery Corporation, Flemington, N. J., a corporation of New Jersey
Original application January 4, 1949, Serial No. 69,151, now Patent No. 2,623,223, dated December 30, 1952. Divided and this application April 25, 1952, Serial No. 284,303
4 Claims. (Cl. 12—1)

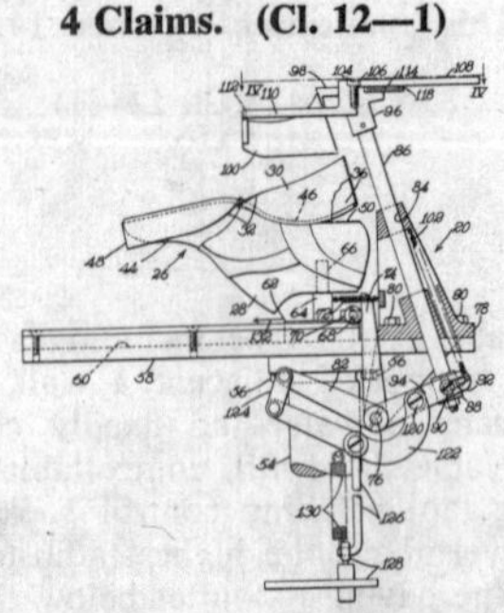

1. In a device for spotting the heel end of a platform unit upon the sock lining of a slip-lasted shoe which is mounted upon a last with the rear portion of its platform cover turned right side out to form a pocket with the sock lining, a support for the shoe on its last mounted for movement in a direction generally lengthwise of the shoe, a former, means for first effecting rela-

tive movement of the support and the former generally heightwise of the shoe to cause the former to be inserted in said pocket and to effect engagement of the former and the heel-seat portion of said sock lining and for thereafter effecting movement of the support generally lengthwise of the shoe to cause the former to shape the rear portion of the platform cover to approximately the shape it is to assume in the finished shoe.

2,716,764
LASTING MACHINES
Fred C. Eastman, Marblehead, Mass., assignor to United Shoe Machinery Corporation, Flemington, N. J., a corporation of New Jersey
Application January 28, 1954, Serial No. 406,629
22 Claims. (Cl. 12—1)

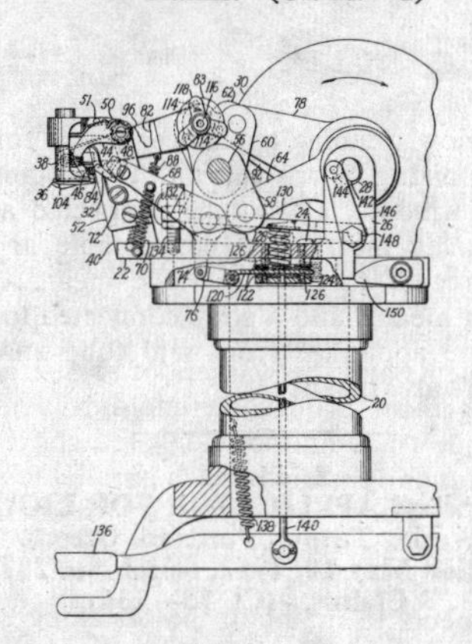

2. In a lasting machine, lasting instrumentalities including a gripper, overdrawing member and wiper respectively operative on adjacent and successive portions of the lasting margin of stock to be lasted, means for moving said instrumentalities heightwise of the shoe and in sequential relation, and means for moving the lasting instrumentalities in unison about an axis extending heightwise of the shoe to feed it, said overdrawing member having a margin-engaging surface adapted to cause each of said successive portions, when released from said gripper, to be slanted toward the operating level of said wiper.

2,716,765
SHEET TRANSFERRING MECHANISMS
Edward Quinn, Saugus, and Napoleon A. Monfils, Haverhill, Mass., assignors to United Shoe Machinery Corporation, Boston, Mass., a corporation of New Jersey
Continuation of application Serial No. 147,452, March 3, 1950. This application October 19, 1954, Serial No. 470,784
23 Claims. (Cl. 12—1)

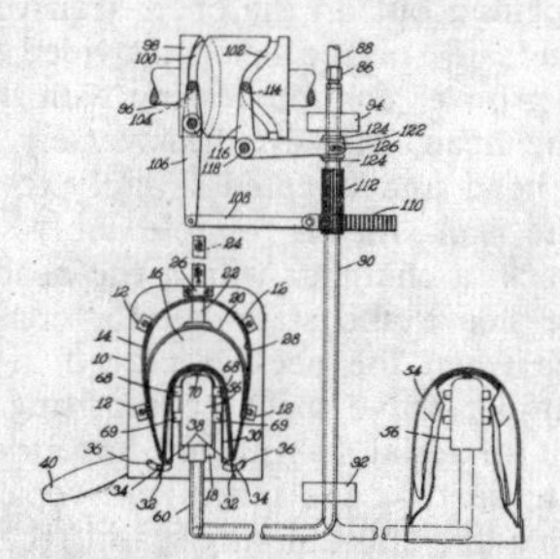

1. In combination, a magazine for last slip papers, a carrier for transferring the leading paper in the magazine to the interior of the heel portion of a supported upper, and suction elements facing in opposite directions for holding the paper in bent condition on the carrier during its transfer from the magazine to the upper.

2,716,766
MACHINES FOR APPLYING PRESSURE TO SHOE BOTTOMS
Helge Gulbrandsen, Beverly, Mass., assignor to United Shoe Machinery Corporation, Flemington, N. J., a corporation of New Jersey
Application August 23, 1952, Serial No. 306,046
9 Claims. (Cl. 12—36)

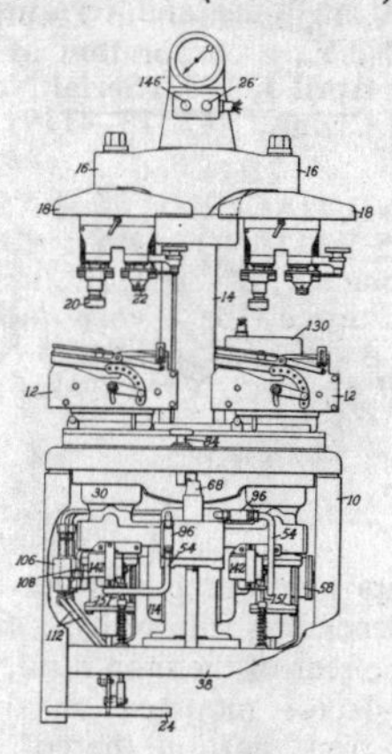

8. A machine for applying pressure to shoe bottoms comprising a pair of operating stations, each station having a pressure applying member and a fluid pressure operated motor connected for moving said pressure applying member to pressure applying position, a pump connected for supplying pressure fluid to each of the fluid pressure operated motors, an electric motor arranged to drive said pump, a control valve associated with each of said fluid pressure operated motors and arranged to control the flow of pressure fluid from said pump to said motors, operator controlled means for individually moving each of said valves to a first position to direct pressure fluid to the associated motor, means for automatically moving each valve to a second position after a predetermined period following its movement into the first position, in which second position pressure fluid is exhausted from the associated motor, operator controlled means for deenergizing said electric motor and for causing movements of said valves into said second positions prior to the expirations of said predetermined periods, and means for thereafter preventing energization of said electric motor and preventing movement of either valve into the first position until the expirations of said predetermined periods.

2,716,767
VEHICLE WHEEL WASHING APPARATUS
Michael Z. Davis, Beverly Hills, Calif.
Application July 27, 1953, Serial No. 370,404
5 Claims. (Cl. 15—21)

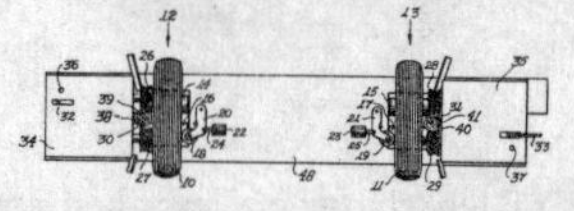

1. Vehicle wheel-washing apparatus, including: a source of power; driving roller means being effectively cooperable to be driven by said power; rotatable driven roller means being spatially related to said driving roller means for support of a vehicle wheel in frictional abutment against both of said roller means whereby said vehicle wheel may be rapidly rotated by said driving roller means; normally stationary brush means being rotatedly and rectilinearly reciprocably movable; rotary force-transmission means cooperably related to said driven roller means and said brush means whereby rotation of said vehicle wheel causing rotation of said driven roller means causes rotation of said brush means; and selectively

operable means for rectilinear reciprocation of said brush means into and out of lateral contact with said vehicle wheel.

2,716,768
WRINGER MOP
Harold H. Schwartz, Port Chester, and Jean S. Tucker, White Plains, N. Y., assignors to Empire Brushes, Inc., Port Chester, N. Y., a corporation of New York
Application April 1, 1952, Serial No. 279,814
7 Claims. (Cl. 15—119)

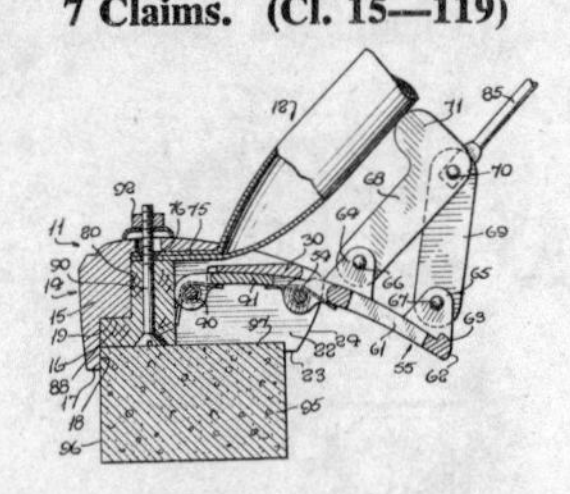

4. A mop comprising a mop head having a front wall, a top wall and depending side walls, a handle extending upwardly from the rear of the mop head, a relatively narrow mounting adaptor mounted on the underside of the head at the front portion thereof, a mop pad of sponge material carried by the underside of the adaptor and projecting below the lower edges of the side walls of the head, the rear portion of the pad extending rearwardly beyond the rear portion of the adaptor, a first hinged member hinged at its forward edge between the side walls of the head directly behind the adaptor and disposed above the exposed rear portion of the pad and mounted for swinging movement between said side walls, a second hinged member hinged at its forward edge to the rear edge of the first hinged member, means for normally holding the hinged members in a raised non-squeezing position, and means for swinging the first hinged member downwardly and forwardly and for swinging the second hinged member downwardly, forwardly and then upwardly to move both hinged members into contact with the rear portion of the pad to squeeze the latter against the front wall of the head and against the adaptor.

2,716,769
MOP FOR VENETIAN BLINDS
Wilton A. Satterfield, New York, N. Y.
Application December 31, 1953, Serial No. 401,494
4 Claims. (Cl. 15—119)

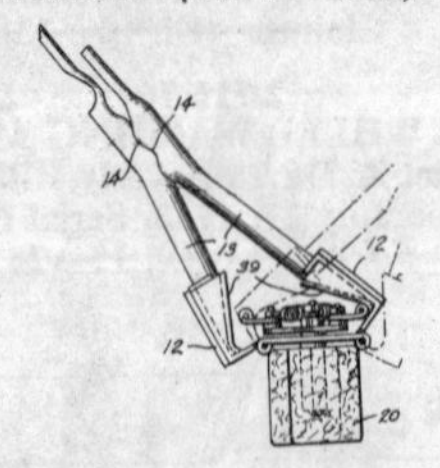

4. A device of the character described comprising an elongated support plate, a plurality of outwardly extending sponges, mounted in parallel spaced relation on one side of said support plate, an L-shaped squeezing plate having an elongated side portion and a narrow base portion hingedly connected to each elongated side of the support plate, a marginal plate having recesses therein adapted to receive side portions of said sponges mounted on each squeezing plate, handle means centrally mounted on each squeezing plate for bringing the plates into squeezing engagement with opposite sides of said sponges and means carried by said support plate operable by said handles for turning said sponges about an axis at right angles to said support plate.

2,716,770
WINDOW CLEANING APPLIANCE
Clarence H. Caldwell and Walter K. Caldwell, Oak Park, Ill.
Application May 27, 1950, Serial No. 164,674
5 Claims. (Cl. 15—126)

1. In a window cleaning appliance, a tank for a cleansing liquid, atomizing means carried by said tank to spray liquid over a window, a flat tube connected at one end to said atomizing means and connectable at the other end to a source of forced air, valve means controlling said atomizing means, and a strap connected to the opposite side of the appliance from said tube and arranged to control said valve means.

2,716,771
CLEANSING APPLICATOR FOR LIQUIDS
Wesley Turner, Portland, Oreg.
Application May 10, 1952, Serial No. 287,157
3 Claims. (Cl. 15—136)

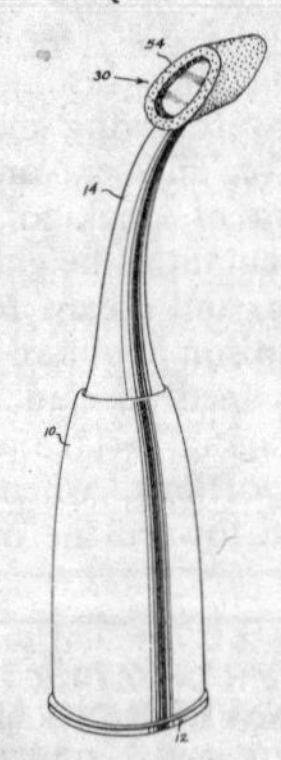

1. A cleansing applicator for liquids comprising a reservoir having walls of resiliently deformable material, a curved neck attached to the reservoir and having therethrough a longitudinal conduit communicating with the reservoir, the conduit terminating at its outer extremity in an orifice opening out on the outer transverse surface of the neck, the said surface being provided with a plurality of radial grooves communicating with the orifice, a liquid-dispensing head, a sleeve transversely penetrating the dispensing head and adapted slidably to receive the outer end of the neck, means for releasably securing the sleeve to the neck, a chamber within the head communicating with the sleeve and stationed for communication with the orifice when the neck is placed in the sleeve, within the chamber valve means comprising a piece of porous resilient material having an impervious surface layer adapted to seat on the outer transverse surface of the neck covering and substantially sealing the orifice, the resilient piece being dimensioned to span the chamber in an axial direction but to leave a passageway for liquid along the sidewalls of the chamber, conduit means axially aligned with the sleeve and extending through the head to the opposite side thereof and communicating with the passageway about the resilient valve member, the con-

duit means opening out into a longitudinally disposed channel extending the length of the head and communicating with a plurality of spaced apart peripheral channels, a pair of flanges one on each end of the head, and a scrubbing jacket made of porous resilient material overlying the head, enclosing the head and being retained thereon by the flanges.

2,716,772
MOTOR VEHICLE WHEEL WASHING ATTACHMENT
Jesse S. Cockrell, Norfolk, Va.
Application July 26, 1954, Serial No. 445,529
3 Claims. (Cl. 15—306)

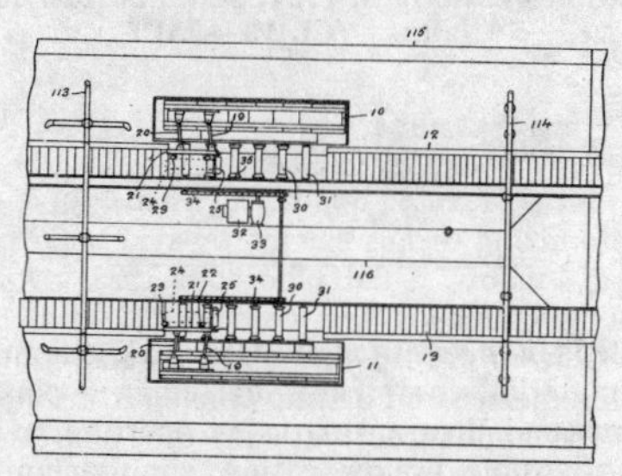

1. In a motor vehicle wheel-cleaning attachment for a mechanical motor vehicle laundry, the combination which comprises spaced parallel tracks positioned to receive wheels of a motor vehicle, auxiliary tracks spaced from outer sides of the former tracks, carriages mounted for longitudinal travel in the auxiliary tracks, swinging arms pivotally mounted on said carriages, rollers positioned to contact side surfaces of wheels on the tracks carried by said swinging arms, rollers pivotally mounted at the ends of the swinging arms and positioned to extend in the paths of wheels of a vehicle traveling on said tracks, resilient means for retaining the pivotally mounted rollers in extended positions, means for releasing the pivotally mounted rollers as carriages upon which the rollers are mounted reach predetermined positions, means for returning the pivotally mounted rollers to positions in paths of wheels traveling on said tracks, and means for returning the carriages to starting positions.

2,716,773
VACUUM CLEANER NOZZLE HAVING PIVOTED CLEANING ELEMENT
Carl E. Meyerhoefer, Brooklyn, N. Y., assignor, by mesne assignments, to Lewyt Corporation, Brooklyn, N. Y., a corporation of New York
Application April 2, 1949, Serial No. 85,103
6 Claims. (Cl. 15—369)

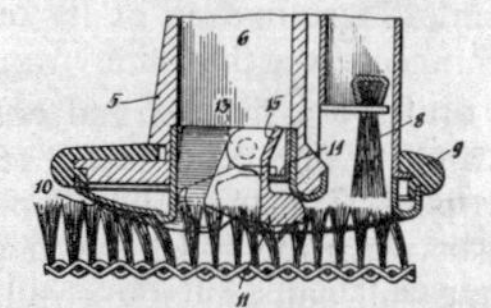

1. In a vacuum cleaner nozzle assembly a comb element to be pivotally supported within a nozzle mouth for rocking movement through an arc of approximately ninety degrees, said element presenting a lower carpet-contacting face defined between longitudinally extending front and rear edges, a row of teeth projecting from said face at a point adjacent said rear edge towards said front edge, a second row of teeth also projecting from said face at a point adjacent said front edge towards said rear edge, the forward ends of said first-named row of teeth being relatively blunt and constituting material engaging and retaining edges, the rear ends of said second row of teeth being relatively sharp and the teeth of said second row being of less height than the teeth of said first-named row at points between the side faces of the latter.

2,716,774
FURNITURE GLIDE
Simon T. Kilmer, Grand Rapids, Mich., assignor to American Seating Company, Grand Rapids, Mich., a corporation of New Jersey
Application October 25, 1952, Serial No. 316,906
1 Claim. (Cl. 16—42)

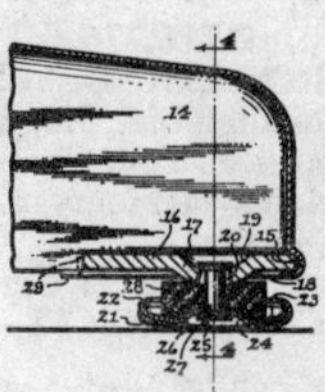

A compensating furniture glide comprising: an attachment plate adapted for attachment to the lower end of a leg of an article of furniture in substantially horizontal disposition, said plate having a depressed embossment whereby the lower surface thereof defines a convex protuberance and whereby the upper surface thereof defines a concave hollow, said embossment having a central aperture therethrough; a floor-contacting member beneath and spaced from said attachment plate; a washer member secured adjacent the upper surface of said floor-contacting member and having an upwardly dished centrally apertured portion forming with the floor-contacting member a compartment; a rivet having a lower head disposed in said compartment freely movably therein and having a shank passing upwardly through the central apertures in said washer and said embossment, and the upper head of said rivet engaging the concave upper surface of said attachment plate, said shank having a smaller diameter than said central apertures to permit universal movement of the rivet relative to both the attachment plate and the floor-contacting member; and an intermediate resilient cushion element squeezed between the washer member and the attachment plate and circumscribing said shank.

2,716,775
HINGE CONSTRUCTION FOR CONVERTIBLE ARTICLE OF FURNITURE
Loy F. Kenimer, Washington, D. C.
Application February 20, 1953, Serial No. 338,081
2 Claims. (Cl. 16—138)

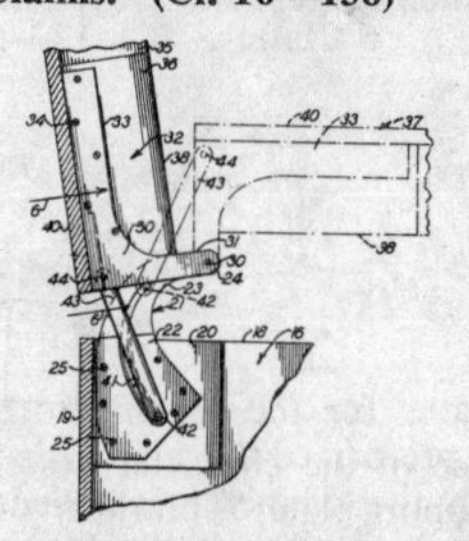

1. For use in a convertible article of furniture of the character described to permit a tiltable back rest to be pivotally manipulated to form alternately a back rest for a seat portion of a chair or a table top overlying the seat portion, a hinge comprising a broadened base portion adapted to be secured to the seat portion and having an elongated arcuately shaped neck portion extending upwardly from said seat portion and disposed substantially at a right angle to said base portion, said base portion and said neck portion having an arcuate slot extending therein, a second hinge member having a right angular

base portion to complement the neck portion of said broadened base portion, a pivotal connection joining the ends of the extending neck portion and said second hinge member, a link member pivotally connected to the base portion of said second member at one end and slidably retained in the arcuate slot of the base portion and the neck portion at the other end of the link member whereby upon movement of the second hinge member the back rest will be positioned alternately at either the horizontal or substantially vertical position.

2,716,776
SHRIMP PROCESSING APPARATUS
Philip A. Streich, Philadelphia, Pa., and Virgil R. Clark, Decatur, and Emmitte P. Tait, Chamblee, Ga., assignors to Tait-Clark-Streich Machinery Corp., Decatur, Ga., a corporation of Georgia
Application January 11, 1952, Serial No. 265,976
10 Claims. (Cl. 17—2)

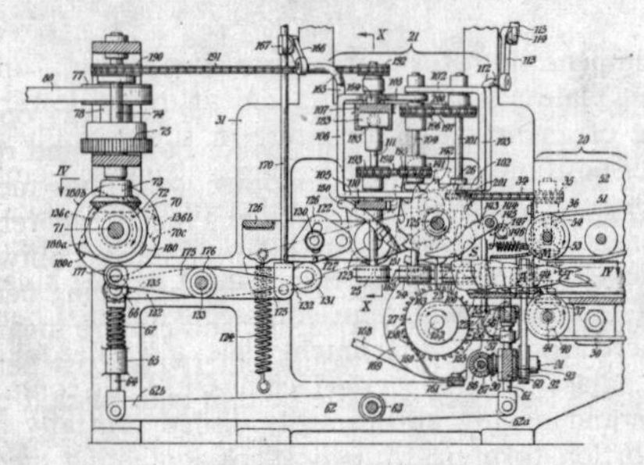

1. Apparatus for removing a part of a shrimp body from a shrimp, comprising conveyor means for said shrimp, remover means above said conveyor means disposed adjacent to the path of conveyance of said shrimp, means for activating said remover means, said remover means being effective when activated to remove said part from said shrimp, control means for inactivating said remover means, and synchronizing means operatively connected to both said conveyor means and said control means, said synchronizing means causing said control means to actuate said remover means in predetermined timed relation to the movement of said shrimp on said conveyor means.

2,716,777
MANUFACTURE OF SHRINKABLE TUBES
Norbert Hagen, Siegburg, Rhineland, Germany
Application August 14, 1951, Serial No. 241,870
6 Claims. (Cl. 18—19)

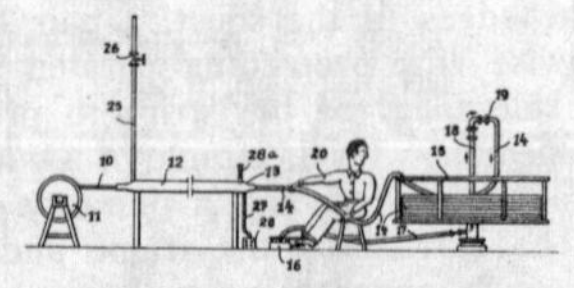

3. Apparatus for the manufacture of heat-shrinkable tubing from preformed thermoplastic tubing of a diameter smaller than the diameter intended for the shrinkable tubing, comprising a longitudinally extending heating and forming device, said device including a forming tube of textile material, adapted for the passage of the tubing therethrough, the forming tube having at its entrance end an internal diameter approximating the external diameter of the starting tubing, and at its exit end an internal diameter approximately equaling the desired external diameter of the shrinkable tubing, and consisting of a widening portion and a cylindrical portion, said widening portion increasing in diameter between the entrance end and a point intermediate the entrance end and the exit end, the cylindrical portion extending between the said point and the exit end, the apparatus further comprising a double-walled heating jacket surrounding the entire length of the forming tube, means to introduce fluid pressure into the tubing to expand same when heated, and means to release said fluid pressure, the inner wall of the heating jacket cylindrically enclosing both portions of the forming tube, thus leaving a free space between itself and said widening section.

2,716,778
METHOD OF MOLDING SEALING STRIPS
Charles H. Beare, Dayton, Ohio, assignor to General Motors Corporation, Detroit, Mich., a corporation of Delaware
Application February 6, 1951, Serial No. 209,624
5 Claims. (Cl. 18—53)

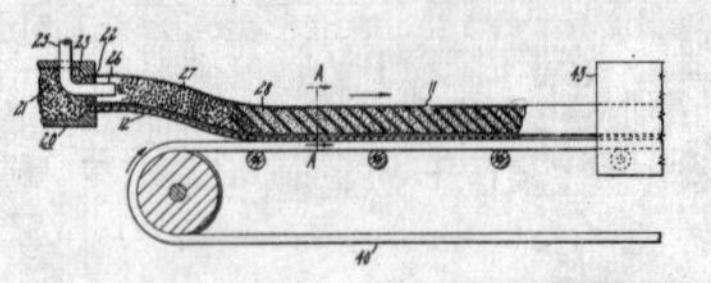

1. The steps in the method of making a cushioning strip of rubber material, comprising: extruding a plastic soft rubber compound thru an extrusion aperture to form a thin-wall collapsible hollow casing, simultaneously and progressively filling said casing with ungelled foamed latex material before said casing collapses by flowing said latex material thereinto thru a conduit passing thru the central portion of said extrusion aperture from the pressure side to the outside of said aperture, progressively gelling said foamed latex filling, after entering said casing and thereby providing a substantial interior support for said thin-wall casing and preventing said casing from collapsing, and subsequently vulcanizing together said casing and filling to form a thin-wall flexible casing filled with spongy rubber.

2,716,779
MEANS FOR DRAFTING TEXTILE FIBERS
William G. Reynolds, St. Petersburg, Fla.
Application November 16, 1950, Serial No. 195,945
3 Claims. (Cl. 19—131)

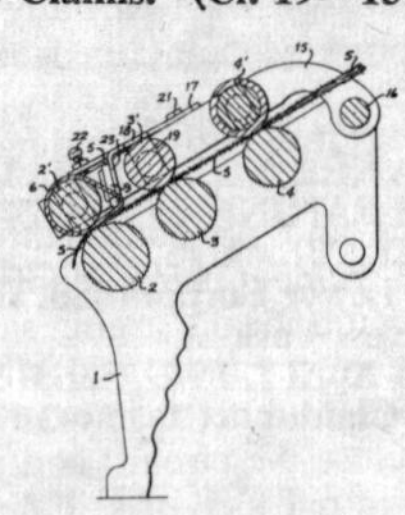

1. Means for drafting a fiber strand comprising, pairs of rolls spaced apart, one pair being restraining rolls and the other terminal drafting rolls adapted to elongate a strand passing between the roll pairs, a guide bar positioned between the roll pairs having a rail lying substantially in a plane parallel to the tops of the bottom rolls of the roll pairs and having upwardly extending spaced guide fingers, and a belt surrounding and driven by one of said terminal rolls and passing around the rail and between the guide fingers of said guide bar, the lower flight of the belt having its end disposed around the rail at a slightly higher elevation with respect to the strand than the end of the flight contacting the terminal rolls whereby light initial contact is made with the strand and maintained with the strand at gradually increasing pressure at the speed of the terminal rolls over an appreciable distance within the drafting zone up to and including the nip between the terminal rolls.

2,716,780
TOP ROLL FOR DRAWING MECHANISMS
Kenneth P. Swanson, Abington, Mass., assignor to Textile Engineering Corporation, Whitman, Mass., a corporation of Massachusetts
Application February 16, 1952, Serial No. 271,903
13 Claims. (Cl. 19—142)

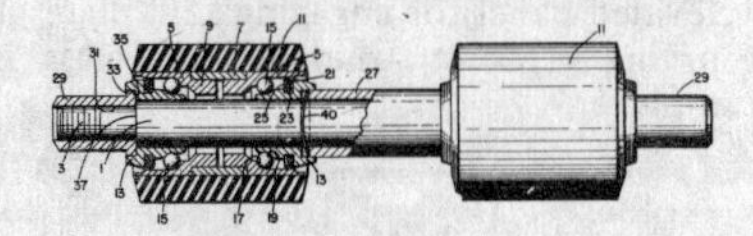

1. A top roll having in combination a shaft, a roll shell, opposed cones sliding on the shaft, bearing balls between the cones and the roll shell, resilient means urging one cone toward the other, means shiftable axially of the shaft to load the resilient means, and stop means on the shaft limiting the extent of shift of the loading means in the loading direction.

2,716,781
SLIVER CONTROLLED STOP MOTION ACTUATING DEVICE
James M. Elliott and Dick R. Green, Raeford, N. C.
Application March 2, 1954, Serial No. 413,587
10 Claims. (Cl. 19—165)

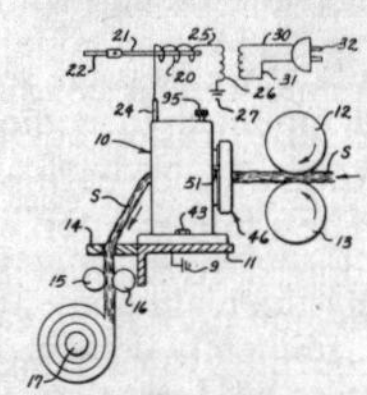

2. An improved textile-sliver-controlled circuit maker and breaker comprising a trumpet having a substantially axial tapered bore therethrough through which the sliver passes and wherein the sliver is withdrawn from the small end of said bore in a direction substantially radially of said bore, a fixed element having a cavity therein of substantially greater dimensions than the trumpet and in which said trumpet is at least partially disposed whereby the trumpet is normally maintained in engagement with one surface of said fixed element by the radial pull of said sliver against the corresponding surface of said tapered bore, an electrode carried by and insulated from said fixed element and projecting into the cavity, said electrode and said trumpet being adapted to be interposed in an electrical circuit, means urging said trumpet toward said electrode whereby, upon said sliver becoming unduly slackened, the trumpet is moved into contact with said electrode to close the circuit, and manually operable means interposed in one side of said circuit for rendering said trumpet ineffective in closing said circuit when it contacts said electrode.

2,716,782
STORABLE INTER-SHELTER CONNECTOR
Harold A. Paulsen, Omaha, Nebr.
Application May 18, 1953, Serial No. 355,426
5 Claims. (Cl. 20—1)

1. A passageway assembly for connecting a first and second shelter, each of said shelters having opposing parallel walls, each having a doorway opening therein, said first shelter having a wall with a recess therein, said recess extending around the sides and top of said doorway opening, a rigid frame member approximately of a shape complemental to the shape of said recess, said rigid frame member having a flat outer surface disposable against the said wall of said second shelter and extending around the doorway thereof, means for supporting said frame against said second shelter wall, a covering member extending around one side of the doorway opening of said first shelter and across the top of said doorway opening and down the other side thereof similarly to said recess, said covering being foldable for fitting in said recess in storage and expandable so as to extend from said recess out to the respective side of said frame at times when said frame is disposed against said second shelter, means for securing the respective ends of said covering to said frame and to a wall of said recess, a floor member disposed under

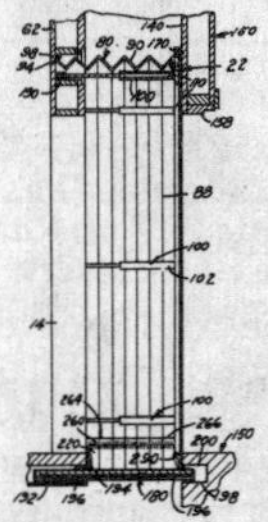

said first shelter in storage beneath the doorway opening portion of said first shelter, means for holding and receiving said floor member during storage, said floor member being slidably received in said holding means whereby the outer end of said floor member can be moved outwardly from its holding means into a position extending between said shelters beneath said covering, and closure means for closing the openings that would otherwise exist between the edges of said floor and the bottom ends of said covering.

2,716,783
DOUBLE WINDOWS
Thomas Gregory Fegan, Hathersage, England
Application November 13, 1952, Serial No. 320,330
Claims priority, application Great Britain November 23, 1951
3 Claims. (Cl. 20—55)

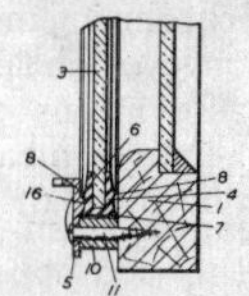

1. A supplementary glazing sheet mounting adapted to convert a single glazed framed window into a double window, said mounting comprising a rectangular supplementary glazing sheet, four separate strips of resilient plastic material with a groove shaped to correspond to the edge-section of the sheet fitted completely round the edges of the sheet, a pair of spaced narrow lengthwise ribs protruding parallel to each other from both side faces of the strips to provide a linear sealing engagement with the frame of a window to which the mounting is applied, and a plurality of swing clips adapted to be carried by the frame and to be swung into compressive engagement with an outer side face of the strips to complete a linear sealing engagement of the ribs on the inner side face with the frame, each clip having a projection adapted to be forced over one rib on the outer side face into the space between the ribs on that face to apply pressure to the body of the strips between the ribs.

2,716,784
SASH AND REGLAZING METHOD THEREFOR
Peter H. Kuyper, Pella, Iowa, assignor to Rolscreen Company, Pella, Iowa, a corporation of Iowa
Application April 2, 1954, Serial No. 420,716
2 Claims. (Cl. 20—56.4)

1. In a window sash construction, a non-metallic frame formed of a number of pieces adapted to be as-

sembled and defining a window opening therethrough, said frame having a rabbet formed substantially completely about its inner periphery and extending in depth in a plane parallel to that of said frame, a pane of window material disposed in said rabbet when the frame is initially assembled, said frame having an elongated shallow groove formed closely adjacent the inner peripheral edge thereof substantially in line with the bottom of the rabbet and extending substantially completely about the

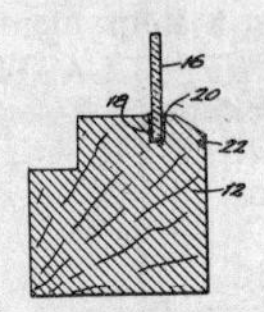

periphery of the frame, said groove extending in depth in a plane substantially perpendicular to that of said rabbet and terminating a substantial distance short of intersection therewith for defining an initially integral stop means capable of being subsequently severed to afford removal of said initially disposed window material, the integral frame thickness between said groove and said rabbet being at least twice the depth of said groove so as to provide substantial material strength for retaining said pane of window material within said frame.

2,716,785
WINDOW SHUTTER AND AWNING
Harvey Schoen, Elgin, Ill.
Application October 21, 1953, Serial No. 387,397
9 Claims. (Cl. 20—57.5)

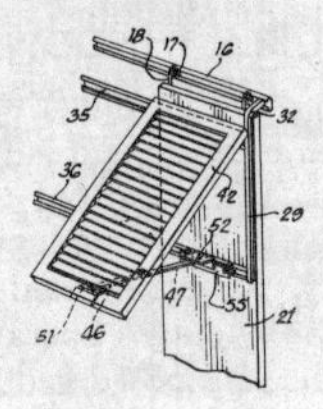

6. An awning apparatus for attachment over sliding sash windows which comprises, an upper guide rail, a lower guide rail parallel to said upper guide rail, a pair of awning elements supported from rollers riding in said upper guide rail, and a foldable bracket associated with each of said awning elements riding in said lower guide rail, each said foldable bracket being pivotally articulated with its awning element.

2,716,786
BRACKET CLIP FOR JALOUSIE WINDOW
John A. Moore, Merrick, N. Y.
Application January 10, 1955, Serial No. 480,883
2 Claims. (Cl. 20—62)

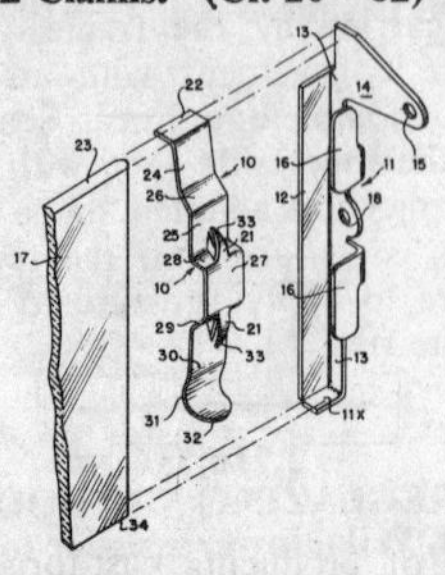

1. A retainer bracket assembly for retaining a pane of jalousie plate glass comprising a bracket piece having a back panel and a side panel disposed at right angles to and integral with said back panel and a pair of spaced-apart platform pieces disposed over said back panel and integral with said side panel, a longitudinal unitary flat leaf spring clip having an elevated top central rectangular panel, a pair of seizure talons integral with said central panel and disposed on opposite sides thereof, and a pressure exerting bottom element disposed on each side of said top elevated panel for engaging said plate glass with constant urging pressure, whereby said glass is locked securely in place.

2,716,787
FLEXIBLE SEALING STRIP
Edward P. Harris, Dayton, Ohio, assignor to General Motors Corporation, Detroit, Mich., a corporation of Delaware
Application December 19, 1951, Serial No. 262,354
1 Claim. (Cl. 20—69)

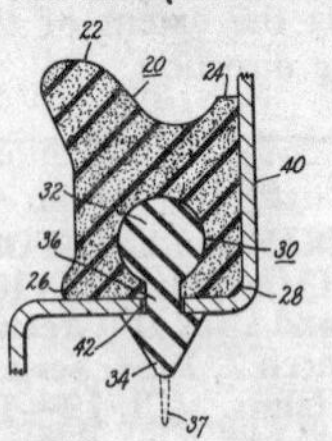

A flexible longitudinally extensible sealing strip adapted to be attached and mounted to a more rigid mounting member having means for mechanically holding a portion of the strip thereto, said strip comprising a resiliently deformable body of rubber-like material of the desired cross section forming the main body of the strip, a longitudinally extending, resiliently deformable part of rubber-like material of greater hardness than the main body portion of the strip and at least partially embedded in and bonded to said body portion, said longitudinally extending, resiliently deformable part being of tubular cross section with respect to that portion thereof that is embedded in the strip and having a longitudinally extending portion exteriorly accessible and integrally formed therewith for providing an interlock with said holding means for holding the strip to the member.

2,716,788
BEADING OR WEATHER STRIP AND METHOD OF MAKING SAME
Harold Burling Naramore, Fairfield, Conn., assignor to Bridgeport Fabrics, Inc., Bridgeport, Conn., a corporation of Connecticut
Application April 5, 1952, Serial No. 280,771
13 Claims. (Cl. 20—69)

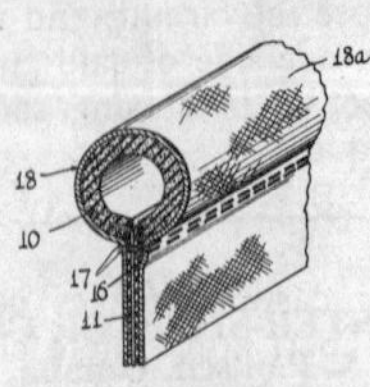

1. A beading or weather strip comprising an elongate member having a mounting flange projecting from one edge and a narrow securing portion projecting from the other edge, said elongate member being bent about a longitudinal axis with the edges in juxtaposition to form a tubular bead with the flange and securing portion disposed outwardly of the bead; and means securing the mounting flange and securing portion together along the juxtaposed edges adjacent the bead to retain the elongate member in tubular form with the mounting flange projecting therefrom.

2,716,789
FOUNDRY CORE FORMING APPARATUS
Ralph P. Davis, Squantum, Mass., assignor to Walworth Company, Boston, Mass., a corporation of Massachusetts
Application September 21, 1951, Serial No. 247,570
4 Claims. (Cl. 22—10)

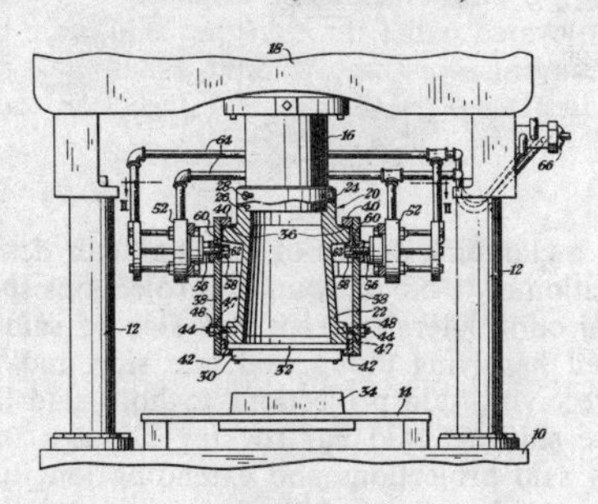

1. An adapter for core-forming apparatus having a discharge nozzle for core material comprising a hollow casing adapted to be secured at one end to the nozzle to overlie core-forming means at the opposite end and to transmit core material from the nozzle to the core-forming means, said casing having a pair of inclined surfaces on opposite sides thereof, a pair of clamping members carried on said opposite sides of said casing and each having a complemental inclined surface for operative engagement with said surfaces on said casing, said members each having a portion projecting beyond said opposite end into overlying relation therewith to receive core-forming means therebetween, and means for moving said members laterally relative to said casing and causing said operative engagement between said inclined surfaces of said casing and said member respectively for maintaining said projecting portions in clamping engagement with the core-forming means.

2,716,790
APPARATUS FOR CASTING METALLIC ARTICLES
Joseph B. Brennan, Cleveland, Ohio
Application May 12, 1951, Serial No. 225,949
6 Claims. (Cl. 22—73)

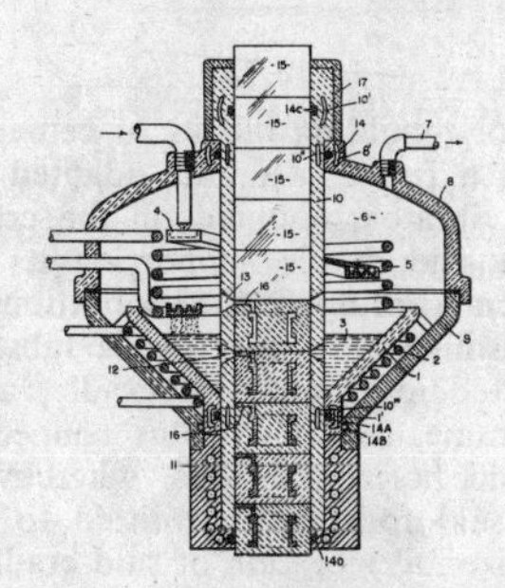

6. Apparatus for casting articles comprising a mold having a cavity, an inert atmosphere chamber containing a receptacle for molten metal with respect to which said mold is adapted to be moved to successively expose the cavity therein to such atmosphere and to communicate the cavity with molten metal in said receptacle for filling said cavity, an upwardly extending guide connected to the upper part of said chamber which embraces said mold and provides a seal against inflow of atmosphere into such chamber as said mold moves through said guide into said chamber, said receptacle being provided with a downwardly extending guide connected to the lower part of said receptacle and aligned with said upwardly extending guide for guiding movement of said mold downwardly and for forming with said mold and metal therein a closure for said receptacle as said mold is thus moved, and means for cooling said mold progressively upwardly from the bottom thereof and thus correspondingly solidifying the metal therein and providing a shrinkhead of molten metal which compensates for shrinking of the metal below.

2,716,791
INVESTMENT CASTING
Eugene L. Schellens, Essex, Conn.
Application April 23, 1951, Serial No. 222,491
10 Claims. (Cl. 22—200)

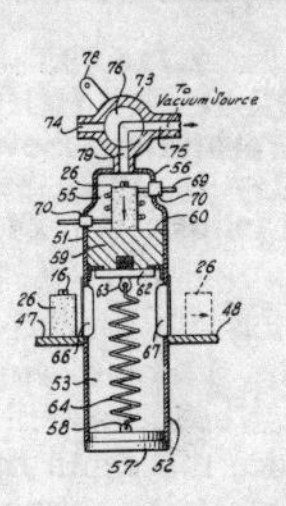

1. The method of producing a solid metallic casting by the use of mold forming investment material, which comprises the steps of, forming into a pattern body having the desired size and shape of an ultimate casting increased by that of a neck formation upstanding therefrom a compacted mixture of heat expansible powdered metal and a thermoplastic binder having a fusing point much lower than that of said metal, embedding said pattern body in flowable settable investment material substantially less expansible than said powdered metal to form therewith a unit in which the main bulk of said pattern body fills said cavity and said neck formation of said pattern body forms and fills a space in said investment material rising from said mold cavity, subjecting said embedded pattern body to increasing heat by stages operative successively to soften said binder while said powdered metal is expanding and later to melt said powdered metal, whereby said pattern body becomes sufficiently plastic prior to the melting of said powdered metal to enable components of said mixture to pass from said mold cavity into said riser space thereabove for accommodating expansive enlargement of said pattern body without subjecting said investment material to fracture by expansive force of said pattern body, cooling the molten metal to form a solid casting, and freeing said casting from the mold.

2,716,792
METHOD OF CAST-FORGING METALS
Karl Kristian Kobs Krøyer, Aarhus, Denmark
Application October 5, 1950, Serial No. 188,578
5 Claims. (Cl. 22—209)

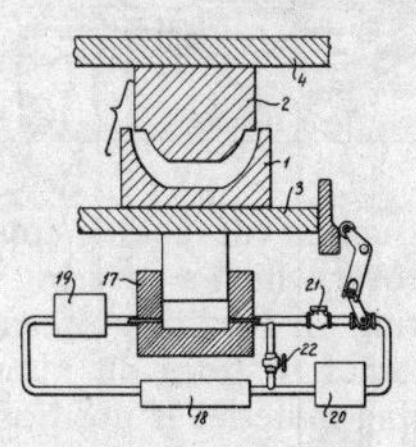

1. A method of producing cast-forged metal objects between an upper and lower die comprising the steps of pouring into the lower die molten metal in a quantity predetermined to be that needed initially between the dies to form the desired article therebetween when the dies are

fully closed, relatively moving the dies toward one another at relatively high speed and low pressure to trap the molten metal between the dies, suddenly altering the closing movement of the dies to a lower speed and a higher pressure at the point in the closing movement of the dies where the volume of the space between the dies is substantially equal to the volume of the aforesaid quantity of molten metal, whereby the closing movement is altered before the dies exert a substantial pressure on the molten metal, and thereafter continuing the closing movement of the dies at lower speed and higher pressure until the closing movement is halted by the resistance offered by the then cast-forged metal between the then fully closed dies.

2,716,793
DRAPERY HOOKS
Samuel Perlmutter, Newton, Mass.
Application May 18, 1951, Serial No. 227,003
3 Claims. (Cl. 24—84)

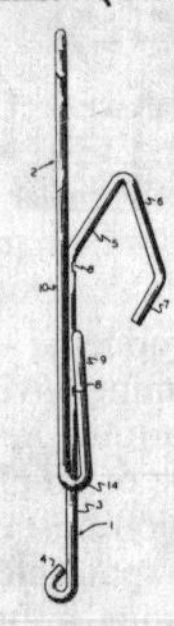

1. A drapery hook of the type described comprising two separable elements, the one formed of a wire having end stem portions extending in the same direction and joined to one another in a reversely bent loop having side sections which are continuations of said wire from said stem portions, the top end of said loop extending across the stems and spaced therefrom a distance less than the thickness of said second separable element, said second separable element comprising a hook with a shank, said shank adapted to engage said first element between said stems and between the stems and the top end of said loop when said shank is in operable position substantially parallel to said end stem portions, said shank having position fixing means engaged by said loop for fixing the position of the second separable element with respect to the first in substantial alignment with the stems thereof.

2,716,794
FABRIC COVERED BUTTON
Ludwig Zelenay, Culver City, Calif.
Application June 29, 1953, Serial No. 364,533
1 Claim. (Cl. 24—113)

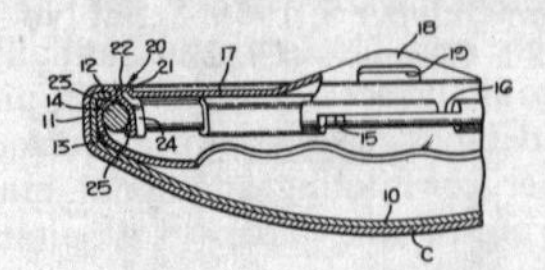

A button construction for making covered buttons comprising a support over which a covering material is placed, said support having an inwardly extending flange about the periphery thereof to form an internal recess within which said covering material is positioned, a split expansion ring partially receivable in said recess to retain the covering material therein, a plate, a U-shaped hinge on the peripheral edge of said plate, and a plurality of spaced apart flanges on said hinge that are urged against said split ring by said hinge when said plate is snapped within the confines of said support.

2,716,795
CLAMPING DEVICE
Charles B. Harker, Rockford, Ill., assignor to Bartelt Engineering Company, Rockford, Ill., a corporation of Illinois
Application January 2, 1952, Serial No. 264,599
5 Claims. (Cl. 24—255)

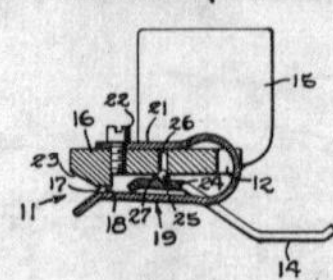

1. In a clamping device of the character described, the combination of, a block, rounded projections spaced apart and projecting laterally from one side of said block, an elongated bar lying along said one side and having recesses receiving said projections to hold said bar against edgewise shifting, said bar rocking about a fulcrum defined by said projections and extending longitudinally of the bar, a U-shaped member of spring material straddling said block transversely of said bar and having a first arm extending over said bar to hold the bar against the block, means for fastening the other arm of said member to the opposite side of said block, the portion of said block opposing said first arm cooperating with the latter to form a clamp, and an operating arm projecting laterally from said bar and movable to rock the bar about its fulcrum and to spread the jaws of said clamp apart.

2,716,796
PRESSES FOR MOULDING PLASTIC SUBSTANCES
William Herbert Smith, East Molesey, England
Application December 18, 1951, Serial No. 262,257
Claims priority, application Great Britain December 21, 1950
6 Claims. (Cl. 25—83)

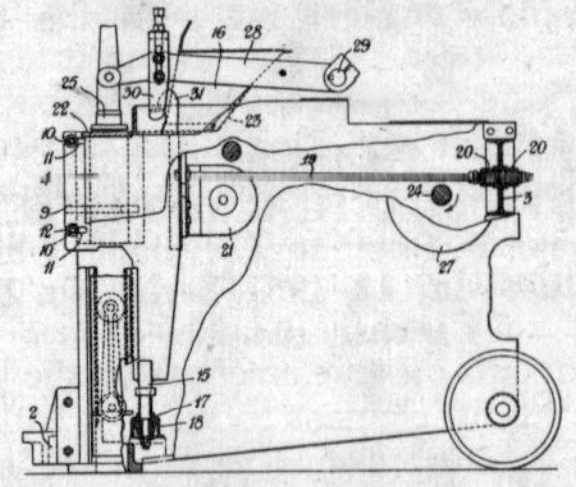

1. A press for moulding blocks of cementitious material comprising a frame, a cradle adapted to support a mould box, a vibrator mounted on the cradle, resilient means beneath said cradle supporting its weight from said frame and absorbing vertical vibrations of said cradle before transmission to said frame, substantially horizontal rods extending from said cradle, and extending through said frame, resilient means connected with and surrounding said horizontal rods, whereby longitudinal movement of said rods is transmitted to said resilient means and horizontal vibration of said cradle is absorbed before transmission to said frame, means for compressing into a block cementitious material fed into the mould box and means for ejecting said block from said mould box.

2,716,797
FABRIC NAPPING APPARATUS
Wilfred N. Hadley, Springfield, Vt., assignor to Parks & Woolson Machine Co., Springfield, Vt., a corporation of Vermont
Application April 2, 1954, Serial No. 420,575
7 Claims. (Cl. 26—29)

1. Fabric napping mechanism comprising two spaced and parallel rolls, a plurality of napping needles disposed in spaced relation around and along the rolls and having hook-like free ends, said ends pointing circumferentially

in one direction on one roll and in the other direction on the other roll, and means including a clutch for driving one roll at a relatively faster peripheral speed from

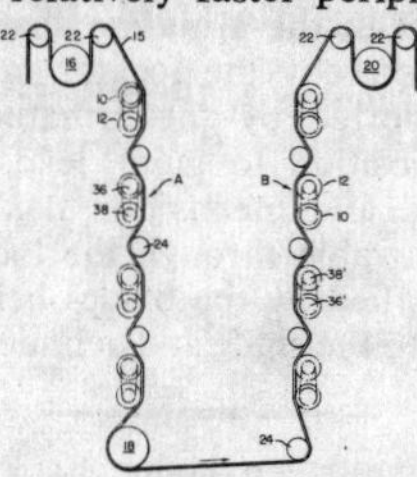

the other roll when the clutch is in one position and the other roll at a relatively faster peripheral speed from said one roll when the clutch is in another position.

2,716,798
SLIDE FASTENER MACHINES
Alexander M. Brown, New Britain, Conn.
Application February 5, 1947, Serial No. 726,624
44 Claims. (Cl. 29—34)

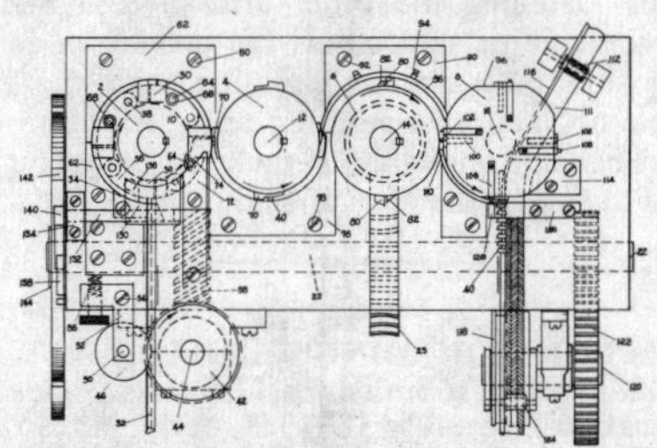

14. A machine for making slide fastener parts comprising a rotating cutter drum, rotating forming drums tangential to each other and one of which is tangential to said cutter drum, and a rotating setting drum tangential to the other of said forming drums, means to drive said drums in synchronism with each other, said cutting drum having a die member on its periphery, a stationary knife member adjacent the periphery of the cutting drum, means to feed a preformed wire past said knife member to said die member whereby said die member at each rotation of the cutting drum cuts off an element from the wire, means to transfer the cut-off element to the first forming drum, means on said forming drums to form the head of the element, means on said setting drum to receive the element from the second forming drum, means to feed a slide fastener tape past said setting drum substantially radially thereof, and means carried by said setting drum to bring an element carried thereby into engagement with such tape and to clamp the legs of such element on the tape.

2,716,799
TOOL HOLDER
William Bader, Birmingham, Mich., assignor to Wesson Multicut Company, Ferndale, Mich., a corporation of Michigan
Application December 6, 1951, Serial No. 260,173
1 Claim. (Cl. 29—96)

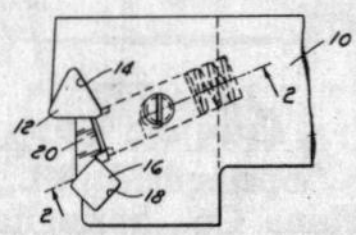

A tool holder comprising a body having two substantially parallel spaced, broached recesses to receive cutter inserts, cutter inserts in said recesses having facing spaced walls and a holder for said inserts comprising a member having a shaft slidable in an opening within said tool body in a direction substantially normal to the axis of the recesses and a shaped portion on the end of said holder having diverging walls on opposite sides to correspond with walls of said recesses and said inserts to contact cutter inserts and force said inserts in opposite directions into said recesses against the tool body.

2,716,800
TOOL HOLDER
William Bader, Ferndale, Mich., assignor to Wesson Multicut Company, Ferndale, Mich., a corporation of Michigan
Application February 25, 1954, Serial No. 412,488
4 Claims. (Cl. 29—96)

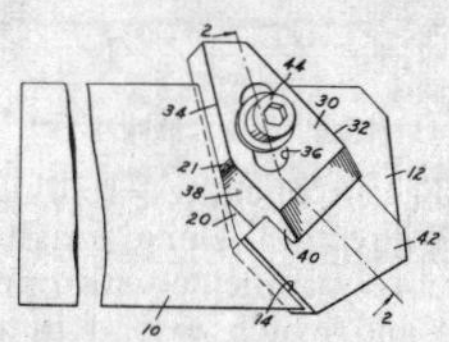

1. A tool holder comprising a body having walls formed thereon converging in the direction of one side of said holder, a recess formed in said body having parallel walls joining said first-named walls, a bit insert to lie in said recess, a clamp block movable along one of said converging walls having a portion to lie behind and a portion to overlie said insert, and means on the other of said converging walls movable transversely of said clamp toward said first converging wall to provide a wedge recess to limit the movement of said clamp block in a rearward direction.

2,716,801
CUTTER
Charles E. Kraus, Rochester, N. Y.
Application June 14, 1952, Serial No. 293,590
2 Claims. (Cl. 29—105)

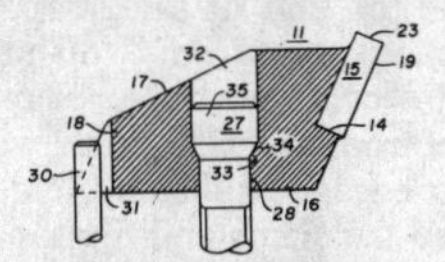

1. In a cutter blade of the type comprising a block of material having opposite flat ends, a flat bottom surface and a top surface generally converging toward said bottom surface in the direction of one of said ends, the latter constituting an abutment surface disposed at an acute included angle to said bottom surface, the other end of said block providing a cutting face projecting from at least one side of said block; a hole extending through said block at right angles to said bottom surface for receiving a fastening member, and said blade having an opening spaced rearwardly from said hole and extending inwardly from said abutment surface and through said block at substantially right angles to said bottom surface for receiving a pin for locating laterally said block, said abutment surface and said fastening member constituting additional means for enabling the securing of said block to a suitable support.

2,716,802
METHOD OF MAKING HEAT EXCHANGE DEVICES
Carl S. Greer, Jr., Albion, Mich., assignor to Tranter Manufacturing, Inc., a corporation of Michigan
Original application October 8, 1951, Serial No. 250,273. Divided and this application October 20, 1952, Serial No. 315,723
2 Claims. (Cl. 29—157.3)

1. The method of forming a heat exchange unit having a tube and having fins applied to said tube, which

comprises forming an elongated flat strip with longitudinally spaced aligned openings elongated in the direction of length of the strip and spaced inwardly from opposite side edges of said strip, notching the strip at opposite sides of the openings and midway between the ends of said openings to provide the openings with aligned extensions, bending the portions of the strip defining the marginal edges of each opening at opposite sides of the extensions laterally outwardly to provide flanges extending at right angles to the strip and having a width less than the depth of the extensions, folding the strip in a laterally inward direction along longitudinally spaced lines located between adjacent openings, folding the strip in a laterally outward direction along longitudinally spaced lines which respectively coincide with opposite side edges of the extensions and which extend in parallel relation to the lines along which said strip is bent in a laterally inward direction to form longitudinally spaced corrugations with aligned open recesses in opposite side walls of each corrugation and to position the flanges in alignment to provide extended bearing surfaces bordering the recesses, positioning a length of tubing in the recesses and securing the length of tubing in seating engagement with the bearing surfaces formed by said flanges.

2,716,803
ARMATURE MAKING APPARATUS
Willard C. Shaw, Anderson, Ind., assignor to General Motors Corporation, Detroit, Mich., a corporation of Delaware
Application November 18, 1948, Serial No. 60,771
4 Claims. (Cl. 29—205)

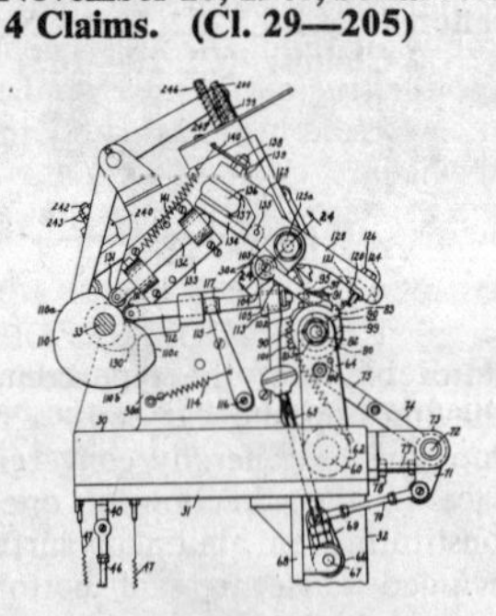

1. In a machine for inserting, into the slots of a toothed armature core assembled with a shaft and a commutator, active coil sides of preformed armature coils, and for staking leads of the coils into notches of bars of the commutator, the combination comprising an armature shaft support including a cylinder having an open recess which receives the end of the armature shaft adjacent to which the commutator is mounted, said recess having a semi-cylindrical portion fitting the armature shaft and a portion extending from the semi-cylindrical portion upwardly to the periphery of the cylinder, said semi-cylindrical portion being located so that the axis thereof is parallel to and above the cylinder axis and at one side of a vertical plane intersecting the cylinder axis, when the armature shaft support is in machine operating position, a hub journalled on the cylinder and including a slot therethrough adapted to register normally with the open end of said recess, a lead locator carried by the hub and having two lead receiving notches, means for locating the hub so that the notches of the locator are substantially vertically above the cylinder axis, means for effecting rotation of the hub to move the slot out of register with the open end of said recess for locking the armature shaft therein and to move the locator to a position such that a plane intersecting the cylinder axis and extending midway between the notches intersects the axis of the semi-cylindrical portion of the armature shaft receiving recess of the cylinder whereby the locator moves close to the commutator supported by the armature shaft and the notches are in position to guide leads into notches of commutator bars, and means for staking the leads including blades movable through the locator notches, the staking forces applied by the blades being symmetrically disposed relative to the shaft and cylinder axes.

2,716,804
METHOD OF MAKING PANELS
Charles Bayard Johnson, Detroit, Mich., assignor to Woodall Industries, Inc., Detroit, Mich., a corporation of Michigan
Continuation of abandoned application Serial No. 64,371, December 9, 1948. This application January 14, 1950, Serial No. 138,621
2 Claims. (Cl. 29—453)

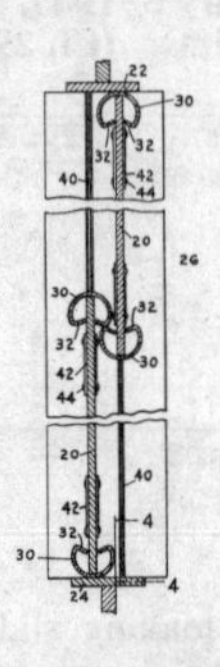

1. That process of fabricating a panel comprising providing a sheet of relatively rigid panel forming material, providing a split tube having opposite resilient edge portions at the split, bending said edge portions correspondingly inwardly of the tube a distance less than half of a diameter of the tube to a position projecting toward each other and spaced apart a distance less than the thickness of the sheet, inserting under pressure along a path perpendicular to the axis of the tube a marginal portion of the panel sheet into the tube between said inturned edge portions spreading said edge portions resistingly apart so that the inturned edge portions of the tube are tensioned by such insertion to grippingly engage the opposite faces of the panel sheet inwardly of the tube and continuing such insertion along said path until the margin of the panel sheet seats against that portion of the side wall of the tube opposite the split.

2,716,805
EXTRUDING INTEGRALLY STIFFENED PANELS
Macdonald S. Reed, Erie, Pa.
Application July 8, 1952, Serial No. 297,666
7 Claims. (Cl. 29—548)

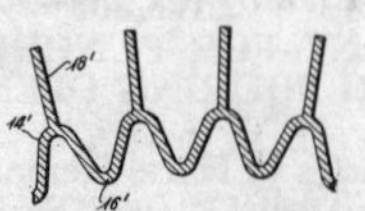

1. A method of forming an integrally stiffened panel comprising, the steps of; extruding a thin corrugated sheet of material having parallel spaced integral ribs on one surface thereof parallel to the corrugations, and thereafter flattening said corrugations to produce a wider and substantially flat panel having spaced parallel integral ribs upstanding from a surface thereof.

2,716,806
MILLING AND STRETCHING INTEGRAL PANELS
Macdonald S. Reed, Erie, Pa.
Application July 8, 1952, Serial No. 297,667
6 Claims. (Cl. 29—548)

1. A method of forming an integrally stiffened panel, comprising the steps of: milling opposite faces of a relatively thick metal plate to provide spaced integral ribs on one side and at least one corrugation between said ribs, and thereafter flattening said corrugations to produce a wider and substantially flat panel of less thickness than said plate having spaced parallel solid integral ribs upstanding from one surface thereof.

2,716,807
CAN OPENER
Arthur Carp, Ferguson, Mo.
Application February 8, 1954, Serial No. 408,931
12 Claims. (Cl. 30—15.5)

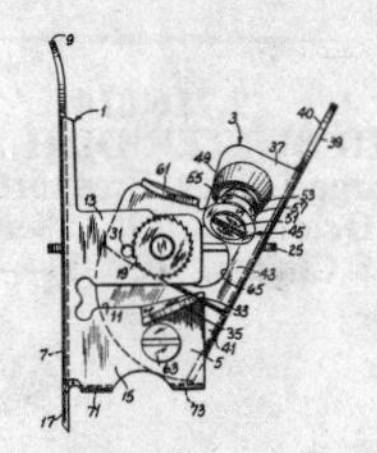

1. A can opener comprising first and second pivoted levers, the first lever having an arm assembly carrying a traction wheel, the second lever carrying a rotary cutter, the first lever being swingable relative to the second from an open to a closed position to dispose the wheel in cooperative relation to the cutter, said arm assembly being adapted for first and second springing movements away from the cutter, the first one of said springing movements being transverse with respect to the pivotal movements of the levers and the second springing movement being substantially parallel to said pivotal movements, the second lever having a first formation adapted for engagement with the arm assembly initially to hold it against said first transverse springing movement as the first lever is swung from open to closed position, the arm assembly being released for said first springing movement when the first lever is in closed position, and the second lever also having a second formation adapted for engagement with the arm assembly to hold it against said second parallel springing movement as the lever is swung from open to closed position, the arm assembly being also released for said second springing movement when the first lever is in closed position.

2,716,808
ATTACHMENT FOR PUNCHING AND FOR CLEARING POURING OPENINGS IN CAN TOPS
Charles C. Hart, Trenton, N. J.
Application April 12, 1955, Serial No. 500,800
3 Claims. (Cl. 30—16)

1. Attachment for an evaporated milk can or other such container, comprising crossed strips of spring material to extend across the top of such a can and having downwardly bent ends to embrace the sides of the can at opposite sides of a diameter of the can whereby to hold on the can against rotational or lateral separating movement, a handle knob secured over said crossed strips at the approximate center of the can structure, an upwardly bowed spring swiveled on said center knob, a conical punch

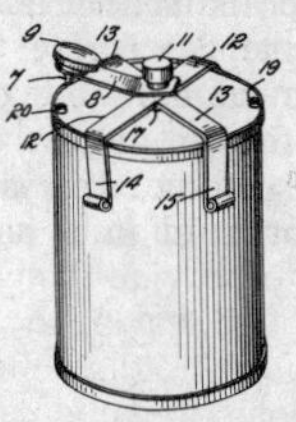

dependent from the free end of said spring and a flat push-button on said free end of the spring over said punch.

2,716,809
HAIR CUTTER
Carl E. Malone, Fort Lauderdale, Fla., assignor of fifty per cent to Hugh M. Sutton, Fort Lauderdale, Fla.
Application February 18, 1954, Serial No. 411,085
6 Claims. (Cl. 30—30)

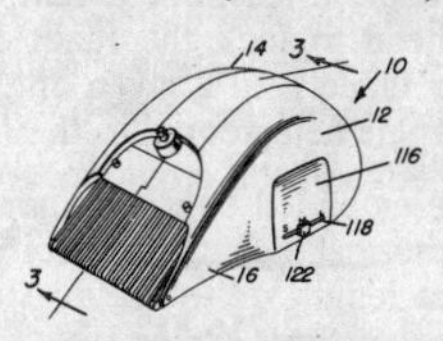

1. A hair cutter comprising a housing, an axle journalled in said housing and having a pair of wheels fixed in spaced relation thereon, a pinion fixed to said axle, a comb member pivotally secured to said housing about an axis extending along the forward edge of its teeth, a blade assembly disposed in overlying relation to and slidably carried on said comb member, and means engaging said pinion and connected to said comb member and said blade assembly for simultaneously imparting a downward swinging motion to said comb member and a forward sliding motion to said blade assembly in response to rotation of said wheels.

2,716,810
PAPERBOOK RAZOR
Walter C. Koval, Los Angeles, Calif.
Application July 6, 1954, Serial No. 441,197
8 Claims. (Cl. 30—47)

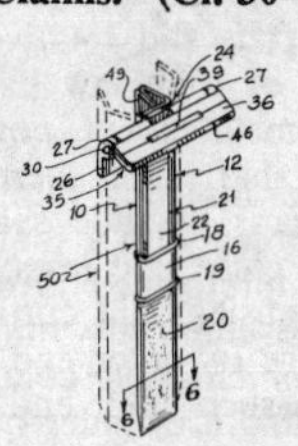

1. A paperbook razor comprising a housing having longitudinal and lateral score lines dividing said housing into central, intermediate and end panels and upper and lower portions; a mounting on the upper central panel having bearings; and a blade holder having pintles adapted to be freely mounted in said bearings.

2,716,811
RAZORS
Larue R. Case, Pasadena, Calif., assignor to Case Company, Pasadena, Calif., a partnership
Application September 2, 1954, Serial No. 453,796
3 Claims. (Cl. 30—47)

1. In a disposable razor, the combination comprising a cover which is adapted to be folded to form a channel-section handle for the razor, a T-shaped razor body hav-

ing a cross-bar with a flat surface and a shank which is attached to the cover, said cross-bar having a top edge lying in a plane which is inclined at a given angle from a plane normal to the flat surface of the cross-bar, a razor head hinged to the cross-bar of the razor body, a guide bar integrally formed as part of the razor head, a razor blade having a cutting edge, said razor blade being attached to the razor head in a position in which the cutting edge is parallel to the guide bar at a given distance from the surface of the guide bar, said guide bar being formed to provide a conduit in the razor head to prevent clogging of the razor, at least one detent integrally formed with the razor head to lock the razor head at an angle so that the razor blade is at a given angle with respect to the shank, said angle being equal to the angle between the plane of the top edge of the cross-bar and the flat surface of the cross-bar.

2,716,812
RIVET STEM CUTTING TOOL
Richard E. Noonan, Bryan, Ohio, assignor to Aro Equipment Corporation, Bryan, Ohio
Application January 2, 1953, Serial No. 329,417
4 Claims. (Cl. 30—94)

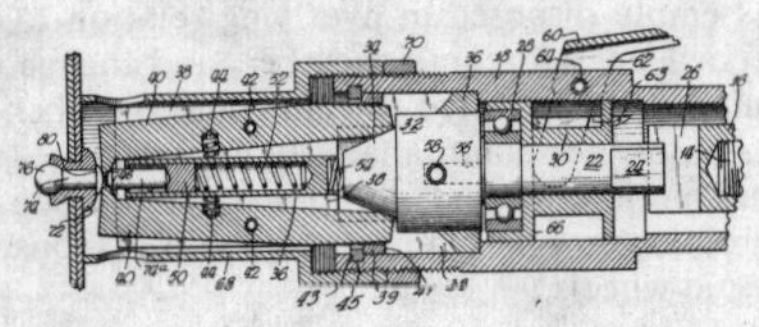

1. In a cutting tool of the character disclosed, a stationary body, a drive shaft rotatable and slidable therein, a jaw carrier rotatable therein and operatively connected with said drive shaft for rotation thereby, jaws pivotally mounted in said jaw carrier, a cone element carried by said drive shaft to spread said jaws at one end thereof when said drive shaft is slid relative to said stationary body, the other end of each jaw having a cutting edge to engage a rivet stem to be cut, power drive means for rotating said drive shaft and thereby said jaws, and manual actuator means to slide said drive shaft when rotating and thereby close said jaws, both of said means when in operation causing said cutting edges to cut the rivet stem and enlarge its outer end against the head of the rivet and burnish such outer end.

2,716,813
EXPLOSIVELY ACTUATED CUTTING TOOL
Lynn E. Smyres, Midland, Mich.
Application August 25, 1953, Serial No. 376,310
2 Claims. (Cl. 30—241)

2. A percussion hot line cutter comprising a housing, a handle connected to an end of said housing, a jaw connected to said housing, a plunger reciprocably arranged in said housing, a blade connected to said plunger and mounted for movement towards and away from said jaw, a pin mounted in said housing, a cylinder rotatably mounted on said pin and provided with a plurality of chambers therein for holding cartridges, there being a cut-out in said housing for the projection therethrough of a portion of said cylinder, said cylinder being provided with a plurality of recesses, a spring member having one end secured to said housing and engaging said recesses, a firing pin reciprocably mounted in said housing, a handle member for moving said firing pin away from the cartridges, resilient means for urging said firing

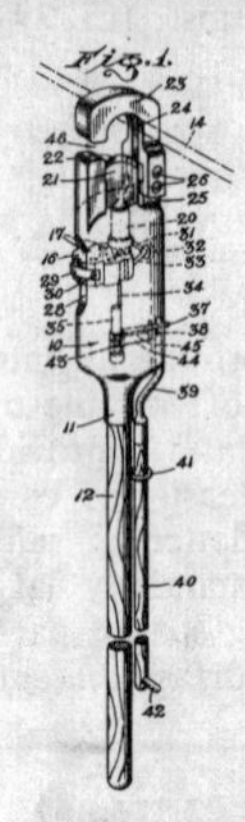

pin into engagement with the cartridges, said housing having passageways therein whereby a portion of the gas from the cartridge is directed onto the recesses in said cylinder to rotate the cylinder as the cartridges are fired.

2,716,814
COMBINED BUTTER DISH AND SLICER
George Lerner, Freeport, N. Y.
Application October 6, 1953, Serial No. 384,424
5 Claims. (Cl. 31—22)

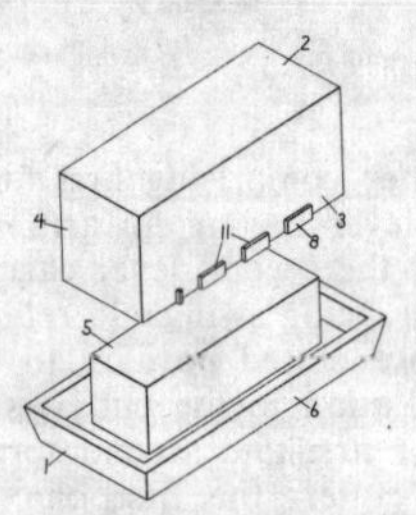

1. A combined butter dish and slicer comprising a butter dish having a cover having two opposite side walls with bottom surfaces provided with grooves opening toward the interior of said cover, and a slicer forming a zig-zag-like strip having portions fittingly inserted in said grooves, being slidable in the same and removable therefrom in downward direction, said slicer having other portions connecting said inserted portions and extending across the interior of said cover.

2,716,815
DENTAL ARTICULATOR AND METHOD
Wayne B. Ford, Burbank, Calif.
Application April 24, 1952, Serial No. 284,175
6 Claims. (Cl. 32—32)

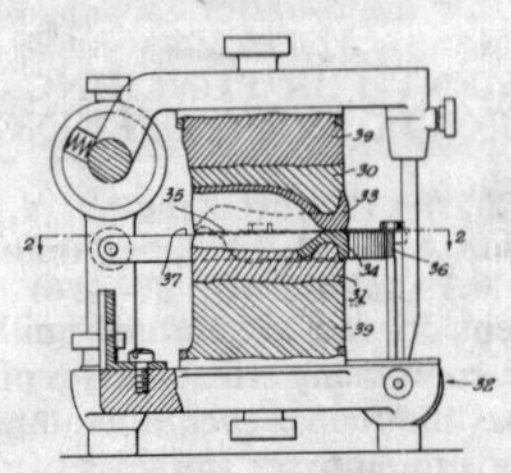

1. A method for locating and setting a mating pair of molar tooth blocks in proper relationship to the gum ridges of a set of plaster models while the latter are posi-

tioned in a dental articulator, said method consisting in establishing the plane of occlusion in the space between said models and positioning a plate on said plane, suspending from said plate said pair of tooth blocks while temporarily connected to and aligned with each other, shifting the connected tooth blocks along the top surface of the plate relative to the gum ridges of the upper and lower plaster model clamping said blocks in place, and filling in molten wax between said models and the tooth blocks of the mating pairs.

2,716,816
DENTAL MULLING RECEIVER FOR AMALGAMS
Blanchard K. Braum, Minneapolis, Minn.
Application August 27, 1954, Serial No. 452,489
3 Claims. (Cl. 32—40)

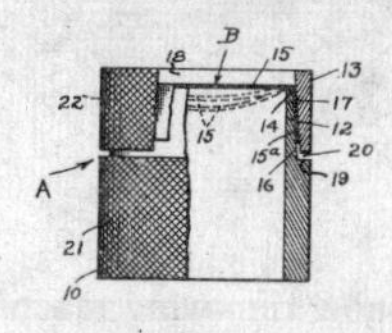

1. In a device of a character described, the combination of; a holder including a frame member having an annular rim thereon and a clamp-ring removably applied to said rim telescopically about the same, a flexible diaphragm removably resting on the lip of said rim with its body portion more or less cupped or bellied within the confines of said lip and with its marginal portion turned down about said rim between the outer surface thereof and the inner surface of the clamp ring, said clamp ring acting to bind such marginal portion of said diaphragm between it and said rim securely to hold said diaphragm in place on the rim.

2,716,817
ADJUSTABLE GARMENT PATTERN POSITIONER
Edna M. Franklin, Oklahoma City, Okla.
Application December 7, 1953, Serial No. 396,430
4 Claims. (Cl. 33—15)

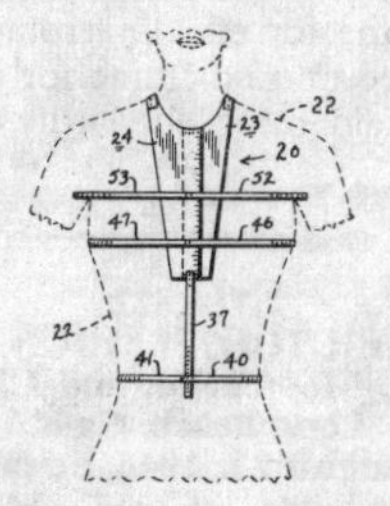

1. An assemblable clothes pattern positioner for gauging and recording specific human body measurements, comprising: a front and back pair of flat flexible elongated indicial neck-sizers, co-operatively and adjustably fastened adjacent their upper ends for encompassing a neck line; a pair of flat flexible longitudinal indicial strips each having one end adjustably fastened adjacent to and depending from the lower end of each said pair of neck-sizers, respectively; three sets of flat flexible longitudinal indicial strips, one set comprising two pairs fastened only at the center front and back to said neck-sizers, respectively, for describing the front and back arm to arm lines, respectively, a second set comprising two pairs fastened only at the center front and back to said neck-sizers, respectively, encompassing a body at the bust line, a third set comprising two pairs encompassing a body at the waist line and vertically and adjustably fastened only at the center front and back to said depending strips, respectively, whereby the neck-sizers and longitudinal strips may be adjusted to fit a particular person's body position measurements; a flat flexible adjustably fastened under-arm seam guide, said guide having oppositely disposed upwardly extending end portions and having arcuately curved inward edges describing an under-arm line; and means for marking the under-arm seam.

2,716,818
WIRE CUT-OFF CONTROL
Lester D. Fitler, Rome, N. Y., assignor to Rome Cable Corporation, Rome, N. Y., a corporation of New York
Application October 17, 1952, Serial No. 315,251
5 Claims. (Cl. 33—131)

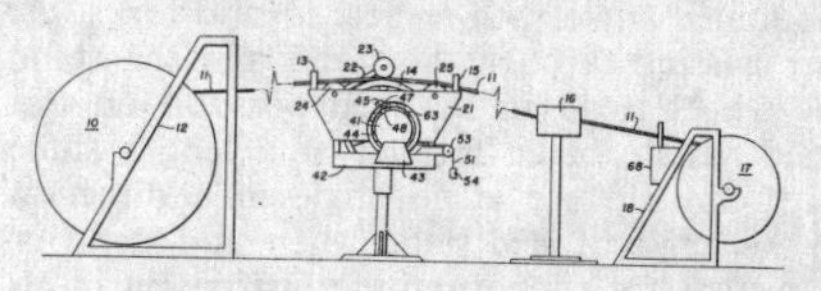

2. A dial plate, an arm movable relative to said dial plate, a first micro-switch, a second micro-switch, means including said dial plate and said arm for adjustably spacing said micro-switches, a contact arm movable between said switches, means actuated by a travelling cable, wire, or filamentary article to measure the length of said article passing said means and to move said contact arm from said second micro-switch to said first micro-switch in synchronism with said measuring means, cut-off means for severing said article and an electric circuit for operating said cut-off means actuated when said contact arm contacts and closes said first micro-switch, clutch means between said means actuated by said travelling article and said contact arm, a second electric circuit actuated by the closing of said first switch to shift said clutch, means operable upon the shifting of said clutch to return said contact arm from contact with said first micro-switch to contact with said second micro-switch and a third electric circuit actuated by the closing of said second micro-switch to open said second electric circuit, and means for returning said clutch means to its original position upon opening of said second circuit thus causing said contact arm again to be moved toward said first switch in synchronism with said means actuated by said travelling article.

2,716,819
MEASURING WHEEL
James D. Staples, Bell, Calif., assignor to Rolatape, Inc., Santa Monica, Calif., a corporation of California
Application November 26, 1952, Serial No. 322,788
6 Claims. (Cl. 33—141)

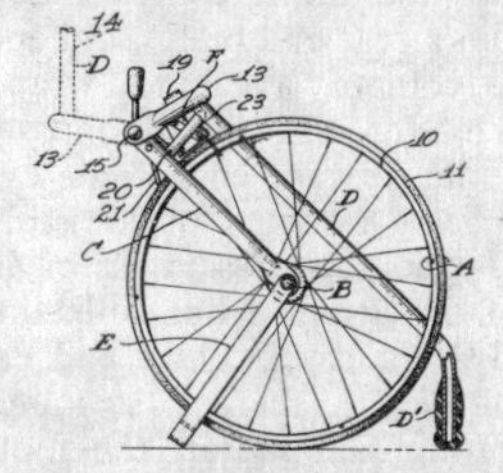

1. In a measuring wheel instrument, a wheel member, an axle on which the wheel member is rotatable, a pair of forks connected to the axle, a cross member joining the forks together at their upper ends, a handle movably joined to said cross member by a hinge connection, said handle providing means by which the wheel member may be rolled over a surface to be measured, means for locking said hinge connection upon movement of the handle into a pre-selected position relative to the cross member, a measurement recording counter supported by said cross member closely adjacent the rim of the wheel member with the recording dial thereof

facing up and visibly exposed, a brake mounted between the wheel forks and positioned with respect to the tread of the wheel member so that it will have a continuous bearing contact therewith, the brake lifting and the wheel member passing therethrough only during the forward rolling movement of said member, the wheel member having a series of projections evenly spaced around its rim, an actuating arm connected to said counter and extending into the path of travel of said projections and rocked in one direction by said projections upon contact therewith for actuating said counter during the forward rolling movement of the wheel member, a stand for supporting the instrument fixed to the wheel axle at a predetermined angle with respect to said forks, the wheel member passing through the stand and the stand clearing the wheel tread in its fixed position on the wheel axle said handle being formed of a length sufficient to extend to the surface at an angle in a direction opposite to the angle of said stand when said stand is angled to the surface for supporting the instrument in an at-rest position.

2,716,820
DRYING APPARATUS
Howard L. Bourner, Nashville, Tenn., assignor to Temco, Inc., Nashville, Tenn., a corporation of Tennessee
Application November 26, 1952, Serial No. 322,776
1 Claim. (Cl. 34—82)

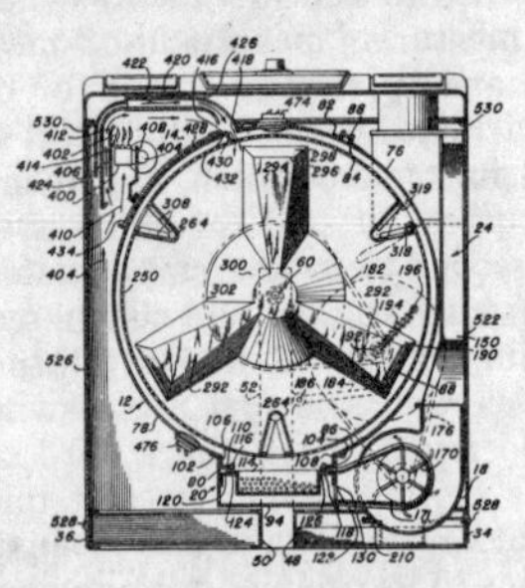

Apparatus of the class described comprising front, rear and side members making up a substantially rectangular base, two cross members running lengthwise of the base between the front and rear base members and spaced apart a predetermined relatively short distance, a column extending upwardly from between the facing sides of the cross channels adjacent the rear base member end, a laundry tumbling tub rotatably mounted on a shaft extending through a bearing mounted in said column, said shaft and bearing extending horizontally, a drum surrounding said tub, the drum comprising a cylindrical section mounted on a rear end plate, said end plate being supported on the column, a cabinet wrap-around supported from said base, a seal ring mounted on said cabinet wrap-around, the forward end of the cylindrical section of said drum being mounted on said seal ring, a drum base section at the bottom of said drum and extending the full length of the drum, a lint collecting drawer insertable into said drum base section and extending substantially the full length thereof, a motor, mounting means for supporting said motor, the motor mounting means being fixed to at least one of the members forming said base, an impeller mounted on a shaft extending from the motor, the motor being positioned parallel to the tub and immediately to one side of the rearward portion of drum base section, a housing surrounding the impeller, a duct leading from the forward end of the drum base section to the impeller housing, a shaft extending from the end of the motor opposite the end on which the impeller is mounted, a pulley on said opposite shaft end, idler speed reducing pulleys of differing diameters supported from said column, a pulley mounted on said tub shaft and fixed thereto at the rear of said column, a first belt running from the motor pulley to the larger of the idler pulleys, a second belt running from the smaller of the idler pulleys to the tub pulley, the arrangements being such that the motor may drive the impeller mounted on the shaft thereof and also the tumbling tub with a minimum of belting.

2,716,821
MOBILE SNOW REMOVING MACHINE
Harry Campbell Grant, Jr., Ridgewood, N. J.
Application March 28, 1951, Serial No. 218,035
7 Claims. (Cl. 37—43)

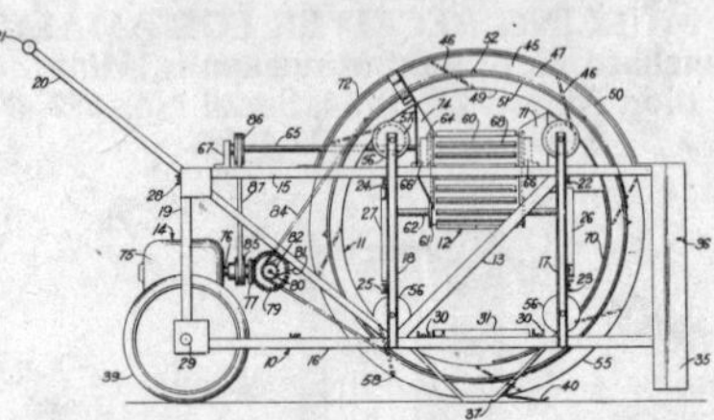

1. In a mobile snow removing machine, the combination of a frame, means on said frame for facilitating moving said frame across a surface from which snow is to be removed, a drum-like hollow rotor having an open end and a ring adjacent each end thereof, a plurality of elongate transversely extending blade-like elements having leading and trailing edges and being constructed and arranged for lifting snow and depositing the same into the interior of said rotor, said rings and blade-like elements having cooperating means for mounting said blade-like elements for swinging movement between said rings, said blade-like elements normally being inclined with respect to the radius of said rotor with the leading edges thereof forwardly disposed in the direction of rotation of said rotor, means adjacent the upper portion of said rotor for effecting radially inward swinging movement of said blade-like elements, bearing means on said frame for supporting said rotor rings for movement about a horizontal axis with the lower portion of said rotor adjacently above the surface from which snow is to be removed, power driven means on said frame for effecting rotation of said rotor, and means mounted on said frame for directing the snow deposited within said rotor outwardly through the open end of said rotor.

2,716,822
DIGGER TOOTH MOUNTING
Ernie L. Launder, Montebello, and Chester C. Hosmer, Long Beach, Calif.
Application August 21, 1946, Serial No. 691,978
5 Claims. (Cl. 37—142)

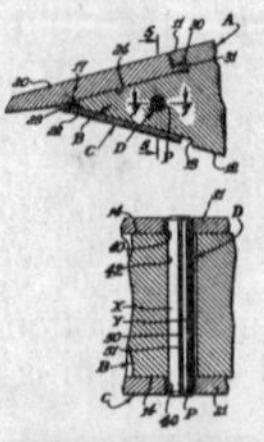

1. A structure of the character described including, a tooth having a transverse hole therethrough, a cap over the tooth having a housing with spaced sides overlying opposite sides of the tooth and provided with holes communicating with the hole in the tooth at the ends thereof, the hole in the tooth and the holes in the housing being intially out of line with each other, and a pin engaged in the passage and opening and including coextensive relatively movable sections and resilient means

between the sections maintaining the sections tight in the hole in the tooth and also tight in the holes in the housing whereby pressure is maintained in opopsite directions on the tooth and cap.

2,716,823
DOZER MOLDBOARD AUXILIARY DEVICE
Henry A. Mullin, Albuquerque, N. Mex.
Application June 30, 1950, Serial No. 171,482
10 Claims. (Cl. 37—145)
(Granted under Title 35, U. S. Code (1952), sec. 266)

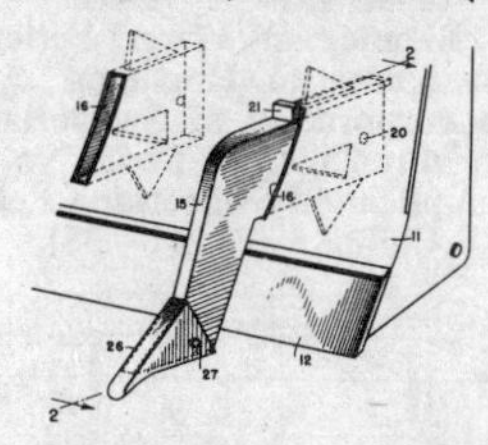

1. An auxiliary device for the moldboard of a dozer, a blade portion having an edge on the bottom of said moldboard, a mounting socket including an upper wall fixed to the rear surface of said moldboard above said blade portion and below the top edge of said moldboard, a bracket, means for securing said bracket in said socket on said moldboard comprising a horizontally disposed wedge between the upper edge of said bracket and said upper wall of said socket, a forwardly and downwardly extending portion on said bracket having a horizontal bottom surface for slidingly engaging the ground surface to be worked, an attachment device removably secured to said forwardly and downwardly extending portion and said bottom surface of said bracket, and an inclined abutment surface on the rear of said forwardly and downwardly extending portion located below said mounting means engaging the front of said blade portion of the moldboard of said dozer.

2,716,824
SLUSHING SCRAPER
Paul R. Francis, Los Angeles, Calif., assignor to Alloy Steel and Metals Company, Los Angeles, Calif., a corporation of California
Application September 8, 1950, Serial No. 183,792
8 Claims. (Cl. 37—147)

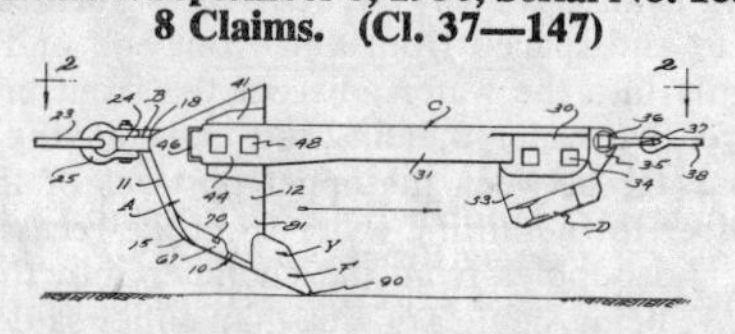

1. In a structure of the character described, a bucket-like body having, a bottom and sides projecting up from the ends of the bottom, blades on the body including corner blades each with a bottom section shielding a part of the bottom and a side section shielding a part of one side, and a retaining means on each blade including a tongue on the blade engaged in a recess in the bottom, and a lock pin holding the tongue in the recess, the pin having a deflected end and the bottom having an opening through which said end of the pin is accessible.

2,716,825
FLATIRON SUPPORT
Frederick W. Kulicke, Jr., Philadelphia County, Pa., assignor to Proctor Electric Company, Philadelphia, Pa., a corporation of Pennsylvania
Application January 15, 1952, Serial No. 266,556
17 Claims. (Cl. 38—79)

1. In a flatiron, a sole plate having an opening in the forward part thereof and also having a pair of laterally spaced openings in the rearward part thereof, a cover secured to said sole plate and forming therewith an enclosure, said cover also having an opening in the forward part thereof generally aligned with the forward opening of said sole plate, a first support means movably mounted within said enclosure including a front leg member associated with the forwardly disposed openings of the sole plate and cover and movable between an upwardly retracted position in which it projects through the cover opening and a downwardly projected position in which it projects through the forward opening of the sole plate and supports the forward part of the flatiron in elevated relation to an ironing surface, means within said enclosure operatively connected to said support

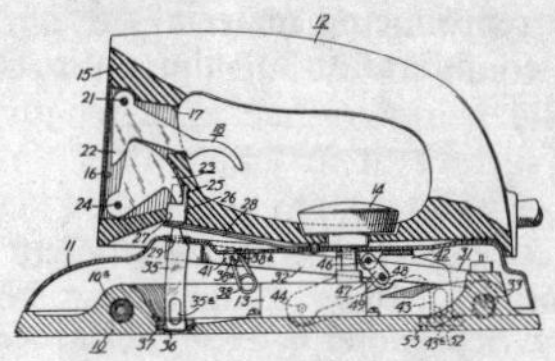

means to retain the same in either of its positions, manually-operable means engageable with the upper end of said leg member to move it from its retracted position toward its projected position, a second support means movably mounted within said enclosure including a pair of rear leg members associated with the rearwardly disposed openings of the sole plate and movable between an upwardly retracted position within said enclosure and a downwardly projected position in which they support the rearward part of the flatiron in elevated relation to the ironing surface, and means within said enclosure operatively interconnecting the two support means to effect actuation of said second support means by said first support means.

2,716,826
APPARATUS FOR REPRODUCING IMAGES
William C. Huebner, Mamaroneck, N. Y., assignor to The Huebner Company, Dayton, Ohio, a corporation of Ohio
Application October 24, 1951, Serial No. 252,966
8 Claims. (Cl. 41—1)

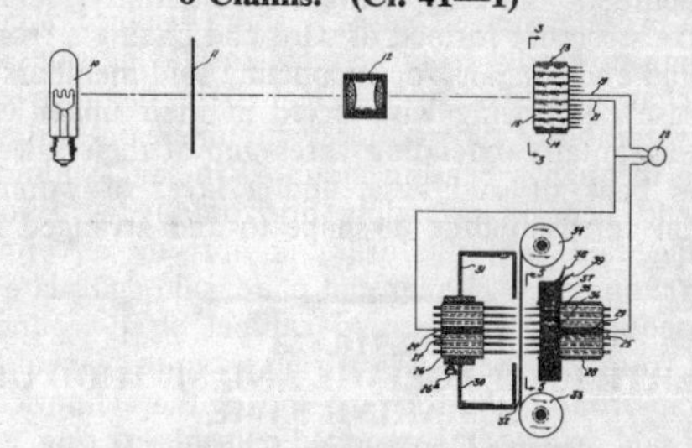

1. An apparatus for reproducing complete images directly and instantly on receiving material which comprises a plurality of closely grouped needle-like discharge elements mounted in a predetermined compact arrangement within a predetermined area, a plurality of similarly closely grouped needle-like attraction elements spaced from the discharge ends of said discharge elements with an atmospheric gap therebetween and with each attraction element correlated to a discharge element and in alignment therewith, an independent electrical circuit connected to each cooperating pair of discharge and attraction elements, means for energizing said circuits selectively to produce electrostatic fields of force between predetermined discharge elements and their correlated attraction elements and including a target having an area correlated to the area within which the elements are grouped and provided with light sensitive electron emit-

ting devices corresponding in number to said circuits and in arrangement to said discharge and attraction elements and dispersed throughout the target area but closely grouped relative to each other, means for positioning receiving material in said atmospheric gap, means for passing clouds of reproducing material through said gap intermediate said receiving material and said discharge elements, and means for projecting simultaneously onto said target area light rays in the configuration of the complete image to be reproduced to illuminate simultaneously certain of said light responsive devices to energize simultaneously certain of said circuits and cause the electrostatic fields of force between the pairs of discharge and attraction elements in said certain circuits instantly to deposit the reproducing material on the receiving material in the configuration of the entire and complete image.

2,716,827
FLORAL DISPLAY PIECE
Leona M. Mixter, Binghamton, N. Y.
Application September 6, 1950, Serial No. 183,440
2 Claims. (Cl. 41—12)

1. A floral piece comprising a plurality of laterally spaced individual members each formed of laterally spaced back wires and laterally spaced front wires defining a letter, a plurality of wires connecting said back wires to each other and to said front wires, the latter being free from connection to each other so that the space within the wires of each member is wholly open at the front of such member for the insertion therein of a penetrable material, horizontally extending wires connected to the back wires of each member to support said members in predetermined relation to spell a word, whereby floral picks are insertable in the penetrable material for the connection of flowers with respect to each member to spell said word in flowers, a horizontal base of substantial area underlying said members and comprising a substantially rectangular tray-like structure formed of wire and having a front, and back and ends, means for supporting said members above said base comprising wires fixed at their upper ends to said horizontally extending wires and at their lower ends to said back of said base, and a layer of water-proof material corresponding in shape to and arranged in said tray.

2,716,828
ARTIFICIAL FOLIAGE AND METHOD OF MAKING SAME
Joseph I. Adler, Jr., Chicago, Ill.
Application February 23, 1954, Serial No. 411,731
5 Claims. (Cl. 41—13)

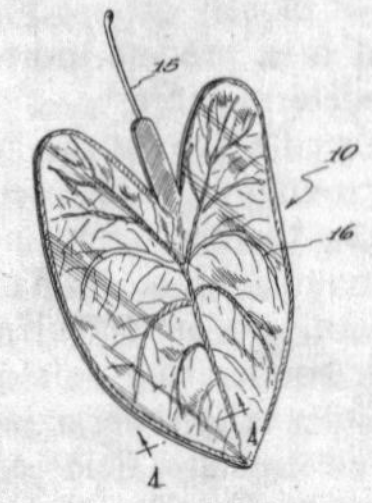

1. An artificial leaf comprising a flexible fabric sheet preformed to simulate the configuration of a natural leaf and having reproduced on each surface thereof a leaf design replicating color, veins and surface texture of a natural leaf, a stem-forming member secured to a surface of said sheet, said sheet having a coating of transparent, non-inflammable plastisol on said surfaces and encasing the stem member, said coating being thin and substantially uniform whereby light reflected from said artificial leaf appears to an observer to have originated from within said leaf in simulation of natural leaf effects.

2,716,829
TIME PIECE DIALS
Pierre Huguenin, Bienne, and Marcel Fetterle, Neuchatel, Switzerland, assignors to Huguenin & Cie, Bienne, Switzerland, a corporation of Switzerland
Application May 28, 1953, Serial No. 358,086
Claims priority, application Switzerland February 5, 1953
8 Claims. (Cl. 41—35)

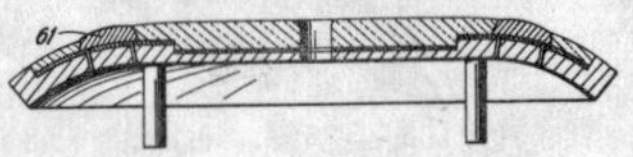

1. Time piece dial, having at least one recess extending at least in the region neighboring its periphery, said recess being provided with a filling of transparent material, characterized by the fact that a decoration is situated on the bottom of the recess and is visible through the material of said filling and that it comprises relief hour signs which are directly in contact with the said filling and which, seen from the front side of the dial, are surrounded by the filling.

2,716,830
FISH LURE
Martin L. Burden, Alexandria, Ind.
Application April 20, 1953, Serial No. 349,900
1 Claim. (Cl. 43—42.06)

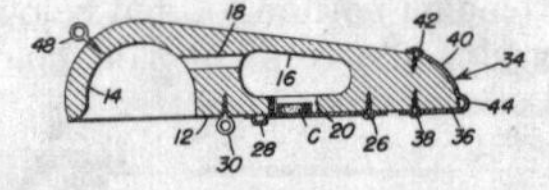

A fish lure adapted to float on the water comprising a buoyant elongated body having closed front and rear ends, said body having a chamber formed therein adapted to contain a chemical capable of generating a gas when mixed with water to form bubbles in said chamber, said body having an upwardly arched recess formed therein adjacent to and spaced from said front end and opening downwardly into the water adjacent the front end of the body to receive water therein, said body having a longitudinal passage between the upper portions of the recess and chamber for admitting water from the recess into the chamber and bubbles from the chamber into the recess, said passage being smaller in area than either said recess or chamber to retard the passage of bubbles from the chamber into the recess and said recess being larger in area than said chamber for collection and expansion of bubbles and gas therein until the gas pressure in said recess raises the front end of the body upwardly out of the water for escape of the bubbles and gas from said recess to produce a sound in the water.

2,716,831
CASTING PLUG
Arthur A. Glass, Freeport, N. Y.
Application August 28, 1953, Serial No. 377,059
2 Claims. (Cl. 43—42.31)

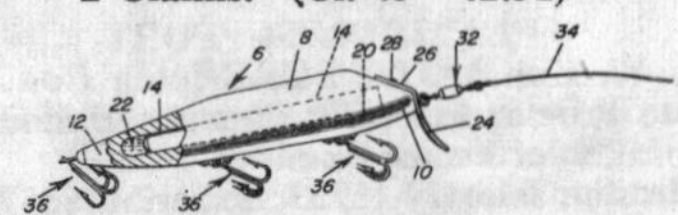

1. A casting plug comprising a hollow body having a chamber portion and having suitably designed head

and tail portions, fishing hooks pivotally mounted on said body, and movable weight means in said chamber portion of said body, said means comprising a rollable mass of mercury of predetermined and required heaviness occupying a position of limited confinement in the tail portion of said chamber at the start of the cast and staying there until the cast is completed and being automatically shiftable to and retaining a restricted position of limited confinement in the head end of said chamber when the plug is being pulled in and thus retrieved, said chamber portion having a cavity at the head portion of said body and said cavity providing a pocket into which the mercury is shifted, and confined during said retrieving step, and deflector plate affixed to the leading end of said head portion and depending to a plane below the plane of the ventral portion.

2,716,832
FISHING LINE SINKER
Raymond J. Minnie III, Vallejo, Calif.
Application April 24, 1953, Serial No. 351,063
7 Claims. (Cl. 43—43.12)
(Granted under Title 35, U. S. Code (1952), sec. 266)

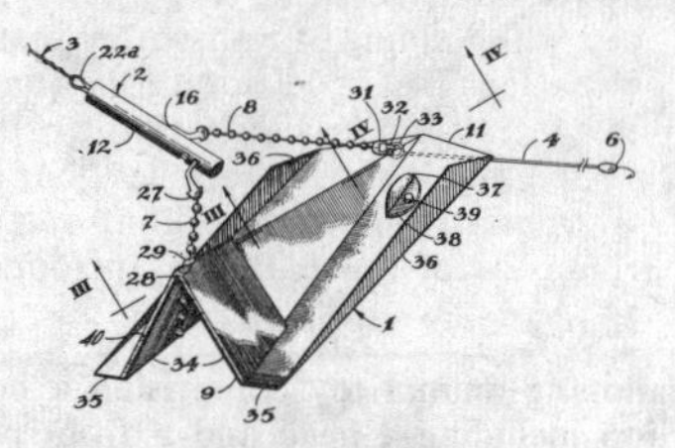

1. Depth-seeking apparatus comprising a vane having a depth-seeking leading end and a surface-seeking trailing end, an elongate casing, means releasably connected to said casing for suspending said leading end from said casing, and means swingably secured to said trailing end for tautly suspending it from said casing, said trailing end normally being carried in an elevated position with respect to said leading end whereby said vane assumes a depth-seeking inclination, and said leading end being adapted upon release by said suspension means to swing to a position rearwardly of said trailing end, whereupon said vane assumes a surface-seeking inclination.

2,716,833
FISHING GEAR
Aldon M. Peterson, Wayzata, Minn.
Application January 25, 1954, Serial No. 405,698
4 Claims. (Cl. 43—44.88)

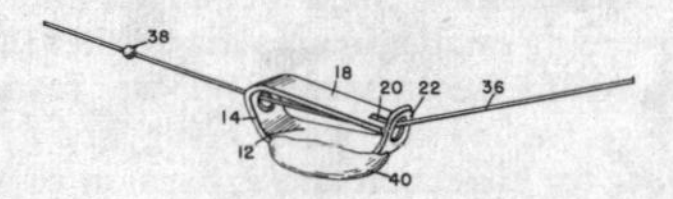

3. A fishing sinker comprising a one piece metal frame of quadrilateral shape, upright end portions thereof being provided with openings adapted to slideably engage a fishing line, and a weight secured to one side of the frame and the other side having a free end that terminates adjacent one of the upright end portions, the free end of the side being provided with a substantially rectangular notch.

2,716,834
BAIT BUCKET
Arthur L. De Bonville, Fitchburg, and Joseph S. Swercewski, Gardner, Mass.
Application May 26, 1953, Serial No. 357,512
2 Claims. (Cl. 43—55)

1. A bait bucket comprising an outer shell including a bottom and a side wall extending upwardly therefrom, said shell being formed of a material of good insulative qualities, the side wall of said shell having a coating thereon throughout its area, over its outer surface, of a heat-reflective paint, and over its inner surface with a moisture-impervious material; an aluminum foil side wall

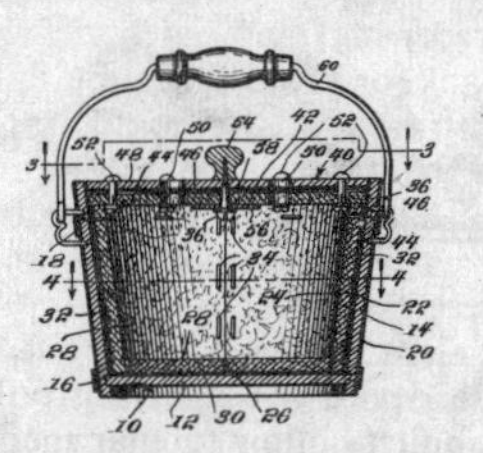

liner overlying the moisture-impervious coating; a bottom liner of aluminum foil separate from said first-named liner overlying the inner surface of the bottom in peripheral contact with one end of the first-named liner; and an inner shell overlying said liners and formed of a material highly pervious to moisture.

2,716,835
ANIMAL TRAP
Clarence J. Siegel, Grand Rapids, Mich.
Application September 17, 1953, Serial No. 380,663
2 Claims. (Cl. 43—95)

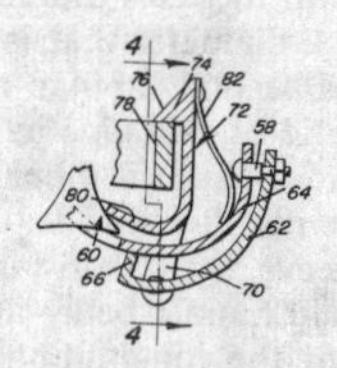

1. In an animal trap having a base provided with upstanding ears at its opposite ends, a pair of jaw members pivotally received between said ears, an arm projecting laterally from the approximate mid-portion of the base and having an upturned free end portion, said jaws being of generally U-shaped configuration with a portion of one of the jaws lying adjacent the free end portion of said laterally projecting arm when the trap jaws are disposed in set position, a spring member associated with said jaws for urging the same to a closed, animal trapping position, and trip mechanism including a pan positioned over the central portion of said base and a trip arm secured to said pan and pivotally secured adjacent the free end of said laterally projecting arm, said trip arm having a cam finger, an improved jaw latch and release mechanism comprising a mounting plate rigidly affixed to the under surface of said free end portion of the laterally projecting arm, said mounting plate having a projection on each of its opposite side edges, a generally J-shaped member comprising an arcuate lower end portion having a pair of depending ears straddling said laterally projecting arm and pivotally mounted on said projections, the upper end of said J-shaped member projecting above the said one of the jaws when the jaws are in set position and being provided with a horizontal latching nose, spring means secured to said J-shaped member and abutting said upturned free end portion of said laterally projecting arm for urging said J-shaped member in such a direction that its nose will overlie said one jaw and retain the same in set position, the arcuate lower end portion of said J-shaped member being in the path of movement of the said cam finger whereby depression of the pan will rotate said cam finger thereby rotating said J-shaped member in opposition to the spring means and release said one jaw.

2,716,836
APPARATUS FOR DISTRIBUTING INSECTICIDES
Fritz Winkelsträter, Wuppertal-Wichlinghausen, Germany, assignor to M. A. Bitzer, Elizabethton, Tenn.
Application August 19, 1950, Serial No. 180,468
1 Claim. (Cl. 43—146)

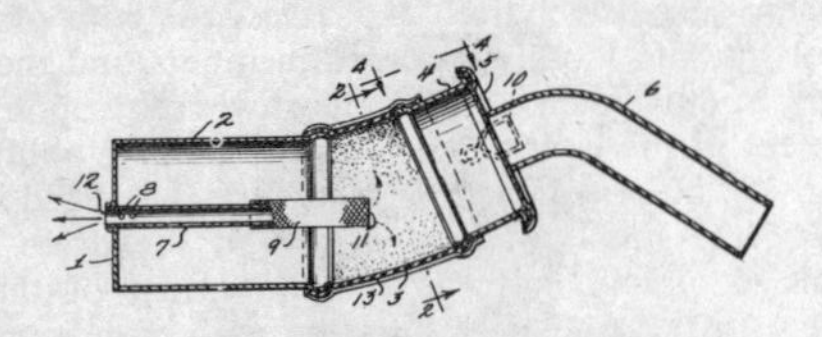

In a portable device for applying pulverulent insecticides to plants, a first hollow tubular member of rigid material open at one end and closed at the other end, a second hollow tubular member of rigid material open at one end and having a bottom at the other end, a cylindrical sleeve of flexible elastic material attached at one end to the open end of the said first mentioned hollow tubular member, said sleeve being attached at its other end to the open end of said second hollow tubular member, said sleeve being of such strength that it is self-sustaining and normally supports said members in spaced relation, said second member being adapted to serve as a container for the pulverulent material, an upwardly inclined handle rigidly secured to said first hollow tubular member, the ends of said sleeve being askew to each other so that said sleeve holds said first member and said second member in spaced relation with the axes thereof out of alignment, a small pipe having a diameter that is small compared to the diameter of said second tubular member, means for fastening one open end of said pipe in an aperture formed in the bottom of said second member so that said pipe is in communication with the outside atmosphere, one of the sides of said sleeve being substantially shorter than the opposite side, said handle being attached to said first hollow member with the longitudinal axis of the handgrip portion of said handle tilted with respect to the longitudinal axis of said first hollow member and extending upwardly toward the longer side of said sleeve, the axis of said handle lying in a plane that passes through both the short side and the long side of said sleeve so that said short side of said sleeve acts as a hinge between said first and said second members while the long side folds upon itself when said second member is flexed with respect to said first member, said pipe having small apertures therethrough near the bottom of said second member through which the pulverulent material is forced out of the device when said first and second members are flexed with respect to each other, said pipe extending upwardly through the pulverulent material in said second member and into said sleeve so that air is drawn into the device through said pipe when the device is unflexed and said self sustaining sleeve assumes its normal self sustained configuration.

2,716,837
TOY BLOW PIPE AND BALL
Carl B. King, Lakin, Kans.
Application February 5, 1953, Serial No. 335,260
5 Claims. (Cl. 46—44)

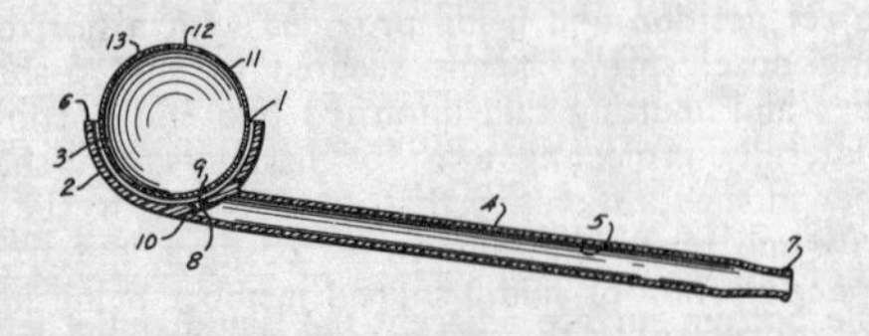

1. A pneumatic toy comprising, a cup having a recess therein, said cup having spaced air passages communicating with the recess adjacent the bottom thereof and spaced from the bottom center, said passages being inclined upwardly toward a point between the bottom center and the upper edge of the opposite side of the cup, and a hollow lightweight thin shell ball in the recess, said ball being spherical and having a plurality of spaced apertures whereby air issuing from the passages will impinge on the lower portion of the ball and cause said ball to rotate and rise to float within the recess and some of the air will enter the apertures in the lower portion of the ball and escape through apertures in the top portion increasing the air pressure at said top portion to stabilize the ball.

2,716,838
GALLOPING ANIMAL TOY
Alessandro Quercetti, Turin, Italy
Application October 23, 1951, Serial No. 252,592
Claims priority, application Italy October 26, 1950
2 Claims. (Cl. 46—105)

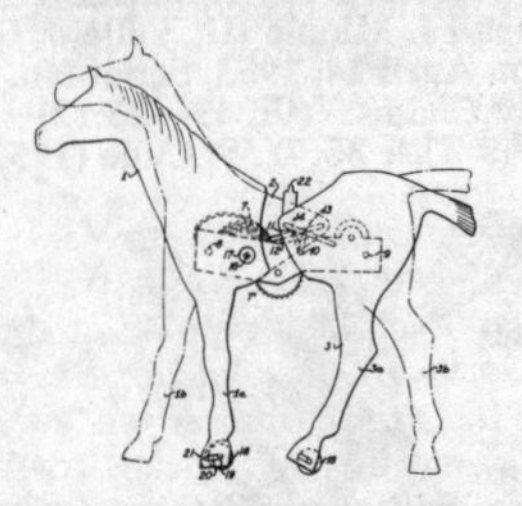

1. A galloping animal toy comprising a hollow front body portion including a head and a front pair of legs, a hollow rear body portion including a rear pair of legs, a motor having a frame substantially wholly enclosed by said front and rear portions, said two body portions being pivoted to said frame at two longitudinally spaced pivots, a motor shaft intermediate said pivots, said shaft having crank ends extending from said frame, a groove in the front body portion opposite one crank, a groove in the rear body portion opposite the second crank, said cranks engaging the respective grooves, a ground engaging roller at the lower end of each leg, and a one-way detent means on each leg for locking the roller upon a backward swing of the leg.

2,716,839
MOVABLE FEATURE TOY FIGURE
George G. Young, Des Moines, Iowa, assignor to Marvin I. Glass, Chicago, Ill.
Application September 25, 1952, Serial No. 311,476
4 Claims. (Cl. 46—135)

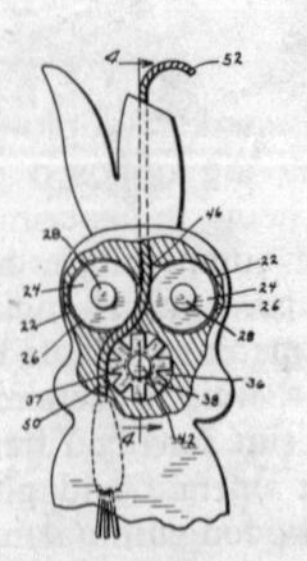

1. In a toy figure comprising a body and head, said head having a face with eye openings and a mouth, a pair of rotatable discs having eyes thereon and a separate rotatable teeth member positioned adjacent said eye openings and said mouth respectively, a flexible cord passing through said head and extending above and below said head and simultaneously engaging the rotatable discs and teeth member and adapted when pulled upwardly or downwardly to rotate same.

2,716,840
MECHANICAL BOXER TOY
Raymond E. Armstrong, Toronto, Ontario, Canada
Application September 30, 1953, Serial No. 383,334
18 Claims. (Cl. 46—142)

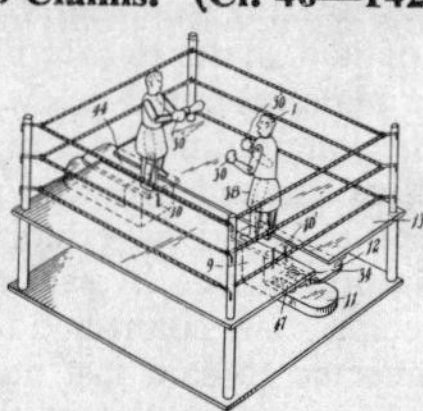

1. An animated toy figure arrangement of the kind described comprising a head member, a torsal frame, flexurably jointed left and right arm members, hinged leg and thigh members, a support beam, and a base therefor, and a mounting platform supplied with a slot through which said support beam loosely projects in upward direction with said base located below the platform, means for hinging the head member for forward and backward motion thereof with respect to said torsal frame, means for hinging the said torsal frame for forward and backward movement thereof with respect to said support beam, means for latching the said torsal frame rigidly to the support beam at a specified attitude of the frame, means sensitive to a backward movement of said head member for unlatching the torsal frame from the said rigid relationship, a solar plexus plate, means for resiliently associating said plate with the said torsal frame, means sensitive to an inward excursion of said solar plexus plate for unlatching the torsal frame from its said rigid relationship with said support beam, resilient means for urging said base toward said platform in such a manner that the base and support beam can be tilted selectively over a predetermined range of angles about a vertical polar axis and can also be lowered along the said axis to a predetermined limit, manually operable means for actuating said flexurably jointed arm members, and manually operable means for tilting and lowering said base.

2,716,841
TOY NOISEMAKER
John A. Loftin, Louisville, Ky.
Application August 14, 1953, Serial No. 374,191
4 Claims. (Cl. 46—191)

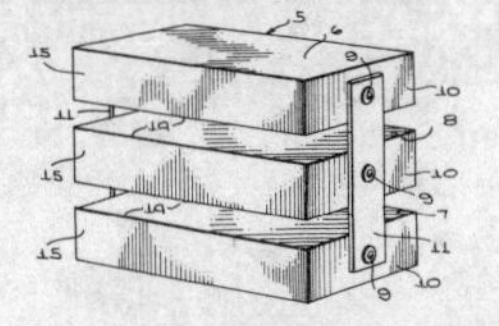

2. In a toy noisemaker, a plurality of generally rectangular blocks having opposed ends and longitudinal edges, said blocks being in superimposed relation to each other, adjacent ones of the blocks being spaced from each other at distances less than the width of the blocks, end bars extending across the opposed ends of the blocks and across the spaces between adjacent blocks, and means pivoting the blocks to said end bars to swing relative to each other on axes extending longitudinally of the blocks and spaced from the longitudinal edges of the blocks.

2,716,842
BRAKE FOR TOY MOTOR VEHICLES
Willy Kellermann, Nurnberg, Germany
Application December 10, 1951, Serial No. 260,833
Claims priority, application Germany June 15, 1951
2 Claims. (Cl. 46—208)

1. A brake for a toy motor vehicle of the type including a clockwork mechanism drive means comprising a rotatable support operably associated with the drive means and rotatably responsive thereto, a plurality of centrifugally influenced members loosely mounted on the support and movable radially outwards of the support under the action of centrifugal force when the support rotates, a movable control member, means mounting the member for movement successively into and across the path of movement of the said loosely mounted members and then into the path of movement of the support whereby upon initial movement of the control member the same is abutted by the loosely mounted members so that the speed of rotation of the support and thus that of the drive means is retarded and upon continued movement of the control member the same passes into the path of movement of the support to stop the rotation of the same and thus to

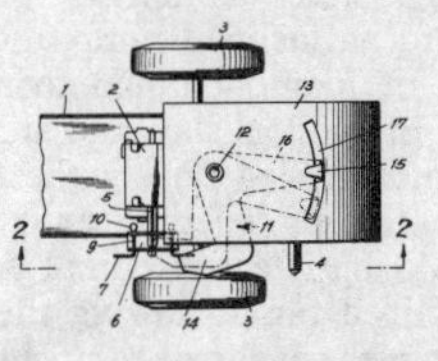

stop the drive means, the drive means to be controlled including a shaft, said rotatable support consisting of a sheet metal member of U-shaped cross-section including spaced flanges, said flanges having aligned apertures at opposite ends of the member and at the mid portion thereof, the apertures at the mid portion mounting said flanges and thus the member on said shaft, said centrifugally influenced members constituting a pair of members each comprising a wire including two limbs bent at right angles to one another, one limb of each wire being loosely mounted in the aligned apertures at the respective ends of the flanges of the support, means preventing disassociation of the said limbs from said support and the other limb of each member being movable radially outward of the support under the influence of centrifugal force.

2,716,843
SLAG FOAMING
John C. K. Stuart, Toronto, Ontario, Canada, assignor, by mesne assignments, to Denman Enterprises, Limited, Hamilton, Ontario, Canada, a corporation of Canada
Application February 28, 1951, Serial No. 213,201
5 Claims. (Cl. 49—1)

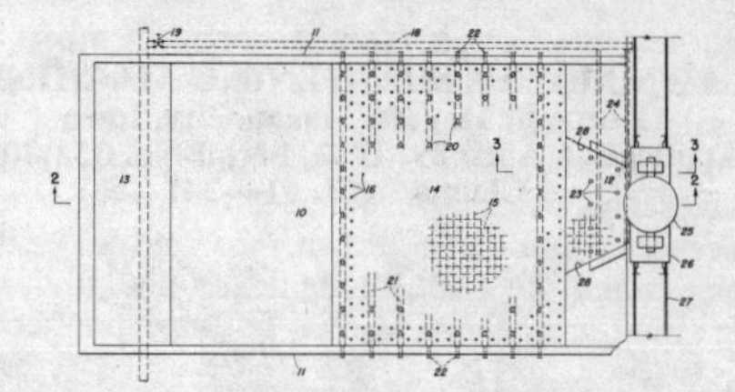

1. Apparatus for use in carrying out an operation of the type in which molten slag is contacted with a liquid capable of vaporizing under the influence of heat contained in the slag and in which vapor produced as a result of contact of the slag and liquid causes swelling of the slag with the production of porous plastic mass of bank slag of substantially greater volume than the volume of the molten slag, which comprises a foaming pit having a solid bottom surface of substantially moisture impervious material and provided throughout a portion of its length with a plurality of liquid outlet jets extending to the surface of the bottom at spaced intervals thereof, an inclined end surface of solid material joined to the bottom surface adjacent the liquid outlet jets and provided with a plurality of similar liquid outlet jets extending to the surface at spaced intervals thereof, means for supplying a controlled amount of liquid under

pressure to said liquid outlet jets, a plurality of porous drain channels formed integral with the solid bottom surface and positioned at spaced intervals adjacent the liquid outlet jets, and means for introducing a mass of molten slag onto the inclined end surface of said pit from an initial pouring head sufficient to spread the slag in a unidirectional flow across the portion of the bottom surface containing the liquid outlet jets and substantially across the remainder of the bottom surface of the pit.

5. The method of treating molten slag by contact with a liquid capable of vaporizing under the influence of heat contained in the slag to cause swelling of the slag with the production of a porous mass of foamed slag having a substantially greater volume than the volume of the molten slag, which comprises pouring a mass of molten slag onto an inclined ramp and providing a continuous supply of foaming liquid from beneath the descending mass of molten slag and directed under pressure into the molten slag during the entire pouring operation from outlet jets opening onto the surface of the inclined ramp to initiate substantial foaming within the slag on the inclined ramp, flowing the foaming mass of molten slag from the inclined ramp by means of gravity flow into a pit having a substantially moisture-impervious bottom surface, initially introducing foaming liquid directly into the pit through outlet jets provided at the surface of the pit bottom to provide a relatively shallow layer of foaming liquid over the entire bottom surface of the pit prior to the introduction of the molten slag therein, progressively flowing the slag across the bottom of the pit under its own momentum in a substantially uni-directional flow preceded by a thin wall of foaming liquid while introducing additional controlled amounts of foaming liquid directly into the slag from the outlet jets provided at the surface of the pit bottom, and relatively gently arresting the motion of the slag flow to permit complete foaming thereof through the action of foaming liquid entrapped beneath and within the mass of slag, arrestment of the motion of the moving mass of slag being effected by permitting the forward portion of the moving mass to mount a relatively gently inclined slope provided at the output end of the pit with at least a portion of the slag and foaming liquid overflowing the slope and foaming on the ground adjacent the pit, thereby avoiding backwash of foaming liquid and slag within the pit.

2,716,844
LAWN MOWER SHARPENING MACHINE
George F. Meyers, Mason City, Iowa
Application April 29, 1953, Serial No. 351,807
4 Claims. (Cl. 51—34)

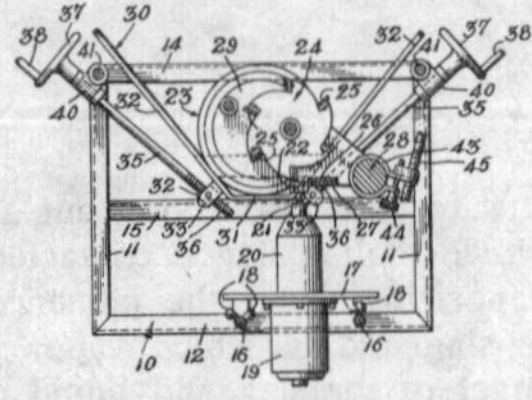

1. A lawn mower sharpener comprising a pair of spaced parallel rectangular frame members each including end pieces and a top and bottom piece, a pair of spaced parallel horizontally disposed cylindrical rods extending between the bottom pieces of said frame members and secured thereto, a pair of spaced parallel horizontally disposed bars extending between the tops of said frame members and secured thereto, a pair of adjustable brackets positioned between said frame members for supporting the wheels of a lawn mower to be sharpened, each of said brackets including a horizontal crosspiece and a pair of arms extending upwardly and outwardly from said crosspiece, a clamp secured to the outer surface of each of said arms adjacent the lower end thereof, a block mounted in each of said clamps and provided with a threaded bore, a shaft having a threaded portion engaging each of said bores, a handle for rotating said shafts, a bushing slidably connecting each of said shafts to said bars, a horizontally disposed movable plate mounted below said brackets and movable along the longitudinal axis of the lawn mower, a plurality of rollers connected to said plate for engaging said rods, an electrically driven grinding wheel mounted on said plate, a motor carried by said plate for actuating said grinding wheel, a shaft connected to one of said bars, and a clamp connected to said last named shaft for engaging the ground-engaging roller of the lawn mower.

2,716,845
LAWN MOWER SHARPENING MACHINE
Richmond Viall, Jr., Providence, Russell I. Peterson, Jr., East Providence, and Roger H. Johnson, Davisville, R. I., assignors to The Graham Manufacturing Co., Inc., East Greenwich, R. I., a corporation of Rhode Island
Application July 17, 1951, Serial No. 237,132
6 Claims. (Cl. 51—48)

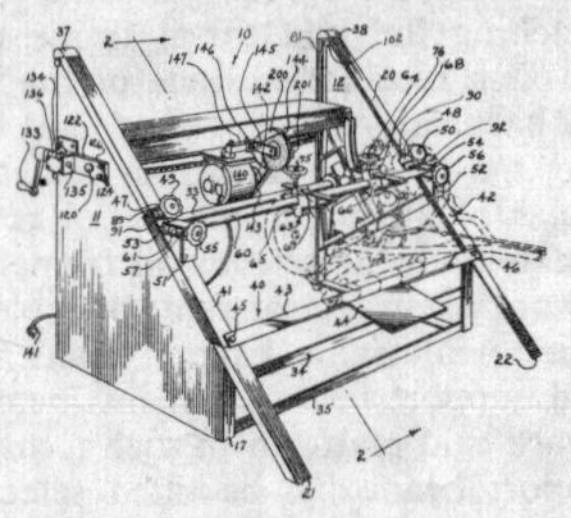

3. A lawn mower sharpening machine having a common frame support which is provided with surfaces inclined to the horizontal, a lawn mower holding frame slidably mounted on said surfaces and provided with clamping means to secure said holding frame to said surfaces and an abrasive wheel having a mounting supported in said common frame for parallel movement before said lawn mower holding frame.

2,716,846
ROTARY KNIFE SHARPENER
Roy J. Gerth, San Fernando, Calif.
Application June 2, 1953, Serial No. 359,109
5 Claims. (Cl. 51—128)

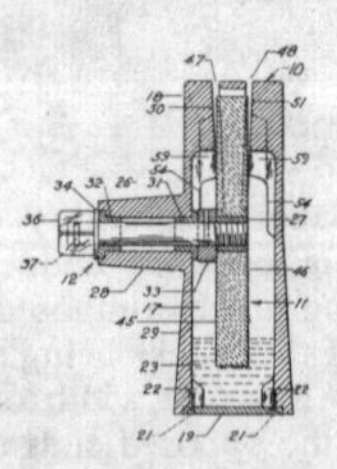

1. In a knife sharpener, a casing having a generally semicylindrical upper wall, a horizontal shaft journaled in said casing substantially axially of said upper wall, a grinding wheel mounted in said casing and axially on said shaft for rotation thereby, said wheel having outer and inner side faces the peripheries of which are inwardly adjacent said upper wall and lie in parallel planes perpendicular to said shaft, outer and inner slots formed in the major portion of said upper wall and lying in substantially vertical planes transverse to said shaft, said vertical planes converging in such a way that adjacent slot ends are respectively more widely spaced and more closely spaced than the spacing between said parallel planes so

that said slots each cross over said peripheries of said side faces, and outer and inner grinding tables respectively provided within said casing and adjacent said slots to guide a knife blade into grinding relationship with said peripheries of said side faces.

2,716,847
HONE FOR SHARPENING SHEARS
James V. Amendola, Kenosha, Wis.
Application April 14, 1953, Serial No. 348,613
2 Claims. (Cl. 51—187)

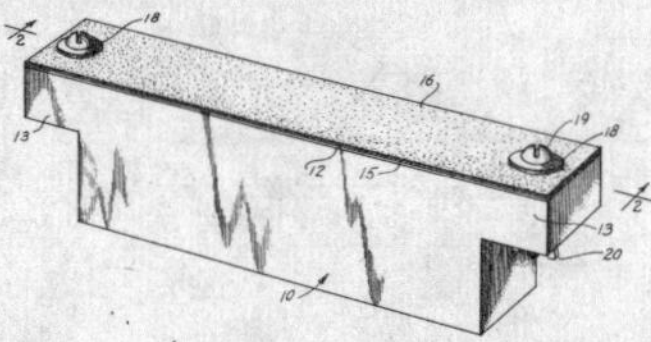

1. A scissor hone comprising an elongated base member of substantial height and having a planiform upper face, a resilient pad on said face, an emery sheet on said resilient pad, fastening means for securing respective ends of said pad and emery sheet to said base member, and each of said fastening means including raised portions projecting above said emery sheet functioning as stops for limiting movement of the blade of a pair of shears during honing.

2,716,848
POLISHING DEVICE
Alfred Schmidt, Nurnberg, Germany
Application July 6, 1954, Serial No. 441,291
15 Claims. (Cl. 51—187)

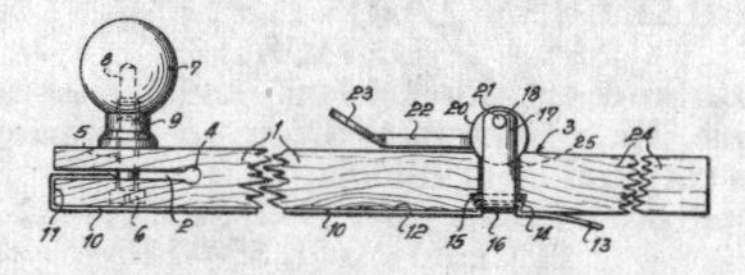

1. A polishing device having in combination a long stick with an upper and lower flat side, a slot on one end of said stick being adapted to receive one or the other end of a strip of thin, flexible material, means for pressing the bodies forming the sides of said slot together, a handle on the slotted end and on said upper flat side fitting in the palm of the human hand, a free part on the other end of said stick, a groove on said lower flat side behind said free end and perpendicular to the longitudinal axis of said stick, and clamping means holding said strip into said groove.

2,716,849
GRINDING AND CLEANING DEVICE
Albert Blumstein, Bronx, N. Y., assignor to Cavitron Corporation, a corporation of New York
Application January 27, 1955, Serial No. 484,343
6 Claims. (Cl. 51—187)

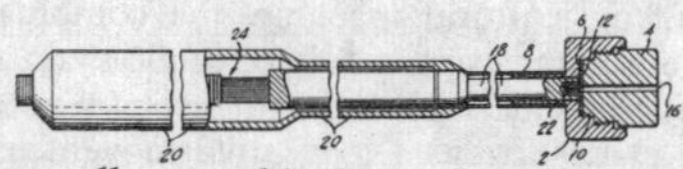

3. A manually operable device for cleaning and maintaining the end surface of an ultrasonic dental tool holder perpendicular to the axis thereof comprising a cylindrical cup member having an axially disposed tubular extension of an internal diameter sufficient to accommodate the cylindrical end of a dental tool holder and a stud member having a transverse plane surface at one end thereof, said stud member being screw threadable into said cup member to bring said transverse plane surface adjacent the inner end of said tubular extension and to hold therebetween an abrasive disc for grinding engagement with the end surface of a tool holder when inserted into said tubular extension.

2,716,850
KIT FOR SHARPENING HYPODERMIC NEEDLES
Charles Reitzes, Atlantic City, N. J.
Application November 10, 1953, Serial No. 391,193
1 Claim. (Cl. 51—211)

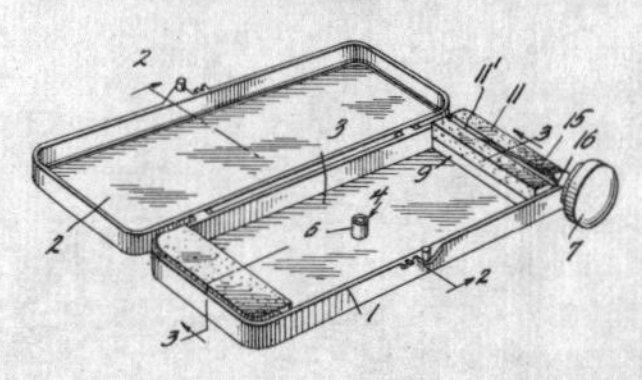

In a sharpening kit for hypodermic needles, a lidded box, a bridge securely fastened in the bottom of the box, said bridge extending throughout the length of the box and extending upwardly and outwardly near its opposite ends to form supporting shelves, an abrasive stone cemented on one of said shelves with its upper surface extending in a plane just above the plane of the upper edge of its adjacent side wall, said abrasive stone having a V-shaped groove extending throughout its greater length, a layer of jeweler's rouge bonded to the opposite self with its upper surface also lying in a plane just slightly above the upper edge of its adjacent side wall, said bridge having a socket in the bottom thereof adapted to support a chuck for conveniently holding a hypodermic needle while the same is being sharpened or polished on the abrasive stone or jeweler's rouge.

2,716,851
CAM-CONTROLLED WORKHOLDER
Carl A. Ulfves, Worcester, Mass.
Application July 23, 1953, Serial No. 369,863
4 Claims. (Cl. 51—232)

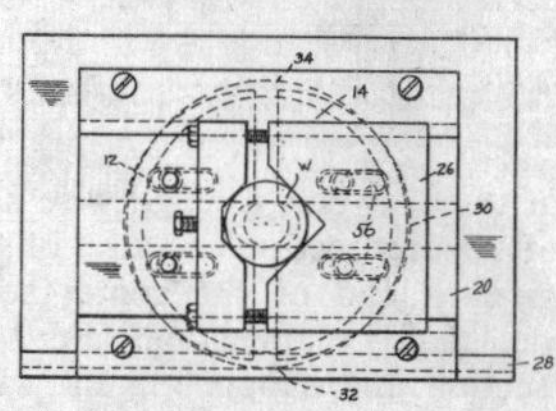

2. The device of the class described comprising a table-like base, an upstanding fence thereon, a semi-circular cam on the base in free sliding engagement therewith for engagement with the fence at its edge and a second cam superposed relative to the first cam and having edges thereof extending beyond certain portions of the edges of the first-named cam for engagement of the extending edges with the fence, and a workholder on the cams free of the fence.

2,716,852
MACHINE FOR CLOSING AND SEALING FILLED BAGS
Harold V. Kindseth, Minneapolis, Minn., assignor to Bemis Bro. Bag Co., Minneapolis, Minn., a corporation of Missouri
Application February 9, 1953, Serial No. 335,818
41 Claims. (Cl. 53—148)

1. In a machine for closing and sealing filled bags, a frame structure, means carried by the frame structure and including a bag supporting element for receiving and transporting a filled bag, power operated mechanism for imparting intermittent movement to the bag supporting ele-

ment to deliver the filled bag successively to a plurality of spaced stations, power operated mechanism at one of said stations and comprising opposed intucking members and cooperating creasing members for forming a top closure for the bag, mechanism located at a different one of said stations for applying adhesive to the intucked and creased top portion of the bag, mechanism at another station for moving the filled bag out of engagement with the bag supporting element, bag closing means engageable with the intucked and creased closure forming portion of the bag to fold the same into closed relationship, and means for simultaneously adjusting the height of the bag top folding and adhesive applying mechanisms and bag closing means relative to the bag supporting element, whereby said mechanisms and bag closing means cooperate to close and seal bags of various heights.

2,716,853
HARROW DRAWBAR
John A. Schulte, Sac City, Iowa
Application January 21, 1953, Serial No. 332,260
4 Claims. (Cl. 55—84)

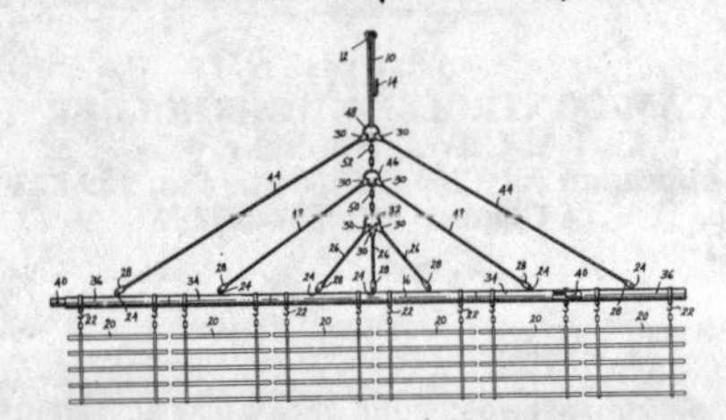

1. In a harrow drawbar, a lengthwise adjustable drawbar formed of a center section and a set of stub sections extending from each end of the center section, coupling means for detachably connecting said stub sections to said center section and supporting the same in endwise alignment, a clevis hitch, a hook formed on said clevis hitch, a first coupling ring engageable with said clevis hitch, a flexible pull member connecting each stub section to said first coupling ring, a second coupling ring, flexible pull members connecting said center section to said second coupling ring, a flexible linkage connecting said rings, said stub section detachable from said center section and arrangeable in tandem relationship thereto, and said second coupling ring engageable with said hook when said tandem alignment is formed.

2,716,854
DRAFT HITCH
Maurice E. Scheibner, Coulee City, Wash.
Application September 8, 1952, Serial No. 308,428
7 Claims. (Cl. 55—89)

1. A draft device for attaching a towing vehicle to a trio of tillage implements such as disk harrows in spread position where the implements traverse parallel paths and are staggered with the middle implement ahead of the outside implements, and in tandem position for transportation, said device comprising a T-shaped tow beam having means at the free end of the stem for attaching it to the towing vehicle, the cross piece of the tow beam having pivots at its ends, implement attaching arms attached to said pivots, said arms extending forwardly and rearwardly from the cross piece and diverging rearwardly from the pivots, wheel frames fixed to the rear ends of said arms, a combination cross brace and tow bar detachably connected to the wheel frames to maintain the arms in their outermost spread position and the tow beam having a tow bar attaching lug and said cross brace, when

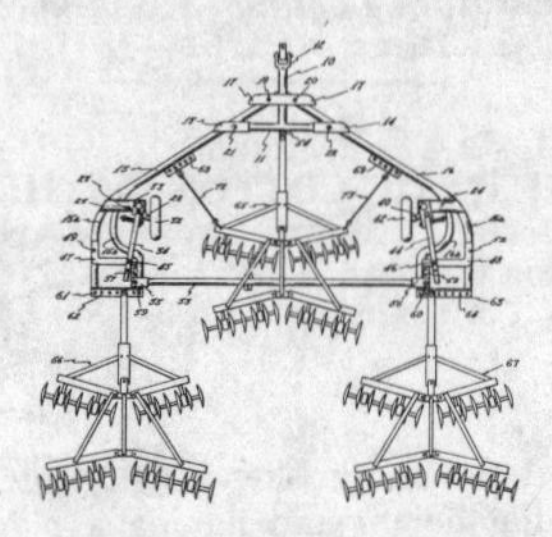

detached from the wheel frames, being attachable to the tow beam for trailing the implements, wheels supporting the wheel frames, the portions of arms extending forwardly from the cross piece and the stem of the tow beam having securing members thereon for securing the said portions of the arms to the stem with the rear portions of the arms in outermost spread position, and implement attaching means on the rear of said wheel frames, operable also to interconnect the wheel frames when the arms are not spread.

2,716,855
PLURAL CONNECTION IMPLEMENT HITCH DEVICE
Knud B. Sorensen, Rock Island, and Walter R. Peirson, Moline, Ill., assignors to Deere & Company, Moline, Ill., a corporation of Illinois
Application January 24, 1952, Serial No. 268,088
11 Claims. (Cl. 55—126)

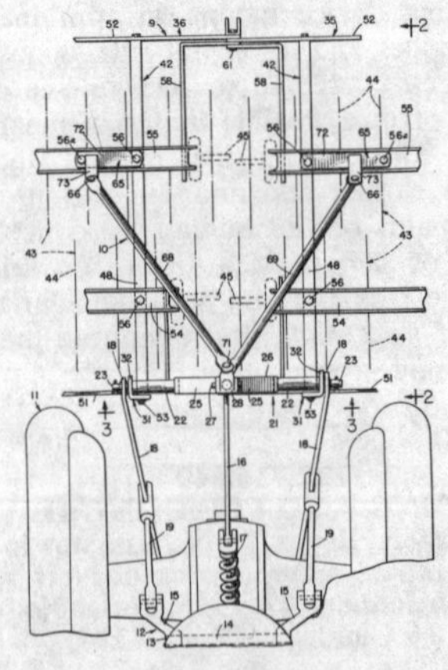

11. In an agricultural machine, the combination with an implement comprising a pair of sections normally disposed in side-by-side relation, and means pivotally interconnecting said sections for relative movement about a generally fore-and-aft extending axis, of a hitch device for connecting said implement with a propelling tractor or the like, comprising a generally upright hitch frame means, pivot means connecting said implement sections on each side of said axis, with said frame means, a pair of suspension links extending generally downwardly, outwardly and rearwardly from said hitch frame means and swingably connected at opposite ends with said hitch frame means and said implement sections for generally

lateral swinging relative thereto, said hitch frame means including a crossbar fixed thereto and downwardly extending brackets fixed to said crossbar and having generally fore-and-aft extending pivot-receiving apertures, pivot means disposed therein for connecting said implement sections to the hitch frame means for pivoting relative thereto about generally fore-and-aft extending axes, and a bracket pivoted to the rear end of each suspension link and attached to the associated implement section above and laterally outwardly of said axis.

2,716,856
PNEUMATIC COTTON GATHERER
Beeler D. Burns, Coolidge, Ariz.
Application October 1, 1952, Serial No. 312,566
3 Claims. (Cl. 56—30)

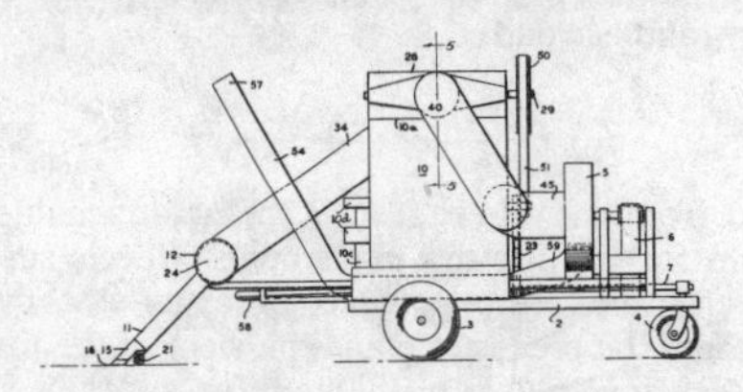

1. A cotton salvaging implement comprising a mobile frame, a mechanically operated cotton cleaner vertically mounted on said frame and having an open upper inlet end and provided with separate outlets for cotton and dirt, a blower having an intake and an exhaust end mounted on the frame, a transverse suction header mounted on one end of the frame, a cotton separator mounted over the cleaner in communication with the upper inlet end, a duct connecting the header to the separator, ground engaging suction pipes carried by the header and terminating in inlet means conforming to the curvature of the spaces between cotton plant rows, said suction pipes being spaced apart on the header to be disposed between side by side rows, a duct connecting the separator to the blower, means operatively mounted in said separator and disposed within the suction line from the blower to the suction pipes for intercepting the cotton and dirt and closing off the inlet end of the cleaner to the suction while permitting the free gravitation of the cotton and dirt into the inlet end of the cleaner, a duct connected to the exhaust end of the blower and connected to the cotton outlet in the cleaner so as to blow the cleaned cotton to a storage receptacle and means for diverting a portion of the air in said duct to the outlet for the dirt, including a duct communicating with said dirt outlet, for moving the dirt from the outlet.

2,716,857
BROOM RAKE STRUCTURE
Francis F. Melvin, Elwood, Ind.
Application October 22, 1954, Serial No. 463,877
3 Claims. (Cl. 56—400.17)

1. A broom rake structure comprising a head having a major flat area across its lateral length; a flange downturned from its front edge and having a plurality of spaced apart slots therethrough adjacent the juncture of the flange with the flat area; a plurality of tines, one tine extending through each of said slots and lying by a rear portion under and against said area; a restricted neck in each tine; a head portion spaced under and along said area across its rear end portion and carried integrally by said area through a rear bend therefrom; an upturned flange along the front of said portion having a free edge notched at regular intervals to fit over said necks; a handle member plate extending over the top side of said head area and secured thereto; a flange downturned from each lateral end of said plate fitting around said bend and extending forwardly under said portion in engagement therewith,

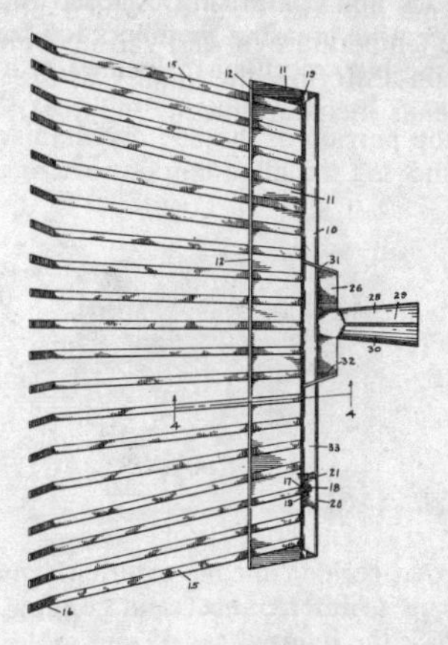

holding said bend against spreading; and a handle receiving member extending rearwardly from said plate.

2,716,858
DETECTOR MECHANISM
Fletcher Schaum, Philadelphia, Pa., assignor to Fletcher Works Incorporated, Philadelphia, Pa., a corporation of Pennsylvania
Application April 30, 1952, Serial No. 285,150
6 Claims. (Cl. 57—80)

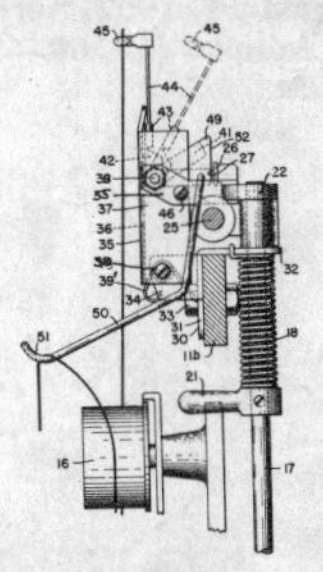

1. Apparatus for detecting yarn break or exhaustion in machines having oscillatory rocker members and yarn feed control mechanism comprising a support, a holder, spring hinge members interposed between said support and said holder, said holder having a front plate portion, a mounting pin in said holder, a limit pin in said holder spaced from said mounting pin, a plurality of drop wire carriers pivotally mounted on said mounting pin and having abutment portions engaging with said limit pin in one position of said carriers, drop wires carried by said carriers having guide eyes each for engagement with a yarn and held by the yarns in positions with said abutment portions spaced with respect to said limit pin, and a movable positioning member for said holder connected to said control mechanism, said abutment portions being engageable by a rocker member in said one position of one of said carriers for moving said holder against the force of said spring hinge members and permitting movement of said positioning member.

2,716,859
HEAT EXCHANGER
William H. Compton, South Euclid, and John M. Duff and Julian K. Smith, Cleveland, Ohio, assignors to The Reliance Electric & Engineering Company
Application January 15, 1953, Serial No. 331,378
11 Claims. (Cl. 57—100)

1. Cooling means for an electric motor driven spinning machine in a room and having a plurality of yarn spin-

ning bobbins, conduit means extending out of said room, said electric motor being located at one end of said machine and the lost heat of which affecting non-uniformly the temperature of said various bobbins and hence affecting non-uniformly the quality of the spun yarn, said cooling means including means for totally enclosing said

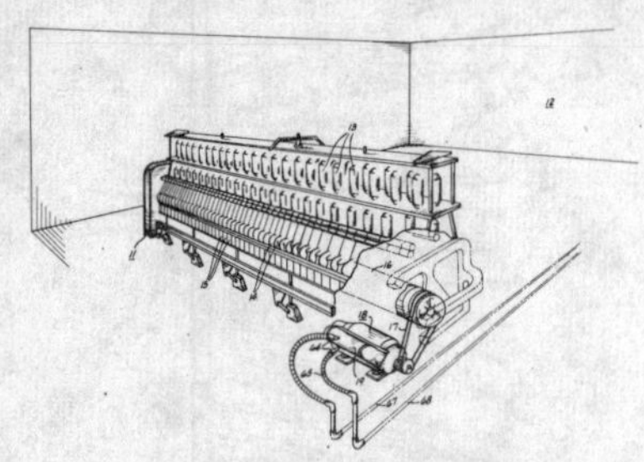

motor, second conduit means externally mounted on said motor, means to interconnect said conduit means, blower means to pass the internal gas of said motor to said second conduit means to pass motor heat out of the room containing said spinning machine to reduce the heat transferred from said motor to said spinning machine to achieve a more uniform quality of spun yarn.

2,716,860
PNEUMATICALLY WOUND DRIVE MECHANISM
John B. McGay and Gilbert B. Clift, Tulsa, Okla., assignors to Rockwell Manufacturing Company, Pittsburgh, Pa., a corporation of Pennsylvania
Application August 21, 1952, Serial No. 305,606
8 Claims. (Cl. 60—7)

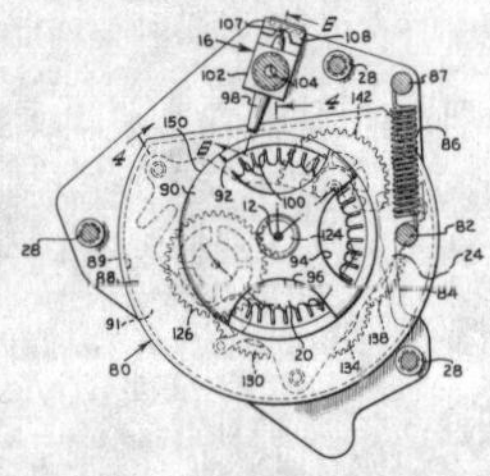

1. In combination with an escapement mechanism and an output shaft controlled thereby; means for imparting drive torque to said escapement mechanism controlled output shaft comprising a turbine rotor, means coupling said output shaft to said rotor, means for directing a stream of fluid under pressure to said turbine rotor, means operable to sheld said turbine rotor from the fluid stream formed by said directing means, and means responsive to variations in the relative rate of rotation of said output shaft and said turbine rotor for controlling the actuation of said shielding means.

2,716,861
PRESSURE ENERGY TRANSLATING AND LIKE DEVICES
James Wallis Goodyear, Polpenwith, Constantine, Falmouth, England
Application May 17, 1949, Serial No. 93,804
Claims priority, application Great Britain May 19, 1948
19 Claims. (Cl. 60—13)

1. In a pressure energy translating device, a rotatable disc having slots equidistantly spaced therearound, a rotor arranged to rotate about an axis approximately at right angles to the axis of rotation of the disc and comprising a rotor body with a convolute vane outstanding therefrom at a gradually varying radial inclination relatively to the axis of the rotor enabling it to mesh with the slots of the disc so that portions of the disc between the slots extend into spaces between the convolutions of the vane to form partitions therein, a casing for said rotor including an inlet chamber and an outlet chamber and an open ended shroud which is located between said chambers and communicates therewith and through which said disc projects to mesh with the rotor, the body of the rotor being of longitudinal curvature corresponding to the peripheral curvature of the disc and running closely adjacent to the periphery of the disc, and the convolute vane being of gradually varying height to cause the vane to extend to a varying extent into the slots of the disc, whilst the shroud is shaped in conformity with the envelope of the vane and is of axially curved non-cylindrical form with its ends substantially coinciding with the ends of said inlet and outlet chambers to permit of smooth and continuous flow of a fluid from the inlet chamber, through the shroud and into the outlet chamber in a direction substantially following the longitudinal contours of the rotor body and shroud.

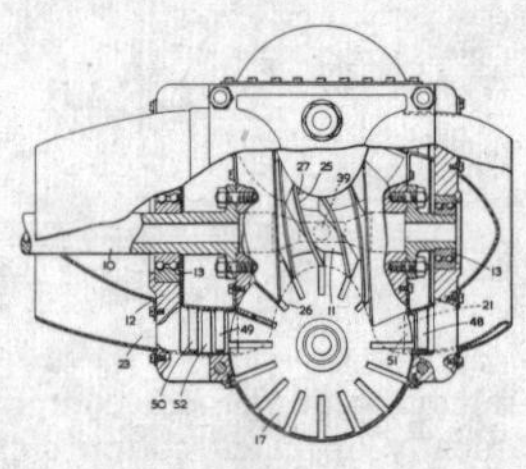

6. A pressure energy translating device according to claim 1 in which the axial contour of the shroud and of the rotor body are such as to provide therebetween a working space for the vane of gradually varying cross sectional area appropriate to compression of a fluid followed by expansion.

7. In a pressure energy translating device according to claim 6, a fuel injection means adapted to deliver fuel to said working space in the region of its minimum cross sectional area.

2,716,862
FUEL REGULATING VALVE MECHANISM
Arnold H. Block, Hackensack, N. J., assignor to Bendix Aviation Corporation, Teterboro, N. J., a corporation of Delaware
Application April 26, 1950, Serial No. 158,170
4 Claims. (Cl. 60—39.28)

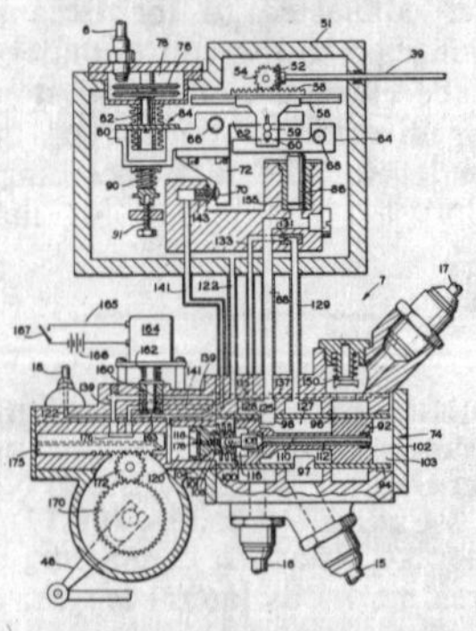

1. For use in regulating the flow of fuel to the combustion chamber of an engine having an air inlet conduit, a conduit for supplying fuel under pressure, and a fuel by-pass conduit; the combination comprising a by-pass valve adjustably positioned for regulating the flow of fuel from said fuel supply conduit to the combustion chamber and to said by-pass conduit, means for controlling the position of said by-pass valve including air inlet pressure responsive means, combustion chamber fuel pressure responsive means, a servo valve mechanism for controlling the position of said by-pass valve under force of said fuel supply pressure, linkage means for

operatively connecting the air inlet pressure responsive means and the combustion chamber fuel pressure responsive means to said servo valve mechanism, and said linkage means having an adjustable fulcrum means for varying the operating relationship between said air inlet pressure responsive means and the combustion chamber fuel pressure responsive means.

2,716,863
CONTINUOUS FLOW AND INTERNAL COMBUSTION ENGINES, AND IN PARTICULAR TURBO-JETS OR TURBO-PROPS
Lucien Reingold, Paris, and Claude Fouré, Becon-Courbevoie, France, assignors to Office National d'Etudes et de Recherches Aeronautiques (O. N. E. R. A.), Chatillon-sous-Bagneux, France, a society of France
Application July 3, 1951, Serial No. 234,968
Claims priority, application France July 4, 1950
7 Claims. (Cl. 60—39.28)

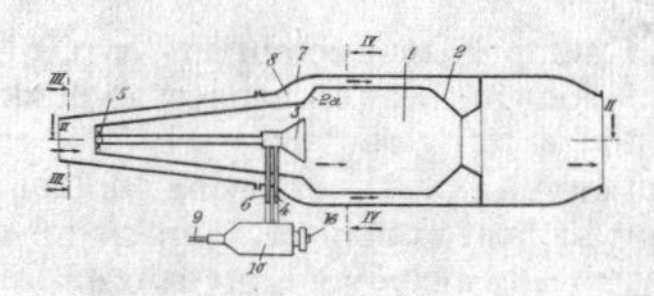

1. A continuous flow and internal combustion engine including at least one combustion chamber, means for feeding a stream of air to the inlet of said chamber, a pilot combustion device for igniting the fuel and stabilizing the flame located in said chamber, a device, located upstream of said pilot device in said chamber, for injecting fuel into a portion of the air stream fed to said chamber, means for feeding fuel to both of said devices, valve means for adjusting the ratio of the amount of fuel fed to said pilot device to the total amount of fuel fed to both of said devices, said last mentioned means being arranged always to supply at least a minimum fuel feed to said pilot device, and means for by-passing, upstream of said fuel injecting means, another portion of the air stream fed to said chamber and reintroducing said by-passed air portion into said chamber downstream of the combustion zone thereof.

2,716,864
CULVERT CLAMP
George H. Hacker, Benson, Minn.
Application August 6, 1951, Serial No. 240,595
1 Claim. (Cl. 61—16)

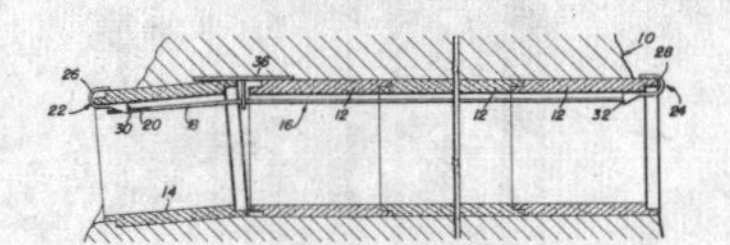

In a culvert, a pair of culvert sections having juxtaposed ends and remote ends, a pair of hook elements engaging over said remote ends and having flanges in said ends extending perpendicularly from shank portions of said hook elements, a connecting rod joining said flanges and extending through said sections with threaded ends, nuts on said threaded ends for abutment against said flanges whereby said hook elements can be clampingly engaged with said remote ends of the culvert sections, a flat plate positioned on the juxtaposed ends of the culvert sections exteriorly thereof and bridging the same and having a U-shaped element extending downwardly therefrom between said juxtaposed ends, said connecting rod engaging and being supported between said juxtaposed ends by said U-shaped element.

2,716,865
REFRIGERATING APPARATUS
Carl A. Stickel, Dayton, Ohio, assignor to General Motors Corporation, Dayton, Ohio, a corporation of Delaware
Application September 19, 1952, Serial No. 310,473
6 Claims. (Cl. 62—4)

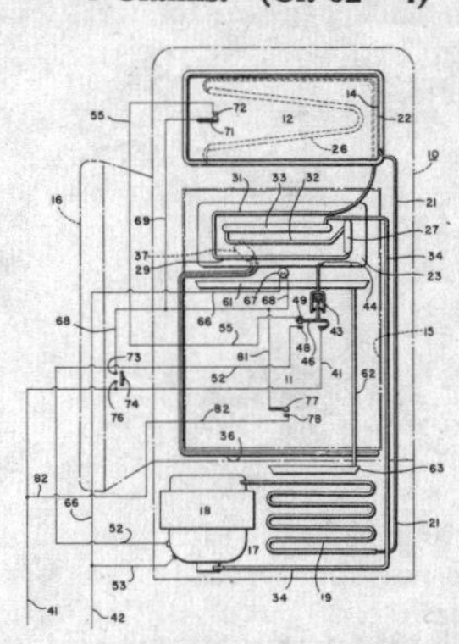

5. A refrigerating apparatus comprising in combination, a cabinet having an unfrozen food storage compartment and a frozen food storage compartment therein, said compartments being insulated from one another and isolated against flow of air therebetween, a closed refrigerating system associated with said cabinet, said system including an evaporator for cooling said frozen food compartment, another evaporator for cooling said unfrozen food compartment and a motor and compressor driven thereby for circulating refrigerant through the evaporators in succession for producing differential temperatures within said compartments, electric means for heating said unfrozen food compartment evaporator to remove frost and ice therefrom, an electric circuit for said motor and said heating means, said circuit including a thermostatic means responsive to the temperature of said frozen food compartment for energizing and deenergizing said heating means, another thermostatic means responsive to a near freezing temperature of air in said unfrozen food compartment for also energizing and deenergizing said heating means independently of said first named thermostatic means, and a thermostatically operated snap switch responsive to the temperature of said unfrozen food compartment evaporator for starting and stopping said motor, said snap switch being interposed in the circuit leading to said heating means and deenergizing said heating means when said motor is started irrespective of the position of said first named thermostatic means.

2,716,866
WATER HEATING SYSTEMS OF THE HEAT PUMP TYPE
William Charles Silva, Chicago, Ill., assignor to General Electric Company, a corporation of New York
Application May 27, 1953, Serial No. 357,767
12 Claims. (Cl. 62—4)

1. In a water heating system including an upstanding hot water storage tank, a cold water inlet connection communicating with the lower portion of said tank, and a hot water outlet connection communicating with the upper portion of said tank; the combination comprising a first refrigerant condenser arranged exteriorly of and in good heat exchange relation with the upper portion of said tank, a second refrigerant condenser arranged exteriorly of and in good heat exchange relation with the lower portion of said tank, a refrigerant compressor operative to compress expanded gaseous refrigerant, first thermal means selectively governed by the temperature of the water in the upper portion of said tank for conducting compressed gaseous refrigerant from said compressor selectively into said first and second condensers, second

thermal means selectively governed by the temperature of the water in the lower portion of said tank for selectively operating said compressor, a refrigerant evaporator, means

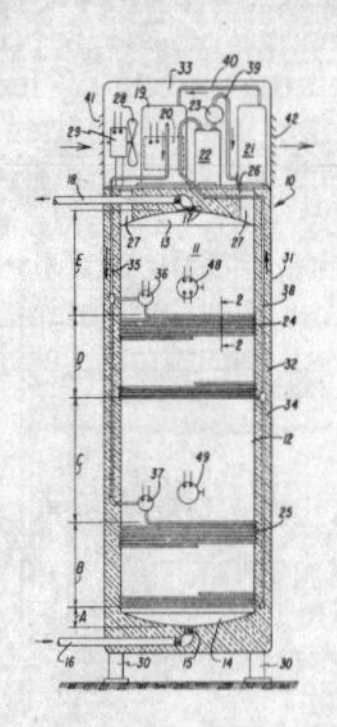

for expanding liquid refrigerant from said first and second condensers into said evaporator, and means for conducting expanded gaseous refrigerant from said evaporator back into said compressor.

2,716,867
REFRIGERATING APPARATUS
James W. Jacobs, Dayton, Ohio, assignor to General Motors Corporation, Detroit, Mich., a corporation of Delaware
Application July 2, 1953, Serial No. 365,592
9 Claims. (Cl. 62—4)

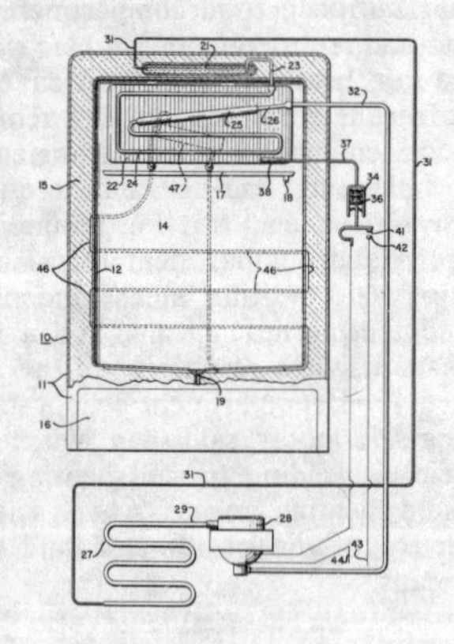

9. In a closed refrigerating system having a single evaporator, refrigerant liquefying means and means for supplying liquid refrigerant from said liquefying means to said evaporator, said last named means including a pressure reducing means, a non-refrigerating flash back chamber interposed in said closed system intermediate said single evaporator and said pressure reducing means, insulating material completely surrounding said flash back chamber to isolate the same from ambient temperatures thereabout, thermostatic means directly responsive to a temperature of said single evaporator below freezing for stopping said refrigerant liquefying means, said thermostatic means also being directly responsive to a temperature of said single evaporator above freezing for starting said refrigerant liquefying means, and said insulated non-refrigerating flash back chamber automatically receiving liquid refrigerant from said single evaporator in response to said thermostatic means stopping said refrigerant liquefying means for permitting the temperature of said evaporator to rise above 32° F.

2,716,868
HEAT PUMP SYSTEMS
Gerald L. Biehn, Needham, Mass., assignor to Westinghouse Electric Corporation, East Pittsburgh, Pa., a corporation of Pennsylvania
Application May 1, 1952, Serial No. 285,413
5 Claims. (Cl. 62—6)

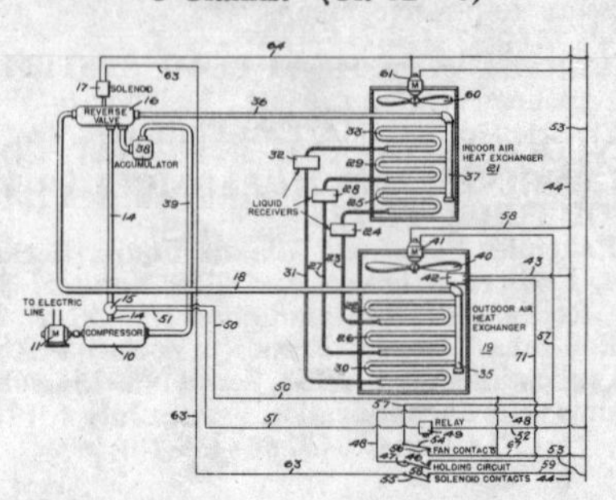

1. A heat pump system comprising a refrigerant compressor, an indoor air heat exchanger, an outdoor air heat exchanger and a refrigerant reversal valve connected in a refrigerant circuit; a fan for moving outdoor air through said outdoor air heat exchanger; a motor for driving said fan; a relay; means including a pressurestat responsive to an increase in pressure drop in the air passing through said outdoor air heat exchanger when it is acting as an evaporator in said circuit and ice forms on the surface of said outdoor air heat exchanger for energizing said relay, and means controlled by the energization of said relay for stopping said fan motor and for actuating said reversal valve to operate said outdoor air heat exchanger as condenser, said relay having holding contacts connected to said pressurestat for maintaining said relay energized when said pressurestat becomes inoperative to energize said relay as a result of the stopping of said fan motor.

2,716,869
FLAKE ICE MAKING MACHINE AND KNIFE THEREFOR
Gerald M. Lees, Chicago, Ill., assignor to Akshun Mfg. Co., Chicago, Ill., a corporation of Illinois
Application April 9, 1951, Serial No. 220,044
10 Claims. (Cl. 62—107)

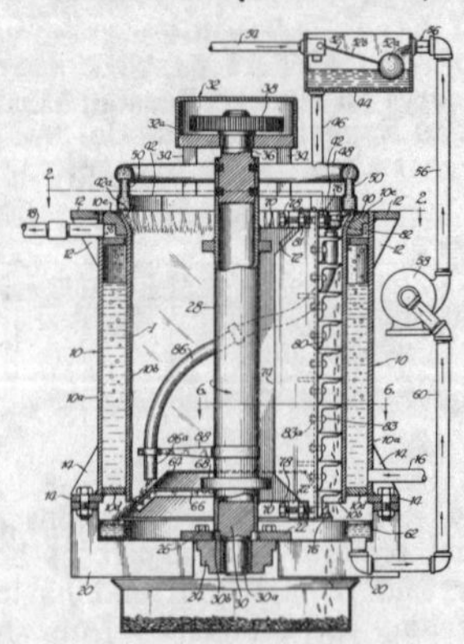

1. An ice making machine adapted to produce dry flake ice without the use of relatively moving water-carrying pipe parts comprising in combination: a vertical drum defining a cylindrical ice forming surface, an annular header non-rotatably disposed above the drum and in registry with the surface, the header having a series of water outlets positioned above the drum to cause circumferentially uniform water flow onto the surface, a rotor within the drum and adapted to rotate about the axis thereof, the rotor having a series of vertically spaced ice cutting knives mounted to engage the ice on the drum

as the rotor turns, and a shield mounted for rotation in unison with the rotor and above the knives to interrupt the flow of water to the drum in advance of the knives to permit the water to form dry ice for engagement by the knives.

2,716,870
REVERSE CYCLE HEAT PUMP SYSTEM
Gerald L. Biehn, Needham, Mass., assignor to Westinghouse Electric Corporation, East Pittsburgh, Pa., a corporation of Pennsylvania
Application April 1, 1953, Serial No. 346,227
8 Claims. (Cl. 62—115)

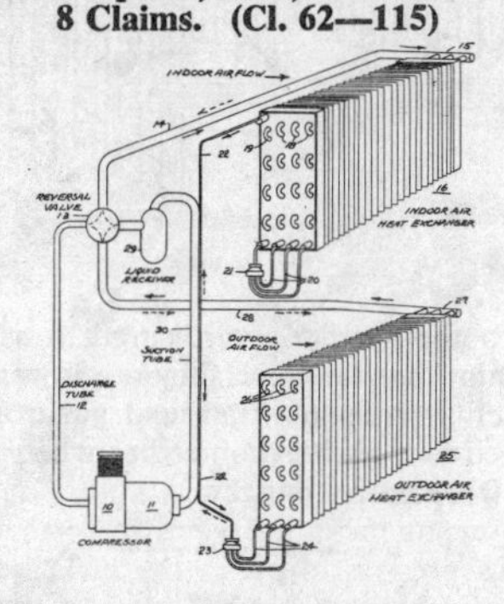

1. A heat pump system comprising a first indoor air heat exchanger, a second indoor air heat exchanger located adjacent and downstream with respect to the flow of indoor air of said first exchanger, an outdoor air heat exchanger, a refrigerant compressor, a reversal valve, means including suction and discharge tubes connecting said valve to said compressor, first refrigerant flow means connecting said outdoor and first indoor exchangers, second refrigerant flow means connecting said valve and second exchanger, third refrigerant flow means connecting said valve and said outdoor exchanger, and fourth refrigerant flow means connecting said first and second exchangers, said valve during air cooling operation supplying refrigerant through said third flow means to said outdoor exchanger from which the refrigerant flows through said first flow means to said first indoor exchanger from which the refrigerant flows through said fourth flow means to said second indoor exchanger from which the refrigerant flows through said second flow means back to said valve, said valve during air heating operation supplying refrigerant through said second flow means to said second indoor exchanger from which the refrigerant flows through said fourth flow means to said first indoor exchanger from which the refrigerant flows through said first flow means to said outdoor exchanger from which the refrigerant flows through said third flow means back to said valve.

2,716,871
OVER ICE CHILLER
Eugene L. Bud Brown, San Jose, Calif.
Application May 20, 1952, Serial No. 288,973
4 Claims. (Cl. 62—146)

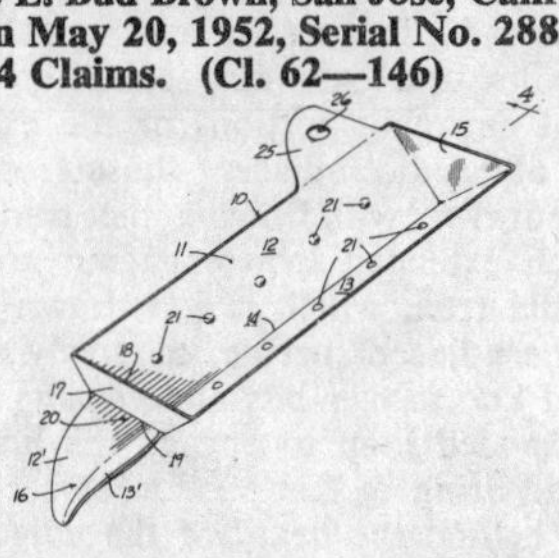

1. A device for chilling liqueur during pouring of the same from a bottle into a small mouthed glass comprising an elongated open trough-like body having a pair of side walls disposed at a substantially right angle to each other and joined at their contiguous edges to provide a valley bottom, a wall at one end of said side walls for closing that end of said trough-like body, a cross bar joining the upper portions of the opposite ends of said side walls for maintaining cubes of ice in said trough-like body and cooperating with the latter to form an opening in said opposite end of said trough-like body, and a spout formed integrally with said side walls and extending beyond said opposite end thereof for directing liqueur poured onto the ice cubes in said trough-like body for continuous flow into a small mouthed glass.

2,716,872
VENTILATED CONTAINER SYSTEM OF TRANSPORTATION
Ellis W. Test, Hinsdale, Ill., assignor to Pullman-Standard Car Manufacturing Company, Chicago, Ill., a corporation of Delaware
Application December 31, 1948, Serial No. 68,688
2 Claims. (Cl. 62—171)

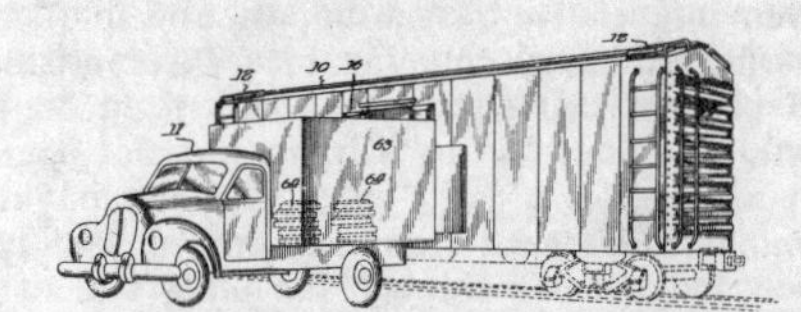

2. A method of maintaining lading in refrigerated condition during transportation thereof, which comprises disposing lading in a ventilated pilferage-preventing container having a heat-insulating base, conveying the container with the lading therein to a point en route in a refrigerated vehicle while refrigerating the lading by subjection to the temperature of the atmosphere of said vehicle, disposing a heat-insulating hood on said base over the laden container and the refrigerated vehicle atmosphere adjacent thereto at said point, transferring to an unrefrigerated vehicle the lading in the container and refrigerating atmosphere enclosed by the hood and base for conveyance from said point, and maintaining the lading under refrigeration by said atmosphere in the hood and base while conveying the lading in the unrefrigerated vehicle.

2,716,873
RESILIENT COUPLING
Robert C. Byers, Kensington, Md., assignor to the United States of America as represented by the Secretary of the Navy
Application February 3, 1953, Serial No. 334,800
1 Claim. (Cl. 64—11)

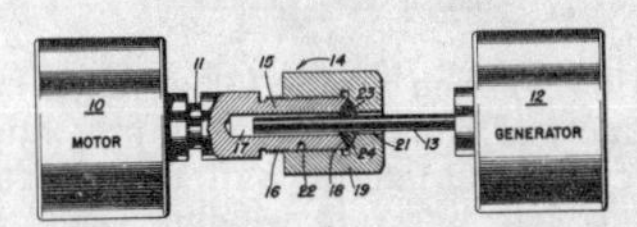

A resilient coupling for operatively connecting a pair of shafts, comprising an externally threaded member connectable to and forming an integral part of one of said shafts when connected thereto, said member having an axial bore loosely receiving a substantial portion of the end of the other of said shafts in spaced relation thereto, and a diverging cone-shaped seat at one end, the axis of said cone-shaped seat and the axis of said first shaft coinciding with one another, a nut-like member having a coaxial bore for loosely admitting the end of said other shaft and a threaded counterbore ending in a cone-shaped seat, said nut-like member threadedly engaging said externally threaded member to bring the two of said cone-shaped seats face to face, and a resilient ring on said other shaft and between said cone-shaped seats, said ring being compressed into tight driving engagement with said externally threaded member and said other shaft whereby rotary motion of constant torque is obtained.

2,716,874
COMBINED SOUND CHANGE-OVER AND VOLUME CONTROL FOR SOUND MOTION PICTURE PROJECTORS
Evans C. Wiley, Cape Girardeau, Mo.
Application May 27, 1952, Serial No. 290,192
1 Claim. (Cl. 64—23)

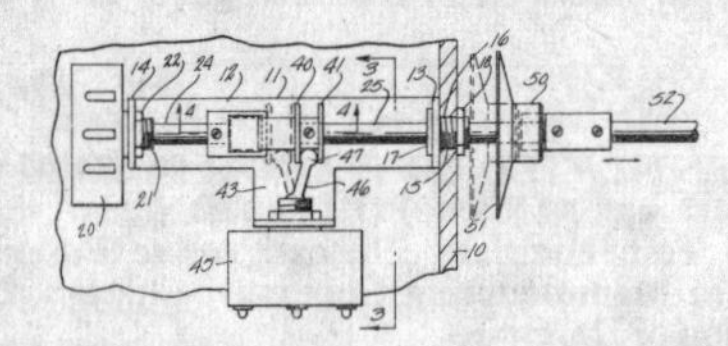

In combination, a support, first and second axially aligned shafts mounted on said support for endwise and rotary movements relative to said support, said shafts having axially inward ends, coupling means comprising a body fixed on said first shaft and a link fixed on said second shaft, said body having a transverse slot opening through opposite sides of said body, said slot having a closed terminal end spaced from one end of said body and an open end opening through the other end of the body, said body having a longitudinal opening communicating with the open end of the slot and the axially inward end of said first shaft being secured in said opening at the open end of the slot, said link having side members positioned at opposite sides of said body and a transverse member extending between said side members and engaging through said slot, said transverse member being shorter lengthwise of said body than the distance between the closed terminal end of the slot and the axially inward end of the first shaft whereby endwise movements of said body relative to said link are limited only by engagement with the axially inward end of the first shaft and the terminal end of the slot, and means on said first shaft for rotating said first shaft relative to the support and moving said first shaft endwise relative to said second shaft and relative to the support.

2,716,875
WATCH WINDING MEANS
Charles Hill, Reno, Nev., and Charles K. Johns, New York, N. Y.
Application March 20, 1953, Serial No. 343,624
5 Claims. (Cl. 64—29)

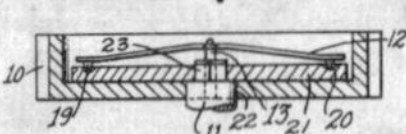

1. A spring winding means for a watch comprising a hollow cylindrical winding member presenting a closed side and a cylindrical interior wall, diametrically opposed slots in said wall, a spring winding arbor passing into said winding member, a rotary plate member fixed to said arbor and presenting detent openings adjacent the periphery thereof, a free flexible disk coextensive with and adjacent said rotary plate member and presenting protuberances for yielding engagement with said openings, and prongs on the periphery of said disk engaging said slots for rotating the disk with said winding member.

2,716,876
APPARATUS FOR KNITTING ELASTIC FABRIC AND METHOD
Julian H. Surratt, Denton, N. C.
Application April 26, 1954, Serial No. 425,500
14 Claims. (Cl. 66—9)

1. In a knitting machine having a circular series of independent needles, a set of stitch cams and means for feeding yarn to the needles; in combination, an auxiliary stitch cam, and means for moving the auxiliary stitch cam into position below at least one of the stitch cams in said set to thereby form an extension to said one of the stitch

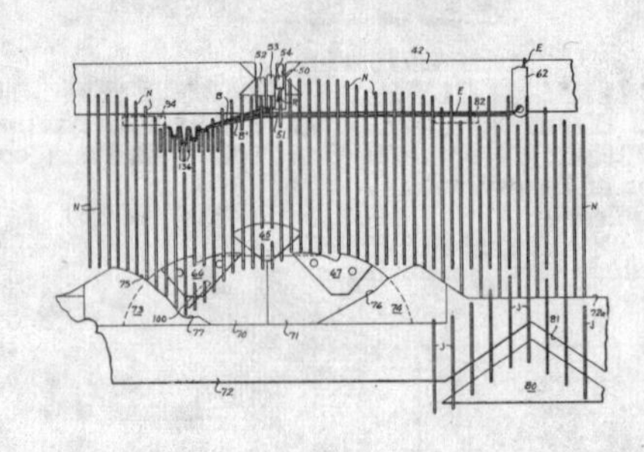

cams whereby the needles are lowered to a substantially lower level than that to which they are lowered when the auxiliary stitch cam is in withdrawn position to thereby form elongated loops with the needles when the auxiliary stitch cam is in operative position.

2,716,877
METHOD OF KNITTING FABRIC
Kurt Willi Wickardt, Liverpool, England, assignor to Hosemaster Machine Company Limited, Liverpool, England, a British company
Original application September 20, 1951, Serial No. 247,396. Divided and this application December 24, 1952, Serial No. 327,774
Claims priority, application Great Britain September 23, 1950
16 Claims. (Cl. 66—96)

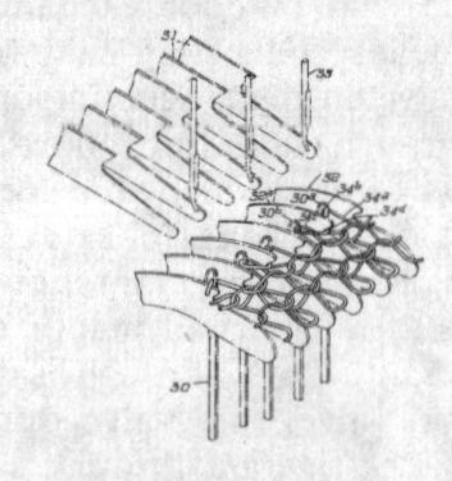

1. A method of knitting a run resisting fabric on a straight bar knitting machine having bearded needles, hooked knock-over bits and transfer points, including after the knock-over operation has been completed, the steps of lowering the transfer points each to engage a loop around a predetermined needle corresponding thereto, displacing the needles and transfer points away from one another and drawing said loops rearwardly of the fabric, moving the transfer points transversely of the fabric through a distance at least equal to the spacing of the needles and extending said loops, lowering the needles and engaging portions of the extended loops in the beards of needles adjacent those from around which they were drawn, withdrawing the transfer points and leaving each extended loop engaging around the shank of the needle from which it was drawn and around the beard of the adjacent needle, relatively approaching the needles and knock-over bits and causing another portion of each extended loop to engage the hook of a knock-over bit, and bringing that portion of each extended loop which was behind the beard of the adjacent needle to a position in front of the shank thereof, inverting the extended portion of each extended loop to form a closed turn around the shank of the adjacent needle.

2,716,878
TWO PART KNITTING MACHINE NEEDLE ASSEMBLY
Joseph L. Morris, Elmhurst, N. Y.
Application November 15, 1952, Serial No. 320,685
8 Claims. (Cl. 66—120)

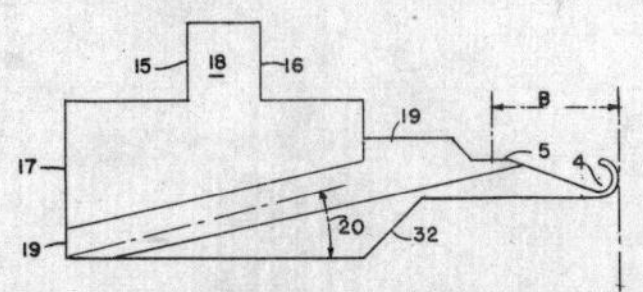

1. A two part knitting needle assembly comprising a knitting hook member having an elongated body portion and a hook-supporting shank extending from one end of the body portion, said body portion having external surfaces adapted to fit for reciprocating sliding movement within and along a needle groove of a knitting machine, and having a straight parallel-walled groove of uniform depth traversing one of said external surfaces, said guide groove being inclined in the direction of reciprocation of said knitting hook member and being generally aligned with the point of said hook, a cast-off member having a point adapted to cooperate with said hook to cast off loops in the knitting operation, said cast-off member having a straight body portion fitting between the walls of said guide groove for sliding movement therein relative to said knitting hook member, the said inclination of the guide groove and the alignment thereof relative to said hook point causing the point of said cast-off member to lie well inwardly of the point of said hook member when remote therefrom, and to move outwardly and over the point of the hook upon relative movement of said cast-off member along said guide groove toward said hook point.

2,716,879
FLAT FILE GAS LIGHTER
John W. Hobing, Cincinnati, Ohio, assignor to Raymond A. Haneberg and Richard H. Haneberg, both of Cincinnati, Ohio
Application October 9, 1953, Serial No. 385,178
1 Claim. (Cl. 67—6.1)

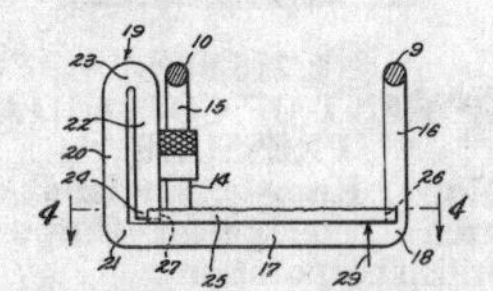

A gas lighter holder having distinct and integral parts and made from a length of resilient and flexible wire stock comprising a pair of relatively movable arms joined together at common ends; a downwardly extending terminal end formed on one arm for mounting a pyrophoric element thereon; the intermediate portion of the other arm being turned downwardly and then laterally to form a fixed, upstanding file holder portion and a transverse base portion, respectively, on the said arm; an inverted U-shaped file clamp element formed on the end portion of the said other arm and having its outer upstanding leg joined to the free end of the base portion by a bend portion, and its inner spring finger spaced from the leg and having its terminal end unconnected with and free to flex laterally adjacent the base portion; a flat file frictionally engaging the pyrophoric element and positioned adjacent to and parallel with the base portion; said file having a notch formed in one end to engage the inner, lowermost end of the upstanding portion; and a notch formed in the opposite end of the file for engagement with the terminal end of the spring finger, said file being releasably held in the holder by the flexure of the spring finger outwardly upon forcible insertion of the file between said finger and the upstanding file holder portion.

2,716,880
PYROPHORIC LIGHTERS
Harry Faulkner, Rochdale, England
Application March 30, 1954, Serial No. 419,869
3 Claims. (Cl. 67—7.1)

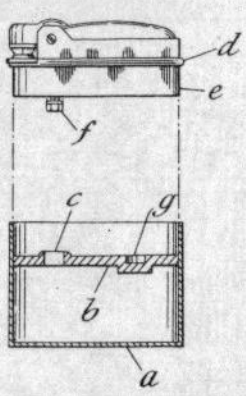

1. A pyrophoric lighter, comprising a casing, open at the top, a transverse partition in the casing dividing it into a lower compartment serving as a receptacle for liquid fuel and an open upper compartment, the partition having one opening offset from its centre through which the upper and lower compartments communicate with one another, a plate, lighter mechanism mounted on the plate, a skirt depending from the plate and adapted to fit into the said upper compartment, absorbent material contained within the skirt, a plug depending from the plate and adapted to close the opening in the transverse partition when the skirt is inserted in the upper compartment in its normal position, and to leave said opening unobstructed when the skirt is removed, turned through 180° and then replaced in the upper compartment, so as to allow fuel contained in the lower compartment to flow through the opening when the lighter is inverted.

2,716,881
PORTABLE LOCK FOR SLIDING PANELS
John R. Terrill, Collingswood, N. J.
Application July 3, 1953, Serial No. 365,903
1 Claim. (Cl. 70—14)

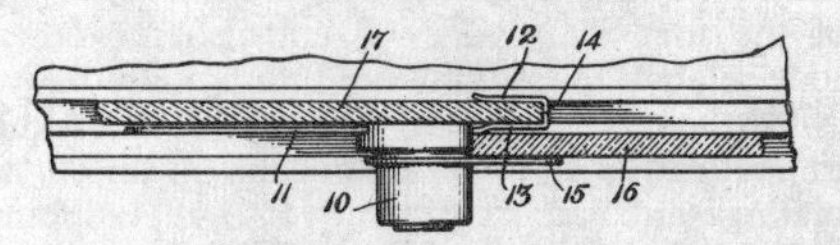

A portable lock for sliding panels comprising a cylindrical body having a transversely disposed bar providing a lock bolt extended through one end, said bar having a U-shaped section on an end extended from the body of the lock, said U-shaped section having a tongue extended from a cross section at one end of the bar, and a plate having an opening therein through which said cylindrical body extends and to which cylindrical body the plate is rigidly secured, said plate being parallel to said bar and laterally spaced from said U-shaped section, said U-shaped section of the bar forming the lock bolt being positioned to receive an edge portion of a panel adapted to extend across the back of the cylindrical body and said plate spaced from the U-shaped section of the bar being positioned to overlie an edge portion of a second panel overlapping the first mentioned panel.

2,716,882
GUARD FOR PROTECTING COIN OPERATED SWITCHES AND LIKE METERING DEVICES
William M. Gill, Kansas City, Mo., and Earl D. Vold, Mission, Kans.
Application June 4, 1954, Serial No. 434,562
6 Claims. (Cl. 70—159)

5. A guard for a coin control switch of the type including a casing having a coin slot and a starting lever projecting from the casing, a door for removal of coins from the casing and having a dial, said guard including a fixed member for seating the casing therein and having an opening to pass said starting lever and to expose the

coin slot, a complementary member cooperating with the fixed member to embrace the casing, a hinge connecting complementary member with the one side of the fixed member and having a portion exposing the dial, flanges

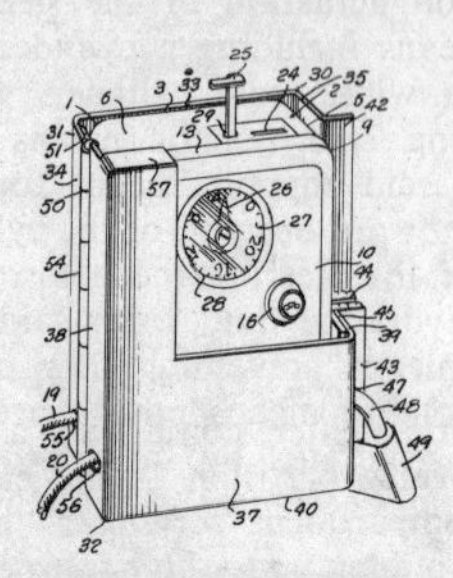

extending outwardly from the opposite side of said members and having registering openings for passing the hasp of a padlock to lock the members together, and means for reinforcing said flanges at the point of said lock.

2,716,883
TESTING MACHINE FOR SHOCK ABSORBER OR THE LIKE
Clark A. Tea, Southfield Township, Mich.
Application June 25, 1952, Serial No. 295,484
5 Claims. (Cl. 73—11)

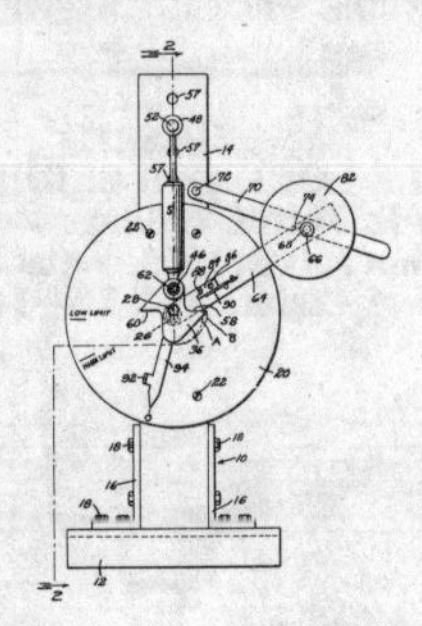

1. A testing machine for a shock absorber or the like comprising, a frame, an arm pivotally mounted on said frame, a weight on said arm at one side of the pivot of said arm, a ratchet having a contact face pivotally carried by said frame on an axis eccentric to the pivot axis of said arm, a support carried by said ratchet and eccentric to the pivot of the contact face for pivotally receiving one portion of the shock absorber, a support carried by said frame for pivotally receiving another portion of the shock absorber, means carried by said arm and in the path of said contact face for engagement with the contact face of said ratchet during only a limited portion of the downward swinging movement of said arm and released from such engagement at the end of the downward swinging movement, and means for indicating the maximum free swinging movement of said arm after the limited portion of the swinging movement of said arm.

2,716,884
GAGE FOR DETERMINING LOOSE AND TIGHT PACKING OF HEAT ABSORBING PLATES
George E. Rosenberg, Wellsville, N. Y., assignor to The Air Preheater Corporation, New York, N. Y., a corporation of New York
Application November 6, 1953, Serial No. 390,653
5 Claims. (Cl. 73—12)

1. Apparatus for determining the relative packing density of plates in a container comprising; a wedge to be driven between parallel plates; a housing loosely supporting said wedge in an upright position with the apex thereof pointing downwardly, and a driving means for subjecting said wedge to a predetermined force whereby

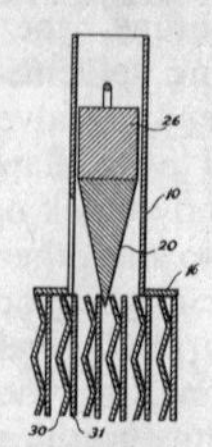

said wedge is driven between plates a distance proportional to the packing density of the plates.

2,716,885
LEAK DETECTOR
Edward Orville Owens and Louie Martin Ezzell, Abbeville, Ala.
Application August 27, 1952, Serial No. 306,544
2 Claims. (Cl. 73—48)

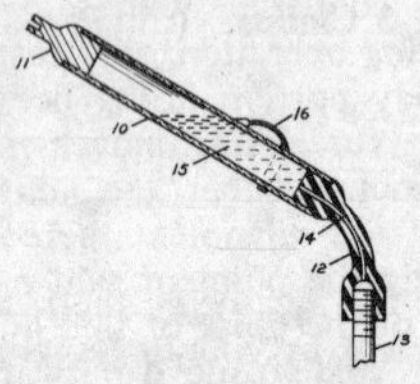

1. A leak detector for valves of pneumatic tires comprising a transparent tube hermetically sealed at one end thereof, a conduit having one end fitted to the other end of said tube, said conduit being provided with a center passage therethrough and being sufficiently large at its free end to receive a valve stem, and liquid partially filling said tube, said liquid having a surface tension sufficiently great to prevent its entering said center passage of said conduit, said center passage being sufficiently large to allow air leaking from said valve stem to pass into said tube.

2,716,886
MEASUREMENT OF COLLOID OSMOTIC PRESSURE
David Stuart Rowe, Knowle, England, assignor to National Research Development Corporation, London, England, a British corporation
Application September 17, 1954, Serial No. 456,743
Claims priority, application Great Britain September 24, 1953
10 Claims. (Cl. 73—53)

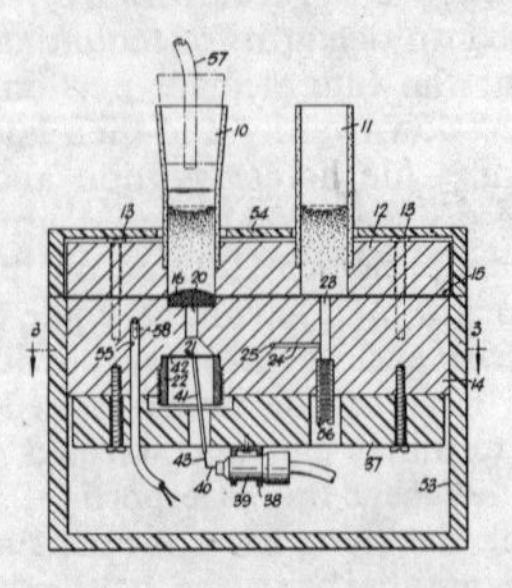

1. An osmometer comprising rigid body and cover blocks rigidly but separably secured together and adapted to clamp between their junction faces a semi-permeable membrane; two spaced upright tubular columns mounted on the cover block, one of said tubes, for containing colloidal solution, being adapted to be connected at its open upper end to a source of pressure including pressure in-

dicating means and at its lower end being open to the junction face of the cover for exposure to one side of the semi-permeable membrane, the other of said tubular columns, for containing solvent, being open to ambient pressure at its upper end; a valve; a compartment within the body block closed by a diaphragm and open to the junction face of the body block opposite to the "solution" column for exposure to the other side of the membrane, there being a passage in the body block connecting the said diaphragm compartment to the valve and there being a further passage connecting the valve to the "solvent" column; sensitive means for detecting small deflections and means for affecting said sensitive detecting means by deflections of said diaphragm.

2,716,887
APPARATUS FOR MEASURING THE RESONANT FREQUENCY OF A VIBRATORY ELEMENT
Roy A. Smith, Sun Valley, Calif., assignor to Collins Radio Company, Cedar Rapids, Iowa, a corporation of Iowa
Application February 25, 1953, Serial No. 338,801
3 Claims. (Cl. 73—69)

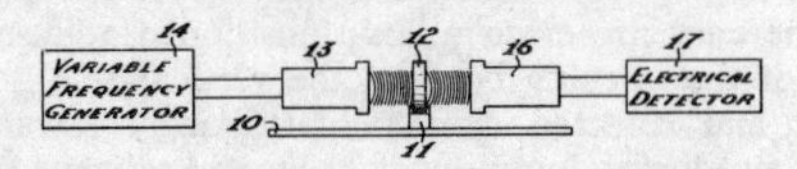

3. Apparatus for measuring the resonant frequency of a vibratory element by coupling energy through the surrounding air comprising an input electrical-mechanical transducer including a vibratory member having a transmitting surface exposed to the surrounding air, a variable frequency oscillator electrically connected to said first transducer for energizing said vibratory member at selectable frequencies, an output mechanical-electrical transducer including a vibratory member having a receiving surface, means for electrically isolating said output transducer from said input transducer, a detector electrically connected to said output transducer for measuring the energy transmitted between said surfaces, said transmitting and receiving surfaces being disposed oppositely in spaced, parallel relation and defining an energy transmission path therebetween, support means for holding said element in said energy transmission path, shield means comprising a plate defining an opening, said plate being positioned between said transmitting and receiving surfaces with said opening aligned with said path, said shield means and said support means having a resonant frequency outside the range of variation of said oscillator, whereby said element is adapted to transfer all the energy as measured by said detector between said transmitting and receiving surfaces.

2,716,888
APPARATUS FOR TESTING POWER STEERING
Ernest J. Svenson, Rockford, Ill.
Application November 6, 1953, Serial No. 390,678
19 Claims. (Cl. 73—116)

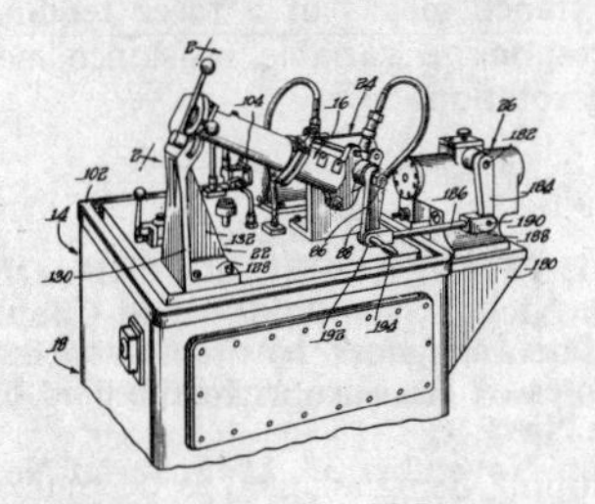

1. A testing apparatus for a steering mechanism, comprising base means, means associated with said base means for supporting a steering mechanism to be tested, hydraulic reactor means for building up a fluid pressure in accordance with torque developed by said steering mechanism upon actuation of the steering mechanism, said reactor means including a cylinder and a relatively shiftable piston within said cylinder, means for operatively connecting said reactor means with a steering mechanism on said support means for relatively shifting said piston toward one end of the cylinder in response to operation of the steering mechanism, conduit means for directing hydraulic fluid from said one end of the cylinder, and means preventing flow of hydraulic fluid through said conduit means until a predetermined pressure has been generated by said piston in response to operation of said steering mechanism.

2,716,889
METHOD OF DETERMINING AND ADJUSTING THE AERODYNAMIC PITCHING MOMENT OF A FULL-SIZED AERODYNAMIC MEMBER AND APPARATUS THEREFOR
Glidden S. Doman, Trumbull, Conn., assignor to Doman Helicopters, Inc., Danbury, Conn., a corporation of Delaware
Application November 25, 1952, Serial No. 322,508
18 Claims. (Cl. 73—147)

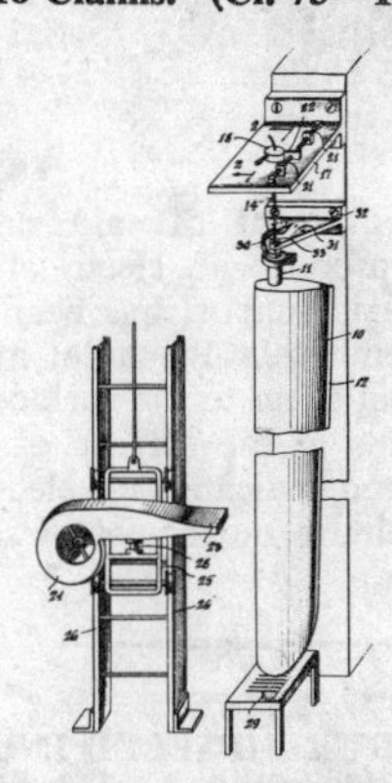

1. A method of testing the aerodynamic pitching moment or coefficient of a full-sized aerodynamic member such as a rotary wing aircraft blade having a pitching moment adjusting element to correct for inaccuracies of airfoil contour comprising supporting the aerodynamic member for movement about the axis of its mean aerodynamic center, creating an air stream having a width of very short span relatively to the length of the aerodynamic member, directing the air stream at the leading edge of the aerodynamic member and successively at a plurality of positions along the length thereof, angularly adjusting one of the elements including the aerodynamic member and air stream to eliminate lift effects of the air foil if such should appear, opposing yieldingly any angular displacement of the aerodynamic member, and measuring the angular displacement of the aerodynamic member at each position.

2,716,890
APPARATUS FOR DETERMINING POINT AT WHICH A PIPE IS STUCK IN A WELL
Philip W. Martin, Huntington Park, Calif.
Application October 3, 1949, Serial No. 119,302
8 Claims. (Cl. 73—151)

1. In a device for determining the point at which a member is stuck in a hole by testing different portions of the member to see whether or not these portions are movable in the hole: a support adapted to be moved into the hole to a position adjacent a selected portion of said member; suspension means for moving said support in

the hole from place to place along said member; engagement means extending laterally with respect to said support for effecting an engagement of said support with a selected portion of said member in the hole which will produce movement of said support by said portion of said member in the hole when the same is moved as the result of change in force applied to the member; an inertia element comprising a mass; yieldable means supporting said inertia element on said support so that it may remain stationary when said support is moved, thereby producing relative movement of said support and said inertia element; means operating in response

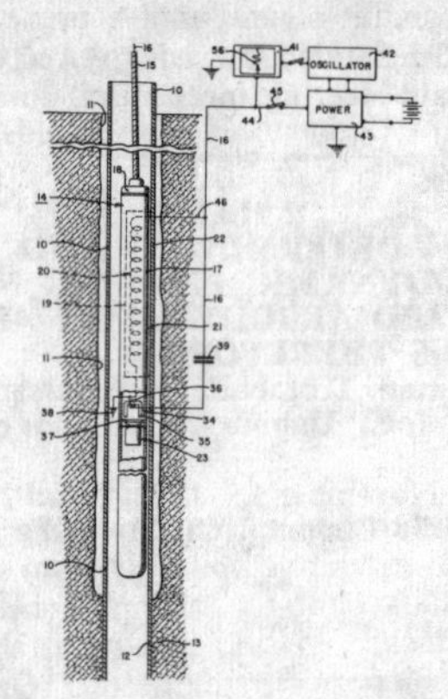

to said relative movement of said support and said inertia element to produce an electrical effect which is an indication that said support has been moved by said portion of said member in the hole; means for transmitting said electrical effect to the surface of the ground; and indicating means at the surface of the ground arranged to receive and indicate said electrical effect and thereby evidence relative movement of said support and said inertia member.

2,716,891
IONIZING TRUE AIRSPEED INDICATOR
Alfred A. Stuart, Ridgewood, N. J., assignor to Bendix Aviation Corporation, Teterboro, N. J., a corporation of Delaware
Original application June 18, 1948, Serial No. 33,390, now Patent No. 2,679,162, dated May 25, 1954. Divided and this application October 1, 1953, Serial No. 383,465
6 Claims. (Cl. 73—194)

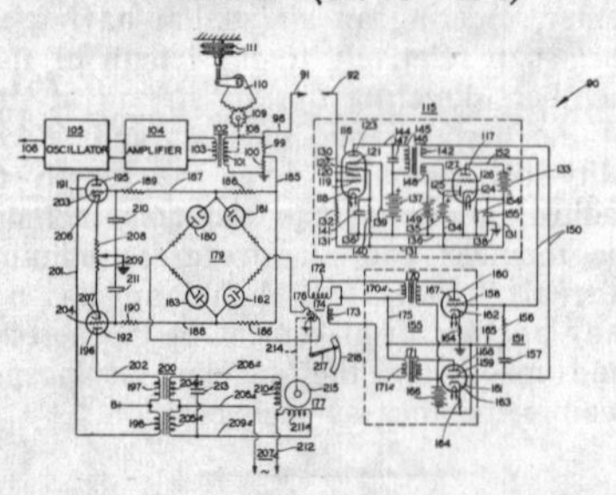

1. A meter for measuring the velocity of a flowing fluid medium comprising an ionizing electrode, a source of periodically varying voltage connected to said electrode of sufficient potential to provide ionization of the fluid medium, a pick-up electrode at a known distance downstream from said ionizing electrode adapted to have a voltage induced therein by the ions carried downstream by the fluid medium, and a phase discriminator connected to said electrodes and responsive to the phase difference between the ionizing and induced voltages to indicate the velocity of the ions between the two electrodes.

2,716,892
WATER GAGE ILLUMINATOR
Walter J. Kinderman, Philadelphia, Pa., assignor to Yarnall-Waring Company, Philadelphia, Pa., a corporation of Pennsylvania
Application May 21, 1951, Serial No. 227,496
4 Claims. (Cl. 73—293)

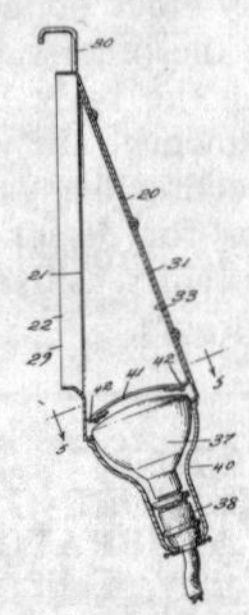

4. A liquid level gage illuminator comprising a body having one vertical side provided with a vertically extending opening, a diagonal reflector behind the vertical side converging upward toward the top of the opening and a mercury arc electric lamp positioned adjacent the bottom of the opening between the vertical side and the reflector and directed diagonally upwardly toward the opening, producing luminous light limited to wave lengths between 400 and 540 millicrons in the optical range.

2,716,893
MEANS AND APPARATUS FOR UTILIZING GYRODYNAMIC ENERGY
Edwin H. Birdsall, San Diego, Calif., assignor, by mesne assignments, to General Dynamics Corporation, a corporation of Delaware
Application October 18, 1949, Serial No. 122,036
25 Claims. (Cl. 74—5)

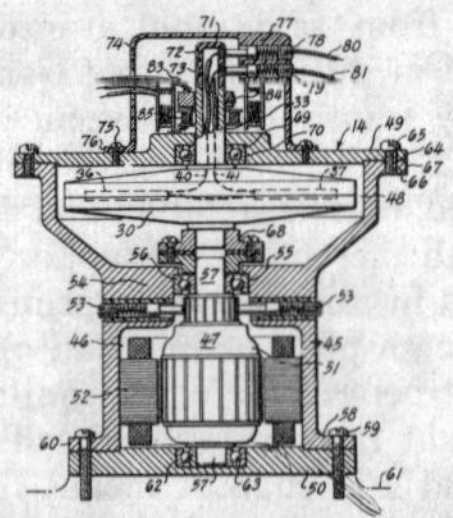

1. Apparatus for utilizing gyrodynamic energy comprising strain-responsive variable resistance means for varying the characteristics of an electric current adapted for rotation around an axis, means for providing continuous rotation to said resistance means, a source of electric current operatively connected to said variable-resistance means, and means operatively connected to said variable-resistance means for apprehending the variations in the character of said electric current resulting from gyrodynamic reactions on said strain-responsive variable resistance means of a force tending to disturb said strain-responsive variable resistance means relative to its axis of rotation.

2,716,894
ROLL AND PITCH ACCELERATION DETECTOR
Nathaniel B. Nichols, Cambridge, and Clinton H. Rider, Boston, Mass., assignors, by mesne assignments, to the United States of America as represented by the Secretary of the Navy
Application November 30, 1945, Serial No. 631,945
3 Claims. (Cl. 74—5.41)

1. For use with a vessel subject to angular motion, a gyro stable element including a gravity reference, a

gyro, and an erecting couple between said gravity reference and gyro to maintain said gravity reference and gyro in alignment during said motion, means producing an alternating voltage having a period equal to that of said angular motion and of a magnitude depending on the angular velocity of said motion, a differentiating

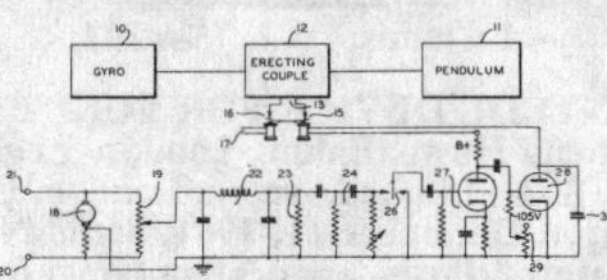

circuit for said voltage for producing a voltage depending on the angular acceleration of said motion, means responsive to the magnitude of said differentiated voltage to disconnect said couple when said differentiated voltage passes beyond a predetermined value and to reconnect said couple when said differentiated voltage returns to said value.

2,716,895
ENGINE STARTING APPARATUS
John E. Antonidis, Anderson, Ind., assignor to General Motors Corporation, Detroit, Mich., a corporation of Delaware
Application August 5, 1952, Serial No. 302,647
7 Claims. (Cl. 74—7)

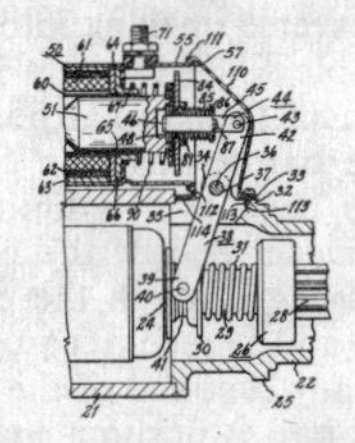

1. In an engine starting apparatus for internal combustion engines the combination comprising, an electric motor; a housing; driving means within the housing for engagement with the engine; a lever pivotally supported on the housing for shifting the driving means into engagement with the engine; a solenoid starting motor switch adapted to be mounted directly on the housing, said switch including a shell open at one end only; a solenoid coil supported within the shell; open at one end only; a solenoid coil supported within the shell; an armature within the coil; a pair of stationary contacts insulatingly supported on the shell adjacent the open end thereof; a link connected with the armature and the lever; a biased movable contact under the control of the armature and adapted to close a circuit between the contacts when the armature is moved by the coil; and resilient means to oppose movement of the armature and to bias the movable contact to an open position, whereby when said

2,716,896
ROTATION LIMIT DEVICE
Lauren F. Beldt and Dean M. Lewis, Cedar Rapids, Iowa, assignors to Collins Radio Company, Cedar Rapids, Iowa, a corporation of Iowa
Application October 8, 1952, Serial No. 313,732
9 Claims. (Cl. 74—10.2)

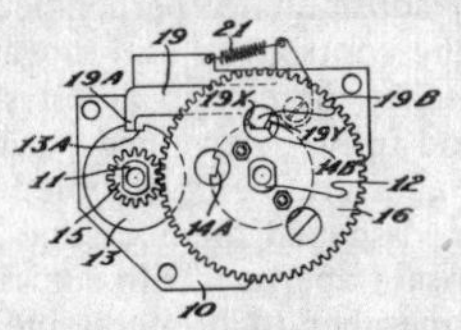

1. Apparatus for limiting the rotation of a pair of shafts comprising, a frame member rotatably supporting said shafts, a first cam mounted on the first shaft, a second cam mounted on the second shaft, said first cam formed with a slot in its periphery, said second cam formed with a pair of slots in its periphery, a pawl pivotally supported on said frame member and formed with a pair of projections, the first projection of said pawl engageable with the first cam and said first projection substantially narrower than the slot formed in said first cam, said second projection engageable with the first and second slots formed in the second cam and said second projection generally V-shaped in form.

2,716,897
DIFFERENTIAL MECHANISM
Arthur H. Wulfsberg, Cedar Rapids, Iowa, assignor to Collins Radio Company, Cedar Rapids, Iowa, a corporation of Iowa
Application November 5, 1953, Serial No. 390,351
4 Claims. (Cl. 74—10.54)

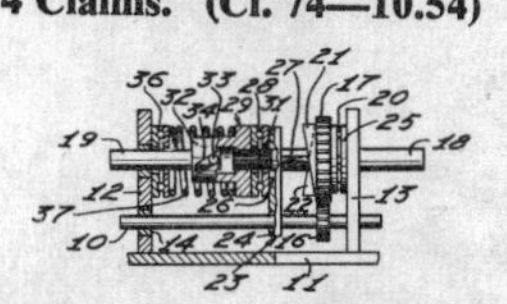

1. A differential mechanism comprising a base plate, a pair of standoffs extending upwardly from said base plate, a first input shaft rotatably supported by the standoffs, a first gear attached to said first input shaft, an ouput shaft rotatably supported by the second standoff, a second gear rotatably supported on said output shaft, a pair of cam surfaces mounted on the face of said second gear, said second gear in mesh with the first gear, a cam-follower member formed with a pair of openings through which the first input and output shafts extend, a pair of cam followers attached to said cam-follower member and in engagement with said cam surfaces, a collar slidably but non-rotatably attached to said output shaft adjacent said cam-follower member and formed with a hollow portion and a pair of transverse slots cut at an angle relative to the longitudinal axis of said output shaft, a second input shaft supported by the first standoff and in axial alignment with the output shaft, a pair of transverse pins attached to the end of the second input shaft and said pins received in the slots formed in said collar, and a spring mounted between the first standoff and said collar and maintained under compression.

2,716,898
VARIABLE SPEED PHONOMOTOR
Saburo Akai, Ohta-ku, Tokyo, Japan
Application December 12, 1951, Serial No. 261,305
Claims priority, application Japan August 9, 1951
3 Claims. (Cl. 74—194)

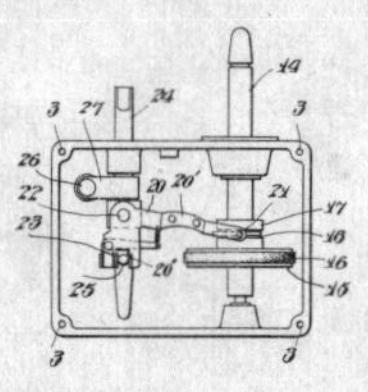

1. A variable speed phonomotor comprising main and auxiliary half cases so formed as to be combined together to form a unit box, the main half case containing a motor having a rotor axle provided with a worm gear, a worm wheel supported to engage the worm gear, and a friction disc wheel mounted to rotate with the worm wheel, and the auxiliary half case containing a rotary shaft, a wheel axially slidable but non-rotatively mounted on the rotary shaft, an operating rod operable from outside, a boss on the slidable wheel, leverage

having pins engaging a groove formed in the boss of said slidable wheel and which has bifurcated ends to engage a stud fixed on the operating rod, said slidable wheel being positioned to frictionally engage said friction disc wheel when the half cases are joined together so that actuation of the leverage through the operating rod varies the engagement point of the sliding wheel with the friction disc wheel to thereby control the speed of rotation of said rotary shaft.

2,716,899
PIVOTAL GEARING
Anthony L. Lado, Rome, N. Y., assignor to Pettibone New York Corporation, Rome, N. Y., a corporation of New York
Application October 24, 1952, Serial No. 316,712
7 Claims. (Cl. 74—385)

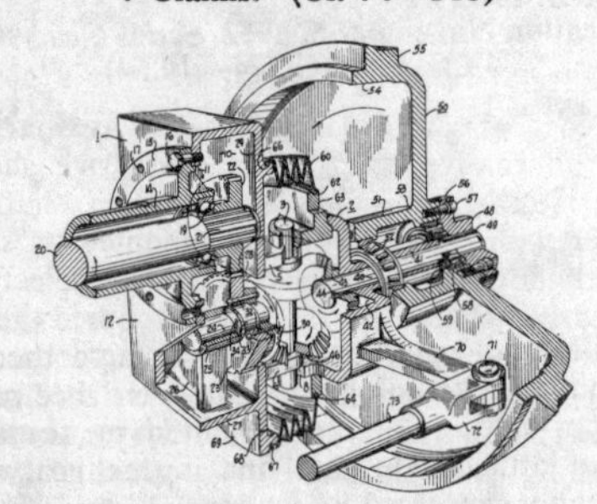

1. A drive comprising a pair of housings, pivot means mounting one housing on the other for relative swinging movement, a drive transmitting element and means mounting such element for rotation about the swing axis of the housings, rotary drive means journaled in one of the housings and engaged with the drive element to rotate the latter, rotary driven means journaled in the other of the housings and engaged with the drive element to be rotated by the latter, a flexible tubular sleeve interposed between the housings and surrounding the pivot means the ends of the sleeve being sealingly engaged with the housings, a hub rotatably mounted on the other of the housings, and means connecting the driven means to the hub to turn the latter.

2,716,900
PRESS
Joseph Barkham, Chicago, Ill., assignor to Chisholm, Boyd & White Company, a corporation of Illinois
Application June 11, 1952, Serial No. 292,796
10 Claims. (Cl. 74—520)

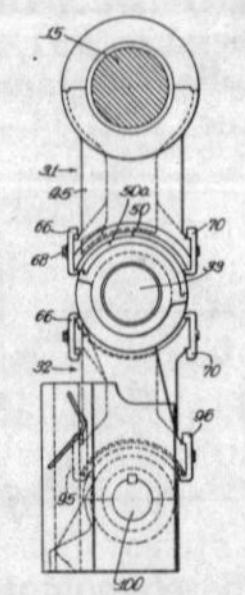

1. In a press having a frame, a vertical toggle construction comprising an upper toggle arm rotatably mounted at its upper end in the frame, a plurality of male bearings on the lower end of the toggle arm, a shaft rotatably secured to the lower end of the toggle arm, a plurality of female bearings on the lower end of the toggle arm, a lower toggle arm rotatably secured at its upper end to said shaft and having a plurality of male bearings at its upper end engaging the female bearings, a plurality of female bearings in said end of the lower toggle arm and engaging the male bearings of the upper toggle arm, a plunger rotatably secured to the bottom of the lower toggle arm, and means for swinging the shaft to move the toggle over dead center to move the plunger.

2,716,901
CONTROL DEVICE FOR FUEL PUMPS
Frank Reginald Howe, Ealing, London, England, assignor to C. A. V. Limited, Acton, London, England
Application December 11, 1951, Serial No. 261,087
Claims priority, application Great Britain December 30, 1950
2 Claims. (Cl. 74—526)

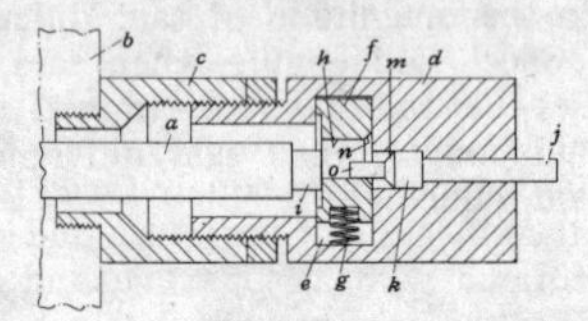

1. In a liquid fuel pump having a slidable bar for regulating the output of the pump, the combination of a stationary hollow body part into one end of which the slidable bar extends, a spring-loaded stop slidable within the hollow body part in a direction at right angles to the direction of movement of the slidable bar, the stop being formed with a transverse opening into which the said end of the bar is movable when the stop is in an appropriate position, but which in the normal position of the stop is inaccessible by the bar so that the extent of movement of the bar into the body part is limited by contact with the stop of the adjacent end of the bar, and a manually slidable plug extending through the end of the body remote from the bar, and shaped to co-operate, when moved inwardly, with the adjacent end of the aperture in the stop to move the latter into a position in which the aperture is accessible by the adjacent end of the bar which upon being moved into the aperture serves to return the plug to its initial position.

2,716,902
MECHANISM CONTROL
Willard C. Skareen, Toledo, Ohio, assignor to The Bingham-Herbrand Corporation, Toledo, Ohio, a corporation of Ohio
Application October 19, 1949, Serial No. 122,175
5 Claims. (Cl. 74—541)

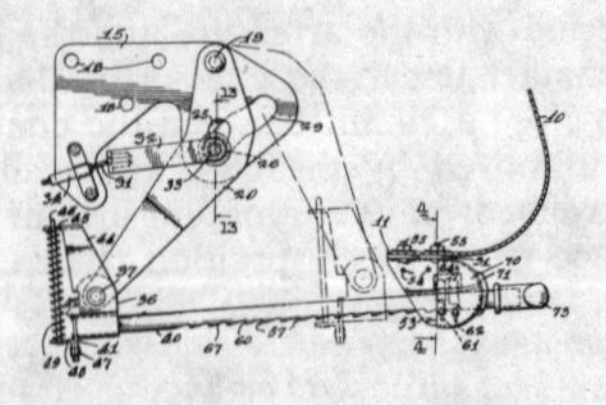

1. A brake actuating mechanism for vehicles comprising a rod-shaped brake control member having an integral ratchet, a relatively fixed part, an aperture in said part through which the rod-shaped member extends, and is slidable for brake actuation, a latch member in the relatively fixed part substantially perpendicular to the longitudinal axis of the control rod and movable for cooperating with the ratchet thereby to arrest sliding movement of the control rod in predetermined positions, the interior surface of the aperture being adapted at least in part to support the rod-shaped member and said member extending through said aperture with sufficient clearance for the rod-shaped member to be nutatable in said aperture about an axis perpendicular to the longitudinal axis of the rod.

2,716,903
STEERING FACILITIES OF AN AUTOMOBILE
Hans Peter Hansen, Dunellen, N. J.
Substituted for application Serial No. 61,500, November 22, 1948. This application July 12, 1954, Serial No. 442,504
5 Claims. (Cl. 74—557)

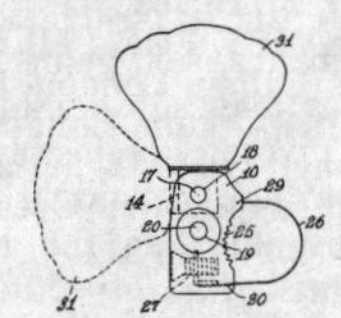

2. In a device of the class described, a housing formed with a cut-away portion at the top thereof, a handle pivotally mounted in said cut-away portion, a knob at one end of said handle, and a square-faced block member integral with the other end of the latter, said housing being in one side formed with an opening, and in alignment with the latter in the opposite side with a square opening, a shank transversely and slidably mounted in the housing and through said openings, said shank being at one end formed with a square-faced block member adapted to engage and cooperate with the square-faced block member of the handle, the square-faced block member of the shank being adapted to move through the square opening in the side wall and beyond the latter in permitting the handle to move from a vertical to a horizontal position, and an extension spring actuating said shank in its movement forth and back.

2,716,904
DAMPER ASSEMBLY
Herbert H. Schuldt, Indianapolis, Ind., assignor to Schwitzer-Cummins Company, Indianapolis, Ind., a corporation
Application April 15, 1953, Serial No. 349,015
5 Claims. (Cl. 74—574)

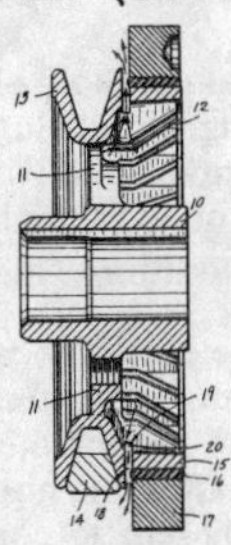

1. A damper assembly comprising a hub adapted for mounting on an engine crank shaft, an accessory drive pulley carried by said hub, a spaced series of spokes interconnecting said hub and pulley, an annular damper unit including an elastic member, said unit being offset and spaced from said pulley to provide a circumferential air passage therebetween, and a series of peripherally spaced radiating vanes interconnecting said damper unit and pulley positioned to pump air through said passage.

2,716,905
SELECTIVE SPEED REDUCTION GEARING
Max Wernli, Schaffhausen, Switzerland, assignor to Georg Fischer Aktiengesellschaft, Schaffhausen, Switzerland
Application October 1, 1951, Serial No. 249,057
Claims priority, application Switzerland October 5, 1950
5 Claims. (Cl. 74—665)

1. Selective speed reduction gearing apparatus comprising a stationary housing, a spindle carried by said housing, a swingable gear housing articulated on said spindle carried by said stationary housing, a drive shaft rotatably supported in said gear housing, a driven shaft and at least one auxiliary driven shaft each rotatably supported in said stationary housing, said drive, driven and auxiliary driven shafts being supported in their respective housings with their axes at equal distances from and spaced along a circle about the axis of said spindle carrying said gear housing on said stationary housing whereby said drive shaft is enabled to be aligned with any of said driven and auxiliary driven shafts by movement of said gear housing on said spindle, coupling

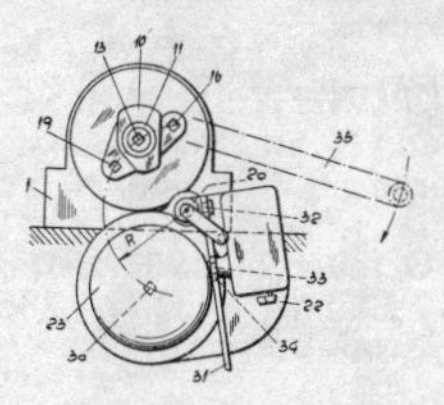

means on each of said drive, driven and auxiliary driven shafts for alternative coupling of said drive shaft with either of said driven or auxiliary driven shafts when aligned therewith, an electromotor mounted on said gear housing, reduction gear means between said electromotor and said drive shaft, an internally toothed gearwheel angularly coupled to said driven shaft, a toothed pinion on said auxiliary driven shaft, said gear wheel and said pinion being mounted in said stationary housing said pinion meshing with said internally toothed gearwheel.

2,716,906
AUTOMATIC TRANSMISSION
Frederick W. Seybold, Westfield, N. J.
Application December 12, 1951, Serial No. 261,249
25 Claims. (Cl. 74—677)

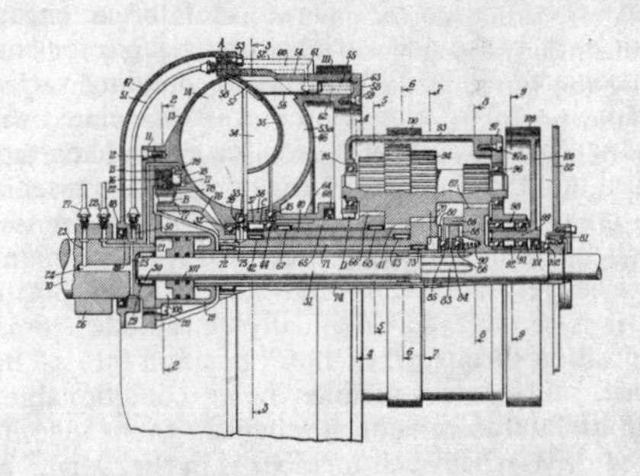

1. In a transmission having driving and driven shafts; a plurality of intermediate shafts arranged in telescoping relation with each other and with said driving and driven shafts, sun gears of respectively different sizes on said intermediate shafts, a sun gear rotatable on and supported by said driven shaft, cluster planet gear means meshing with said sun gears, a frame rotatable on the axis of said shafts and rotatably supporting said cluster planet gear means, brake means selectively engageable for holding said frame against rotation in one direction, an overrunning clutch between said frame and brake means for permitting rotation of said frame in a direction opposite to the said one direction, means for coupling said driving shaft first with the one of said intermediate shafts having the smallest sun gear thereon and then with the other of said intermediate shafts having the largest sun gear thereon, clutch means on a third of the sun gears on said intermediate shafts which is intermediate the size of the said other two sun gears on said intermediate shafts, clutch means on said sun gear rotatable on said driven shaft, and clutch means movable on and drivingly connected to said driven shaft engageable with the said clutch means on said sun gears for coupling said driven shaft to the sun gear rotatable on said driven shaft or to the said third sun gear.

2,716,907
CONTROL FOR STEERING-BY-DRIVING MECHANISM
William O. Bechman and Harry A. Land, Chicago, and David B. Baker and William W. Henning, Riverside, Ill., assignors to International Harvester Company, a corporation of New Jersey
Application April 19, 1950, Serial No. 156,764
10 Claims. (Cl. 74—720.5)

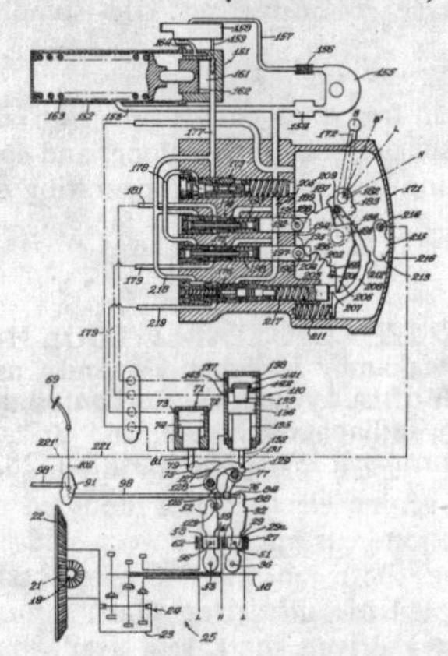

1. For use in the control of vehicle steering-by-driving apparatus including driven shafts respectively drivingly connected with propelling means at opposite sides of the vehicle, brakes for said shafts, and change-speed power transmission mechanisms selectively operable for transmitting driving force respectively to said driven shafts to drive the same at relatively high or low speeds or operable to neutral position to transmit no driving force to said shafts: the combination of separate shaft brake engaging means for each brake and variably energizable for engaging its associated brake to cause corresponding variation in its resistance to rotation of the shaft associated therewith; and a separate control means for each change-speed mechanism and the shaft brake engaging means associated with the brake for the shaft driven by such transmission, each control means including a control member manipulatable selectively between "High" and "Low" positions wherein it is settable and manually retractible from the "High" position through the "Low" position into a "Braking" range, said control member being conditionable for operation of its transmission mechanism to arrange it for high speed drive when such member is in the "High" position, for low speed drive when such member is in the "Low" position, and in neutral when such control member is in the "Braking" range, and each shaft brake engaging means being operable to incur shaft rotation resistance of a magnitude correlated with the degree of retraction of the control member therefor into its "Braking" range.

2,716,908
AUTOMATIC GRINDER FOR CIRCULAR SAWS
Roland O. Lundberg, Lincolnwood, Ill.
Application July 19, 1954, Serial No. 444,104
7 Claims. (Cl. 76—40)

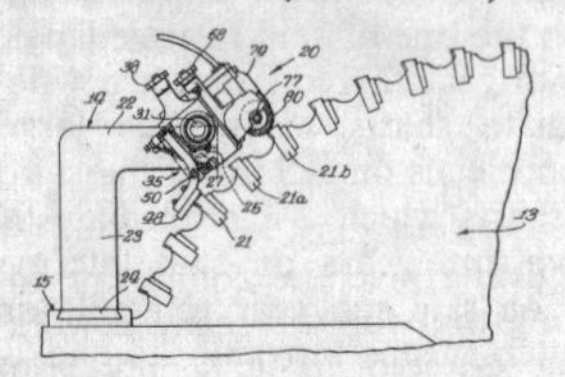

6. In a saw grinder of the class described for sharpening the teeth of circular metal cutting saw blades, the combination comprising, a main body casting attachable to a rigid support system for operating in adjacency to the radial periphery of the blade to be sharpened, means for supporting the casting pivotally about an axis transverse to the plane of the saw blade, a guide means connected to the casting on one side of its pivotal axis, a grinding wheel supported on said casting and at a position located on the opposite side of its pivotal axis from said guide means, said guide means and grinding wheel being disposed in substantially coplanar relationship, means for pivotally adjusting the said guide means with respect to the rotational axis of the saw blade, and means for adjusting the radial positioning of said grinding wheel with respect to the rotational axis of the saw blade independently of the guide means; the arrangement being such that said guide means is normally disposed in contact with the outer peripheral edge of one tooth of said saw blade while the grinding wheel engages and sharpens a succeeding tooth of the saw blade to the end that a tooth sharpened by said grinding wheel acts as a limit means for determining the radial positioning of the guide means with respect to the rotational axis of the saw blade, the guide means in turn determining the radial disposition of the grinding wheel with respect to the axis of the saw blade so as to result in the uniform sharpening of the saw's teeth.

2,716,909
METHOD OF MAKING A REFLECTOR MOLD
Richard Rupert, Independence, Mo.
Application February 9, 1952, Serial No. 270,861
6 Claims. (Cl. 76—107)

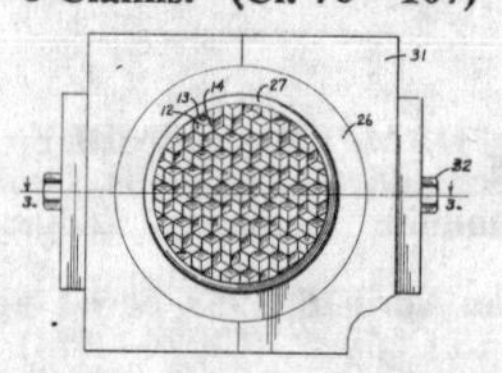

1. The process of making mold elements of the character described comprising, producing a male hob with locating faces and a pyramidal end with the surfaces thereof arranged in a circuit around the axis of the hob, said surfaces intersecting in a point on said axis and in accurate relation to the locating faces of said hob, polishing the surfaces, aligning the axes and locating faces of the hob and an element of formable material in a confined passage with the pyramidal end of the hob adjacent an end of the element, and pressing the hob into the element and forming the material therein to conform to the confined passage and the pyramidal end of the hob to provide a female mold element with an axial recess having mold forming surfaces sloping inwardly toward the axis of the element and arranged in a circuit therearound, the movement of the hob into the formable material effecting a moving engagement of the polished surfaces of the hob with the mold forming surfaces in the axial recess to accurately smooth said mold forming surfaces and provide a polish thereon.

2,716,910
INSULATING HOOKED PLIERS FOR THE HANDLING OF FUSES
Roger Jean Guerinet, Malakoff, France, assignor to Societe Stapfer & Cie, s. a. r. l., Paris, France, a body corporate of France
Application March 20, 1952, Serial No. 277,586
Claims priority, application France March 22, 1951
5 Claims. (Cl. 81—3.8)

1. A tool for handling cartridge fuses comprising in combination a hollow handle provided with an aperture at one of its ends, a first blade-shaped element fixed in flat position in said hollow handle and freely projecting at

its outer end through said aperture, a second blade-shaped element disposed in flat position above the first one and having one of its ends projecting through said aperture, a substantial semi-cylindrical hook-shaped extension on said projecting end of the second blade-shaped element, having its axis parallel to and transversally disposed with respect to said first blade-shaped element, a slide fixed to

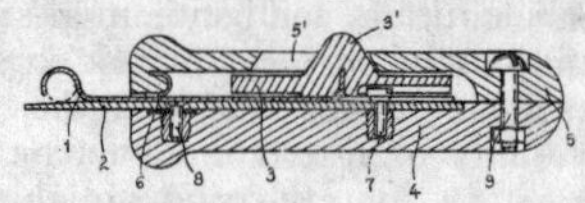

the other end of said second blade-shaped element and disposed within the hollow handle, a guide-slot in the handle and a thumb-piece on said slide projecting through the guide-slot, whereby the said slide may be pushed forwards to have the hook located beyond the outer end of the first blade-shaped element and rearwards to bring the hook above the said first blade-shaped element.

2,716,911
OPEN-CENTER VISE
Gustavus K. Focke, East Stroudsburg, Pa.
Application October 1, 1954, Serial No. 459,621
2 Claims. (Cl. 81—23)

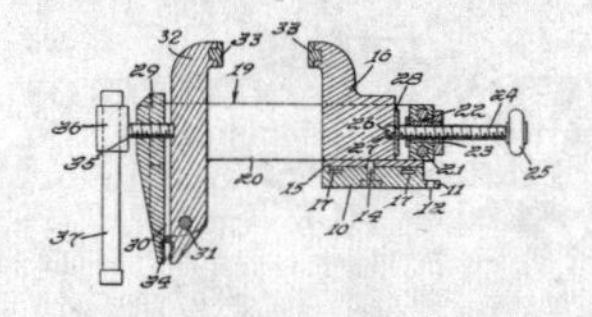

1. In an open-center vise, a base provided with a plurality of apertured ears adapted to be secured to a supporting structure, a swivel pin extending upwardly from said base, a horizontally disposed plate engaging said pin and rotatably mounted on said base, a first jaw secured to said plate, there being a pair of spaced parallel grooves in the sides of said jaw, an open frame including a pair of spaced parallel and elongate side bars slidably engaging said grooves, an end member extending between a corresponding end of each of said bars and secured thereto, said end member being provided with an opening, a sleeve nut seated in said opening so that its axis extends parallel to elongate axes of said parallel bars, a screw member threaded through said sleeve nut, said jaw having a recess in a wall confronting said end member for rotatably receiving an end of said screw member, said screw member having an annular groove therein, a collar secured to said wall of said jaw and having a portion engaging said annular groove, a bracket extending between a corresponding opposite end of each of said side bars and secured thereto, a portion of said bracket extending down below said side bars, said bracket including a pair of spaced parallel side walls, a pin supported by and extending between said side walls, a second jaw pivotally mounted on said pin, a spring member interposed between the lower end of said bracket and said second jaw to bias said second jaw to its non-clamping position, and manually operable means for pivoting said second jaw against a work piece to be clamped between said first and second jaws.

2,716,912
LEVER-AND-LINK FEED, SLIDABLE JAW WRENCH
Charles Leslie Maitland, Jasper, Ontario, Canada
Application December 9, 1952, Serial No. 324,900
4 Claims. (Cl. 81—86)

1. In a wrench, an elongate member including a jaw, said member being formed with a toothed rack, a second jaw movable on said member toward and away from said first jaw, a pivoted operating handle bodily movable with said second jaw, a pair of pawls pivoted to said handle at longitudinally spaced points therealong

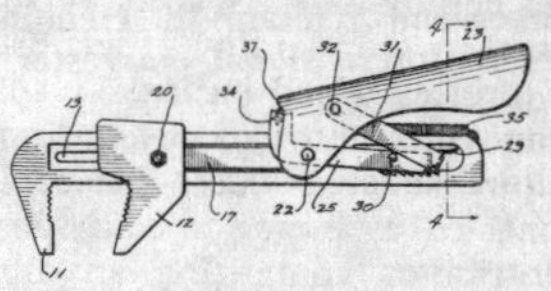

and having their free ends engageable with said rack at longitudinally spaced points therealong, and spring means resiliently holding said pawls in co-operating engagement with said rack.

2,716,913
UNIVERSALLY ADJUSTABLE TOOL HOLDER
Herbert H. Leerkamp, Indianapolis, Ind., assignor, by mesne assignments, to John A. Rockwood and Herbert H. Leerkamp, Indianapolis, Ind.
Application April 19, 1952, Serial No. 283,193
3 Claims. (Cl. 82—12)

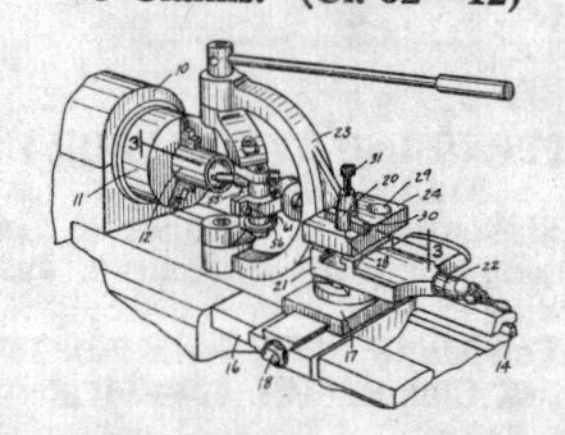

1. A tool holding apparatus for lathes adapted to move the cutting surface of a tool about a center on a fixed radius comprising a mounting yoke adapted to be mounted on the tool post of an engine lathe, a swing yoke pivotally mounted between the ends of said mounting yoke, a post eccentrically mounted on said swing yoke, a double clamp, one portion of which is pivotally connected to said post for rotational movement thereon, an arm slidably and rotatably mounted in the other portion of said double clamp, a primary tool holder pivotally mounted at one end of said arm for angular movement with respect thereto, and a secondary tool holder slidably and rotatably mounted in said primary tool holder.

2,716,914
SMOKE SUPPRESSING SYSTEM FOR USE WITH MACHINE TOOLS
Reginald J. S. Pigott, Pittsburgh, Pa., assignor to Gulf Research & Development Company, Pittsburgh, Pa., a corporation of Delaware
Application February 20, 1952, Serial No. 272,670
2 Claims. (Cl. 82—34)

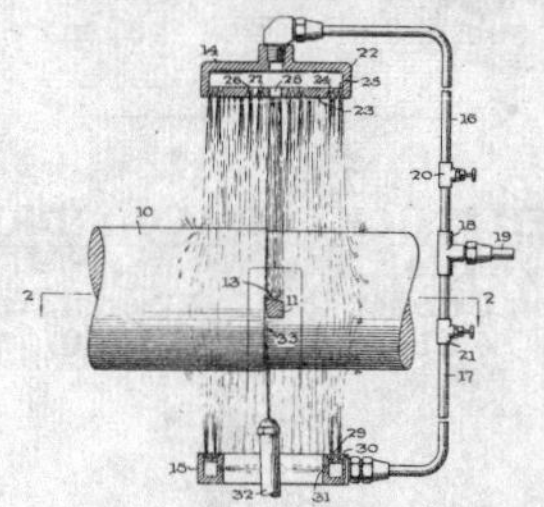

1. A metal cutting machine comprising a work piece support, a cutting tool for traversing a work piece held in the support, a source of liquid supply, an upper liquid distributor positioned above the area where the tool engages the work and connected by piping to said source of supply, said upper distributor comprising an enlarged head having a face plate formed with a central opening

overlying the point of cut and small, closely spaced orifices adjacent its periphery, whereby a heavy stream of liquid discharges through said central opening onto the chip formed in cutting and a spaced, surrounding liquid curtain is formed by the plurality of jets discharged from the peripheral orifices which encloses a space about the point of cut and thereabove, a lower liquid distributor positioned below the area where the tool engages the work and connected by piping to said source of supply, said lower distributor comprising an annular conduit formed with a series of closely spaced orifices directed upwardly toward the work and having in its inner annular wall a series of orifices discharging radially inward, whereby an enveloping curtain of spray will be discharged upwardly to merge with the descending spray from above, the liquid-walled chamber so formed by the upper and lower spray jets being closed at its bottom by the radially directed jets, whereby smoke produced in the cutting operation will be confined and absorbed by contact with the liquid sprays and chips from the cutting operation are removable at bottom without disruption of the seal provided by the radial jets.

2,716,915
SMOKE SUPPRESSING SYSTEM FOR USE WITH MACHINE TOOLS
Albert Biber, Pittsburgh, Pa., assignor to Gulf Research & Development Company, Pittsburgh, Pa., a corporation of Delaware
Application May 9, 1952, Serial No. 287,059
2 Claims. (Cl. 82—34)

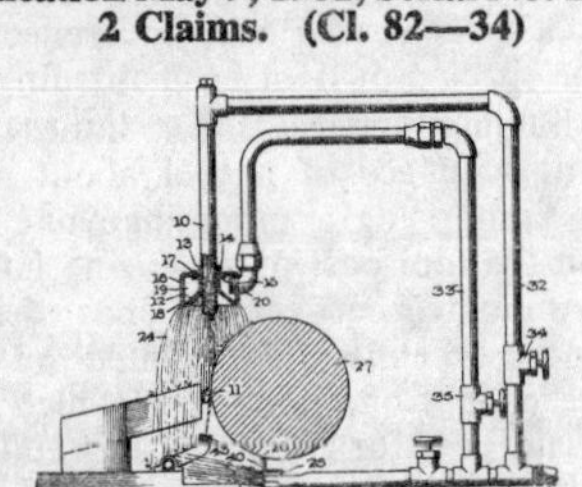

1. In combination with a metal-cutting machine having a cutting tool, a liquid distributor positioned above the cutting tool and workpiece engaged thereby for discharging a curtain of liquid downwardly to surround the point of cut and spaced therefrom to provide a continuous liquid surface of large area within which smoke generated in the cutting operation is confined, a second liquid distributor extending beneath the work piece and comprising an elongated head provided with conduit means for supplying liquid thereto, said head being formed with an upwardly opening slot and communicating slots angularly related thereto at the ends thereof whereby liquid is discharged from said head in sheets which substantially box the space beneath the point of cut.

2,716,916
REED BLOCK FOR ACCORDIONS
Angelo Piana, Newark, N. J.
Application October 23, 1952, Serial No. 316,416
4 Claims. (Cl. 84—360)

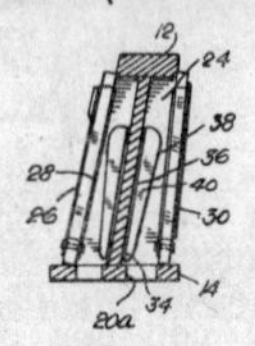

1. A reed block for accordions, comprising a longitudinal partition, a top wall on said longitudinal partition, a bottom wall on said longitudinal partition, a plurality of spaced transverse partitions on said longitudinal partition, on both sides thereof, said transverse partitions being connected to said top and bottom walls to form air chambers between said transverse partitions and between said top and bottom walls, a plurality of reeds extending across said air chambers on the outside edges of said transverse partitions and between said top and bottom walls, a plurality of openings formed in said bottom wall, one for each air chamber and communicating therewith, and a plurality of inserts mounted on said longitudinal partition, one for each said air chamber, each said insert being made of vibratory material and being provided with pads between the insert and the longitudinal partition to enable said insert to vibrate under the influence of the reed vibrations which would be generated by the passage of air in either direction through the openings in the bottom wall, the air chambers and the reeds.

2,716,917
TONE MODIFIER FOR MUSICAL INSTRUMENTS
Frederick J. Troppe, Joliet, Ill.
Application February 24, 1953, Serial No. 338,297
3 Claims. (Cl. 84—400)

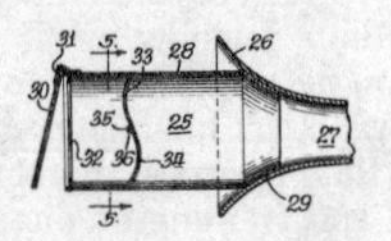

1. In an accordion having a bellows, a reed-mounting panel adjacent one end of the bellows, and having means for selectively controlling the passage of air from the bellows through the reed-mounting panel including a keyboard disposed adjacent the latter, an improved resonating chamber for enclosing the side of the reed-mounting panel opposite said bellows and adjacent said keyboard, said chamber comprising a generally shallow housing having a plurality of openings along the outer surface thereof, a hollow cylindrical tube surrounding each of said openings and extending outwardly from said housing, the axis of each of said tubes intersecting the plane of said keyboard, and valve means disposed in covering relation to the outer end of each of said tubes for controlling the passage of sound from within the housing, said valve means comprising a lid having an edge portion thereof hingedly mounted at the outer end of said tube for swinging movement to and from a position closing the opening at said outer end, all of said lids being similarly hingedly connected to their respective tube at a position intermediate the transverse horizontal and vertical axes and above the transverse horizontal axis of said tube, as determined when the accordion is in its upright operative position with the bellows expandable in a generally horizontal direction, whereby a generally lateral swinging movement of the accordion will effect a substantially simultaneous movement of said lids relative to said tubes through generally parallel paths.

2,716,918
SLIDE VIEWER
John R. Miles, Chicago, Ill., assignor to Michael S. Wolk, Chicago, Ill.
Original application August 25, 1950, Serial No. 181,553, now Patent No. 2,621,993, dated December 16, 1952. Divided and this application October 23, 1952, Serial No. 317,257
3 Claims. (Cl. 88—1)

1. A portable viewer for slides such as is adapted to be formed by plastic molding and which comprises a principal body for enclosing a rectangular space, an in-

clined mirror disposed in said space to reflect rays from a top direction of said body toward a front direction thereof, a front wall to said body having a viewing aperture, a lens over said aperture, the top of said body having a light projection aperture therein, said body including a peripheral frame surrounding and above the projection aperture, said frame having a pair of side approach channelways opening at the side of the viewer for the reception of a light diffusion slide, and a pair of side approach channelways opening at the side of the viewer for the reception of picture slides between the

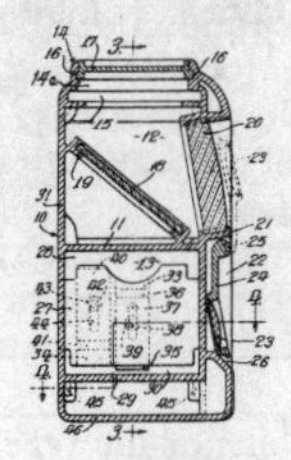

light diffusion slide and the light projecting aperture, the frame being narrower at the place of opening of the first channelway than the spacing between the second channelways so that a straight slide inserted in the second channelways projects beyond the frame at the first channelway to facilitate gripping of the slide for removal, the side of said frame opposite the one in which said second channelways open having a slide back stop to limit the extent of slide insertion thereby and to facilitate the rapid insertion and framing of picture slides for successive exposures.

2,716,919
PICTURE PROJECTING SYSTEMS AND SCREEN THEREFOR
Ernest Gordon Beard, Willoughby, New South Wales, Australia
Application December 3, 1948, Serial No. 63,317
Claims priority, application Australia December 15, 1947
5 Claims. (Cl. 88—16.6)

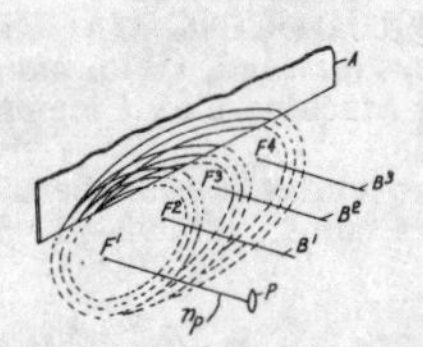

1. A system for stereoscopically viewing a pair of projected images comprising a planar screen having an optically polished surface defined between a pair of side edges and a pair of upper and lower end edges, said surface being provided with a set of spaced grooves lying along elliptical paths in the plane of said screen, said grooves extending uninterruptedly between said side edges and being confocal on two foci positioned in the plane of said screen and in an area extending beyond said end edges and defined by said side edges, the cross-section of said grooves being substantially constant and corresponding at least in part to a circular arc, and stereoscopic means positioned to project a pair of images on said screen, one of said images to be viewed by the right eye of an observer and the other of said images to be viewed by the left eye of said observer, said projecting means being spaced from the plane of said screen and positioned at a location beyond the end edges of the latter, whereby light rays from each of said images are bent by said grooves so that rays from one of said images are received by one eye of said observer while they are not received by the other eye of said observer.

2,716,920
DEVICE FOR TAKING MOTION PICTURES IN RELIEF
Raymond Henri Denis Rosier, Paris, France, assignor to Societe d'Optique et de Mecanique de Haute Precision, Paris, France, a corporation of France
Application December 12, 1951, Serial No. 261,340
Claims priority, application France December 13, 1950
1 Claim. (Cl. 88—16.6)

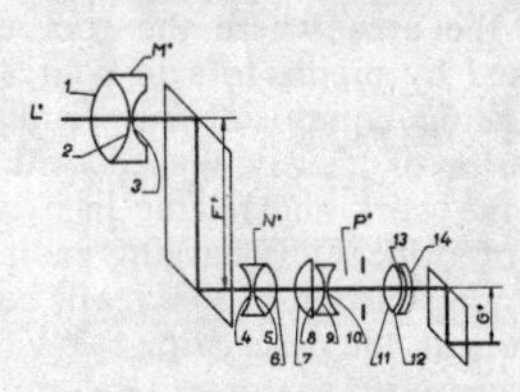

A stereoscopic photographic device adapted to replace the normal single objective of an ordinary camera and to obtain two images of the object photographed juxtaposed on the same film with a spacing between their geometrical centers equal to one-half the width of the normal single image, comprising a pair of objectives having parallel axes spaced apart a distance intermediate the stereoscopic spacing and the spacing of the two images on the film, a pair of rhombic prisms at the rear of said objectives for reducing the spacing between parallel light beams issuing from the objectives to that of the images on the film, and an afocal lens system in front of each objective providing a field of the same size as that of the normal single objective, said afocal lens systems including lenses having stereoscopically spaced axes and reflecting means for reducing the spacing of light beams issuing from said lenses from the stereoscopic spacing to the spacing of the axes of said pair of objective.

2,716,921
INTERMITTENT FILM FEEDING MECHANISM
Herbert Edward Holman, West Drayton, and George Charles Newton, Anerley, London, England, assignors to Electric & Musical Industries Limited, Hayes, England, a company of Great Britain
Application December 4, 1951, Serial No. 259,754
Claims priority, application Great Britain December 8, 1950
9 Claims. (Cl. 88—18.4)

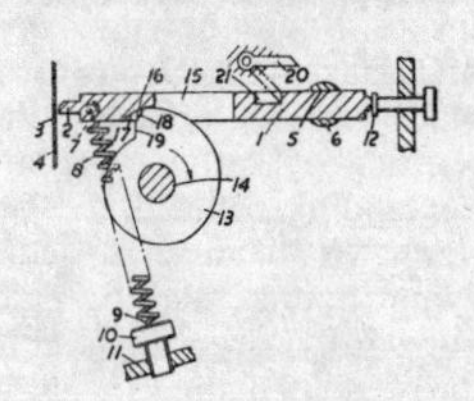

1. A mechanism for intermittently moving films, comprising an arm, means mounting said arm for pivotal and axial movement, a claw carried by said arm, a rotatably mounted cam co-operatively associated with said arm to impart movement thereto, energy storage means coupled to said arm, said cam having an ascending surface to impart movement in one direction to said arm against an opposing force of said storage means to store energy in said storage means, said ascending surface terminating in a first descending surface sloping away from said ascending surface in a direction opposite to the direction of rotation of said cam to move said arm towards the film to engage said claw with said film under the action of the energy stored in said storage means, said cam having a second descending surface of increased slope compared with said first descending surface to enable said stored energy to effect a rapid advance of said film after engagement of said claw with said film.

2,716,922
SOLENOID OPERATED MAGAZINE SLIDE PROJECTOR
Theron W. Stephens, Bakersfield, Calif.
Application April 15, 1953, Serial No. 348,854
2 Claims. (Cl. 88—28)

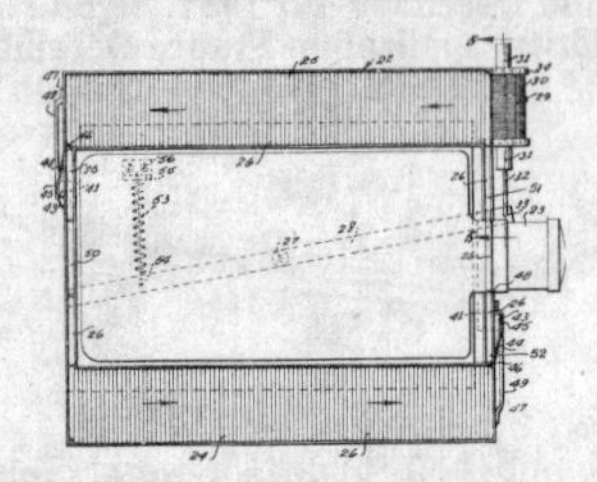

2. An automatic slide projector comprising a base, a housing supported above said base, a lens assembly extending forwardly from said housing, a slide feeding trough arranged on one side of said housing, a slide receiving container arranged on the opposite side of said housing, there being a passageway interconnecting the rear ends of said trough and container together, there being a passageway interconnecting the front ends of said trough and container together, means for automatically moving the slides through said passageways and past said lens assembly, said means comprising a lever pivotally mounted in said base, a solenoid mounted on the front of said housing and including a movable core, a rod connecting said core to said lever, a bar arranged transversely with respect to each end of said lever and secured thereto, an arm extending at right angles with respect to said bar and secured thereto, a pair of fingers projecting from each of said arms and each provided with a lug for engagement with said slides, there being a pair of slots communicating with said passageways for the slidable projection therethrough of said lugs, a bracket secured to said base, resilient means connected to said bracket and to said lever, and a time switch electrically connected to said solenoid.

2,716,923
FIRING MECHANISM FOR A RIFLE
Alonzo F. Gaidos, Redwood City, Calif.
Original application February 1, 1950, Serial No. 141,801. Divided and this application December 8, 1952, Serial No. 324,857
4 Claims. (Cl. 89—140)
(Granted under Title 35, U. S. Code (1952), sec. 266)

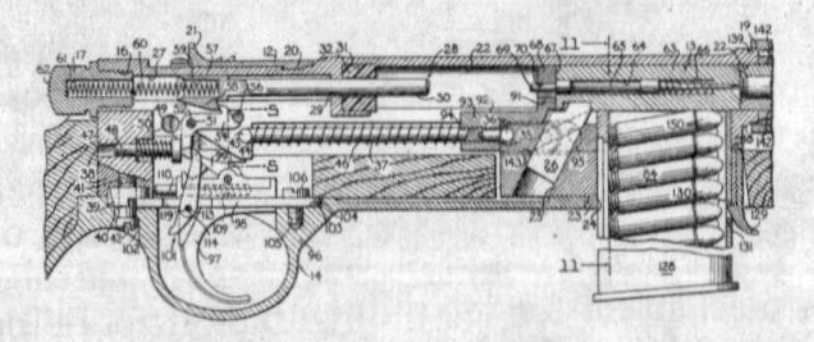

1. In an automatic firearm, a receiver, a trigger guard, a trigger slidably mounted in said trigger guard having a first and second firing position, a spring-biased semi-automatic sear pivotally mounted on a transverse pin in said receiver, a spring-biased automatic sear pivotally mounted in juxtaposition with said semi-automatic sear on said pin, a spring-biased striker slidably mounted in said receiver, said striker having a downwardly depending lug engageable by both of said sears, pivotal means on said trigger for releasing both of said sears from said striker in said first firing position to obtain semi-automatic firing, an actuator bar slidably mounted in said receiver for reciprocating movement therein whereby said automatic sear is disengaged from the striker engaging position as said bar moves rearwardly, and means on said trigger for holding said semi-automatic sear out of engagement from said striker in said second firing position whereby the cyclic release of said striker by said actuating bar effects automatic firing.

2,716,924
COPYING MACHINES
Frederick William Whitehead, Bristol, England, assignor to The Bristol Aeroplane Company Limited, Bristol, England, a British company
Application April 4, 1952, Serial No. 280,549
Claims priority, application Great Britain April 12, 1951
17 Claims. (Cl. 90—13)

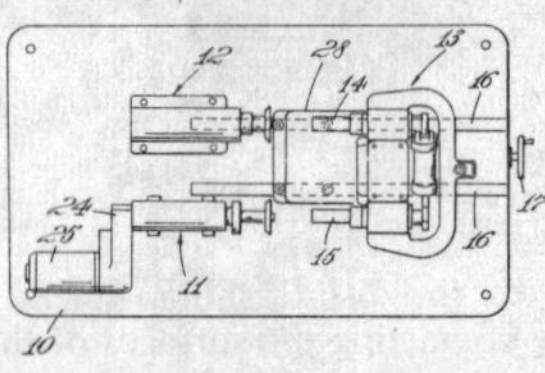

1. A copying machine of the type set forth comprising a base, a guide member carried from the base and fixedly located relatively to the base to make local engagement with the master, a tool carried from the base and fixedly located relatively to the base similarly to engage the workpiece, means for continuously oscillating the master and workpiece together about parallel axes through the cutting range, said parallel axes lying generally in the plane containing the long axes of the master and workpiece, means for supporting the master and workpiece for pivotal movement about a common axis normal to the axes of oscillation to move both said axes towards or away from the guide member and tool, the arrangement being that oscillating movements of the master resulting in pivotal movements thereof produce corresponding pivotal movements of the workpiece, and means for moving the master and workpiece relatively to the guide member and tool respectively whereby the master and workpiece are traversed across the guide member and tool generally in the direction of the axes of oscillation.

2,716,925
REPRODUCING MACHINE
Erwin G. Roehm, Norwood, Ohio, assignor to The Cincinnati Milling Machine Co., Cincinnati, Ohio, a corporation of Ohio
Application November 6, 1952, Serial No. 319,117
6 Claims. (Cl. 90—13.5)

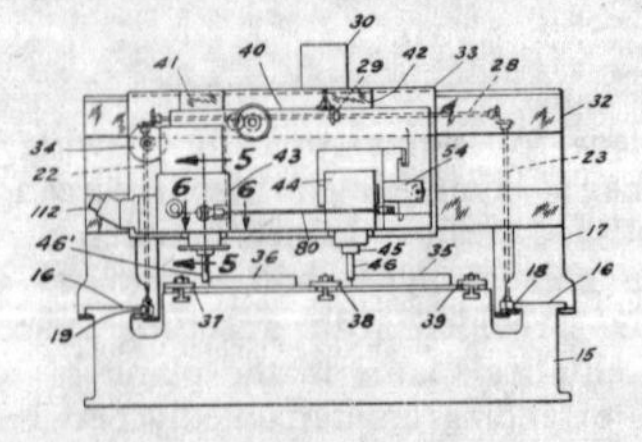

1. A reproducing machine including a carriage, a tracer support and a tool support mounted on said carriage, a tracer finger mounted on the tracer support for displacement relative thereto, a first power means for moving the carriage in the angularly related directions, control connections between the tracer and power means for determining the operation of the power means in accordance with the deflections of the tracer with respect to its support, a second power means coupled with the tool support to move said tool support relative to the carriage in two angularly relative directions, and additional control connections between the tracer and said second power means for activating said second power means to move the tool support relative to the carriage in accordance with the displacement of the tracer with respect to its support.

2,716,926
APPARATUS FOR TREATING WOOD PULP
Bruce Armstrong, Saginaw, Mich., assignor to Jackson and Church Company, Saginaw, Mich., a corporation of Michigan
Application January 2, 1951, Serial No. 204,039
8 Claims. (Cl. 92—26)

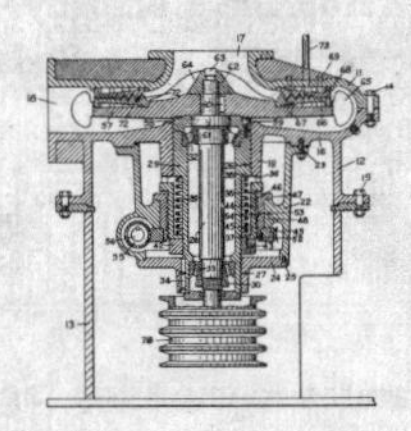

2. In a device for disintegrating fibrous material and simultaneously contacting the surfaces of the fibers thereof with a treating fluid, and including a fixed upper disk and a rotating lower disk, the improvement comprising: a supporting frame; a vertically disposed shaft; a tubular member surrounding said shaft, supporting same rotatably with respect thereto and holding said shaft against relative axial movement with respect thereto, said tubular member having an external, circumferential, downwardly facing shoulder near its upper end; a pair of spaced, elongated, vertically aligned and concentric bearings supporting said tubular member for axial movement with respect to said frame, said shoulder being positioned immediately below the uppermost of said bearings; an annular ring supported by said supporting frame, concentric with said shaft and axially positioned between said bearings; an elevating sleeve rotatably supported on said annular ring and having internal threading; an annular elevator screw having external threading engaging the threading of said elevating sleeve and supported thereby, an internally extending annular flange at the lower end of said elevator screw; means on said tubular member engaging said elevator screw for preventing relative rotative movement therebetween but permitting axial movement therebetween; a coil spring surrounding said tubular member, resting on said annular flange and bearing against said shoulder for resiliently urging said tubular member upwardly; a worm wheel surrounding said tubular member and affixed to said elevating sleeve for rotation therewith; a worm and means supporting same in operative engagement with said worm wheel; manually operable means for rotating said worm.

2,716,927
FOURDRINIER
Sylvester C. Sullivan, Hudson Falls, N. Y.
Application June 3, 1953, Serial No. 359,283
6 Claims. (Cl. 92—44)

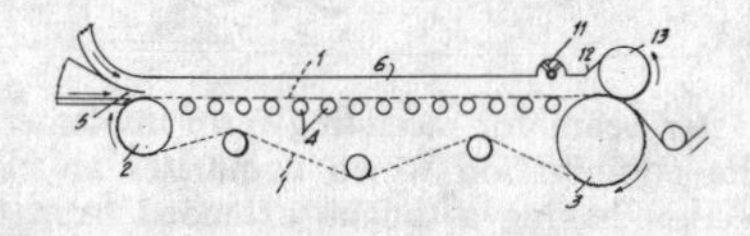

5. In a Fourdrinier machine comprising a wire and a slice, the combination with means enclosing the upper side of said wire from the slice throughout a zone within which a web of paper on said wire is substantially completely formed; of means for maintaining a flow of air through said enclosing means in the same direction as said wire moves; a movable damper at the end of said enclosing means for the discharge of said air therethrough; and means controlled by the air pressure within said enclosing means for moving said damper to maintain a substantially constant air pressure within said enclosing means.

2,716,928
THUMB HOLE PUNCH
Harry Keller, Philadelphia, Pa.
Application June 2, 1950, Serial No. 165,742
2 Claims. (Cl. 93—58.5)

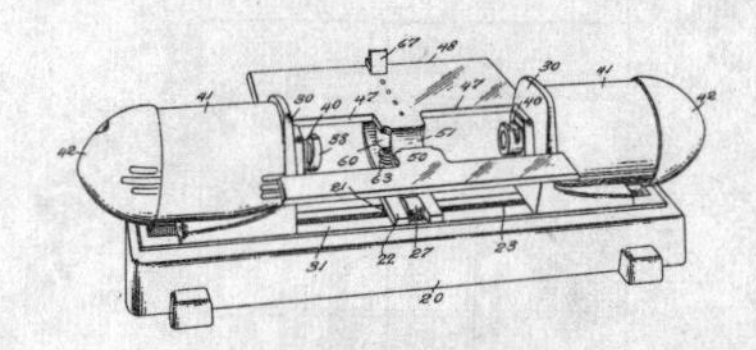

1. In a thumb hole punch, a work table having opposed recesses, a pivotal mounting for the work table at one side of the recesses, spring means urging the work table into limiting upper position, an electric switch positioned beneath the work table to be closed thereby, opposed solenoids mounted in line with the work table recesses and operatively electrically connected to the electric switch, means for adjusting the solenoids toward and away from one another while moving the solenoids in the recesses of the work table, the solenoids having armatures, punches in line with the solenoids and supported on the armatures and upwardly slotted dies cooperating with the punches, the slots of the dies extending to a position on a level with the depressed position of the work table.

2,716,929
PHOTOGRAPHIC ROLL HOLDER
Clarence Elwood Smith, Rochester, N. Y., assignor to Graflex, Inc., Rochester, N. Y., a corporation of Delaware
Original application July 25, 1949, Serial No. 106,628, now Patent No. 2,588,054, dated March 4, 1952. Divided and this application May 26, 1951, Serial No. 228,488
6 Claims. (Cl. 95—34)

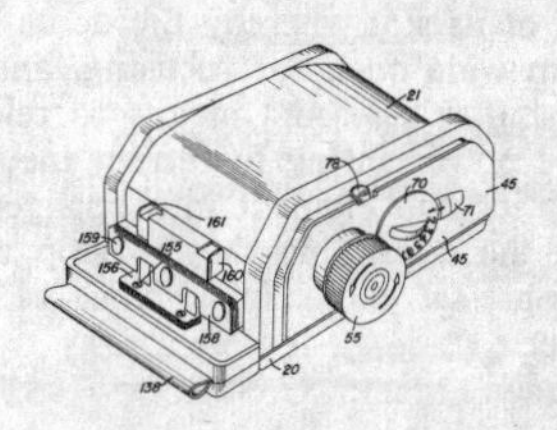

5. In a photographic film holder, a base having a central exposure opening extending therethrough from the front of the base to the back thereof, said base having a slot therein between the front and back registering with the exposure opening and extending to one side of the base to receive a dark slide, a film spool carriage removably mounted on the base, rotatable supports on the carriage for a supply spool and a take-up spool, said carriage having a bottom plate constituting a pressure pad to seat the film against the exposure opening and over which film is fed from the supply spool to the take-up spool, a pair of rails projecting from the base along the opposite, parallel sides of the exposure opening to support the film in its movement from the supply spool to the take-up spool, a second pair of rails formed on the base parallel to the first pair of rails and projecting further from said base than said first pair of rails to form a seat for the carriage and provide accurate clearance for the passage of film, a cover hingedly connected at one end to the base, and latch means for securing the cover to the base in closed position to hold the carriage on the base, said base and carriage, and said cover and carriage having interengaging portions for preventing light leakage when the cover is closed over the carriage and secured to the base.

2,716,930
AUTOMATIC LENS DIAPHRAGM CONTROL
Samuel Marson, New York, N. Y.
Application October 27, 1952, Serial No. 317,096
1 Claim. (Cl. 95—64)

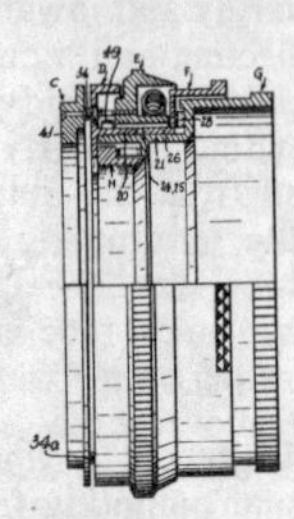

A detachable lens-diaphragm control device for use selectively with all complete conventional lens mounts of single lens reflex cameras, which comprises a ring fixedly positioned relative to a lens mount, a diaphragm-actuating ring operatively connected to the diaphragm ring of a camera lens, an adjustable pre-setting ring having an abutment and settable to stop said auxiliary ring at desired rotative position relative to the fixed ring, an extensible power spring connecting said actuating ring and presetting ring to rotate together as a unit while presetting, said auxiliary ring being movable to open the diaphragm wide and place the power spring under tension, a stop on said last-mentioned ring cooperating with said abutment, said fixed ring having slots arcuate in form, pins on said actuating ring and extending through said slots, a plate connected to said pins and rotatable therewith, means on said plate to connect the diaphragm ring of the camera lens thereto, clutch teeth on said fixed ring, cooperating clutch teeth on said adjustable ring, said last ring slidable axially to engage and disengage said clutch, spring means engaging said adjustable ring to hold the clutch teeth engaged, a latch spring on said fixed ring and cooperating with one of said pins to latch it at one end of its slot whereby the actuating ring holds the diaphragm wide open for focusing, and means operated by the shutter releasing means to release the latch and allow the power spring to retract the actuating ring back to bring its stop element into engagement with the abutment on the adjustable ring to stop the diaphragm down to the pre-set value, in advance of the operation of the shutter.

2,716,931
DAYLIGHT DEVELOPING TANK
Michael Lesjak, deceased, late of Ichenhausen, near Augsburg, Germany, by Babette Viktoria Lesjak, administratrix, Goggingen, Germany
Application August 23, 1952, Serial No. 305,946
Claims priority, application Germany September 20, 1951
8 Claims. (Cl. 95—90.5)

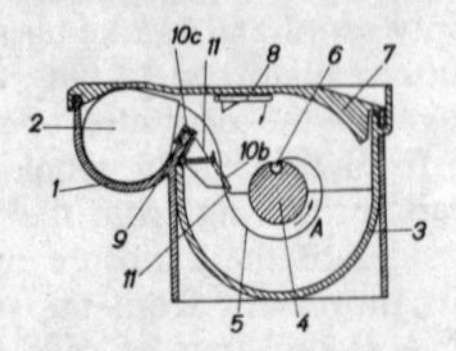

1. In a daylight developing tank for developing light-sensitive film strips and including an enclosed, covered tank divided into a pair of chambers separated by a partition wall and having a rotatable film strip receiving core member in one chamber and a storage member for said film strips in the other chamber, the improvement which comprises an automatic stationary, film strip catching device positioned intermediate said two members and including structure for catching and retaining the trailing end of a film strip as it is unwound from the storage member to the core member, said device being located adjacent to but below the normal path of movement of said film strip between said members.

2,716,932
ROOT SEPARATOR
Henry Grounds, Dallas, Tex.
Application December 29, 1952, Serial No. 328,461
3 Claims. (Cl. 97—10)

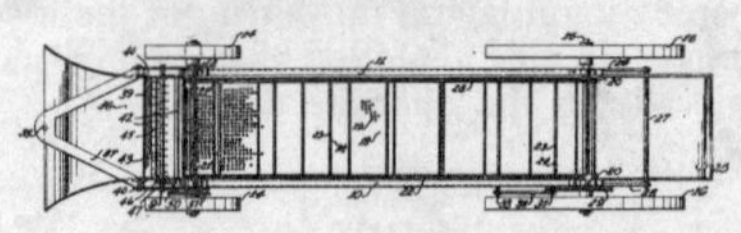

1. A root separator comprising an inclined frame including a first and second spaced parallel beam each having a U-shaped cross section, spaced parallel front and rear axles journaled on said frame, front and rear ground-engaging wheels mounted on said axles, a platform reciprocably supported by said beams and provided with a plurality of spaced perforations, an inclined scraper extending downwardly from the front end of said frame, there being a slot in said first beam, a block secured to said platform and arranged contiguous to said slot, a securing element extending through said slot and connected to said block, a first gear wheel arranged contiguous to said rear axle, a link having one end pivotally connected to said securing element and its other end eccentrically connected to said gear wheel, a second gear wheel mounted on said rear axle, and meshing with said first gear wheel, sprockets mounted on said front and rear axles, a pair of endless chains trained over said sprockets, a plurality of spaced parallel bars extending between said pair of chains and secured thereto, a plurality of spaced parallel fingers projecting from each of said bars, a hopper mounted on the front end of said frame for receiving material from said scraper, spaced parallel shafts extending through said hopper and operatively connected to said front axle, and staggered teeth projecting from said shafts.

2,716,933
MASTER IMPLEMENT AND TOOL CARRYING BEAM
Roy F. Smith, Silt, Colo.
Application May 20, 1952, Serial No. 288,857
1 Claim. (Cl. 97—26)

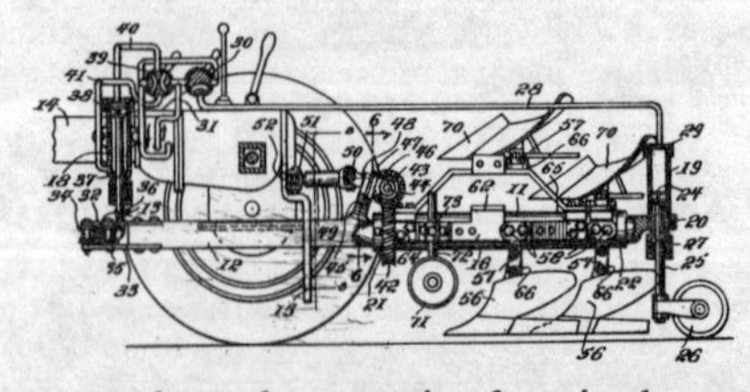

In a master beam for mounting farm implements on a tractor, the combination which comprises an elongated tubular section having a tongue extended from the forward end and having a hub, the axis of which is vertically positioned on the trailing end, a sleeve having flat upper and side surfaces rotatably mounted on said tubular section, said sleeve having threaded bolt-receiving openings in said flat surfaces and also having a worm gear on the forward end, a bracket having a bar parallel to and spaced from the sleeve adapted to be bolted to the sleeve through a flange on a diagonally disposed section at the forward end and also through a flange on a right angularly disposed section at the opposite end, said bracket being adapted to have a farm implement attached thereto, means for connecting the forward end of the tongue to a tractor, a worm mounted on the tongue and positioned in meshing relation with said worm gear, means for, selectively, actuating said worm by the power take-off of the tractor or with a hand crank, a wheel pivotally mounted in the hub on the trailing end of said cylindrical section of the beam, a yoke having a horizontally disposed lower section mounted on the tractor housing, rollers rotatably mounted on the tongue and positioned to travel on the lower portion of said horizontally disposed lower section, and a fork mounted on the tractor housing and positioned to receive said tongue, said fork being positioned between the yoke and forward end of the tubular section of the beam whereby upon lateral movement of the forward end of the tongue in one direction the trailing end thereof moves laterally in the opposite direction.

2,716,934
CULTIVATOR SHOVEL
Leroy A. Demorest, Ashley, Ohio, assignor to Cultiguard Shovel Company, a corporation of Ohio
Application September 1, 1950, Serial No. 182,732
2 Claims. (Cl. 97—204)

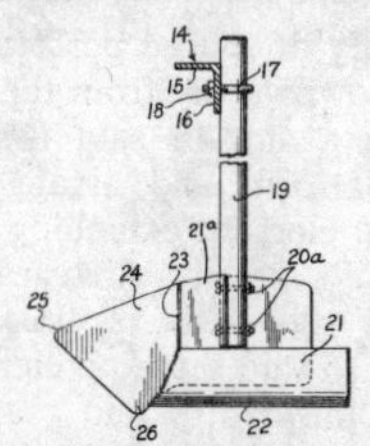

1. A soil-penetrating, unitary cultivating and hilling shovel device comprising a flat, vertically arranged blade member terminating at one end in a forwardly pointed extremity; a generally vertically arranged, shield-forming, plate-like body rigidly secured at its forward end to one side of said blade member intermediate the ends of the latter and extending laterally outwardly from said one side of said blade member in acutely angular relation thereto, said body being provided at its lower edge with an elongated, downwardly inclined cultivator wing disposed in laterally offset and acutely angular relation to said blade member and terminating forwardly in a sharpened cutting edge, said body also being provided at its rearward end with an angularly related attachment web extending inwardly toward and rigidly secured to said one side of said blade member, the attachment web of said body being disposed in substantially perpendicular relation to said blade member and provided with bolt-receiving openings for securing said shovel device to a shank of a cultivator frame, and said attachment web being shielded against soil accumulation during soil penetration by said plate-like body.

2,716,935
CULTIVATOR SHOVEL
Leroy A. Demorest, Ashley, Ohio, assignor to Cultiguard Shovel Company, a corporation of Ohio
Application October 18, 1952, Serial No. 315,563
7 Claims. (Cl. 97—204)

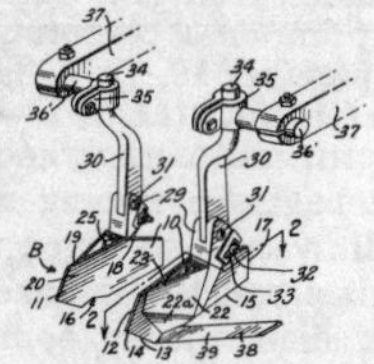

1. A unitary cultivating and hilling shovel for tractor-operated farm cultivators, comprising: a rigid body structure including a vertically disposed longitudinally extending plate-like blade having flat inner and outer surfaces, said blade being formed at the front end thereof with a soil-penetrating nose, the latter including upper and lower reversely extending relatively angular edges, the diverging rear parts of the latter terminating rearwardly in the upper and lower edges of the blade body, the lower edge of the blade at the rear terminating end of the nose being formed with an upwardly and inwardly directed soil-passing recess, a soil-deflecting and mounting plate forming a component part of said body, said plate being rigid with said blade and extending angularly and rearwardly in a longitudinal direction from the outer face thereof above said recess, said plate terminating rearwardly in a perforated shank-attaching web, the latter having an inner edge joined with the outer face of the blade, and an elongated soil-penetrating cutter wing constituting a component part of the shovel body structure and formed rigidly with and projecting from the outer face of the body portion of said blade beneath said plate, the upper surface of said wing being inclined with respect to the horizontal throughout the full length thereof.

2,716,936
APPARATUS FOR ROASTING COFFEE
Joseph L. Kopf, Maplewood, N. J., assignor to Jabez Burns & Sons, Inc., New York, N. Y., a corporation of New York
Application January 17, 1951, Serial No. 206,400
3 Claims. (Cl. 99—236)

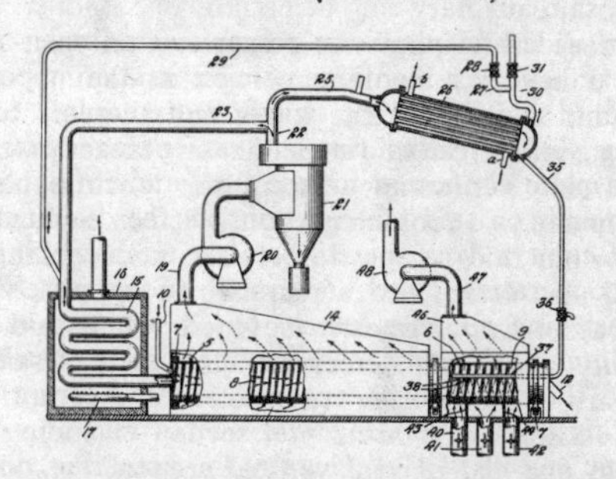

1. In a coffee roasting system means for continuously conveying a stream of coffee through a roasting chamber and a cooling chamber in sequence comprising coaxial perforate drums so that a continuous stream of coffee is roasted, cooled and delivered from the cooling chamber, means introducing heated gases into the interior of the roasting chamber at the entrance end thereof, means collecting the gases passing outward through the perforate drum constituting the roasting chamber, means receiving gases from said collecting means and recirculating them through a heater and said gas introducing means leading to the entrance end of said roasting chamber to roast the coffee therein, and means for continuously cooling and discharging coffee from said cooling chamber.

2,716,937
COFFEE MAKER
Cesare A. Milano, Atlantic City, N. J.
Application December 16, 1952, Serial No. 326,233
3 Claims. (Cl. 99—306)

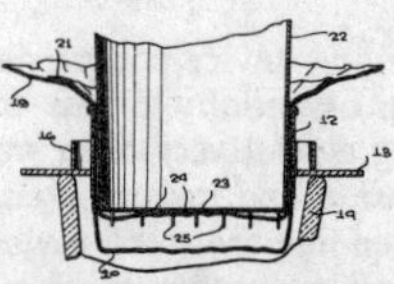

1. A coffee making apparatus comprising a coffee cup, an annular collar member, means on said collar member engageable with the rim of the cup, a first sheet of disposable porous flexible material disposed in and extend-

ing outwardly over the rim of said collar member, said sheet being arranged to receive ground coffee, a second sheet of disposable porous flexible material disposed in and extending outwardly over said first sheet, a cup member shaped to fit inside said collar member, said cup member being tightly received inside said first and second sheets, said cup member having an apertured bottom wall, and depending projections on said bottom wall adapted to deform said second sheet to facilitate the flow of hot water therethrough.

2,716,938
BASKET FILTER FOR DEEP FAT FRYING APPARATUS
Harry L. Smith, Lake Waccamaw, N. C.
Application May 8, 1951, Serial No. 225,208
3 Claims. (Cl. 99—408)

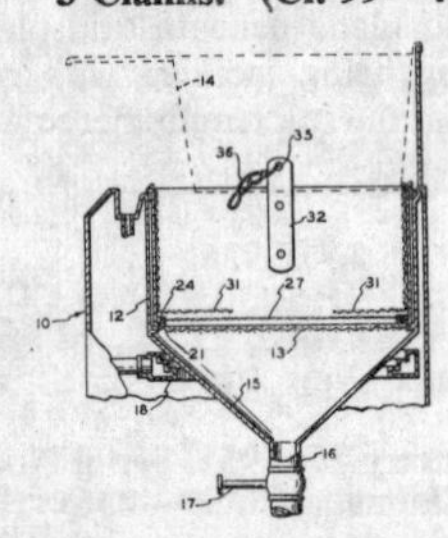

1. A deep fat frying device comprising an open-topped receptacle having a cooking portion at the upper end thereof and a sump at the lower end thereof, conduit means extending from said sump, heating means externally encompassing said sump adjacent the juncture between said sump and said cooking portion, a screen element positioned within said receptacle between said cooking portion and said sump, and a plurality of oppositely positioned basket-support means mounted within said cooking portion above said screen element, each of said support means being of mesh construction and comprising a vertical section and a horizontal section extending laterally from one end of said vertical section, the opposite end of said vertical section being provided with a hooked portion releasably clamped over an upper edge portion of said receptacle, said horizontal section being vertically spaced from said screen.

2,716,939
BASKET STRAINER FOR PORTABLE DEEP FAT FRYING APPARATUS
Harry L. Smith, Lake Waccamaw, N. C.
Application December 10, 1951, Serial No. 260,849
3 Claims. (Cl. 99—408)

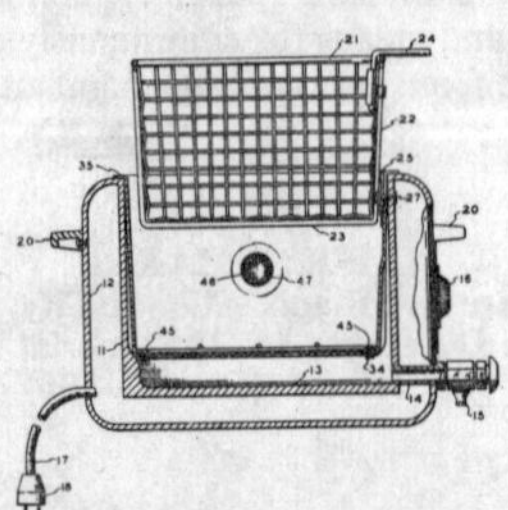

1. A deep-fat frying device comprising a double-walled housing consisting essentially of an outer shell and an inner shell defining a relatively deep well having substantially imperforate side and bottom walls and an open top end, a conduit extending from the lower side wall portion of said well through said outer shell, a drain means connected to said conduit externally of said outer shell, an internal shoulder provided at the lower side wall portion of said well in spaced relationship to the bottom wall thereof, a relatively deep receptacle in said well, said receptacle having a substantially imperforate side wall area and open top and bottom ends, said open top end of said receptacle being defined by a laterally outwardly extending flange portion, and said open bottom end of said receptacle being defined by a laterally inwardly extending flange portion, a screen supported by said inwardly extending flange portion and closing said open bottom end of said receptacle, said receptacle being supported at the bottom end thereof by the internal shoulder of said well and, at the top end thereof, by said outwardly extending flange portion engaging said housing, spacer bars provided on the internal surface of the screen closing said open bottom end of said receptacle, screened openings in the side wall area of said receptacle, and a relatively deep basket insertable into said receptacle from the open end thereof and adapted to rest on said spacer bars.

2,716,940
BORDER STRIP DESIGN MACHINE
Walter O. Kemper, Shreveport, La.
Application September 26, 1951, Serial No. 248,375
6 Claims. (Cl. 101—6)

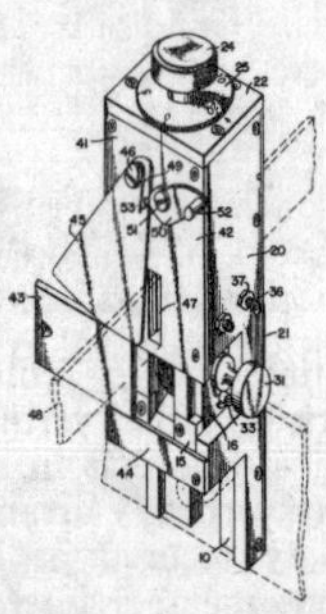

1. In a device of the class described, a stationary inner housing having upper and lower ends, a yoke encircling said inner housing and slidably mounted with respect to said inner housing and having upper and lower ends positioned respectively most nearly adjacent the upper and lower ends of said inner housing, spring means between the lower end of said yoke and said inner housing normally urging the upper end of said yoke toward the upper end of said inner housing, an outer housing surrounding said inner housing and said yoke and provided with a top cover plate and a front plate and a back plate, a back-up wheel rotatably mounted upon a shaft carried by said inner housing, a circular die rotatably mounted upon a shaft mounted in said yoke, and gauging means connected to said yoke for raising and lowering said yoke with respect to said inner housing, said front plate and said back plate being apertured for the passage of border strip through said device between said die and said back-up wheel.

2,716,941
MACHINES FOR MARKING CYLINDRICAL ARTICLES
Frederick A. Hattman, Hubbard, Ohio, assignor to United States Steel Corporation, a corporation of New Jersey
Application September 15, 1950, Serial No. 185,047
6 Claims. (Cl. 101—40)

1. A machine for marking cylindrical articles comprising a support having a sloping surface along which an article can roll, a shaft mounted on said support for rotation on an axis parallel to the axes of rotation of the articles, a pair of diametrically opposed marking heads fixed to said shaft, one of said marking heads occupying a position in which an article on said surface can roll across it and receive a marking, the other of said marking heads occupying a position for recoating with marking

fluid, drive means for rotating said shaft in half revolution steps and thereby reversing said marking heads, an electric circuit including a switch on said sloping surface adapted to be actuated by an article passing thereover for operating said drive means, rotating said shaft and thus reversing said marking heads, a second shaft rotatably mounted on said support and being parallel to and below said first named shaft, arms fixed to said second shaft, a coating transfer roll rotatably mounted on said arms, means occu-

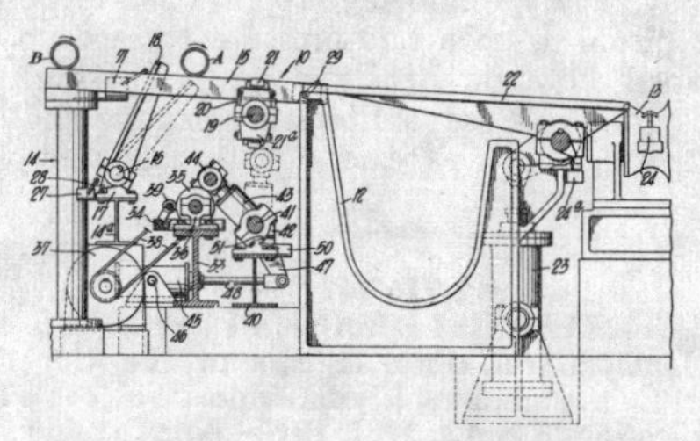

pying a fixed position for applying marking fluid to said roll, said arms normally holding said roll in a position to receive marking fluid from said last named means and clearing said marking heads, and additional drive means operatively connected with said circuit for rotating said second shaft first in one direction and then in the other after each reversal of said marking heads and thus wiping said roll across the marking head which is in its recoating position and then returning said roll to its normal position.

2,716,942
PRINTING MACHINE
Ernest A. Timson and Charles Hillingdon Dickinson, Kettering, England
Application January 24, 1951, Serial No. 207,492
Claims priority, application Great Britain January 24, 1950
7 Claims. (Cl. 101—144)

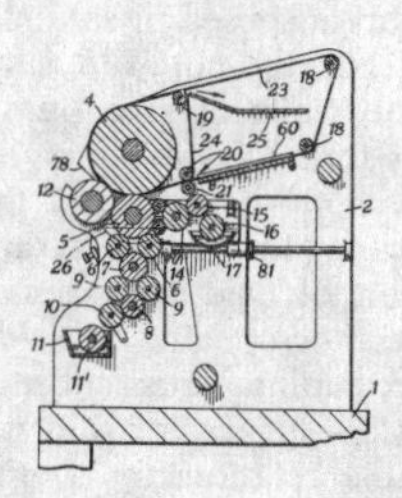

2. A test printing machine comprising an impression cylinder, eccentric bearings mounting said impression cylinder in the machine, at least one countercyclinder, cam means for moving said impression cylinder into printing engagement with said countercylinder, a tiltable table, sheet conveyor tapes arranged to pass through grooves in said table and through the printing zone, a lever system for causing tilting of said table to bring a deposited sheet on to said tapes, cam means for actuating said lever system to initiate the sheet feed, ink transfer form rolls removably mounted in the machine, cam means for automatically causing the moving of said form rolls into inking and non-inking position, a removable offset cylinder arranged to contact the impression and printing cylinders, cam means for moving said offset cylinder into operating contact with the printing cylinder, an assembly of water transfer rolls movable into and out of an operative position, all said cam means being disposed on a common cam shaft to become temporarily operative in a timed sequence, and clutch means controlling the drive to said cam shaft.

2,716,943
LIQUID METAL HIGH PRESSURE PUMP
Leonard V. Vandenberg, Sharon Springs, N. Y., assignor to the United States of America as represented by the United States Atomic Energy Commission
Application January 16, 1953, Serial No. 331,653
3 Claims. (Cl. 103—1)

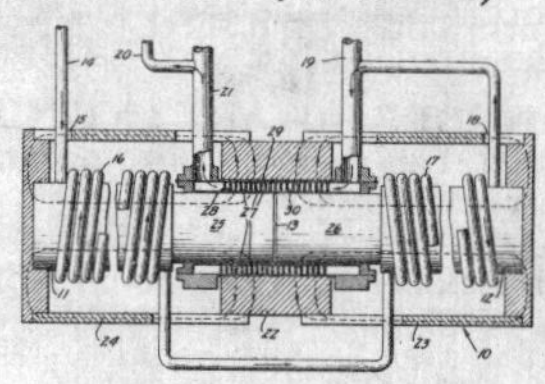

1. A high pressure electromagnetic pump for liquid conductors comprising a pump inlet and outlet, an annular pumping chamber communicating with the inlet and outlet, fins dividing the chamber into a plurality of spiral channels, means for passing an armature current through the liquid conductor, and a pair of electromagnets arranged in spaced end to end relationship, said electromagnets being connected in series with each other and in series with the armature current, said electromagnets being arranged to establish a radial magnetic field and positioned to interact with the armature current whereby a pressure is created on the liquid conductor in the direction of the spiral channels.

2,716,944
MECHANISM FOR PUMPING A LIQUID AND A LUBRICANT SIMULTANEOUSLY
Walter Ferris, Milwaukee, Wis., assignor to The Oilgear Company, Milwaukee, Wis., a corporation of Wisconsin
Application May 24, 1954, Serial No. 431,660
20 Claims. (Cl. 103—2)

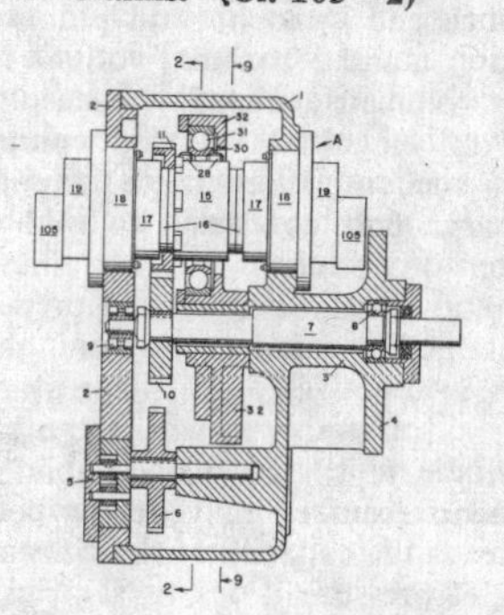

1. In a pumping mechanism, the combination of a rotatable cylinder barrel having arranged therein a plurality of main cylinders and a plurality of auxiliary cylinders, a main cylinder port communicating with each of said main cylinders and an auxiliary cylinder port communicating with each of said auxiliary cylinders, pistons fitted in said cylinders, means for effecting reciprocation of said pistons in response to rotation of said cylinder barrel, a valve engaging said cylinder barrel and having a main inlet port and a main discharge port arranged therein in the path of said main cylinder ports, means for supplying a first liquid to said main inlet port, means for conducting liquid from said main discharge port to a point of use, said valve also having two interconnected low pressure auxiliary ports arranged therein at opposite sides of said main inlet port and two interconnected high pressure auxiliary ports arranged therein at opposite sides of said main discharge port with one low pressure port and one high pressure port arranged in the path of said auxiliary cylinder ports, means for supplying a second liquid to said low pressure auxiliary ports at a pressure

somewhat higher than the pressure of said first liquid, and means for conducting liquid from a high pressure auxiliary port including means for resisting the flow of liquid therethrough to thereby cause the pressure in said high pressure auxiliary ports to exceed the pressure in said main discharge port.

2,716,945
VARIABLE STROKE ROTARY CYLINDER PUMP
Frank G. Presnell, Hollywood, Calif., assignor to Bendix Aviation Corporation, North Hollywood, Calif., a corporation of Delaware
Application October 17, 1952, Serial No. 315,269
4 Claims. (Cl. 103—4)

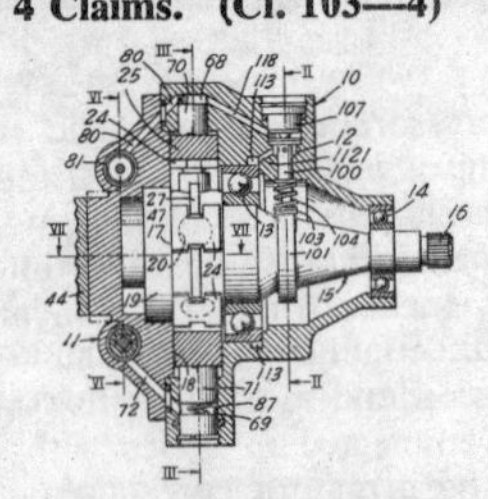

1. A variable volume pump assembly for supplying fluid to a system having a small continuous demand and a relatively large intermittent demand comprising: means defining inlet and outlet passages; a large capacity variable volume main pump unit having a suction port and a discharge port and having a control element movable to vary its pumping action from zero to maximum; means connecting said suction port to said inlet passage; a check valve for passing fluid from said discharge port to said outlet passage and preventing reverse flow; means responsive to pressure in said outlet passage for actuating said control element to reduced said pumping action to zero in response to a predetermined pressure and to incite said pumping action at a lesser pressure; a small constant volume auxiliary pump unit having volumetric capacity exceeding but of the same order of magnitude as said continuous demand and having a suction port connected to said inlet passage and a discharge port connected to said outlet passage; and common drive means for simultaneously driving both said pump units; whereby said predetermined pressure is maintained by said auxiliary unit and said variable volume unit is deactivated and depressurized when the demand on said assembly is within the capacity of said constant volume unit, and the variable volume unit is activated and pressurized only during periods when the demand exceeds the capacity of the constant volume unit and the outlet pressure drops to said lesser pressure.

2,716,946
HYDRAULIC CONTROL SYSTEM
James A. Hardy, Indianapolis, Ind., assignor to Schwitzer-Cummins Company, Indianapolis, Ind., a corporation
Application October 14, 1952, Serial No. 314,682
10 Claims. (Cl. 103—37)

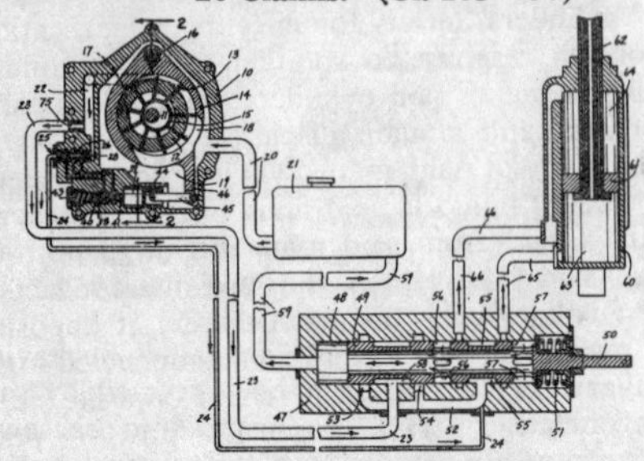

1. In a hydraulic control system, the combination with a variable displacement pump having a fluid inlet and discharge outlet, and a fluid displacement varying member urged toward its non-pumping position by the fluid pressure in said outlet, of a fluid displacement control cylinder and piston connected with said pump member, a fluid pressure passage leading from said outlet to said cylinder, a relief passage leading from said cylinder, a flow control valve interposed in said pressure passage comprising a valve piston having a passage leading from said outlet to a service line, and a spring urging said piston in a direction to open said pressure passage, said piston passage and spring being proportioned to position said valve piston to open said pressure passage up to a predetermined flow through said piston passage and cause said piston to close said pressure passage and open said relief passage upon a greater flow through said piston passage.

2,716,947
REVERSIBLE VANE PUMP
Anton J. Janik, Elyria, Ohio, assignor to The Ridge Tool Company, Elyria, Ohio
Application May 12, 1950, Serial No. 161,660
3 Claims. (Cl. 103—138)

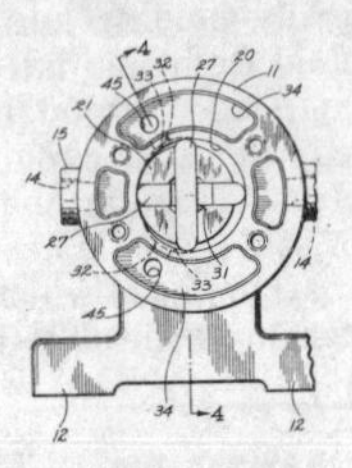

1. A rotary pump comprising a housing and a cylindrical pump chamber formed therein, a rotor rotatably disposed eccentrically in said chamber, a cylindrical chamber formed centrally in said rotor and a pair of elongated diametrical slots formed in said rotor and opening into said cylindrical chamber in said rotor and dividing said rotor into four equal segments, a pair of soft resilient vanes slidably disposed in said slots, each of said vanes having diametrically opposite surfaces in sliding engagement with the walls of said pump chamber, one of said vanes being formed with an opening through its broad side at the center thereof, and the other of said vanes having a necked down portion at its center, said necked down portion of said vane member being positioned within said center opening of said first mentioned vane member and extending across substantially the entire width thereof, the side surfaces of said necked down portion being in slidable relationship to the edges of said central opening in said first mentioned vane member, and thereby providing support for said vane members within the cylindrical chamber of said rotor, the sides of said vane members being in slidable engagement with the sides of said segments and providing support therefor, said vane members having a portion extending outward from said rotor a distance less than the thickness of said vanes, and arranged to engage the walls of the cylindrical pump chamber.

2,716,948
ROTARY PUMP
Ernest A. Cuny, Miami Beach, Fla.
Application August 21, 1953, Serial No. 375,713
4 Claims. (Cl. 103—161)

1. A rotary pump comprising a pump casing having a cylindrical pump chamber, a cylindrical rotor ring fitted to and journaled in said cylindrical pump chamber, said rotor ring having radial pump cylinder passages therethrough, a shaft journaled in said pump casing on an axis eccentric to the axis of the cylinder pump chamber, a cylindrical bearing land on said shaft within

the pump chamber, pistons in said cylinder passages in the rotor ring, said pistons having extensions at the inner ends of the same bearing on said land and provided with laterally projecting wrist pin lugs, an annular retainer concentric with the shaft and engaging over the outer edges of said wrist pin lugs to hold the pistons in substantially concentric relation to the shaft, said piston extensions having flat sides, means engaged with the flat sides of said piston extensions for holding the pistons against turning in the cylinder passages and for retaining the wrist pin lugs in cooperative engagement with the bearing land and retainer ring, a driving element on the shaft having a sliding rocking driving connection with the rotor ring and inlet and outlet ports for the cylindrical pump chamber disposed at opposite sides of a diameter intersecting the eccentrically related centers of the pump chamber and shaft.

2,716,949
APPARATUS FOR SERVICING VEHICLES HAVING A BODY AND A REMOVABLE UNDERCARRIAGE
John O. Converse, Minneapolis, Minn., assignor to Nichols Engineering Company, an Illinois corporation
Application January 17, 1951, Serial No. 206,476
15 Claims. (Cl. 104—32)

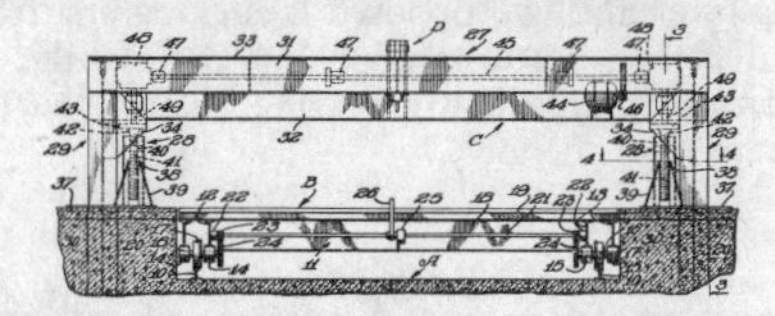

9. A vehicle hoist for use in combination with a shallow pit transfer table of the character described, comprising: a pair of lifting beams spanning said pit at an equal height and spaced apart to receive a vehicle between them with one of its trucks on said transfer table; an integral upright stabilizing column at each end of each of said beams, each of said stabilizing columns extending into a closely fitting well in the floor adjacent the pit to provide lateral stability for said beams, a screw support for said beams adjacent each stabilizing column; mechanism for simultaneously rotating said screw supports to move said beams vertically in unison; and a vehicle supporting arm on each lifting beam, each supporting arm being movable between a retracted position with its inner end substantially flush with the inner face of the beam and a lifting position projecting beneath the body of a vehicle located between said beams.

2,716,950
PIE MARKER
Mary A. Johnston, Huntsville, Mo.
Application February 17, 1953, Serial No. 337,358
1 Claim. (Cl. 107—47)

A pie marker comprising: a vertical, cylindrical housing having a plurality of circumferentially spaced, longitudinal slots therein, a slidable longitudinal guide shaft mounted centrally in the housing and projecting from both ends thereof, one end portion of said shaft for insertion in a pie to be marked, a handle on the other end of the shaft, a plurality of pivoted marking blades mounted radially in the housing around the shaft and swingable through the slots to operative and inoperative positions, means on the housing for releasably securing the blades in either position, and means on the shaft for elevating the housing for disengaging the blades from the pie, the last-named means including a collar fixed on the shaft,

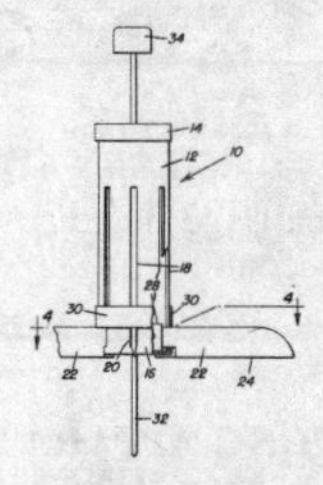

said collar being engageable with the free end portions of the blade for swinging same outwardly through the slots from inoperative position when released by said securing means, and a coil spring on the shaft engaged with the collar and the housing for yieldingly urging said housing and said shaft in opposite directions relative to each other.

2,716,951
RUBBISH BURNER
Frank William Ver Haigh, St. Paul, Minn.
Application November 4, 1949, Serial No. 125,420
1 Claim. (Cl. 110—18)

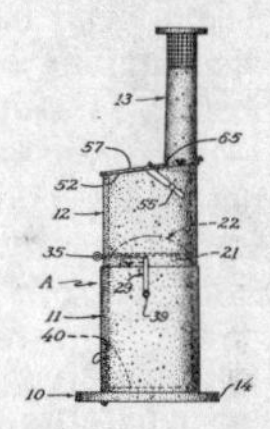

A cylindrical rubbish burner comprising a separable assembly consisting of an upper burner section, a lower burner section and a chimney, the said lower burner section consisting of a cylindrical base provided with a centrally positioned cylindrical raised portion, an open ended cylinder wall having a rectangular opening in its lower half portion and being seated on said base about said raised portion, the said cylinder wall having a ventilated door connected thereto and hinged therewith to fit said rectangular opening, an inner peripheral shoulder and groove in the upper rim edge of said cylinder wall, said upper burner section comprising a second open ended cylinder will mounted on said shoulder and seated in said groove, a movable grate slidably mounted on a support positioned at the base of said second cylinder wall, handle means operatively associated with said grate to tip said grate into an operable and inoperable position, a hinged cover means closing substantially over one-half of the upper open end of said second cylinder wall, and a chimney section supported on the remaining upper end portion of said second cylinder wall.

2,716,952
SPIRAL SEWING MACHINE
Vittorio Pasquini and Vincenzo Guerrini, Pavia, Italy, assignors to Vittorio Necchi S. p. A., Pavia, Italy
Application June 16, 1953, Serial No. 362,098
Claims priority, application Italy June 16, 1952
8 Claims. (Cl. 112—2)

1. Spiral sewing means for a sewing machine having work feeding means comprising a body rotated by said work feeding means carrying the material to be sewn, a first carriage movably mounted with translatory motion towards or away from the sewing point supporting said

body, a pinion mounted on said body, a second carriage movable angularly with respect to the movement of said

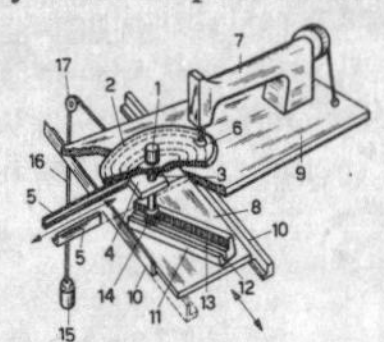

first carriage and a rack meshing with said pinion mounted on said second carriage diagonally with respect to the movement of said second carriage.

2,716,953
ZIG-ZAG STITCHING APPARATUS FOR SEWING MACHINES
Walter Heinrich Daniel Langhein, Hamburg-Volksdorf, Germany, assignor to Ingolf Felix Friedrich Emil Pauls, Basel, Switzerland
Application October 28, 1952, Serial No. 317,312
Claims priority, application Germany August 18, 1952
9 Claims. (Cl. 112—158)

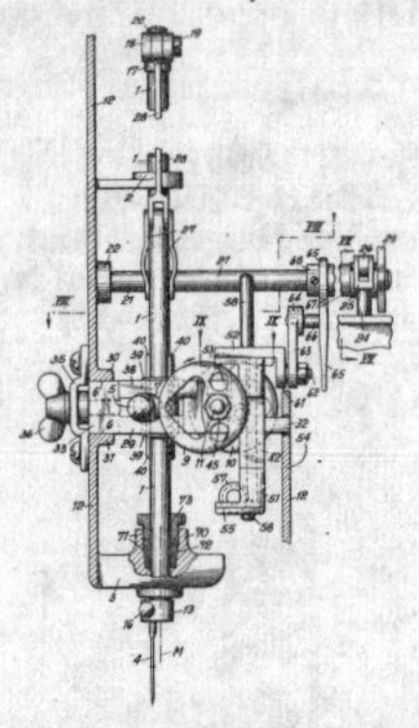

1. A sewing machine apparatus for producing a zig-zag stitching, comprising, in combination, support means; a needle carrying rod mounted on said support means for reciprocating movement thereon; a needle, adapted to pass through the material to be sewn, located to one side of and being parallel to the axis of said rod and extending therefrom; first drive means operatively connected to said rod for reciprocating the same along the axis thereof; and second drive means operatively connected to said rod for reciprocating the same about said axis thereof, said second drive means comprising a lever connected to said rod, extending transversely to the same, and having opposite arms respectively extending from opposite sides of said rod; and a pair of staggered eccentrics turnably mounted on said support means and respectively located opposite said arms of said lever to successively engage the same for reciprocating said rod about said axis thereof.

2,716,954
BLIND STITCHING ATTACHMENT FOR SEWING MACHINES
William C. Haines, La Salle, Ill.
Application August 6, 1952, Serial No. 302,910
25 Claims. (Cl. 112—176)

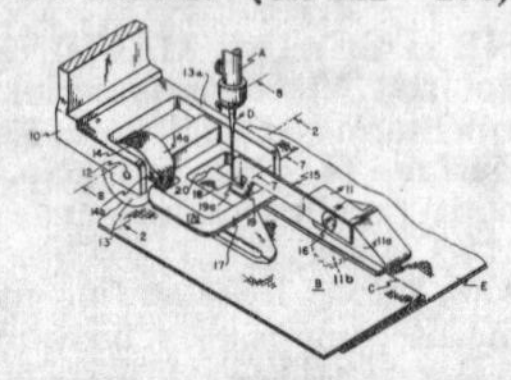

1. An attachment for sewing machines of the type having a reciprocating needle comprising a presser foot having an opening and a working point therein, a guiding groove on said foot adjacent the working point for guiding a folded edge of material past the needle but at a position laterally thereof, pressing means on said foot for periodically so pressing said folded edge at a point closely adjacent the needle as to project same into the path of the needle, and actuating means for said pressing means.

2,716,955
FEED ADJUSTING MEANS FOR SEWING MACHINES
Robert W. Stewart, Greenwich, Conn., assignor to The Singer Manufacturing Company, Elizabeth, N. J., a corporation of New Jersey
Application June 3, 1953, Serial No. 359,282
4 Claims. (Cl. 112—210)

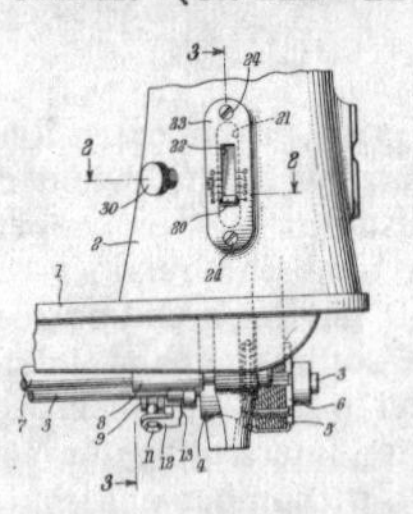

1. In a sewing machine, a frame provided with a slot having parallel opposed side walls, an adjusting-lever extending through said slot and pivotally mounted to said frame on an axis normal to a plane parallel to the opposed side walls of the slot whereby said lever may be swung longitudinally of said slot, and means for frictionally maintaining said adjusting-lever in adjusted position relative to said slot comprising, a U-shaped element including a base having an aperture therethrough for receiving said adjusting-lever and two opposed legs extending from said base and arranged substantially parallel to the axis of said aperture for resiliently bearing against the opposed walls of said slot.

2,716,956
LOOP-TAKERS FOR SEWING MACHINES
Frank Parry, Trumbull, Conn., assignor to The Singer Manufacturing Company, Elizabeth, N. J., a corporation of New Jersey
Application July 27, 1953, Serial No. 370,407
4 Claims. (Cl. 112—230)

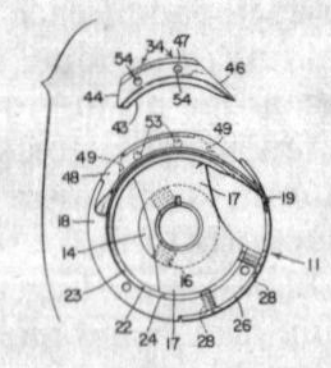

3. A circularly moving loop-taker for lock-stitch sewing machines comprising a thread-case carrier having a peripheral bearing rib; a cup-shaped body adapted to receive said thread-case carrier; a cylindrical side-wall formed as part of said body, said side-wall being provided on its inner surface with a circular raceway having a base and a side-wall, said side-wall also having a substantially crescent shaped circumferential slot extending radially therethrough and a pair of holes extending parallel to the axis of said hub and opening into said slot; a substantially crescent shaped gib of a size such as to fill the slot in said side-wall, and said gib having a pair of threaded holes; a circular flange formed on the inner edge of said gib and overhanging a portion of said circular raceway thus acting to hold the bearing rib of said thread-case carrier in said race-

way; and a pair of threaded screws passing through the holes in said body-wall and being threaded into the holes in said gib.

2,716,957
TORPEDO PATTERN RUNNING SETTING DEVICE
Raymond C. Kent, Jr., and Roland G. Daudelin, Silver Spring, Md., assignors to the United States of America as represented by the Secretary of the Navy
Application November 19, 1952, Serial No. 321,532
2 Claims. (Cl. 114—23)
(Granted under Title 35, U. S. Code (1952), sec. 266)

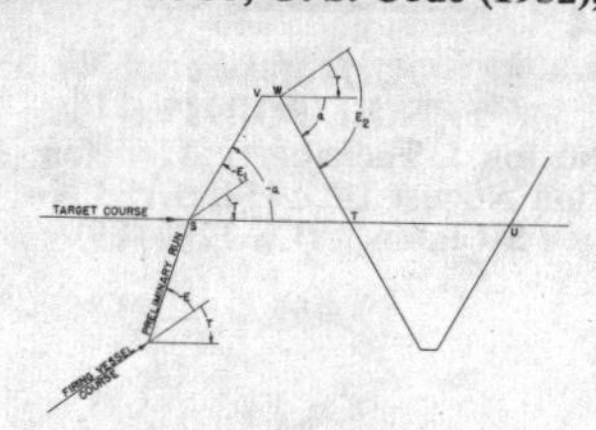

1. The combination with a steering controller of a torpedo wherein said steering controller is of a character comprising means having a first cam means for controlling the direction and length of the preliminary torpedo run after launching, cam follower means, and means including at least two pattern setting course deviate control cams so related to said first cam and said cam follower means as to steer a zig-zag path of travel for said torpedo according to a predetermined presetting; of means disengageably connectible to said torpedo and prior to launching to provide predetermined pattern setting movement of said course deviate cams whereby said follower alters the torpedo travel path following initial run and in correlation with the relative bearing of same with respect to target travel, said last named means including means for providing a first output drive for a first positional input setting of a first of said pattern setting control cams prior to launching, means connected to said last named means for indicating a condition of correlation of said setting with the direction of preliminary run of said torpedo, means providing a second output drive for setting into said steering controller at a second input thereof for the second of said pattern deviate control cams a pattern deviation positioning movement indicative of the algebraic sum of said first setting and the angular deviation of said pattern angle taken with respect to the prior run thereof and the relative bearing of said torpedo, and means connected to said drive means for indicating said pattern setting in correlation with said initial setting, said pattern setting means including differential gearing interposed between the first and second drive outputs thereof.

2,716,958
FLUID CARGO BARGE TANK ASSEMBLY
Clarence W. Brandon, Tallahassee, Fla., assignor, by direct and mesne assignments, of fourteen and one-sixth per cent to N. A. Hardin, fourteen and one-sixth per cent to Hazel H. Wright, and fourteen and one-sixth per cent to Catherine H. Newton, all of Forsyth, Ga., and fifteen per cent to Harvey B. Jacobson, Washington, D. C.
Application February 4, 1949, Serial No. 74,546
13 Claims. (Cl. 114—74)

1. A barge for the transportation of fluids comprising a pair of fluid storage tanks, means for securing said storage tanks in side by side relation for forming the fluid cargo carrying hull of a barge, and end assembly adapted to constitute the prow or the stern of a barge, means for directly securing said end assembly to one end of said storage tanks in embracing relation thereto to form with said storage tanks an end of the barge, an expansion tank, means establishing communication between said expansion tank and one of said storage tanks for fluid flow therebetween, means mounting said ex-

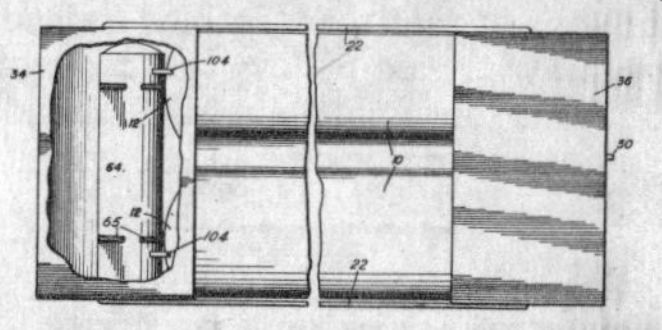

pansion tank in said end assembly in bracing relation thereto to thereby constitute a reinforcing member thereof, means directly securing said expansion tank to said storage tanks for comprising a reinforcing means therefor.

2,716,959
NET GATE FOR MARINE HARBORS
George E. Betts, Jr., and Elmer L. Blekfeld, Falls Church, Va., and Arthur T. McCanner, Jr., Washington, D. C., assignors to the United States of America as represented by the Secretary of the Navy
Application May 17, 1954, Serial No. 430,465
4 Claims. (Cl. 114—241)
(Granted under Title 35, U. S. Code (1952), sec. 266)

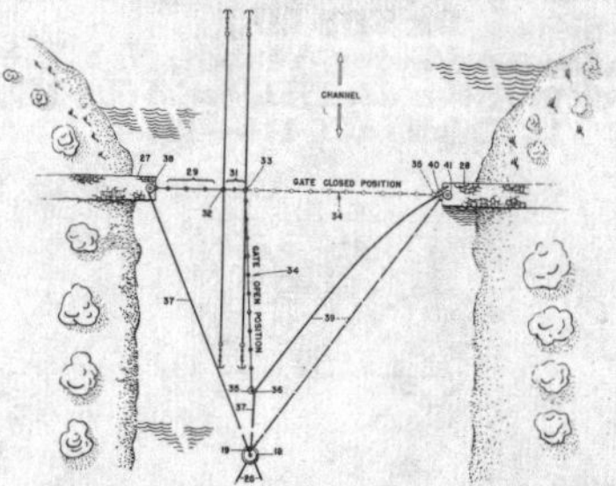

1. A swing type aqueous barrier for closing a shipping channel to surface and undersurface traffic comprising a nonrigid net of a length to span said channel, said net including a hinge end and a free end, a plurality of spaced floats supporting the net including a free end float, means anchoring the hinge end along one side of the channel, an opening cable and a closing cable connected to the free end, means at the same side of the channel as said hinge end but laterally spaced therefrom a distance greater than the net length for drawing in and paying out said opening cable such that said net is maintained in substantially a straight line while being opened or closed, and means at the side of the channel opposite said anchoring means for drawing in and paying out said closing cable.

2,716,960
OUTBOARD MOTOR, HOIST, AND GUIDE
Forest H. McCumber, Port Clinton, Ohio
Application December 30, 1953, Serial No. 401,206
3 Claims. (Cl. 115—41)

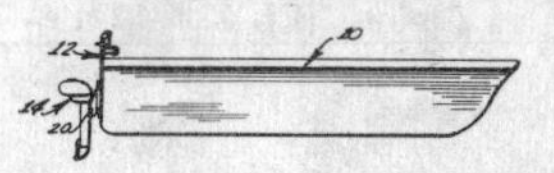

1. In an outboard motor mounting, a frame including a pair of spaced parallel vertically disposed tubes each provided with an outer longitudinally extending slot, a crosspiece extending between the lower ends of said tubes and secured thereto, a bracket adjustably connected to said tubes and including a body portion, a pair of spaced arms, and inwardly extending fingers extending into said slots, a pair of spaced parallel bars connected to said bracket, a first shaft extending through said bars, a drum mounted on said shaft, cable means trained over said

drum and adapted to be connected to an outboard motor, a pair of spaced parallel ears extending from said tubes, a second shaft journaled between said pair of ears, a second drum mounted on said second shaft, pulleys supported by the tops of said tubes, cables trained over said second drum and over said pulleys and connected to said bracket, and means for maintaining said second drum immobile in its adjusted positions.

2,716,961
WATER MARKING DEVICE
Theodore B. Manheim, Philadelphia, and Joseph L. Castelli, Ardmore, Pa.
No Drawing. Application December 29, 1952, Serial No. 328,536
3 Claims. (Cl. 116—124)
(Granted under Title 35, U. S. Code (1952), sec. 266)

1. A marker for air-sea survival usage comprising a rigid, porous resinous foam body having a high degree of cell interconnection, said body being impregnated with a water-soluble dye, said dye being adapted to readily go into solution and flow out of said body when said body is immersed in water.

2,716,962
TUNING DEVICE FOR BROADCAST RECEIVERS OR THE LIKE
Siegmund Loewe, Yonkers, N. Y.
Application October 23, 1951, Serial No. 252,737
11 Claims. (Cl. 116—124.4)

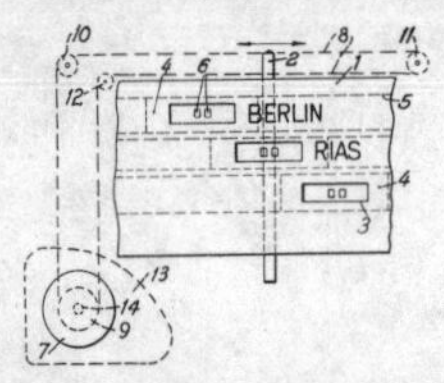

1. In a radio receiving set a tuning indicator device comprising in combination a station dial plate carrying the station names and supplied with slots, each of which is associated with one station name, insert pieces adjustably mounted behind said slots in said dial plate, tuning means for turning the radio set by means of a tuning knob, a pointer movable along said dial plate by operation of said tuning knob, said tuning knob being coupled by a suitable drive with said tuning means and simultaneously with said pointer, thus designating the position of the tuning means on the dial plate, said insert pieces being slidably mounted on said dial plate and provided with means adapted to shift said insert pieces with regard to said dial plate by hand from outside through said slots, thus enabling a readjustment of the exact tuning of the set to single stations to be received.

2,716,963
BLOTTING DEVICE
Robert F. Howe, Leominster, Mass.
Application August 4, 1953, Serial No. 372,320
1 Claim. (Cl. 120—24)

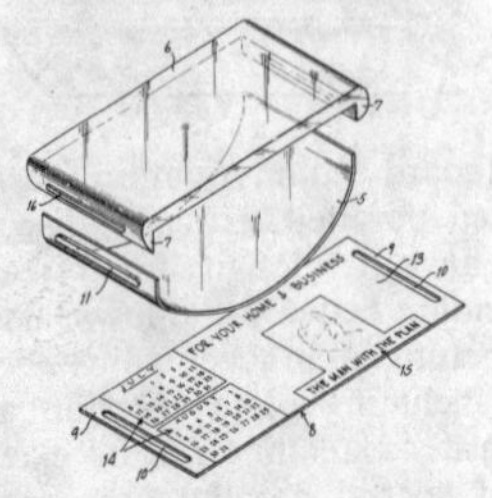

A blotting device comprising a curved resilient transparent plastic body, semi-oval in longitudinal section and having elongated transverse protuberances along its end edges, a blotter positioned around the exterior of the curved body, said blotter having elongated transverse slots along its end edges for receiving and interlocking with the protuberances on said curved body, said blotter having a blotting surface on one side and a glazed surface on the other side, and a flat rectangular-shaped transparent plastic cover plate having end flanges with transverse slots therein for detachably receiving and interlocking with the protuberances of the body and for clamping the blotter between the body and plate, said blotter having indicia on its glazed surface visible through the cover plate.

2,716,964
PENCIL POINTER
Middleton J. Tackaberry, Houston, Tex.
Application August 16, 1954, Serial No. 449,896
5 Claims. (Cl. 120—89)

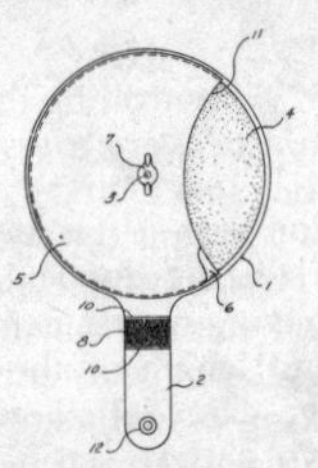

1. In a pencil pointer, a round body member, a handle integral therewith, an axial port in said body member, and a bolt mounted in said port and extending laterally therefrom, a series of abrasive discs mounted on said bolt and a shield mounted on said bolt and covering the faces of said discs, a portion of said shield being cut away to expose a portion of the surface of the topmost disc.

2,716,965
FLUID PRESSURE ACTUATED DEVICE HAVING A NUMBER OF PREDETERMINED POSITIONS
Paul Klamp, Detroit, Mich., assignor to Mechanical Handling Systems Inc., Detroit, Mich., a corporation of Michigan
Application December 3, 1951, Serial No. 259,666
12 Claims. (Cl. 121—38)

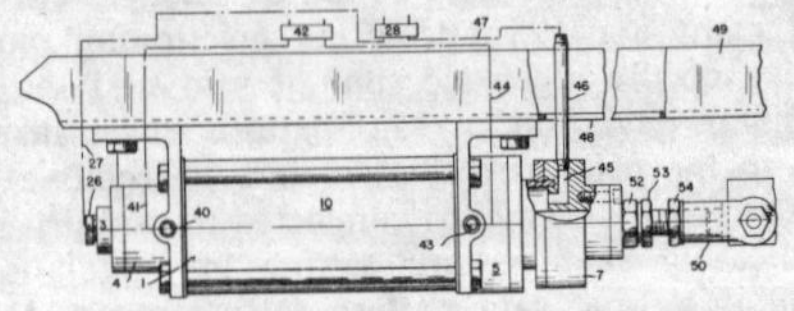

1. A fluid pressure actuated device comprising a cylinder having front and rear heads at its ends, a combined cylinder and piston reciprocable within said cylinder and having at its ends a rear head inside and a front head outside said cylinder, said inside head being provided with a reduced axially extending portion abutting said first mentioned rear head in the rearward position of said combined cylinder and piston and having a port, said first mentioned rear head having outer and inner members with said outer member provided with a port in communication with said cylinder radially outwardly of said reduced portion and said inner member adjustably mounted on said outer member and forming an adjustable abutment for said reduced portion, said inner member having a port in communication with the port in said reduced portion, a valve in the inner member port, means on said reduced portion for holding said valve open in the rearward position of said combined cylinder and piston, said front heads of said cylinder and combined cylinder and piston being provided with ports, and a piston reciprocable within said combined cylinder and piston.

2,716,966
HYDRAULIC RAM AND CONTROL THEREFOR
Clarence A. Hubert, Chicago, and Joseph F. Ziskal, Brookfield, Ill., assignors to International Harvester Company, a corporation of New Jersey
Continuation of abandoned application Serial No. 689,732, August 10, 1946. This application June 18, 1952, Serial No. 294,394
5 Claims. (Cl. 121—40)

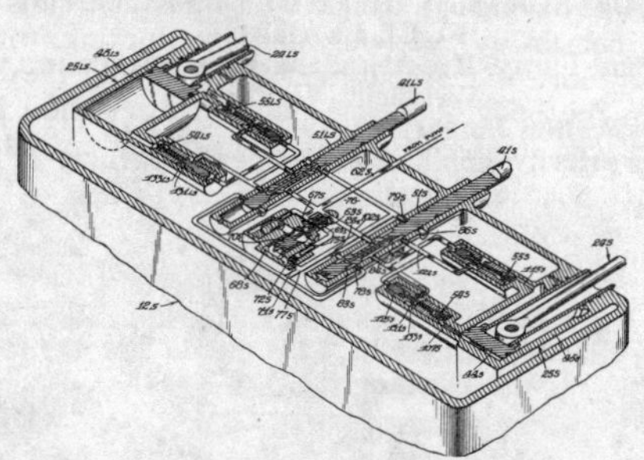

1. In apparatus for controlling the delivery of operating fluid from a constant delivery pressure source thereof to a double-acting piston and cylinder device and from said device to a place of exhaust, a control valve including a bore and a spool type control element axially adjustable therein, said control element having axially spaced intermediate enlargements and end enlargements cooperating with said bore to form a fluid admission chamber between opposed ends of said intermediate enlargements and to form exhaust chambers respectively between the opposite ends of said intermediate enlargements and said end enlargements, a pair of check valves, motor means for said check valves, a pair of reversible flow passages having ports respectively covered by said intermediate enlargements when the control element is in a neutral adjustment, said passages leading respectively through said check valves to respective ends of the cylinder of said device and said check valves being openable by the pressure of fluid in their respective passages tending to flow to the cylinder, said passages also leading to said motor means and the latter being operable under the influence of pressure fluid from each passage to open the check valve of the other passages, a fluid admission passage communicative between said source and said admission chamber, a pressure regulator unit in communication with said admission passage to by-pass the fluid therefrom when subjected to the pressure in such admission passage and thereby incur a low pressure condition in such passage and said unit being operable to terminate said by-pass when subjected to a diminished pressure and thereby incur a high pressure condition in the admission passage, a restricted flow capacity passage between said pressure regulator unit and the admission passage for subjecting said unit to said pressure, and a fluid diversion passage leading from said unit to divert the restricted passage fluid therefrom and incur the diminished pressure subjection, said diversion passage having ports respectively covered by the end enlargements of the valve element while it is in neutral to prevent fluid diversion through such passage, said diversion passage ports being alternately registered with the associated of the exhaust chambers by opposite movement of the control valve element from neutral and the reversible flow passage ports being concurrently alternately registered with the admission chamber and their associated exhaust chamber.

2,716,967
ARCHLESS DOOR FRAMES FOR OPEN-HEARTH FURNACES
Harry A. Shiflet, Greenock, Pa., assignor to United States Steel Corporation, a corporation of New Jersey
Application April 6, 1951, Serial No. 219,636
5 Claims. (Cl. 122—499)

2. In an open-hearth furnace including an arched roof and a horizontal skewback channel supporting the roof on one side of the furnace, the combination therewith of an archless water-cooled door frame including a body portion in flatwise engagement with the web of said channel and spaced legs depending from the body portion, said body portion having an inwardly projecting nose adapted to bear up against the bottom flange of said

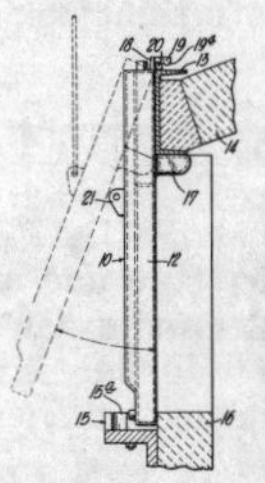

channel, a pivot bearing on top of one of said frame and channel and hook means on the other whereby said nose clears said bottom flange on outward tilting of the frame about said bearing, said hook means having a sloping surface engaging said bearing and exerting a wedging action tending to raise the frame on movement of the body portion thereby away from said channel.

2,716,968
SPLIT FURNACE AND TUBE SUPPORT
Frederic O. Hess and Kurt W. Fleischer, Philadelphia, Pa., assignors to Selas Corporation of America, Philadelphia, Pa., a corporation of Pennsylvania
Original application April 23, 1947, Serial No. 743,358, now Patent No. 2,606,536, dated August 12, 1952. Divided and this application December 26, 1951, Serial No. 263,247
6 Claims. (Cl. 122—510)

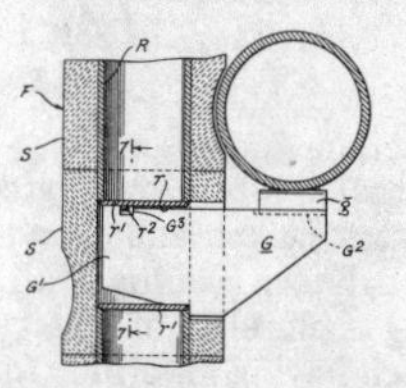

1. A supporting structure for superposed horizontal fluid heater tubes and adapted to withstand high temperatures and comprising a vertically disposed metallic pipe adapted to be cooled by the flow of a fluid through its bore and formed with horizontally extending hollow pockets having metallic walls which extend diametrically through the pipe in the same direction at different levels but do not close the pipe bore, said pockets being rigidly attached to the pipe at opposite sides thereof where the pocket extends through the pipe, and adjacent pockets being open at opposite sides of the pipe, and brackets of heat resistant material each having a portion extending into and removably anchored in a corresponding one of said pockets and having a tube supporting portion external to and extending away from said pipe.

2,716,969
AIR- OR LIQUID-COOLED CYLINDER HEAD FOR INTERNAL COMBUSTION ENGINES
Heinrich Lang, Munich, Germany, assignor to Durex, S. A., Geneva, Switzerland, a corporation of Switzerland
Application July 19, 1952, Serial No. 299,836
Claims priority, application Switzerland July 25, 1951
5 Claims. (Cl. 123—41.16)

1. Air or liquid cooled head for cylinders of internal combustion engines comprising a cylinder head casting including inlet and exhaust ports; an at least approximately cylindrical recess in said casting substantially form-

ing an extension of the cylinder bore; an insert made of heat resistant metal permanently seated in said recess and filling up the same at least substantially to the brim thereof; a combustion chamber within said insert communi-

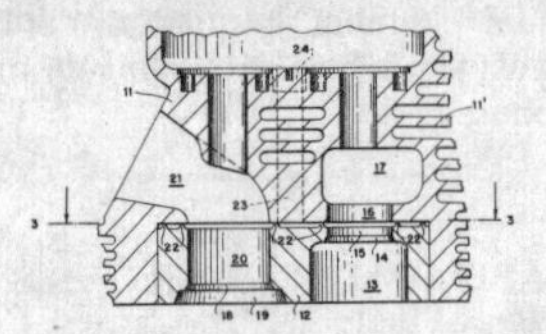

cating with the cylinder bore and arranged eccentrically to the axis thereof; inlet and exhaust ducts in said insert forming extensions of said inlet and exhaust ports, respectively; and valve seats at the terminal duct openings facing the cylinder bore.

2,716,970
MEANS FOR DIRECTING WATER IN THE CYLINDER HEAD OF AN INTERNAL COMBUSTION ENGINE
Ralph Joseph King, Peoria, and Earl Duane Eyman, Canton, Ill., assignors to Caterpillar Tractor Co., Peoria, Ill., a corporation of California
Application December 14, 1953, Serial No. 397,871
3 Claims. (Cl. 123—41.73)

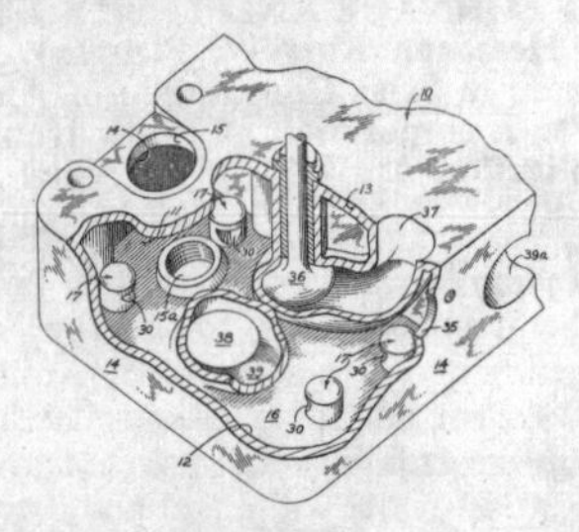

1. In an internal-combustion engine, a cylinder head containing a water jacket and a cylinder sealing wall, means on said cylinder sealing wall within said water jacket comprising a plurality of water directing members, said members each having an orifice forming an elongated throat in its peripheral wall, said orifice having a baffle extending inwardly of the member to direct the flow of coolant into a fan shaped pattern against the cylinder sealing wall adjacent the combustion areas of said engine.

2,716,971
FREE PISTON ENGINE
Allen H. Sykes, Dover, Tenn.
Application April 23, 1953, Serial No. 350,714
5 Claims. (Cl. 123—46)

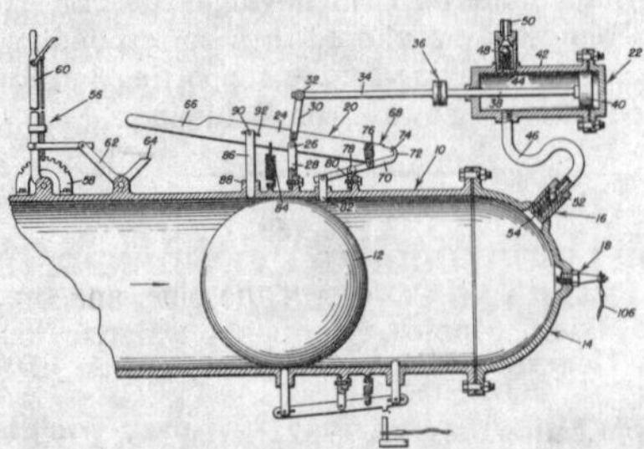

1. In an internal combustion engine having a wall defining a cylinder and a piston reciprocable therein, a fuel injecting nozzle communicating with said cylinder, and means including a member disposed within the path of movement of said piston for injecting fuel through said nozzle, said means also including a second member disposed within the path of movement of said piston when said means has effected the fuel injection, said second member being effective to position the means for subsequent fuel injection.

2,716,972
LUBRICATION OF ENGINE VALVES BY FUEL LEAKAGE
Paul Farny, Ilvesheim, and Ernst Weidmann, Weinheim, Germany
Application January 29, 1953, Serial No. 333,872
Claims priority, application Germany February 4, 1952
6 Claims. (Cl. 123—90)

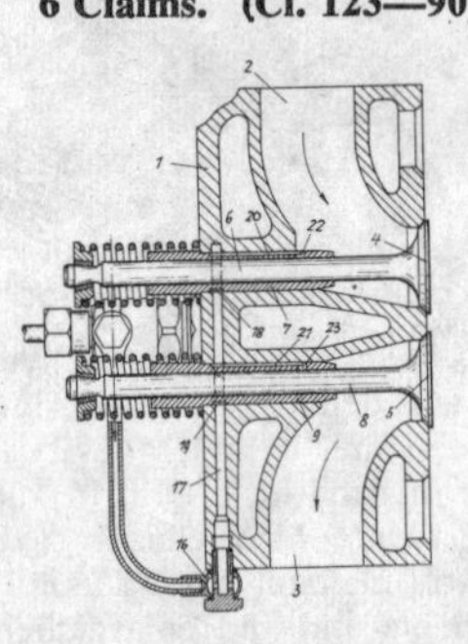

1. A valve-controlled internal combustion engine with fuel injection, comprising in combination at least one injection valve lubricated by a lubricating liquid leaking partly from said injection valve, at least one valve having a valve stem, means for lubricating at least said valve stem of said valve by the leaking part of the liquid lubricating said injection valve, and an oil leakage pipe leading from said injection valve to said valve stem.

2,716,973
BALL THROWING MACHINE
Paul Francis Desi, Brooklyn, N. Y.
Application September 4, 1952, Serial No. 307,832
8 Claims. (Cl. 124—1)

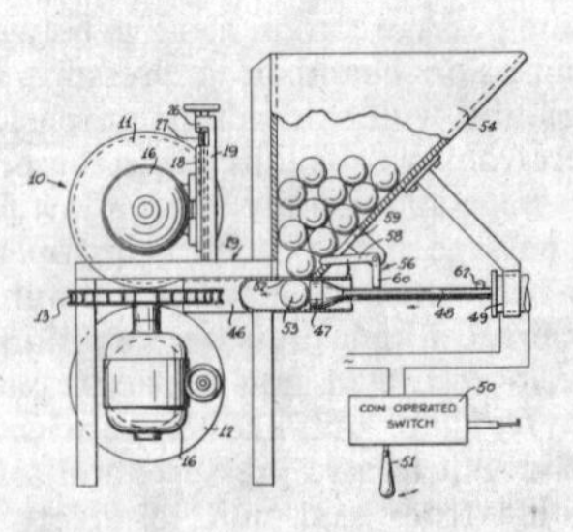

1. A ball throwing machine comprising two pairs of circular ball throwing members, means supporting said members for rotation about axes lying in a single plane with one pair of said members rotating in a first plane and the other pair of said members rotating in a second plane disposed at right angles to said first plane and intersecting the latter along a line between said first pair of members so that the confronting peripheral edge portions of said two pairs of throwing members define a tangential passageway for the reception of the balls to be thrown, said first pair of members constraining a thrown ball to movement in said second plane and said second pair of members constraining a thrown ball to movement in said first plane, means for effecting the continuous rotation of said throwing members to achieve movement of said confronting peripheral edge portions in one direction, and means for delivering balls one at a time to said passageway in said one direction of movement.

2,716,974
GAS FURNACE
George O. Wray and Fred E. Wilson, Milwaukee, Wis.
Application January 30, 1952, Serial No. 269,024
2 Claims. (Cl. 126—109)

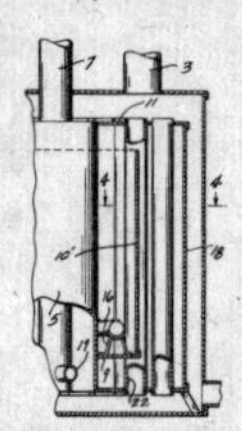

1. A furnace comprising a combustion chamber having an annular inner wall and an annular outer wall surrounding said annular inner wall and spaced therefrom and having a main bottom and a main top wall extending from said annular inner wall to said annular outer wall, said combustion chamber having an annular baffle spaced from said main top wall and main bottom wall and having an auxiliary annular bottom wall spaced from said main bottom wall and extending from said inner wall to the lower end of said baffle, said furnace having a centrally located discharge chamber surrounded by said combustion chamber and having a closed bottom and an outer wall spaced from the inner annular wall of said combustion chamber thereby providing an annular auxiliary chamber, said annular auxiliary chamber being open top and bottom for the upward passage of air to be heated, an annular burner located inwardly of said baffle and above said auxiliary annular bottom wall, said baffle and said inner wall defining the primary combustion chamber and said baffle allowing the products of combustion to pass downwardly between said baffle and said outer wall, means located below the baffle and placing the lower portion of said combustion chamber in communication with said centrally located discharge chamber, and an outer shell surrounding said combustion chamber and spaced therefrom and having air inlet and air outlet means, said baffle having a plurality of spaced air pipes forming a unitary portion therewith and passing through the combustion chamber and opening exteriorly of the combustion chamber at points above and below the combustion chamber and allowing air to pass upwardly from a point below said combustion chamber and to discharge at a point above said combustion chamber.

2,716,975
COMBUSTION TYPE AIR HEATER FOR DRYING PURPOSES
Danal W. Johnston, Piqua, Ohio, assignor to Hartzell Industries, Inc., Piqua, Ohio, a corporation of Ohio
Application November 14, 1951, Serial No. 256,191
8 Claims. (Cl. 126—110)

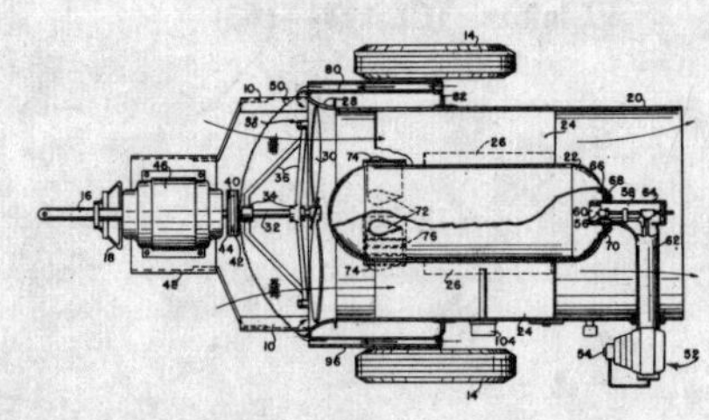

1. In a combustion type air heater, an outer shell at both ends, a fan at one end of the shell for blowing air therethrough, a hollow heater cartridge mounted on the axis of said shell between the ends thereof and having burner means at the end opposite the fan for generating heat within the cartridge, exhaust conduits extending downwardly from the fan end of said cartridge to the outer periphery of said shell, an orifice ring on the said one end of said shell around the fan, a chamber for exhaust gases surrounding said shell at the fan end and communicating with said exhaust conduits for receiving exhaust gases from said exhaust conduits and one end of said chamber being formed by said orifice ring, and exhaust stack means opening from said chamber to the atmosphere.

2,716,976
HEATER APPLICABLE TO FLOOR AND WALL
Thomas Pinatelli, Los Angeles, Calif.
Application June 11, 1951, Serial No. 230,870
6 Claims. (Cl. 126—116)

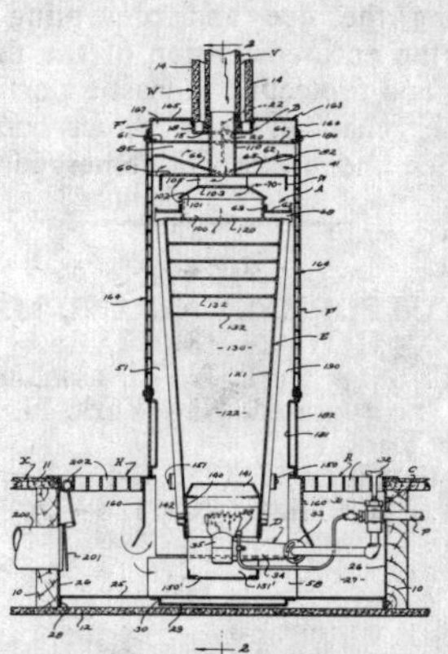

1. A heater applicable to a floor and a wall projecting up from the floor and having spaced studs and a vent therein, including, an upwardly opening box-shaped base adapted to be arranged within the confines of the floor, a single air ducting body structure adapted to be arranged in the wall and having an elongate air collecting head structure extending horizontally and adapted to be positioned between the studs of the wall, said body structure having vertical end duct portions communicating with and depending from the ends of the head and opening into the base, said body structure including a gas discharging pipe projecting through said head and engageable with the vent, the body structure having side openings adapted to open at each side of the wall, a core structure in the body structure defining a combustion chamber and opening downwardly into the base and having heat transferring side members respectively facing said side openings, a burner in the base at the open lower end of the core structure, a discharge pipe projecting from the upper end of the core structure toward said gas discharging pipe, and a pair of shells, each having side sections detachably secured to the body structure directly outward of and in spaced relation to the respective side members of the core structure and each having openings to pass the air heated from the side members of the core structure, each end duct portion of the body structure including, a flat vertically disposed outer wall, a flat vertically disposed inner wall spaced from the outer wall, side walls between the vertical edges of and joining the inner and outer walls, and a partition between the walls dividing the end portion into two vertically disposed air conducting passageways open at the upper and lower ends of the body structure.

2,716,977
HEATER FOR LIVESTOCK WATERING TANKS
Rudolph Otis Loyles and John W. Walker, Wichita, Kans.
Application January 9, 1953, Serial No. 330,413
2 Claims. (Cl. 126—360)

1. In an immersion heater which has a liquid tight burner chamber, a burner therein adapted to operate at a level below the surface of the liquid to be heated, and a heat transfer flue communicating with the chamber

to receive heat from the burner and projecting exteriorly of the chamber into the liquid to be heated, the improvement which comprises a protective liquid tight jacket encasing a portion of the flue from the burner chamber

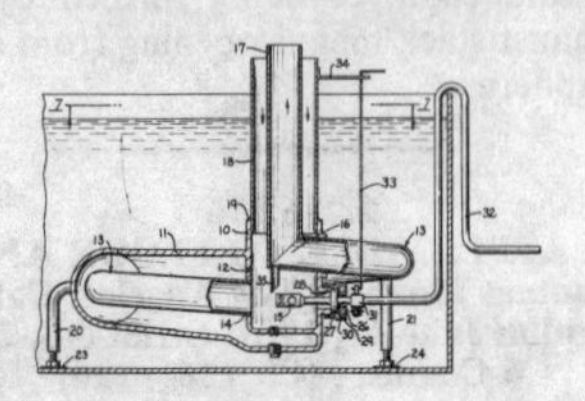

outward along the flue and preventing direct liquid contact with the encased portion of the flue, said jacket also encasing and protecting at least a portion of the wall of said burner chamber from direct contact with the liquid in which the heater is submerged

2,716,978
APPARATUS FOR TESTING AND MEASURING HUMAN REFLEXES
Ugo Torricelli, New York, N. Y., assignor to Torricelli Creations, Incorporated, New York, N. Y., a corporation of New York
Application June 28, 1952, Serial No. 296,165
20 Claims. (Cl. 128—2)

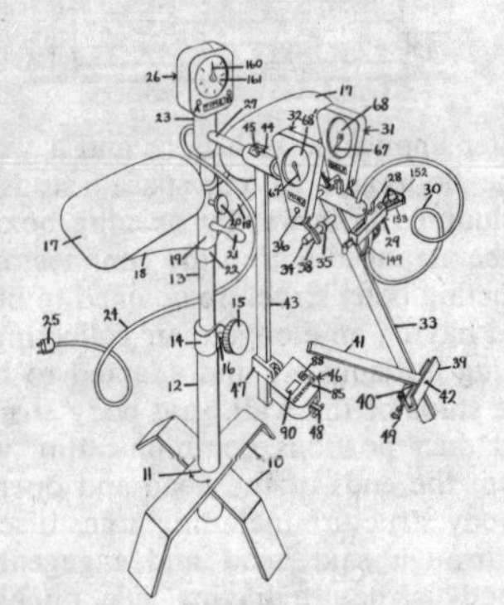

1. In an apparatus for testing and measuring human reflexes, a vertical support, a horizontal arm carried thereby, spaced angle and force reaction gauges mounted on said arm, a pendulum pivotally connected near its upper end to said horizontal arm between said gauges and a limb-engaging bar assembly on the lower end of said pendulum and connections between said pendulum and said gauges for actuating the same upon movement of said pendulum in response to movement of said limb-engaging bar assembly.

2,716,979
METHOD OF DETECTION AND ELECTRIC DETECTOR OF ACUPUNCTURE AND IGNIPUNCTURE POINTS
Pierre Pouret, Chatellerault, France
Application July 31, 1951, Serial No. 239,515
Claims priority, application France August 3, 1950
13 Claims. (Cl. 128—2.1)

7. Apparatus for locating particular cutaneous tissue points for the practice of ignipuncture and acupuncture comprising a condenser, a source of saw-tooth-form pulsating potential, means connecting said circuit to said condenser to charge the condenser, a discharge circuit comprising a plate electrode adapted to be held in contact with an area of said cutaneous tissue, a contact point adapted to be brought into contact successively with selected points on said cutaneous tissue, a discharge circuit coupling said electrodes respectively to opposite terminals of said condenser, and indicating means connected to said discharge circuit to indicate changes in the re-

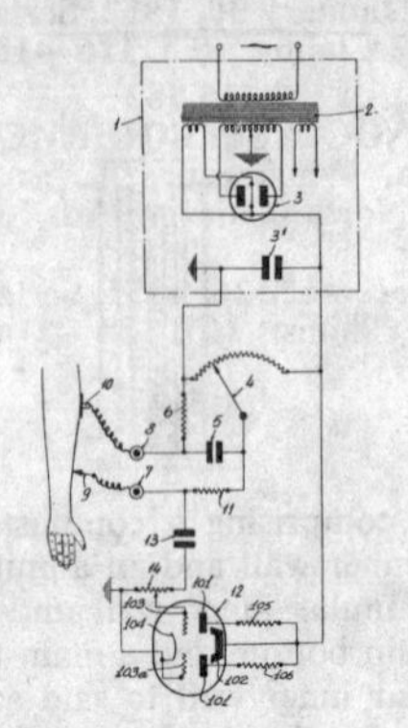

sistance of said cutaneous tissue when said contact point is brought into contact therewith.

2,716,980
EXERCISERS FOR SUB-NORMAL MUSCLES
William Bierman, New York, N. Y.
Application November 3, 1951, Serial No. 254,690
6 Claims. (Cl. 128—44)

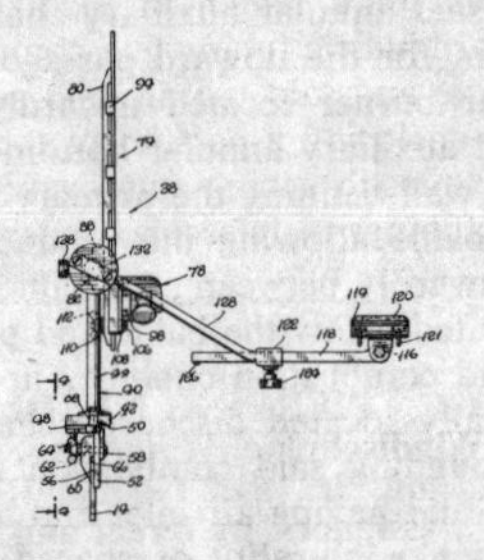

6. An exerciser comprising a first platform for limb extremities, a second platform for limb extremities, means defining an axis about which the first platform is rockable, means defining an axis about which the second platform is rockable, a member carrying said last mentioned means, means detachably securing said member to the first platform, and means carried by the member for supporting a limb, an extremity of which is adapted to be located on the second platform.

2,716,981
FACIAL MASK
Agnes Murray More, Laguna Beach, Calif.
Application August 12, 1954, Serial No. 449,463
4 Claims. (Cl. 128—163)

1. A facial mask to be worn by a woman underneath a hair dryer, comprising: a main body formed with a pair of eye openings, and a single opening for the nose and mouth; a pair of flaps depending from the lower side portions of said main body for covering the front neck portion of said woman, said main body and flaps including a first layer of heat-resistant paper and a second layer of absorbent paper secured to the rear of said

first layer, said second layer being impregnated with a skin cleansing material; and means to secure said main body to said woman's head.

2,716,982
VENOCLYSIS EQUIPMENT

George R. Ryan, Waukegan, Ill., assignor to Abbott Laboratories, North Chicago, Ill., a corporation of Illinois

Application December 26, 1951, Serial No. 263,326
8 Claims. (Cl. 128—214)

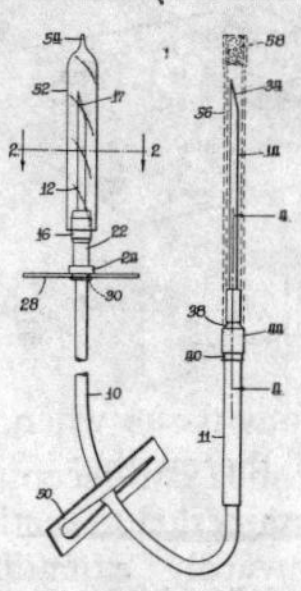

5. A venipuncture needle adapted for use in a disposable venoclysis set comprising, in combination: a length of needle stock having a beveled sharp point at one end thereof; the other end of said needle telescoping inside an end of a length of flexible, plastic tubing; a length of hard plastic tubing telescoping over the juncture of said needle and said tubing; and permanently deformable means enclosing a portion of the telescoping needle and flexible plastic tubing immovably securing the said needle and tubings in telescoped position.

2,716,983
PIERCING NEEDLE

Edward F. Windischman, Waterbury, Conn., and William Lionel Hartop, Jr., and George Richard Ryan, Waukegan, Ill., assignors, by direct and mesne assignments, to Abbott Laboratories, Chicago, Ill., a corporation of Illinois

Application October 8, 1952, Serial No. 313,723
18 Claims. (Cl. 128—221)

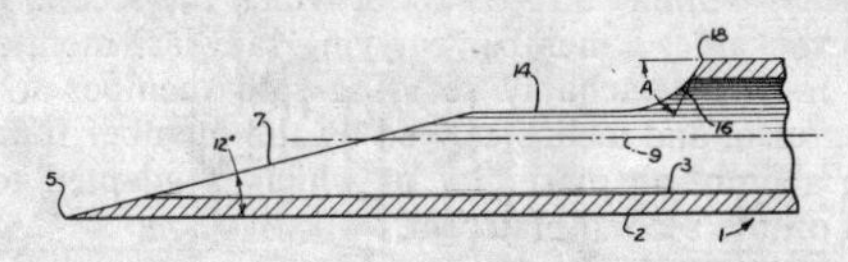

1. A needle useful in the therapeutic arts comprising a cannula through which a fluid-like material is conducted having a piercing point at one end of said cannula formed by an oblique bevel surface extending rearwardly across the cannula from said point, a ledge surface extending rearwardly along opposite side walls of the cannula from the upper ends of the said bevel surface a distance less than half the length of the said cannula, and upwardly inclined surfaces extending along the cannula wall from the trailing edges of the said ledge surface defining a heel surface which presents a relatively blunt surface area to the material being penetrated thereby.

2,716,984
FEMININE HYGIENE DISPENSER AND METHOD

Harold G. Davis, Denver, Colo.

Application April 14, 1952, Serial No. 282,099
13 Claims. (Cl. 128—225)

1. A dispenser comprising the combination of a container adapted to contain and dispense a substance, a removable cover for said container, a separate storage compartment in said container and extending into the container for substantially its entire depth, said storage compartment being attached to and removable with said cover, and a soluble gas-creating substance in said compartment adapted to contact said container substance

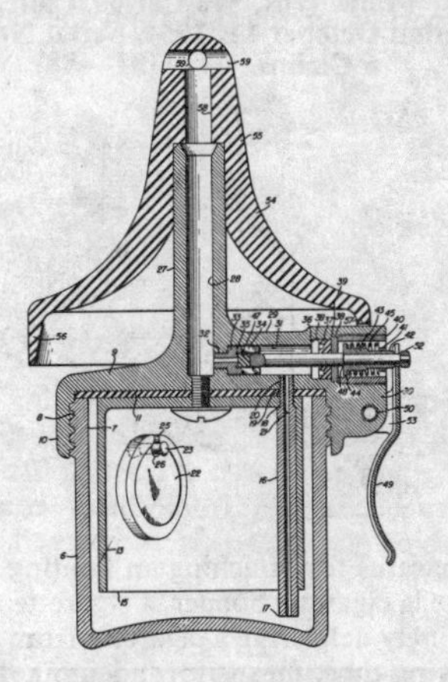

only when said container is operated whereby to cause said gas creating substance to contact said container substance and create sufficient gas pressure to dispense the substance from the container.

2,716,985
COMBINATION LOOSE LEAF NOTEBOOK AND PORTABLE DESK

Harold H. Wolf, Chicago, Ill., assignor to Major Leather Goods Mfg. Co., Chicago, Ill., a corporation of Illinois

Application February 24, 1955, Serial No. 490,189
1 Claim. (Cl. 129—1)

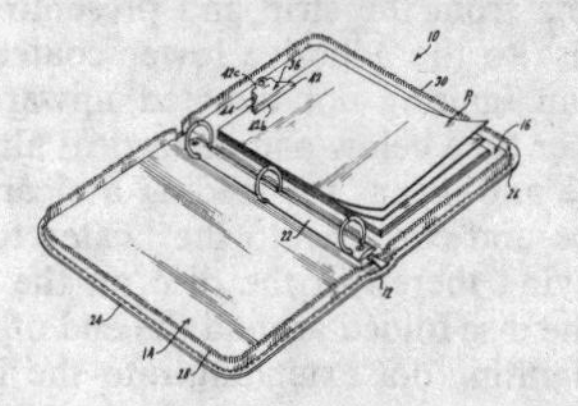

An enclosed portable desk comprising a cover of box-like configuration having a binder panel and front and back panels hinged along inner side edges to said binder panel, a loose leaf binder mechanism extending longitudinally of said binder panel marginal closure strips connected to the outer side edges and upper and lower edges of said panels for completing the box-like configuration of said cover, slide fastener mechanisms connected to said marginal closure strips for releasably closing said cover in said box-like configuration, at least said back panel including facing and backing members, a hidden rigid reinforcing member interposed between said facing and backing members and of an area substantially coextensive with said facing and backing members, and clip means operatively disposed in relation to the inner face of said back panel for holding one or more sheets of paper on said back panel serving as a portable desk, said clip means extending across said back panel adjacent to its upper edge and including an attachment bracket, securing means extending through said facing member and anchored only on said reinforcing member for mounting said attachment bracket on said back panel, said securing means being covered by said backing member whereby said securing means is hidden from view when said cover is viewed from its outer side, a jaw swingably supported in said attachment bracket and having a contact edge engaging said backing member, and a spring yieldably urging said contact edge against said backing member.

2,716,986
CAPPING MACHINE FOR SELF-LIGHTING CIGARETTES
Frank Witt, San Carlos, Calif.
Application October 11, 1954, Serial No. 461,363
6 Claims. (Cl. 131—88)

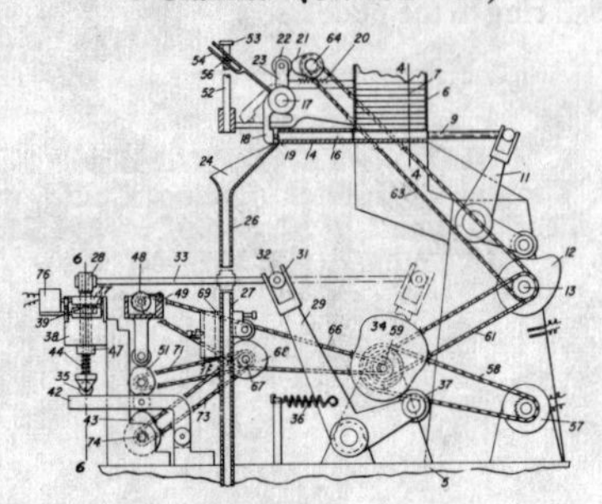

1. In apparatus for attaching an igniting cap to an end of a cigarette, a cigarette hopper, a cigarette receiving tube, means for singly delivering a cigarette from the hopper into the receiving tube, means for indenting the leading end of the cigarette, a guide tube receiving the indented cigarette, a translatable tubular carrier normally positioned below the guide tube, a driven adhesive coated roller in the path of the carrier, means for raising and lowering said roller for intermittent rolling contact with the lower end of the carrier in one direction of movement of the carrier thereover to deposit adhesive on the lower end of the carrier, a die positioned beyond the roller with the carrier movable to a position over the die, means for feeding a strip having spaced igniting dots on opposite faces to said die with the igniting dots arranged in opposed pairs and for punching a disc including an opposed pair of igniting dots from the strip and presenting the upper side of the disc to the adhesive lower coated end of the carrier with an igniting dot directed upwardly into the carrier, a heater tube below and in vertical alignment with the guide tube and means for feeding a cigarette through the guide tube and carrier into the heater tube with the cigarette carrying therewith the disc on the carrier with the edges of the disc folded around the end of the cigarette and with an igniting dot extending into the indented end of the cigarettes.

2,716,987
CIGARETTE TRAP
John S. Keefe, Brooklyn, N. Y.
Application January 11, 1954, Serial No. 403,261
5 Claims. (Cl. 131—237)

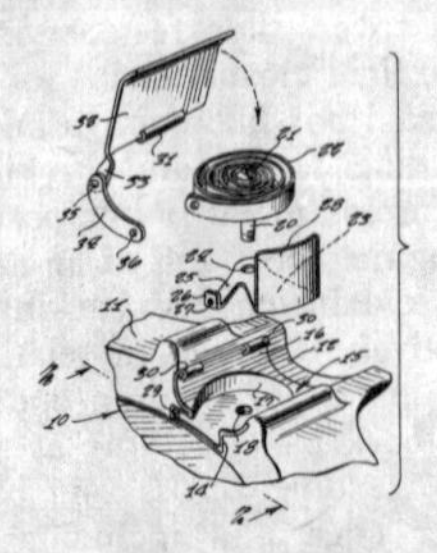

1. An ash tray having a peripheral wall provided with a cigarette receiving groove, a spiral bi-metallic element disposed at the base of the groove, said element having its inner end fixed therewithin and its outer end adapted for movement, a cover plate pivotally connected at one side of said groove and adapted to move downwardly on to a cigarette disposed therewith, means interconnecting the free end of said bi-metallic element and said cover whereby to lower the cover as a cigarette burns on the said bi-metallic element, and end plate means in operative engagement with said free bi-metallic element end adapted to close off the inner end of the groove.

2,716,988
SMOKER'S ACCESSORY
Millard F. Comstock, Fort Erie, Ontario, Canada
Application October 30, 1953, Serial No. 389,435
4 Claims. (Cl. 131—242)

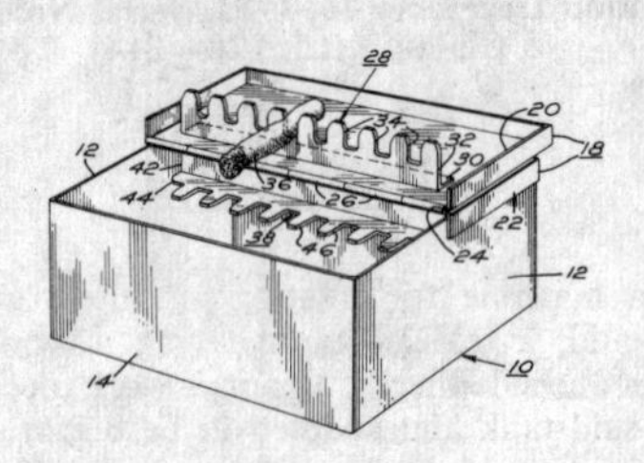

3. A cigarette butt and ash receptacle comprising a generally open top container, duo-section cover portions having laterally downwardly extending marginal rim portions adapted to snugly embrace the upper edge portions of said container when in closed position, said cover portions being hinged at their abutting edges, a cigarette holder rack attached to the inner face of one of said cover portions and being substantially coextensive with and spaced slightly inwardly from the inside edge thereof, and a rack clearing means depending from the lower face of the other of said cover portions and being substantially coextensive with and spaced slightly inwardly from the inside edge thereof; said racks having a plurality of spaced lugs adapted to intermesh with each other when said cover portions are in closed position.

2,716,989
APPARATUS FOR TREATING METAL PARTS
Clarence L. Joy, Detroit, Mich., assignor to Holcroft & Company, Detroit, Mich., a corporation of Michigan
Application June 5, 1950, Serial No. 166,287
12 Claims. (Cl. 134—66)

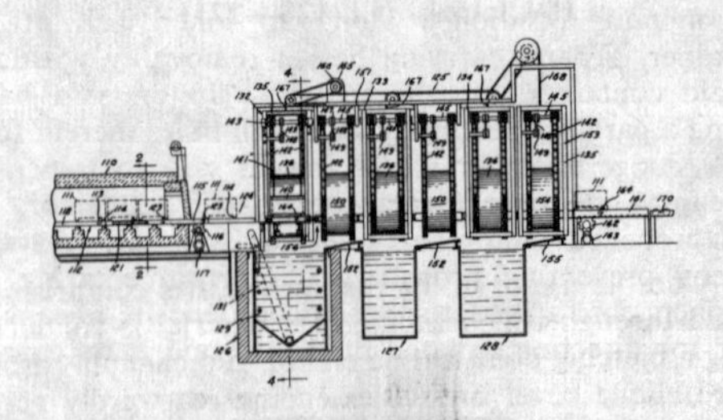

1. Apparatus for treating metal articles comprising a receptacle containing a bath and open at the top, an elevator movable into and out of the bath through the open top of the receptacle, a drum supported on the elevator for rotation about an axis extending generally perpendicular to the path of travel of the elevator and movable by the elevator into and out of the bath, a second drum rotatably supported to one side of the path of travel of the elevator in a position to register with the first drum in the raised position of the elevator, said drums having the opposite ends open and having means for supporting articles therein, means for transferring articles from the first drum into the second drum when said drums are in registration comprising a pair of slides respectively supported in the drums for sliding movement axially of the drums and having coupling portions between the drums movable into interlocking engagement upon movement of the elevator to its raised position, and means for rotating the drums when the drums are out of registration.

2,716,990
TREATING MACHINE FOR TREATING AIR FILTERS AND THE LIKE WITH A LIQUID
Samuel W. Katz and Nicholas T. Salmon, San Antonio, Tex.
Application July 10, 1950, Serial No. 172,878
3 Claims. (Cl. 134—155)

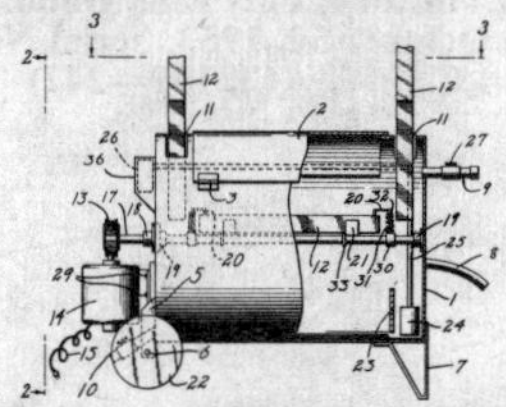

1. In a machine for treating air filters and the like with a liquid, a treating tank having a work admitting opening in one side thereof to admit work pieces to the interior of said tank and adapted to be disposed upwardly during operation of the machine, a closure for said opening, a shaft rotatably mounted in said tank, means for rotating said shaft, a work support secured to and extending transversely of said shaft, and clamps slidably movable longitudinally of the shaft and work support for clamping a work piece against said work support.

2,716,991
CLEANING APPARATUS
Gene Brenfleck, St. Louis, Mo.
Application January 24, 1955, Serial No. 483,611
4 Claims. (Cl. 134—182)

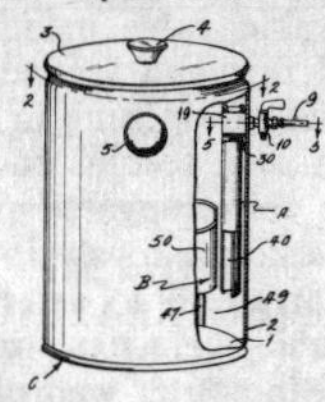

1. A cleaning apparatus comprising a container adapted to hold a cleaning fluid, a fluid propelled motor and agitator positioned along an inner wall of said container, a parts carrying basket removably positioned in said container opposite to said agitator, said basket having a parts supporting surface with holes therein forming a sludge chamber therebelow in said container and a cleaning chamber thereabove in said container, said fluid propelled motor being sealed from said chambers thereby preventing propelling fluid from entering said chambers, and a deflector positioned opposite to said agitator for directing agitated fluid into said parts carrying basket.

2,716,992
SUPPORTING MEANS FOR TENTS
Arthur E. Campfield and Walter Davis, New York, N. Y., assignors to Arthur E. Campfield, Inc., New York, N. Y., a corporation of New York
Application October 30, 1953, Serial No. 389,403
7 Claims. (Cl. 135—3)

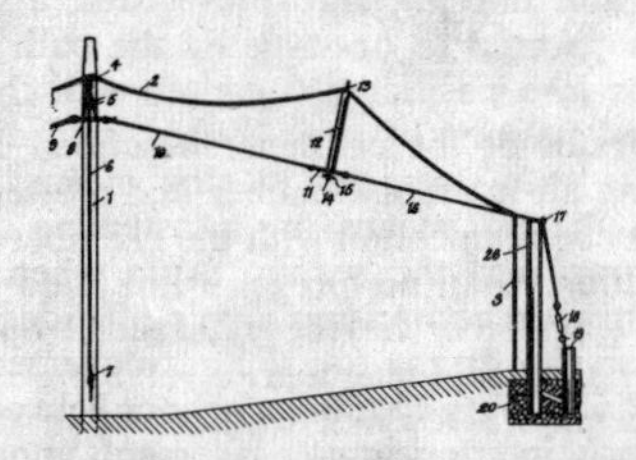

3. A supporting means for tents comprising, a tent post, a bail ring surrounding the post, a tent prop located at a distance from the post and having its upper end supporting a part of the top of the tent cover, a plate on the lower end of the prop, an upright at a distance outwardly of the post, a tensioned cable extending from the plate to the upright and a tensioned cable extending from the bail ring to the plate.

2,716,993
FOLDING TENT FRAME
Thomas H. Codrick, Encino, Calif.
Application June 2, 1952, Serial No. 291,215
10 Claims. (Cl. 135—4)

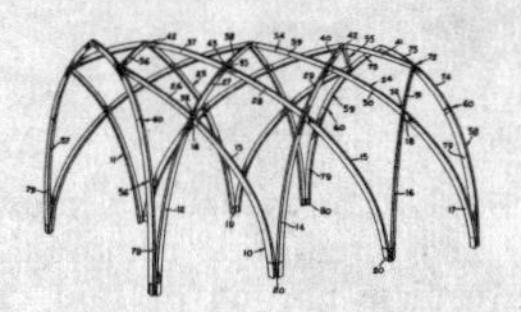

5. A collapsible tent frame comprising two half-dome type tent frames, each of said tent frames being collapsible, means detachably connecting the half-dome tent frames to each other so as to form a dome-type tent frame, each of said half-dome type tent frames comprising a lower set of arcuate members hingedly connected together so as to be capable of assuming positions in side by side relationship or spaced positions, and an upper set of arcuate members hingedly connected together to assume collapsed positions in side by side relationship or spaced positions with respect to each other, and means for detachably connecting the lower ends of the members of the upper set to the upper ends of the members of the lower set.

2,716,994
COMBINED UMBRELLA AND POCKET BAG CONSTRUCTION
Ugo Torricelli, New York, N. Y., assignor to Torricelli Creations, Incorporated, New York, N. Y., a corporation of New York
Application December 14, 1950, Serial No. 200,785
6 Claims. (Cl. 135—33)

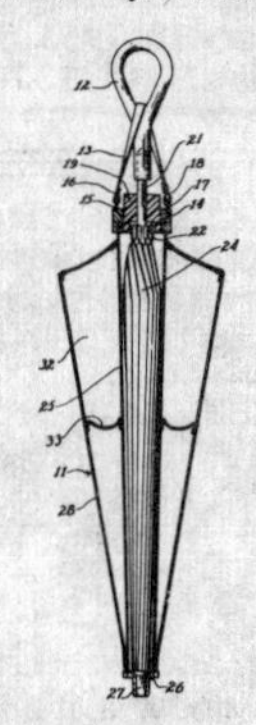

1. A combined umbrella and pocket bag construction comprising an umbrella part having a central shaft and an apertured handle, a pocket bag part having a central tubular, umbrella-receiving sleeve into which the umbrella part is insertable and from which it is removable and a plurality of circumferentially spaced, radially arranged external pockets of generally conical shape secured to said central sleeve and provided at their upper ends with openable closures, means for detachably coupling the upper portion of the central sleeve of the pocket bag part to the shaft of the umbrella part and means connected to said first means and extending through said handle for detachably coupling the pocket bag part and the umbrella part.

2,716,995
VALVE FOR REVERSIBLE FLUID PUMP
Everett L. Baugh and De Loss D. Wallace, Dayton, Ohio, assignors to General Motors Corporation, Detroit, Mich., a corporation of Delaware
Application September 23, 1950, Serial No. 186,396
3 Claims. (Cl. 137—87)

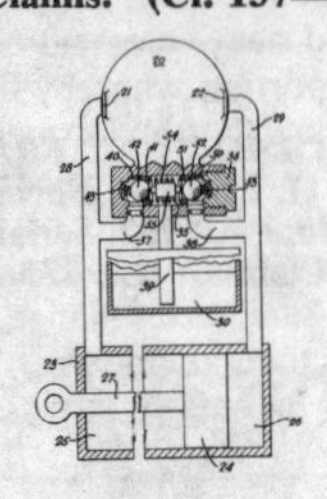

3. A control valve consisting of a housing providing a valve chamber and three ports communicating with the ends and intermediate portion of said chamber respectively; two oppositely disposed check valves in said chamber normally shutting off communication between the two end ports and the intermediate port, each valve consisting of a sleeve slidable in the chamber and providing a valve seat and a ball valve resiliently urged upon the valve seat in the sleeve for closing the sleeve, each sleeve having opening means to connect the port thereadjacent with the interior of the sleeve; a spring interposed between said sleeves for urging each sleeve toward and against a respective end of the valve chamber; and a floating pin between the balls of said valves and normally disengaged from the said valves, said pin being engageable by either valve to engage and open one valve in response to unitary movement of the other valve and to thereafter open said other valve all in response to increasing fluid pressure thereagainst first to a predetermined value and thence above the said value respectively.

2,716,996
DISPENSER FOR CONCENTRATES AND THE LIKE
Carl C. Bauerlein, Mukwonago, Wis., assignor to The Dole Valve Company, Chicago, Ill., a corporation of Illinois
Application February 10, 1953, Serial No. 336,120
9 Claims. (Cl. 137—98)

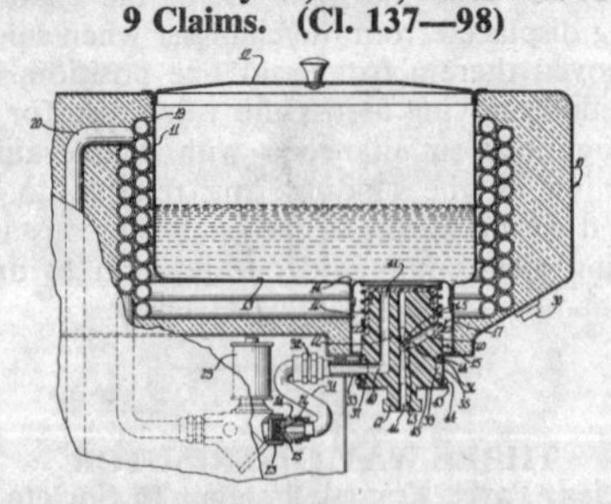

1. A proportioning and dispensing valve comprising a proportioning valve body, a fluid passage leading into said valve body from a wall thereof, and opening through the top of said proportioning valve body, a mixing passageway leading through said body and opening to each end thereof, a concentrate passageway leading into said body from a wall thereof and communicating with said mixing passageway, and a valve plunger having a wall extending about said body and a top extending across said wall and closing the top thereof and movable along the wall of said valve body to close said concentrate passageway and block the passage of concentrate into said body, the side wall and top of said valve plunger also defining a flow passageway from said fluid passage to said mixing passageway and opening said concentrate passageway by the pressure of fluid passing through said fluid passage opening through the top of said valve body.

2,716,997
QUICK RELEASE TIMING VALVE
Robert R. Crookston, Houston, Tex., assignor, by mesne assignments, to Esso Research and Engineering Company, Elizabeth, N. J., a corporation of Delaware
Application November 28, 1949, Serial No. 129,731
6 Claims. (Cl. 137—102)

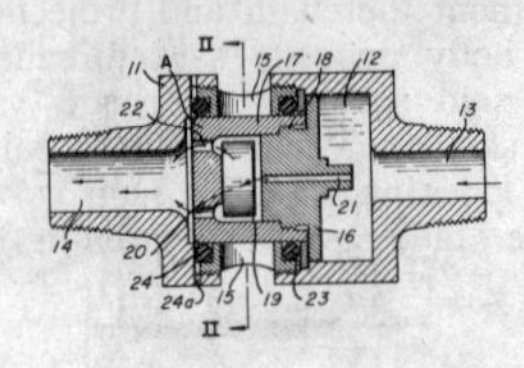

1. A quick release valve consisting of a valve body having an elongated passage with a cylindrical wall with an axial inlet port connecting to one end, an axial delivery port connecting to its other end, at least one radial exhaust port piercing the wall at a point between said inlet and said delivery port with a total effective exhaust port area at least a multiple of the total effective inlet port area, a first stop shoulder and a second stop shoulder spaced away from said first stop shoulder along the longitudinal axis of the valve body, a first circular sealing ring groove in the cylindrical wall of the body between the inlet port and the exhaust port and a second circular sealing ring groove in the cylindrical wall of the body between the exhaust port and the delivery port with a vent communicating the second circular sealing ring groove with an exterior surface of the body, first and second circular sealing rings arranged in the first and second grooves respectively, a movable valve member with a cylindrical outer surface and first and second end surfaces perpendicular thereto with no opening in the outer wall surface between said first and second end surfaces, said valve member being in the elongated passage with its outer cylindrical surface continuously in contact with the first sealing ring and arranged to assume a delivery position with its second end surface resting against the second stop shoulder and the outer wall of the valve member in sealing contact with the second as well as with the first sealing ring, said valve member being longitudinally movable along the axis of the valve body from said delivery position to an exhaust position where its second end surface is clear of the exhaust port and its first end surface is in contact with said first stop shoulder.

2,716,998
COMBINED TIRE INFLATING CHUCK, DEFLATOR AND BLOW GUN
Joseph J. Knasko, Ford City, Pa.
Application March 2, 1949, Serial No. 79,247
4 Claims. (Cl. 137—231)

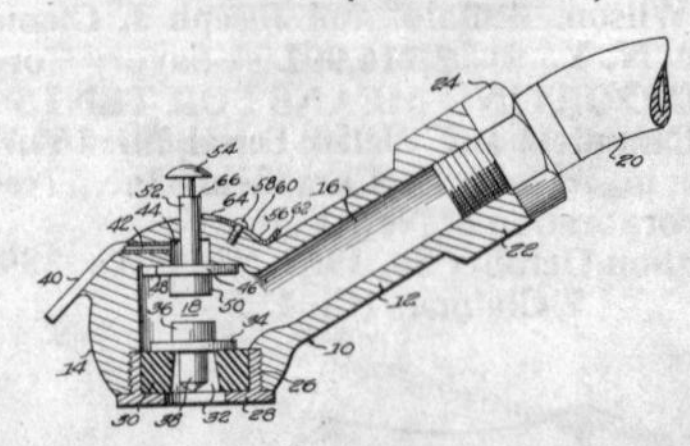

1. Apparatus of the character described comprising a body having an internal chamber and a laterally extending duct in communication with the chamber and adapted for connection to an air pressure line, said body at one side of said chamber having a threaded opening larger in diameter than said chamber, a retaining ring threaded in such opening, a resilient washer within said retaining ring having an axial opening therethrough, a valve normally seated against the inner surface of said washer and provided with an axial projecting valve depresser in

the opening in said washer, a valve seat in said body at the other side of said chamber, a second valve engaging said valve seat, there being a space in said body adjacent said second valve normally closed by such valve to communication with said chamber, said body being provided with a laterally extending passage communicating with said space, and a valve stem carried by said second valve in axial alignment therewith and projecting through the side of said body opposite said threaded opening for operation of said second valve, said valves and said washer being smaller in diameter than said chamber and said threaded opening whereby such valves and washer are insertable into position through such threaded opening.

2,716,999
NOZZLE AND ASSOCIATED DEVICES FOR FUELING AND DEFUELING TANKS
Everett H. Badger, Jr., Carl P. Dahl, and John W. Overbeke, Cleveland, Ohio, assignors to The Parker Appliance Company, Cleveland, Ohio, a corporation of Ohio
Application May 1, 1950, Serial No. 159,346
40 Claims. (Cl. 137—235)

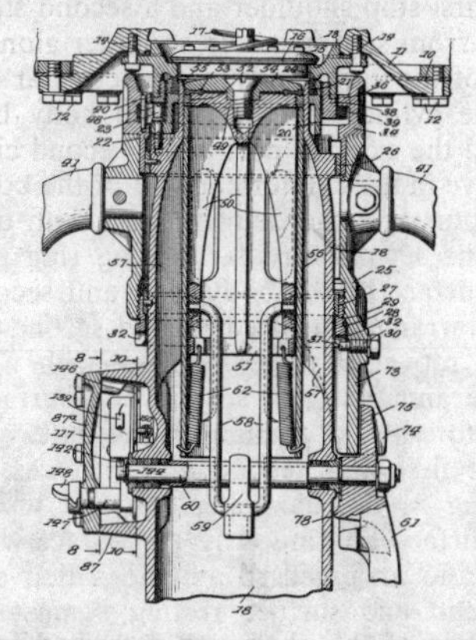

1. A fueling system for a tank having a filling opening controlled by a spring closed valve comprising a nozzle, means for securing the nozzle to the tank, a manually operated valve disposed within the nozzle for closing the same and movable into the tank for opening the tank valve and exposing ports leading from the nozzle into the tank, means within the nozzle for closing said ports while the nozzle is in extended position, and means controlled by a predetermined condition of the fuel in the tank for causing the operation of said port closing means.

2,717,000
CALF FEEDER NIPPLE AND VALVE ASSEMBLY
Ashley F. Wilson, Buffalo, and Joseph J. Clement, Williamsville, N. Y., assignors to Lisk-Savory Corporation, Buffalo, N. Y.
Application May 27, 1950, Serial No. 164,726
2 Claims. (Cl. 137—343)

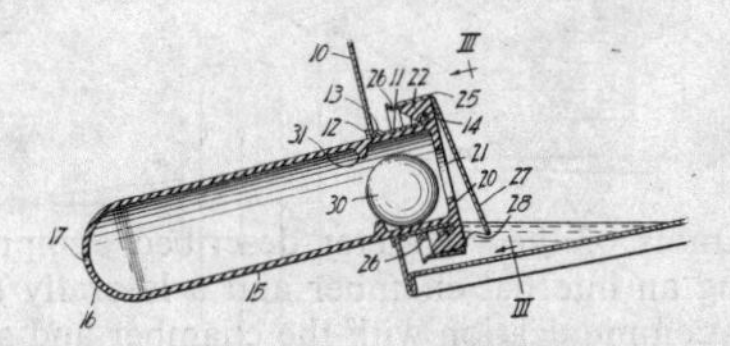

1. Attachment means for detachably securing the end of a flexible tubular member in an aperture in a wall member, said attachment comprising a rigid sleeve element disposed in said aperture and extending inwardly of said wall member and having a peripheral bead at its inner end, a flexible tubular member having radially inwardly and outwardly directed marginal flanges at one end and internal annular protuberances spaced axially from said inwardly directed flange, said outwardly directed flange having an annular channel for receiving said sleeve element bead, and a cap element having a rigid annular portion engaging peripherally about said outwardly directed flange to cause the walls of said annular channel to flexibly embrace said sleeve element bead, and an opening therein at its lower portion whereby fluid from within the wall member enters the tubular member only adjacent the bottom edge of said cap element, and a member of circular cross section adapted to be forced through said inwardly directed flange into said tubular member and adapted to engage said latter flange to control flow of fluid through said tubular member and limited in its outward movement by said protuberances.

2,717,001
VALVE SEAT
Ainslie Perrault, Tulsa, Okla.
Application July 15, 1950, Serial No. 174,051
1 Claim. (Cl. 137—514)

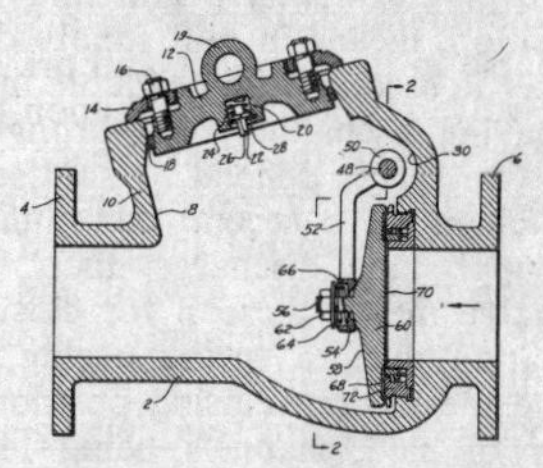

In a valve seat for a safety valve having a pivotable disc capable of rapid automatic opening and closing and comprising ring means secured in the valve body and an annular chamber in said ring means, said chamber being closed at its inner end and open at its outer end, and a freely moveable landing ring disposed in said chamber and adapted to extend outwardly through the open end of the chamber beyond the ring means in one position thereof for contact by the valve disc, said chamber being of greater width than the thickness of said landing ring to define with said landing ring a restricted passageway for fluid into and out of the chamber, said fluid being displaced from the chamber when said landing ring is moved therein from said one position, and said displaced fluid moving across the valve disc for cleaning the seating area simultaneous with a hydraulic cushioning action of the landing ring relative to the ring means, and spring means in the chamber constantly urging the ring means into said one position to draw fluid into the chamber through said passageway.

2,717,002
THREE-WAY DISTRIBUTOR
Rene Lucien, Paris, France, assignor to Societe d'Inventions Aeronautiques et Mecaniques S. I. A. M., Fribourg, Switzerland, a corporation of Switzerland
Application August 13, 1952, Serial No. 304,075
Claims priority, application France February 14, 1952
2 Claims. (Cl. 137—620)

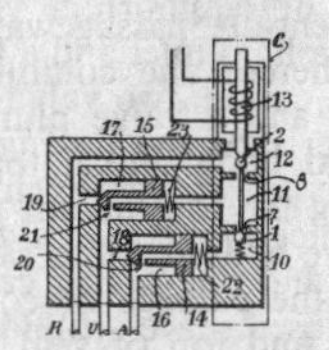

1. The combination, in a single body member, comprising a three-way hydraulic distributor including a body member having a supply duct, a discharge duct

and an output duct therein and comprising apertured walls in said body member defining a central one and two outlying aligned chambers respectively connected to said ducts, a guide inside said central chamber, a piston slidable in said guide, valve stems secured to each end of said piston, valve heads on said stems positioned to seat on said apertured walls in said outlying chambers, a spring urging said piston in one direction, actuating means comprising a plunger positioned to act upon said piston in opposition to said spring, and a return spring positioned to act in opposition to actuating movement of said plunger; and a further hydraulic distributor of greater output comprising, in said body member, three ducts respectively for supply, output and discharge common to said three-way hydraulic distributor and said further hydraulic distributor, means in said output duct defining a first chamber, a first piston slidable in said first chamber and having a channel therethrough communicating with the two spaces defined by said piston in said first chamber, a first communication duct between said supply and output ducts, said first piston having valve means thereon for closing said first communication duct, a spring urging said first piston and valve means into closing position, means in said supply duct defining a second chamber, a second piston slidable in said second chamber and having a channel therethrough communicating with the two spaces defined by said second piston in said second chamber, a second communication duct between said supply and discharge ducts, said second piston having valve means thereon for closing said second communication duct, a spring urging said second piston and valve means into closing position, the cross-section of flow of said channels being less than the useful cross-section of flow of the apertures in said walls.

2,717,003
DOUBLE BEAT OR EQUILIBRIUM VALVES
James Bertram Jay, Albert Joseph White, and Joseph Anthony Hunt, Redditch, England, assignors to The Chloride Electrical Storage Company Limited, near Manchester, England
Application March 13, 1951, Serial No. 215,314
Claims priority, application Great Britain April 21, 1950
2 Claims. (Cl. 137—625.34)

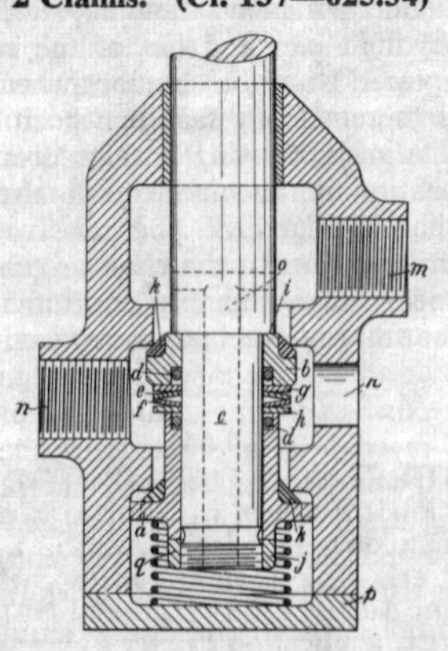

1. A double beat or equilibrium valve comprising a casing provided with a fluid inlet and a fluid outlet branch, a valve spindle in the casing, two valve heads each having an outer and an inner side and having a good sliding fit on the spindle, a passageway through the valve spindle establishing communication between the outer sides of the two valve heads and one of the casing branches, the casing having an enclosed space around the adjacent inner sides of the valve heads, the casing having enclosed spaces adjacent the outer side of each of the two valve heads, one branch on the casing leading to the first named enclosed space and the other branch leading directly to one of the second named enclosed spaces and indirectly through the passageway in the spindle to the other of the second named enclosed spaces, a cylindrical recess in each valve head, a packing ring in the said recess making contact with the valve head and the spindle, at least one Belleville washer between the adjacent inner sides of the valve heads, a flat ring between each valve head and the adjacent Belleville washer, one valve head engaging a shoulder on the spindle, two valve seats on the casing upon which the seating surfaces of the valve heads can seat simultaneously, and an adjusting nut carried by the spindle for moving one valve head towards the other and compressing the Belleville washer.

2,717,004
MULTIPLE VALVE
John S. Page, Long Beach, Calif., assignor to Page Oil Tools Inc., Long Beach, Calif., a corporation of California
Application April 18, 1949, Serial No. 88,211
12 Claims. (Cl. 137—637)

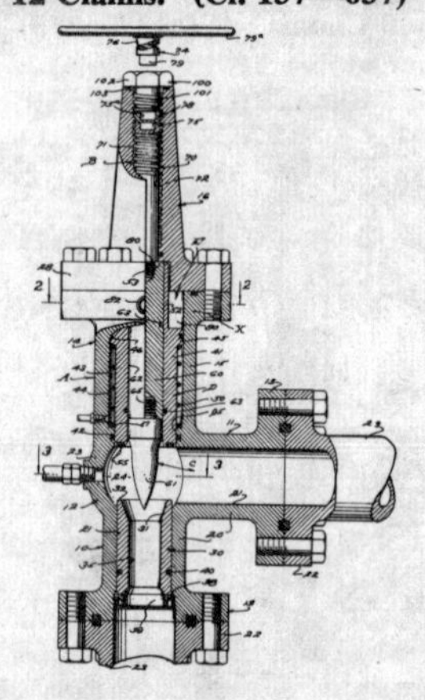

3. A structure of the character described including, a body having angularly related ducts in communication with each other and having a lateral extension, two valves carried by the extension and shiftable relative to each other to each control flow through the body from one duct to the other, operating means for the valves operating the valves independently of each other, and a seat member detachable from the body and carried in one of the ducts of the body and cooperating with both of the valves to seal therewith.

2,717,005
PROCESS OF WEAVING
Frank W. E. Hoeselbarth, Carlisle, Pa., assignor to C. H. Masland & Sons, Carlisle, Pa., a corporation of Pennsylvania
Application June 19, 1950, Serial No. 168,960
6 Claims. (Cl. 139—39)

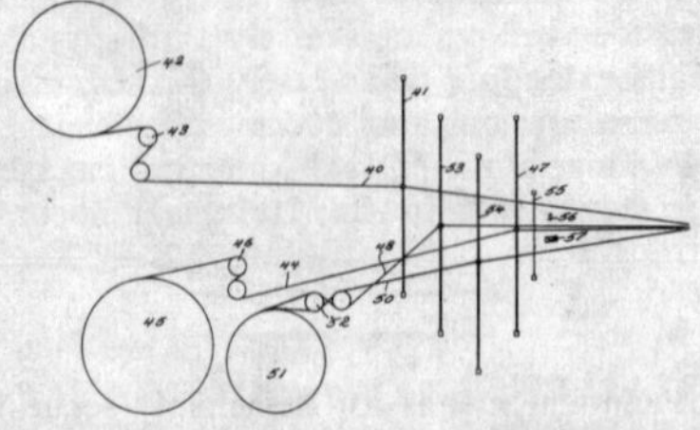

1. The process of weaving a pile carpet, which comprises interweaving at least one pile warp, stuffer warp ends, binder warp ends and wefts into a close woven fabric having in excess of seven rows of pile projections per inch, alternately raising the same pile warp in weaving

over a straight non-cutting pile wire and then over a pile wire having high and low portions within the shed and having a cutter at the far end, withdrawing the wires while cutting the pile warp by the wire having the cutter, and during withdrawal of the wires pulling down the uncut loops formed over the straight wire by the wire having high and low portions within the shed, and restricting such pull-down by virtue of the tightness of weave resulting from the above close weaving and thereby building up stress in the fabric to cause the tufts formed at both sides by cutting the loops through the action of the wire having the cutter to have high and low portions in the same transverse row, whereby both the uncut loops and both sets of cut tufts have high and low portions in the same transverse row.

2,717,006
SWEEPSTICK FOR LOOMS AND SELF-ALIGNING BEARING THEREFOR
Edward V. Dardani, Fairfield, Conn., assignor to The Heim Company, Fairfield, Conn., a corporation of Connecticut
Application October 31, 1952, Serial No. 317,894
1 Claim. (Cl. 139—151)

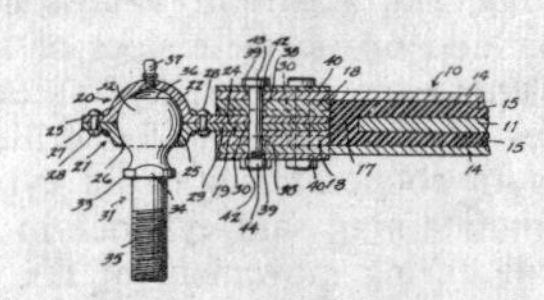

In combination, a sweepstick for looms, said sweepstick consisting of an elongated bar of rectangular cross-section having relatively narrow horizontal top and bottom surfaces and relatively wide vertical side surfaces and having an open end slot in one end centrally of and parallel to said side surfaces and extending between said top and bottom surfaces, and a bearing assembly connected to an end of said bar, comprising a pair of complementary socket-forming members, each having a vertical rigid web portion disposed longitudinally outwardly beyond the end of said bar and having a flat inner surface, said inner flat surfaces being in contact with each other in a vertical plane coincident to the longitudinal central vertical plane of said bar, means rigidly joining said web portions together with said flat inner surfaces in contact, each of said web portions having a socket portion complementary one to the other to form a spherical socket having its center point coincident to both the central horizontal plane of said bar and said central vertical plane thereof, at least one of said socket-forming portions having a central stud-receiving opening having its axis normal to said central vertical plane, and each of said web portions having an integral rigid co-planar tongue extension extending inwardly beyond the end of said bar in parallel relation to said side surfaces, said extensions having their flat inner surfaces in contact and being engaged within said slot of said bar with their outer surfaces in contact with the side surfaces of said slot, horizontal clamping bolt means extending transversely through said bar and said tongue extensions and securing said bearing assembly to said bar, and a ball end stud member engaged for free swiveling movement in said spherical socket and extended transversely outwardly through said stud-receiving opening.

2,717,007
ADJUSTABLE REED CAP AND SHUTTLE GUARD
William M. Battles, Anniston, Ala.
Application May 27, 1952, Serial No. 290,215
2 Claims. (Cl. 139—192)

1. In a loom construction including lay swords and an adjustable reed cap comprising top and bottom rails with the top rail thereof being mounted on the lay swords, the improvement comprising; means fastening said top and bottom rails for limited relative lateral movement including said top and bottom rails having registering vertical bores therethrough, the bores in one of said rails being larger than the bores in the other of said rails,

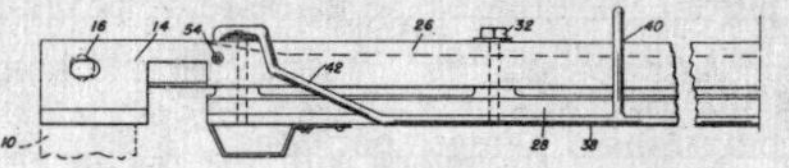

fastening elements extending through said bores adjustably frictionally clamping the top and bottom rails against one another, said fastening elements extending loosely through the larger bores enabling limited lateral adjustment of the rails with respect to one another.

2,717,008
ELECTRIC STOPPING MECHANISM FOR SHIFTING SHUTTLE BOX LOOMS
Charles W. Moss, Canton, Mass.
Application May 18, 1953, Serial No. 355,603
15 Claims. (Cl. 139—336)

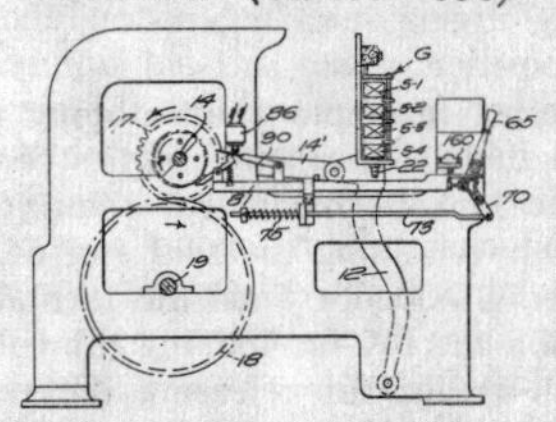

1. In a loom having shiftable shuttle boxes, box shifting mechanism, and loom stopping means including a knock-off mechanism, control mechanism for actuating said knock-off mechanism comprising abutment means associated with and rotatable by one of the operating shafts of the loom, a longitudinally shiftable bar member situated in-between said abutment means and a movable element of said knock-off mechanism and arranged to move said element and actuate said knock-off mechanism upon shifting movement of said bar member in a direction away from said abutment means, a finger movably mounted on said bar and having a portion extending well beyond the end of the bar and normally positioned with its tip close to but out of the rotational path of the abutments carried by said abutment means when said bar is in its initial starting position from which it is shifted for actuating the knock-off mechanism, a solenoid associated with said finger and operative upon being energized to move the finger and cause its tip to enter the rotational path of the abutments of said abutment means and engage an abutment thereof whereby said bar will be pushed thereby and shifted longitudinally away from its initial starting position to move said movable element of said knock-off mechanism and actuate the latter to effect loom stoppage, and an electric switch associated with the box shifting mechanism of the loom and normally inoperative during normal shifting of the shuttle boxes but operable when the boxes are unable to shift to cause energization of said solenoid and bring about the shifting movement of said bar accompanied by actuation of the knock-off mechanism resulting in loom stoppage.

2,717,009
APPARATUS FOR KNOTTING AND LOOPING CABLES FOR EARTH ANCHORS
Claude E. Grimes, Vicksburg, Miss.
Application December 16, 1954, Serial No. 475,851
19 Claims. (Cl. 140—2)
(Granted under Title 35, U. S. Code (1952), sec. 266)

1. In apparatus for providing substantially flat earth-anchors with flexible lead cables, the anchors having a central aperture for receiving a length of the cable, the

improvements which comprise a foundation structure and spaced means mounted on the foundation structure for providing an earth-anchor with the cable and preparing the cable for service, the said spaced means including a cable-clamping box containing means for frictionally clamping a cable passing through the cable-clamping box, supporting instrumentalities for receiving and holding an earth-anchor plate provided with a cable-receiving central aperture therein in alignment with the frictional clamping means for enabling the lead end of the cable to be passed through the said central aperture, movable carriage instrumentalities for stressing the cable a predetermined amount while the cable is held in the frictional

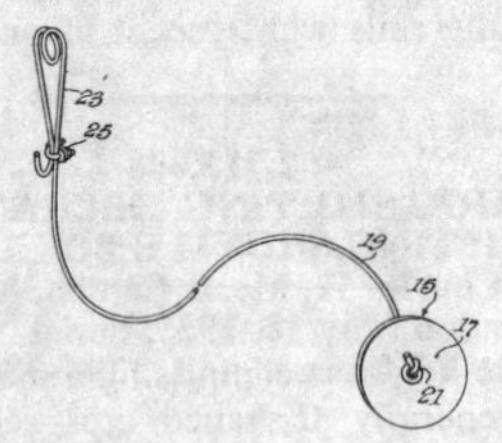

clamping means, turntable cable-looping mechanism on the movable carriage instrumentalities for receiving the lead end of the cable, means for rotating the turntable mechanism while holding the said end of the cable, the said cable being looped around the turntable mechanism during rotation thereof for looping the cable adjacent to the said end, means for releasing the resulting looped cable from the turntable mechanism, and means for actuating the movable carriage instrumentalities in a direction away from the frictional securing means and anchor-supporting instrumentalities for imparting a predetermined tension to the cable while it is held in the frictional securing means.

2,717,010
TRACTOR SUPPORTED POWER DRIVEN SWINGING CIRCULAR SAW
David M. Adams, Mount Crawford, Va.
Application August 14, 1952, Serial No. 304,383
7 Claims. (Cl. 143—43)

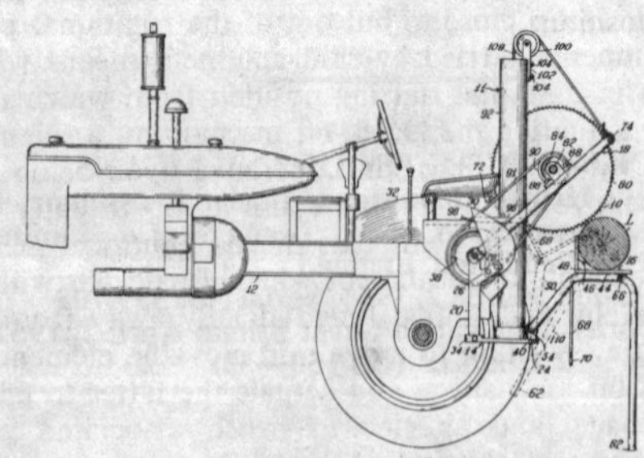

1. A tractor mounted power driven saw comprising draw bar means for rigid attachment to a tractor, a log supporting table pivotally carried by said draw bar means for vertical tilting movement into and out of log supporting position, a disc saw supporting frame pivotally carried by said draw bar means and rotatably carrying a disc saw for movement vertically toward and away from said table, the pivotal axis of the saw supporting frame being common with the axis of rotation of the power take-off pulley of the tractor, means for interconnecting said disc saw with the power take-off pulley of the tractor for driving the same, a tube mounted vertically upon said draw bar means and open at its top, a sheave rotatably mounted at the top of said tube, a counterweight slidable vertically in said tube, and a cable trained about said sheave and fastened at one end to said counterweight and at its other end to the disc saw supporting frame.

2,717,011
MACHINE FOR DRILLING WINDOW SHADE ROLLERS
George A. Phinn, Pasadena, Calif.
Application July 6, 1953, Serial No. 366,135
8 Claims. (Cl. 144—93)

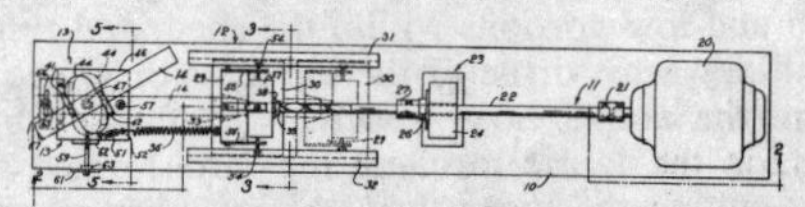

1. A machine for use in extending shade rollers, which comprises a base, a drill connected at its shank end to a motor, slide means mounted on said base, a cylinder-supporting member provided on said slide means for movement longitudinally of said drill, said cylinder-supporting member having a conical opening formed therethrough adjacent the cutting end of said drill, said conical opening being in axial alignment with said drill and having its relatively large diameter base portion remote from said motor, a pivot member spaced from said cylinder-supporting member on the opposite side thereof from said motor, said pivot member being pivotable about an axis normal to and intersecting the axis of said drill and said conical opening, and a pair of V-jaws mounted on said pivot member and on opposite sides of said pivot member axis for engagement at axially spaced locations with opposite sides of a cylinder to be drilled, said V-jaws being shaped to cooperate with said cylinder-supporting member in maintaining said cylinder in axial alignment with said drill during movement of the cylinder toward said drill to form an axial end bore therein.

2,717,012
WOOD SLICING MACHINE
Gilbert D. Schneider, Lexington, Ohio, assignor to Schneider Machine Co., Lexington, Ohio, a corporation of Ohio
Application February 2, 1953, Serial No. 334,607
9 Claims. (Cl. 144—162)

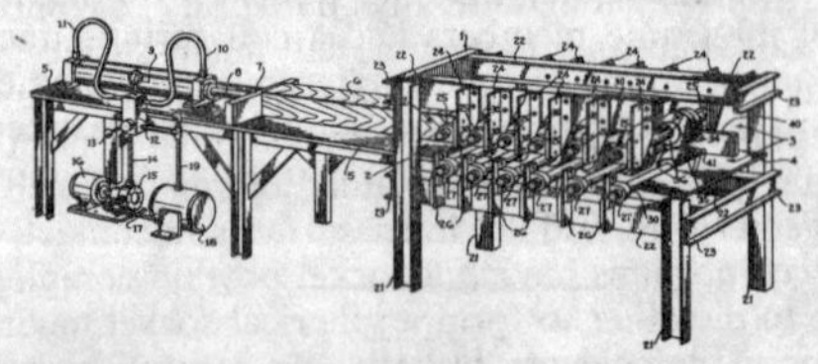

1. Apparatus for slicing lumber comprising a framework having axially aligned feed and discharge zones therein for the lumber being sliced; a first stepped series of knife-edged elements of progressively increasing size disposed in the stretch between the feed and discharge zones of the framework in series arrangement with each other in a single vertical plane paralleling the longitudinal axis of the framework each of which elements in turn cuts its way into the lumber as the lumber traverses said stretch; a second stepped series of knife-edged elements of progressively increasing size disposed beneath the first series, in the above-mentioned single vertical plane, the cutting edges of the knife-edged elements of said second series approaching by increments the cutting edges of the knife-edged elements of said first series; and power means for positively driving the knife-edged elements of both the upper series and the lower series.

2,717,013
SPIRAL GUIDE FOR A MACHINE TOOL
Benjamin R. Van Zwalenburg, Grand Rapids, Mich.
Application November 28, 1951, Serial No. 258,651
4 Claims. (Cl. 144—253)

1. In combination with a machine having a frame supporting a tool holder having an axis about which said

holder and a tool are adapted to rotate, and a base defining a supporting surface disposed axially forwardly of said tool holder and transversely of and about said axis;

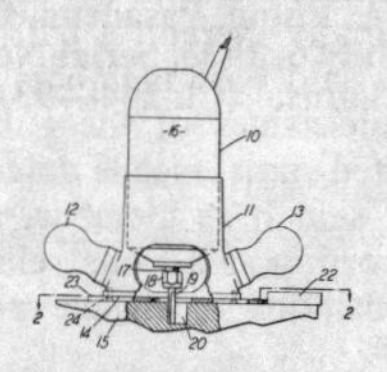

a guide member mounted on said base, said guide member having a guiding periphery extending radially beyond said base and including a plurality of points disposed at unequal radial distances from said axis.

2,717,014
POCKET ASH RECEIVER
John A. Corn, Indianapolis, Ind.
Application May 24, 1954, Serial No. 431,655
2 Claims. (Cl. 150—4)

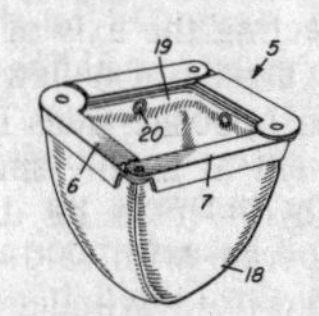

1. A pouch comprising a plurality of rigid frame members pivoted at their end portions in overlapping relation to each other, the frame members being swingable from an open position with the frame members disposed substantially at right angles to each other to a closed position with the frame members parallel to each other, said frame members being of inverted channel shape in cross section to provide an inner downwardly extending flange and an outer downwardly extending flange, a flexible pouch member attached at its mouth to the outer flanges of the frame members, a lining attached to the inner flanges of the frame members and a coil spring attached to a pair of the frame members adjacent the pivoted ends thereof and movable inwardly and outwardly past center of the pivot to tensionally oppose opening and closing movement of the frame.

2,717,015
BAG
Morris J. Berry, Queens County, N. Y.
Application April 10, 1952, Serial No. 281,527
2 Claims. (Cl. 150—34)

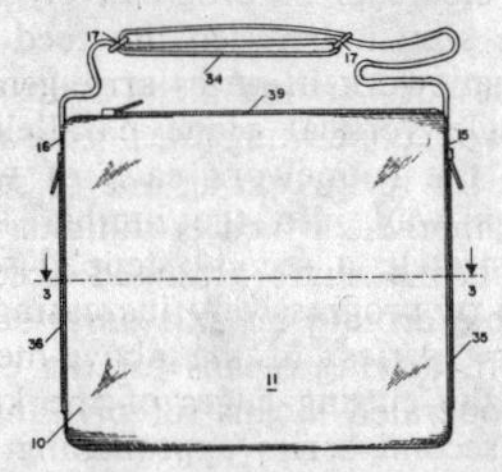

1. A formula bag comprising a substantially rectangular container having a top, bottom, side and end walls, an opening formed in each of the end walls of said bag, a thermally insulated chamber defined by an inner and outer wall of flexible sheet material and a layer of heat insulating material interposed therebetween disposed within said container adjacent each end wall thereof, a longitudinally extending opening provided in each of said chambers, the edges of the material defining each of said chambers, along the openings formed therein, being secured to the end walls of the container along each side of each of the openings in said end walls, a lining member disposed within said bag and substantially co-extensive with the interior thereof, said lining member defining a compartment in the interior of said bag, an opening formed in the top wall of said container providing communication to the interior of said lining compartment.

2,717,016
KEY HOLDER
John Chester Molyneux, Jamestown, N. Y.
Application September 16, 1954, Serial No. 456,472
3 Claims. (Cl. 150—40)

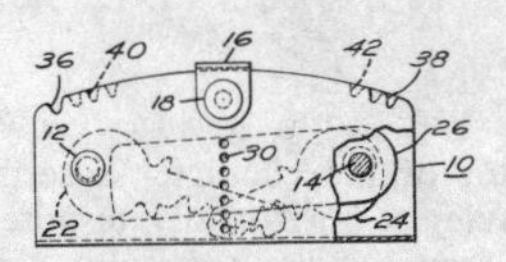

1. A flexible key case comprising a main body portion folded into generally U-shaped cross-sectional form, a plurality of post members fitted in the opposite end portions of said main body portion, a plurality of keys pivotally connected to said post members inwardly of the main body portion for swinging movements outwardly and inwardly of said key case, the post members being spaced so that the ends of the keys mounted thereon will overlap when in stored position inwardly of the case, snap fastener means detachably connected to the opposed open ends of said key case main body portion, means providing a predetermined number of spaced shoulders in said main body portion in fixed positional orientation adjacent each of said keys, said main body portion having its side wall portions weakened or scored substantially medially and crosswise thereof whereby the ends of the main body portion may be flexed relative to each other to form a substantially medial bottom ridge portion in raised position upwardly between said side wall portions incidental to such flexing to eject the ends of said key members outwardly of said main body portion for easy identification thereof by sight or touch of said shoulders when under adverse lighting conditions and to permit selective removal into operative position.

2,717,017
PACKAGING METHODS AND MEANS
John Feasey, Hemel Hempstead, England, assignor, by mesne assignments, to R. A. Brand & Co. Limited, Bridge Mills, Pendleton, England
Application March 2, 1953, Serial No. 339,708
Claims priority, application Great Britain April 2, 1952
2 Claims. (Cl. 150—52)

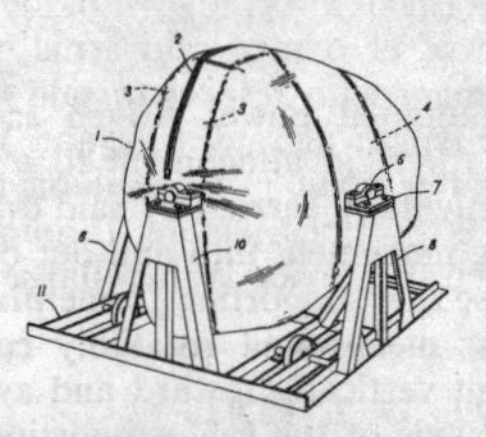

1. A sealable protective plastic cover for packaging an article having trunnions by which the article is to be supported, said cover conforming to the general shape of the article and having at least one sealable opening for receiving the article and being provided, for each supporting trunnion of the article, with a reinforcing element for transmitting support for the weight of the article to said trunnion, each reinforcing element being at least in part included within the plastic skin of the cover.

2,717,018
PLUMBING FIXTURE PROTECTOR
Donald D. Wagner, El Monte, Calif.
Application October 9, 1953, Serial No. 385,147
5 Claims. (Cl. 150—52)

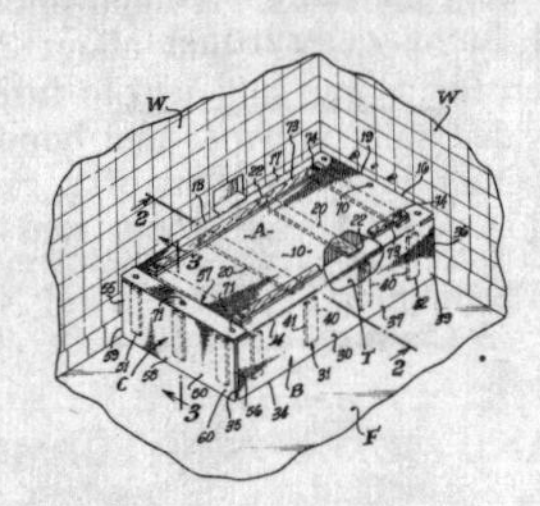

5. A protector of the character described, applicable to a fixture having a horizontally disposed top, a vertically disposed front, and a vertically disposed end, including, a top panel coextensive with and covering the top, battens extending across and reinforcing the top, flaps along the margins of the top, a front panel depending from the top coextensive with and covering the front, battens extending across and reinforcing the front, flaps along the margins of the front, an end panel depending from the top coextensive with and covering the end, battens extending across and reinforcing the end, flaps along the margins of the end, and projections depending from the top, and engageable with the fixture to retain the protector on the fixture.

2,717,019
TRACTION DEVICE FOR VEHICLE WHEELS
Hornsby Sewell Baldwin, Perth Amboy, N. J.
Application November 13, 1953, Serial No. 391,896
3 Claims. (Cl. 152—226)

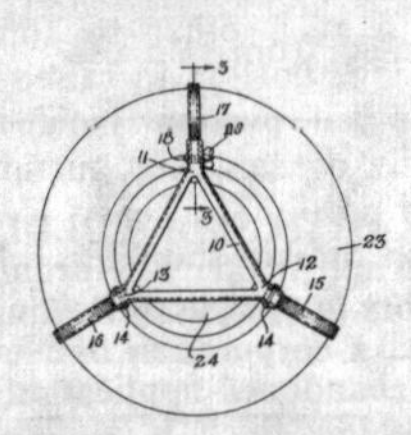

1. A collapsible traction device for attachment to vehicle wheels comprising a rigid frame of isoceles triangular shape, the frame encompassing an opening sufficiently large for the purpose hereafter described, an arm pivotally mounted at each apex portion of the frame for movement transverse to the plane of the frame, a tire-engaging hook formed integrally with the outer end of each arm, the pivot of one of said arms being mounted for lateral movement with respect to said frame to regulate the effective length of said arm, and means for fastening the said latter pivot in place upon the frame, the tire-engaging hook portions passing through the opening of the frame when the device is in collapsed condition.

2,717,020
VEHICLE BODY AND FRAME STRAIGHTENING APPARATUS
George L. Dobias, Mount Pleasant, Mich.
Application June 1, 1954, Serial No. 433,389
3 Claims. (Cl. 153—32)

1. A straightening apparatus of the character described comprising a pair of elongated runways, cross braces disposed beneath and secured to the ends of said runways and supporting the runways in elevated substantially parallel positions, a base member extending transversely beneath said runways, means slidably connecting said base member to bottom portions of the runways for supporting the base member beneath the runways for sliding movement longitudinally thereof, a pair of uprights fixed to and raising from said base member, said uprights being disposed on outer sides of the runways, a pair of fluid pressure actuated rams, means detachably connected to and rising from said base member, said uprights befor adjustably supporting said rams relatively to the uprights in desired positions to engage parts of a vehicle

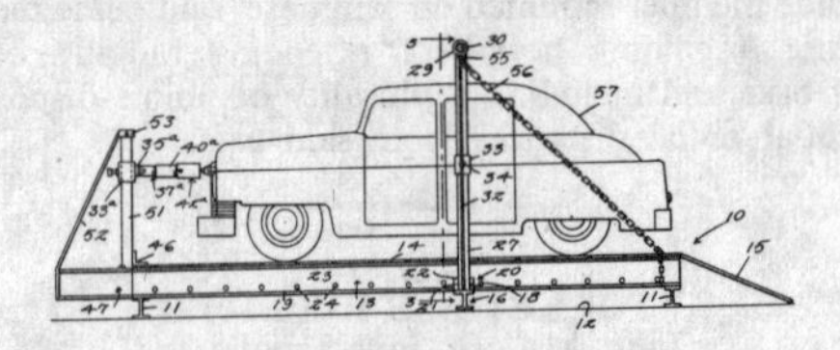

mounted on the runways between the uprights, a second base member disposed between corresponding ends of said runways and slidably supported on the cross brace supporting said runway ends for sliding movement of said second base member transversely of the runways, means retaining said second base member against swinging movement relatively to the runways, an upright fixed to and rising from said second base member and disposed beyond said runway ends, a third fluid pressure actuated ram, and means detachably and adjustably supporting said third ram on said last mentioned upright.

2,717,021
MACHINE FOR BENDING METAL STRIPS
Uno Bernhard Rylander, Arsta, Stockholm, Sweden, assignor to Svenska Skofabrikantforeningen, Stockholm, Sweden, a corporation of Sweden
Application December 16, 1952, Serial No. 326,201
3 Claims. (Cl. 153—48)

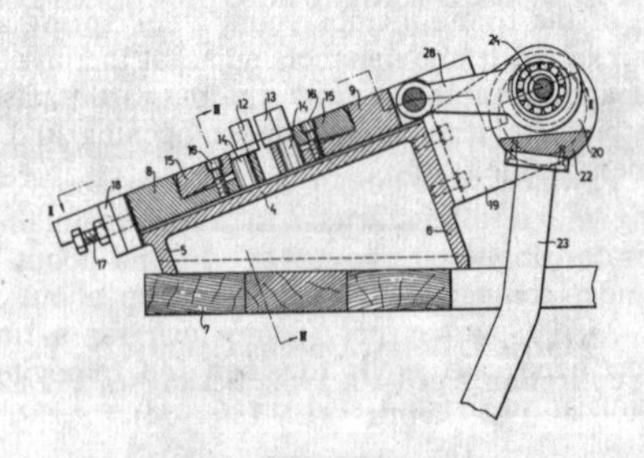

1. A machine for bending metal strips, comprising in combination two cooperating dies, eccentric means including a shaft for reciprocating one of said dies toward and from the other die which is stationary, bearings for the eccentric means shaft arranged eccentrically in a support pivotable around an axis parallel to said eccentric means shaft, driving means for the eccentric means and manually operated means for pivoting said support.

2,717,022
MACHINE FOR APPLYING TREAD MATERIAL TO TIRE CASINGS
Arnold Duerksen, Lodi, Calif., assignor to Super Mold Corporation of California, Lodi, Calif., a corporation of California
Application December 28, 1953, Serial No. 400,651
16 Claims. (Cl. 154—9)

1. In a machine for applying a strip of tread material to a tire casing, a rigid supporting structure, means on the structure to support a tire in an upstanding position and

including a roller unit to engage a tire on the inside at the top directly under the adjacent tread bearing portion of the

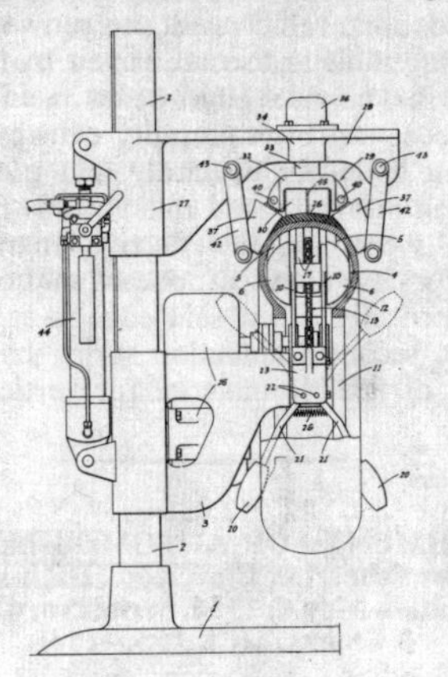

casing in supporting relation, and means to drive the roller unit whereby to rotate the tire.

2,717,023
METHOD FOR FORMING AN O-RING AND O-RING FORMED BY SAID METHOD
Alexander C. Hetherington, Mountainside, N. J., assignor to The M. W. Kellogg Company, Jersey City, N. J., a corporation of Delaware
Application June 29, 1951, Serial No. 234,284
20 Claims. (Cl. 154—33.1)

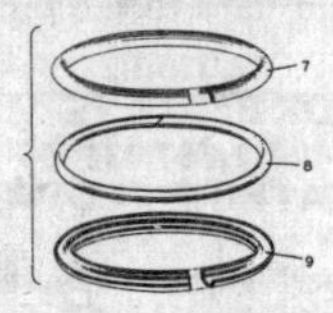

1. A method for forming an O ring which comprises: forming an O ring core comprising a non-metallic resilient solid material; thereafter covering said core with a complementary pair of one-half O ring shells comprising a solid polymer of trifluorochloroethylene having an N. S. T. value from about 220° C. to about 350° C.; heating the resulting article in a die maintained at a temperature between about 415° F. and about 625° F. and under compacting pressure between about 500 and about 25,000 pounds per square inch for a time sufficient to bond said shells together and removing the O ring thus formed from said die.

2,717,024
GASKET AND METHOD OF FORMING SAME
Ulrich Jelinek, Springfield, N. J., assignor to The M. W. Kellogg Company, Jersey City, N. J., a corporation of Delaware
Application September 17, 1952, Serial No. 310,051
16 Claims. (Cl. 154—33.1)

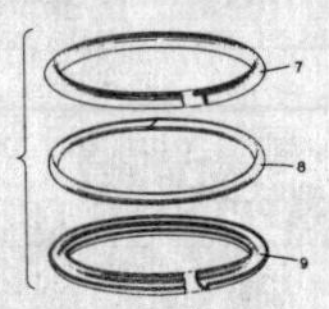

1. A method for forming a gasket comprising: forming a relatively large core comprising a non-metallic resilient solid material; compressing a mass comprising a finely divided solid polymer of trifluorochloroethylene having a no strength temperature value between about 220° C. and about 350° C. at a temperature substantially below 415° F. in the form of toroidal half-shells comprising an adherent unfused mass of said finely divided solid polymer; cover said core with a complementary pair of said toroidal half-shells; heating the resulting article in a die maintained at a temperature between about 415° F. and about 625° F. and under compacting pressure between about 500 and about 25,000 pounds per square inch for a time sufficient to fuse said mass of said finely divided solid polymer and bond said toroidal half-shells together into the form of a gasket having a seamless homogeneous sheath about said core; and removing the gasket thus formed from said die.

2,717,025
GASKET AND METHOD FOR FORMING SAME
Ulrich Jelinek, Springfield, N. J., assignor to The M. W. Kellogg Company, Jersey City, N. J., a corporation of Delaware
Application December 19, 1952, Serial No. 326,911
19 Claims. (Cl. 154—33.1)

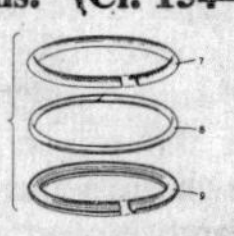

1. A method for forming a gasket comprising: forming a core comprising a non-metallic resilient solid material; compressing a mass comprising a finely divided solid polymer of tetrafluoroethylene at a temperature below about 700° F. in the form of toroidal half-shells comprising an adherent unfused mass of said finely divided solid polymer; covering said core with a complementary pair of said toroidal half-shells; sintering the resulting article in a die at a temperature between about 700° F. and about 800° F. to fuse said mass of said finely divided solid polymer and bond said toroidal half-shells together into the form of a gasket having a seamless homogeneous sheath about said core; and removing the gasket thus formed from said die.

2,717,026
CHAIR WITH SELF-FOLDING SEAT
Alfred C. Hoven and Walter E. Nordmark, Grand Rapids, Mich., assignors to American Seating Company, Grand Rapids, Mich., a corporation of New Jersey
Substituted for abandoned application Serial No. 46,742, August 30, 1948, which is a division of application Serial No. 267,776, January 23, 1952, now Patent No. 2,705,526, dated April 5, 1955. This application August 13, 1954, Serial No. 449,814
1 Claim. (Cl. 155—85)

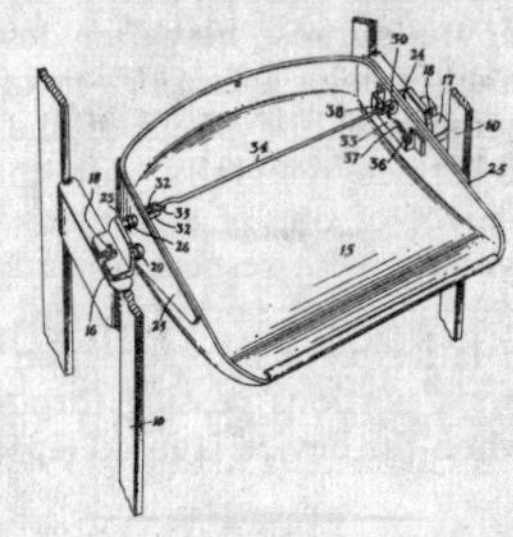

A chair structure comprising, in combination: spaced upright chair supporting standards; mounting members on said standards respectively and having mutually inwardly extending trunnions; a chair seat comprising an upholstered element mounted upon an enclosed lower pan element journalled at its opposite sides on said trunnions respectively for swinging movement to a lowered, substantially horizontal position for seat occupancy and to an upwardly tilted position between the standards; and a torque spring comprising a substantially straight wire disposed within the interior of said pan element and having one end thereof non-turnably connected to one of said trunnions and the other end thereof connected to the pan member turnably therewith adjacent the other of said trunnions, whereby the seat is normally urged to an upwardly tilted position.

2,717,027
ADJUSTABLE HIGHCHAIR
Ralph H. Thatcher, Claremont, Calif.
Application July 19, 1954, Serial No. 444,109
14 Claims. (Cl. 155—115)

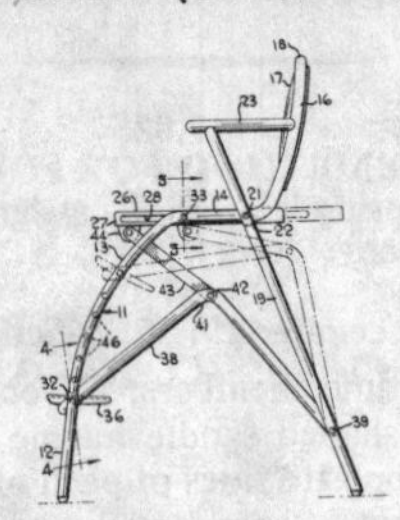

1. In an adjustable chair, a frame, a backrest mounted on the top of said frame, a seat, means connecting said seat and said frame for movement of said seat relative to said frame, a footrest moveable relative to said frame from a lowered to a raised position, and means connecting said footrest and said seat to move said seat rearwardly as said footrest is raised.

2,717,028
FOLDING PICNIC TABLE
Joseph Villemure, Newberry, Mich.
Application October 12, 1954, Serial No. 461,744
5 Claims. (Cl. 155—124)

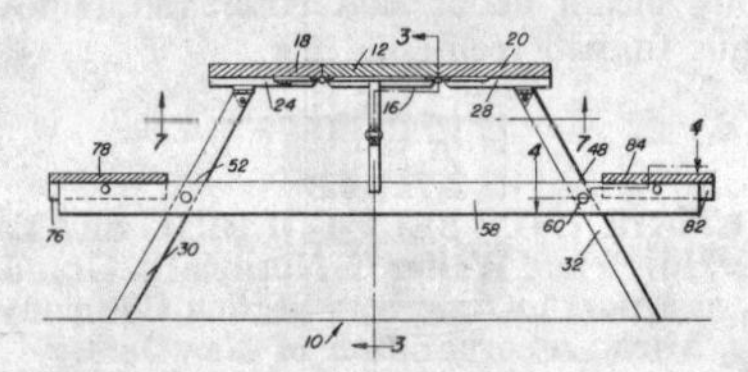

1. A folding picnic table comprising a central table top member, side table top members hingedly secured to said central table top member, pairs of legs hingedly secured to said side table top members, said legs being grooved, side support members received in said grooves and detachably secured to said legs to hold said legs in an extended position, and seats carried by said side support members spaced downwardly and outwardly away from said side table top members.

2,717,029
SPRING CONSTRUCTION FOR FURNITURE
Samuel S. Jonas, Los Angeles, Calif.
Application June 20, 1952, Serial No. 294,581
3 Claims. (Cl. 155—179)

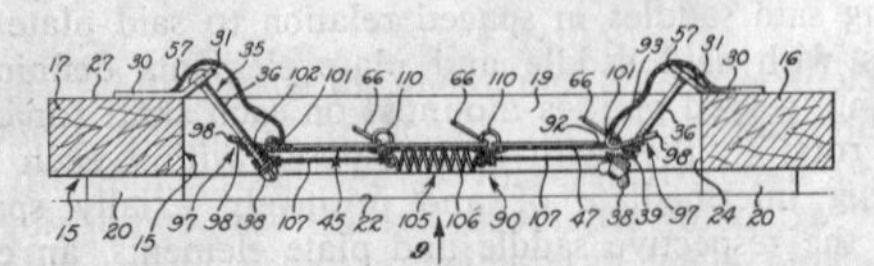

1. In a spring construction for furniture, the combination of: a frame including front and back rails operatively connected to each other to define an opening therebetween; a first foundation spring bank mounted on said rails to define the edges of said construction; a bodily movable sheet of flexible material disposed in said opening with opposed edges thereof being spaced inwardly from said front and back rails, respectively; rigid reinforcing means extending along and secured to each of said opposed edges, each of said opposed edges being movably supported from its adjacent rail by a plurality of supporting members extending obliquely upwardly and outwardly from said reinforcing means to said rails; means for selectively and individually changing the effective length of said supporting members along at least one of said opposed edges whereby to vertically adjust said edge in said oblique direction; and a second foundation spring bank supported on said sheet of flexible material for vertical movement therewith.

2,717,030
ATTACHMENT FOR AUTOMOBILE SEATS
Aloysius Schuszler, Cleveland Heights, Ohio
Application May 4, 1954, Serial No. 427,612
3 Claims. (Cl. 155—188)

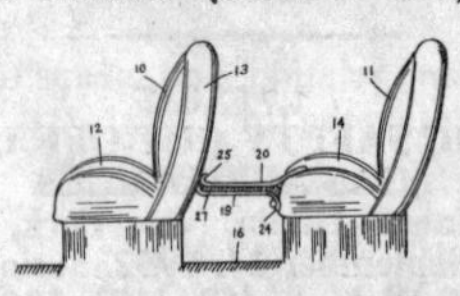

1. A device of the character described comprising, in combination with a vehicle having front and rear seats, a pair of dependent flaps affixed to the forward portion of the rear seat, a pair of dependent flaps affixed to the rear of the front seat, and means for connecting the free edges of each of said flaps to the other of the respective seats.

2,717,031
INJECTOR OPERATED WELDING AND CUTTING BURNER
Felix Heinrich Hubert Damm, Dusseldorf, Germany, assignor of forty per cent to G. Seemann, Coffeyville, Kans.
Application October 17, 1950, Serial No. 190,597
6 Claims. (Cl. 158—27.4)

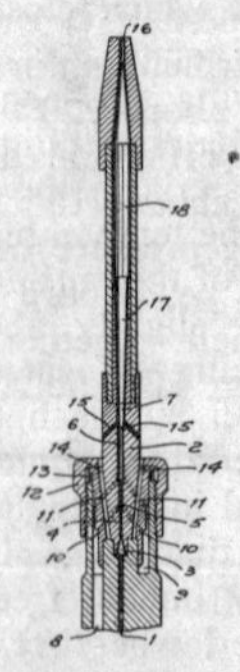

2. In a mixer, a tubular member forming a first chamber, an injector having one tubular end extending into one end of the said tubular member, said tubular end having diminishing inner diameter in rearward direction, a plurality of aligned first borings provided in the said injector in order to set a plurality of stages in the said injector, the said succeeding first borings having smaller diameters in rearward direction, the rearmost of the said first borings terminating into a second conically shaped boring expanding in rearward direction and being aligned with the said first borings, the rear end of the said injector forming an annular recess, the said injector forming an annular shoulder facing forwardly, a plurality of third borings inclined in forward direction and communicating each of the said first borings with the periphery of the said injector, a handle, one portion of the said handle be-

ing received by the said annular recess formed at the rear end of the said injector, another portion of the said handle surrounding the said injector at least for a part of its length, sealing means disposed between the said injector and the surrounding portion of the said handle, the outer wall of the said injector being partly spaced apart from the surrounding portion of the handle to form a first chamber, an oxygen feed line disposed in the said handle and communicating with the said second boring of the injector, a fuel gas feed line in the said handle communicating with the said first chamber, at least the most forward disposed third borings of the injector communicating with the atmosphere and the other of the said third borings of the injector communicating with the said first chamber, a plurality of fourth borings in the said injector extending from the said annular recess disposed at the rear end of the injector and communicating with the atmosphere, in order to provide airing means for any leakage of oxygen at the connection between the handle and the injector, and means for securing the said injector to the handle.

2,717,032
AUTOMATIC SAFETY CUT-OFF DEVICE FOR GAS APPLIANCES
Eugene A. Dupin, New York, N. Y.
Application September 24, 1952, Serial No. 311,292
2 Claims. (Cl. 158—132)

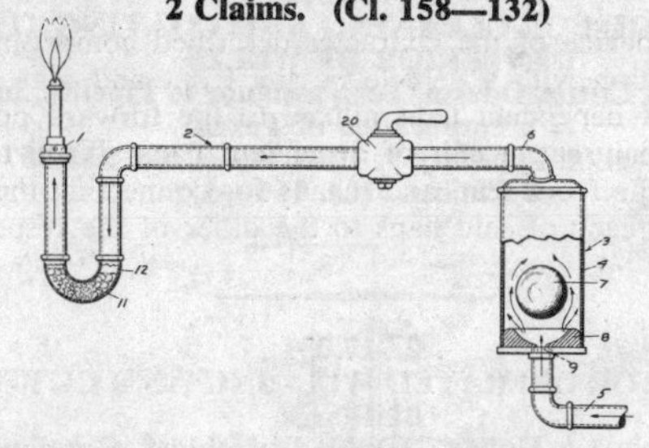

2. An automatic safety cut-off for gas appliances comprising a burner, a supply pipe leading to the burner, an automtically closing valve device in the supply pipe which is held open by the gas therein when it is under normal pressure, but which closes automatically when the gas in the supply pipe has a subnormal pressure, said supply pipe having a U-bend closely adjacent the burner and sealing material filling the bight of the U-bend, said sealing material being in solid form at room temperatures within the temperature range resulting from the usual fluctuations of the outside temperature, but having a melting point slightly above such temperature range, said U-bend being so located relative to the burner that the heat given off by the lighted burner will maintain the sealing material in liquid form, the amount of liquid sealing material in the bight and the specific gravity thereof being such that the gas in the supply pipe under normal pressure may bubble freely therethrough in its passage to the burner, whereby when the gas pressure falls the valve device will automatically close thus extinguishing the burner, and the sealing material will solidify because of the absence of heat from the burner, and thereby effectually seal the U-bend of the pipe.

2,717,033
FOLDING FLEXIBLE PARTITION
Donald M. Breslow, Beverly Hills, and Orland S. Schesvold, Los Angeles, Calif., assignors, by mesne assignments, to Donald M. Breslow, Los Angeles, Calif., doing business as The Curtition Co.
Application February 11, 1950, Serial No. 143,714
5 Claims. (Cl. 160—84)

1. A closure for dividing a room into a plurality of areas, comprising: a horizontal track secured adjacent the ceiling of said room; roller means engaging said track; an expansible-retractable frame supported on said roller means, said frame comprising an upper frame member having a plurality of hinge leaves and a lower frame member having an equal number of hinge leaves, each of said hinge leaves having stop means to limit the opening movement of said hinge leaves to substantially 90°, said hinge leaves when opened comprising a plurality of corrugations; a plurality of stiffeners attached to and extending between the correspondingly positioned hinge

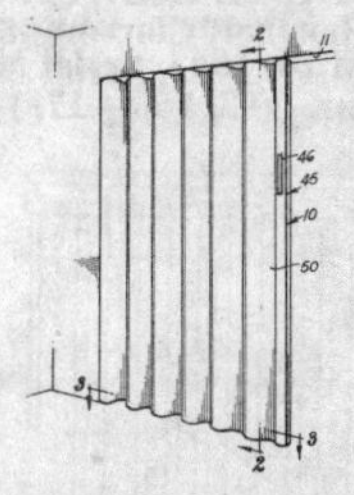

leaves of said frame members; a web-supporting means comprising a U-shaped saddle having a pair of legs extending on the opposite sides of each of said hinge leaves; and a pair of webs attached at the opposite ends thereof to said saddle legs on opposite sides of said frame, said webs being free from attachment intermediate their attachment to said saddle legs; a plurality of said hinge leaves being attached to said roller means for horizontal movement of said closure along said track.

2,717,034
VERTICAL BLIND CONSTRUCTION
William F. Sharpe, Minneapolis, Minn.
Application October 20, 1953, Serial No. 387,138
1 Claim. (Cl. 160—176)

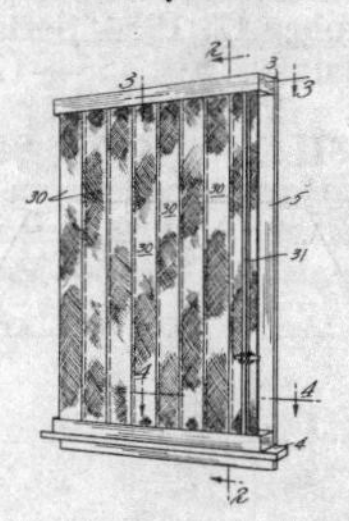

A support and operating structure for a window blind having vertically disposed parallel slats, said support and operating structure comprising an elongated U-shaped channel adapted to be secured horizontally to a window adjacent its upper or lower frame section, the bottom of said channel being provided with aligned longitudinally spaced inturned sleeve openings, vertical shafts projecting inwardly through each of said openings, laterally spaced hook elements on the outwardly projecting ends of said shafts adapted to receive said slats, plate elements having an aperture therein being positioned within said channel with an inturned sleeve being snugly received in each aperture, said plate elements each terminating in a bearing flange, non-rotative saddle elements mounted on said shafts within the channel, spacing means positively retaining said saddles in spaced relation to said plate elements with said saddle and plate elements defining a housing, geared pinions mounted on each shaft for common rotation therewith and disposed within each said housing intermediate of and positively axially spaced from the respective saddle and plate elements, an elongated rack disposed intermediate the geared pinions and said bearing flanges in meshing engagement with the pinions and sliding frictional engagement with the bearing flanges, said axial spacing of said pinions to the saddle elements and plate elements permitting limited floating movement of the pinions when actuated by said rack and resilient means interposed between the inwardly projecting ends of the shafts and the respective saddles biasing said hook elements toward the bottom of the channel.

2,717,035
ATTACHING MEANS FOR VERTICAL VENETIAN BLINDS
Fred A. Groth, Chicago, Ill., assignor to Fred A. Groth and Joseph D. Friedman, Chicago, Ill.
Application August 26, 1953, Serial No. 376,635
4 Claims. (Cl. 160—178)

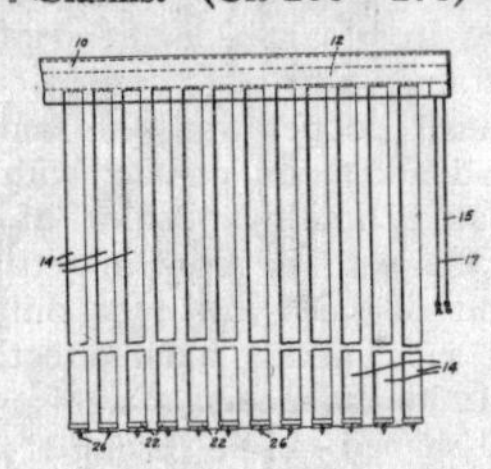

1. The combination with vertical Venetian blinds of the type having a cornice and a plurality of parallel vertical slats hung suspended from said cornice of stabilizer means connecting the lower ends of said suspended slats together, said means including pivotally connected links connected to the lower ends of said slats, each link having a medial opening therethrough and a cap connected to the ends of each of said slats and having a stud extending through an opening in a link, the ends of said studs being in the nature of hollow spherical bulbs and of a larger size than the link opening, said studs being compressible to permit drawing of the bulbous ends of the same through said link openings.

2,717,036
AUTOMOBILE WINDOW SCREEN
Estelle F. Harris, Providence, R. I.
Application November 4, 1952, Serial No. 318,633
2 Claims. (Cl. 160—354)

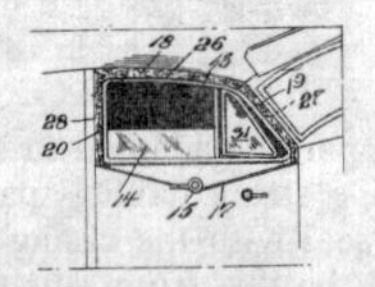

A screen for an automobile door having an opening therein and a projection from its inner surface below said opening comprising a flexible open mesh fabric of a size to extend over the outer face of the door with its top and side edges extending about the upper and side edges and across the entire thickness of the door and over the inner surface of the marginal edge of the door, a pocket formed along said upper and side edges by doubling the edge of the fabric back upon itself, an elastic cord in said pocket of a length to extend below said projection when under tension to draw and hold the said edges taut over said opening, and means attached to the lower edge of the fabric to secure the lower edge to the outer surface of the door below the door opening.

2,717,037
WINDER AND SHEET SEPARATOR
John E. Goodwillie, Beloit, Wis., assignor to Beloit Iron Works, Beloit, Wis., a corporation of Wisconsin
Application May 3, 1950, Serial No. 159,826
5 Claims. (Cl. 164—65)

1. In a slitting, separating, and winding assembly for fibrous web material moving in a given line of travel including superimposed overlapping slitter disks adapted to receive a fibrous sheet therethrough for slitting the sheet into a plurality of relatively narrow strips and a winding mechanism positioned thereafter in said line of web travel having separate coaxial individual cores for each receiving a strip, the improvement of a separator roll positioned in the line of web travel between the winding mechanism and the slitter disks having a helical screw periphery for guiding the strips laterally outward from

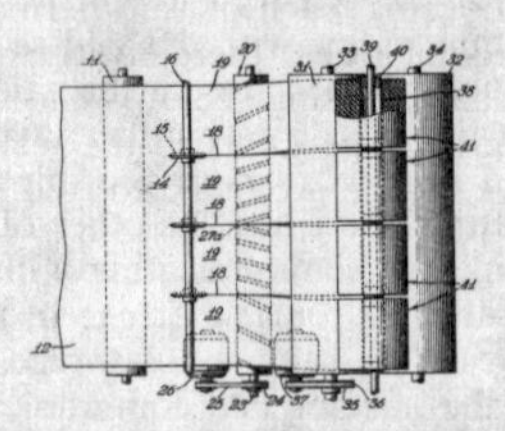

the center of the fibrous web to separate the strips, and means for driving said separator roll independently of said winding and slitting apparatus and at a speed different from the speed of travel of the fibrous web, said separator roll extending laterally substantially the distance between the outer ends of the outermost cores of the winding mechanism.

2,717,038
METHOD FOR CLEANING AND COATING THE INTERIOR OF WELLS
Arvel C. Curtis, Odessa, Tex., assignor to Pipelife, Inc., a corporation of Texas
Application February 9, 1953, Serial No. 335,857
8 Claims. (Cl. 166—1)

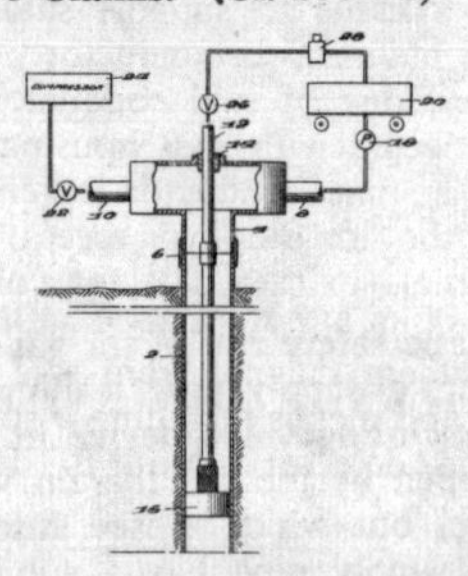

1. Method of coating the inside of a substantially vertical pipe containing another pipe of smaller cross-section therein which comprises blocking off the larger pipe shortly below the lower end of the smaller pipe, filling the interior of both pipes with liquid coating material, and introducing gas into the upper end of the larger pipe under sufficient pressure to force the coating material upward through the smaller pipe until substantially all the coating material is removed from the larger pipe.

2,717,039
DETECTOR DEVICE FOR EXPLORING FERROMAGNETIC STRUCTURE IN WELL BORES
William R. Gieske, Fullerton, Calif., assignor to The Ford Alexander Corporation, Whittier, Calif., a corporation of California
Application September 2, 1952, Serial No. 307,492
2 Claims. (Cl. 166—65)

2. In a device of the character described for lowering into a well through a drilling string that includes fishing tools or the like for the purpose of detecting changes in configuration of ferromagnetic structure along the length of the well bore, the combination of: a cylindrical tubular housing of an outside diameter of the order of magnitude of one and one-half inches; a core member of soft ferromagnetic material positioned in said housing coaxially thereof, said core member having a diameter of the order of magnitude of one-third of the inside diameter of the tubular housing; a detector coil wound

on said core to substantially fill the annular space in the housing around the core; a pair of permanent magnets in said housing positioned in end-to-end contact with the opposite ends of said core, said magnets being of substantially the same diameter as said coil whereby said core, coil and two permanent magnets form an assembly substantially completely filling a longitudinal portion of said housing, each of said permanent magnets having a longitudinal recess therein; an explosive charge supported by said housing below said assembly for detonation to loosen pipe joints in the well; a detonator for said explosive charge grounded to said housing below

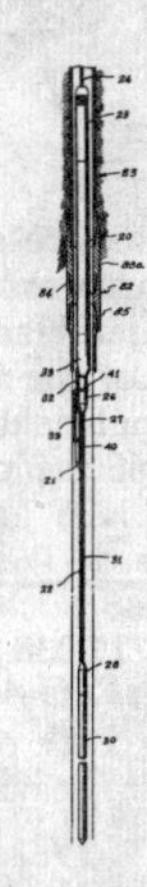

said assembly; a cable to support said housing in the well, said cable having two conductor portions insulated from each other, one of said conductor portions being electrically connected with said housing; a detector circuit including a first conductor extending from said cable through the longitudinal recess of one of said permanent magnets to said detector coil, and a second conductor extending through the longitudinal recess of the other permanent magnet from said detector coil to said detonator; and means including a source of E. M. F. to energize said detector circuit to detonate said explosive charge.

2,717,040
WELLPOINTS
Harold Philip Sidney Paish, Meadowside, Limpsfield, England, assignor to Henry Sykes Limited, London, England, a British company
Application January 8, 1952, Serial No. 265,376
Claims priority, application Great Britain January 9, 1951
2 Claims. (Cl. 166—157)

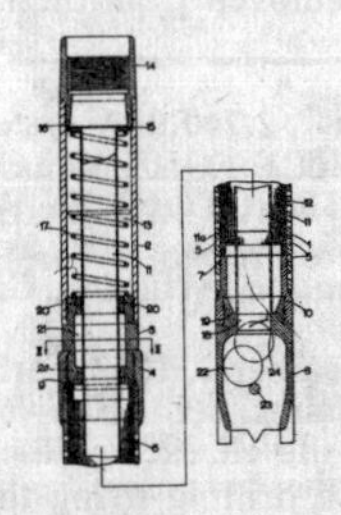

1. A wellpoint comprising an elongated casing having an annular seat adjacent one end thereof, a perforated strainer tube arranged coaxially with respect to said casing, a connector member arranged for interengaging said casing with one end of said strainer tube, a jetting head disposed at the opposite end of said strainer tube and defining a passageway communicating with said strainer tube, a solid-walled flow tube having a first open end and a second open end and arranged for axial displacement between two operative limiting positions with respect to said jetting head and within said casing and said strainer tube, an annular flange connected to said flow tube adjacent said first end and projecting between said flow tube and said casing interiorly of the latter for cooperative abutment against said annular seat in one of said operative limiting positions of said flow tube, resilient means urging said annular flange to abut against said annular seat and to space said second end of said flow tube from said passageway of said jetting head in said one limiting operative position, thereby permitting passage of water from said second end of said flow tube through said strainer tube, said second end of said flow tube forming means cooperating with said passageway of said jetting head and upon contact with same defining the other operative limiting position of said flow tube for restricting the path of flow of jetting water from said second end of said flow tube only through said jetting head in substantially axial direction thereof.

2,717,041
CROSSOVER APPARATUS FOR DUAL PRODUCTION IN OIL WELLS
Cicero C. Brown, Houston, Tex.
Application October 16, 1952, Serial No. 315,161
13 Claims. (Cl. 166—186)

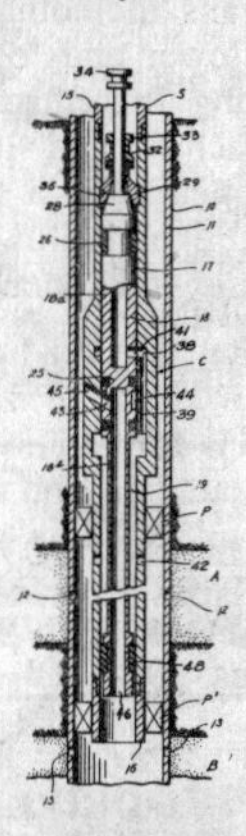

1. A crossover apparatus adapted to be lowered into a well casing on a pipe string having spaced well packers thereon adapted to seal with the casing for dual production in oil wells and the like from a first production zone between said packers and a second production zone below said packers, comprising a tubular housing connected in said pipe string, an inner conductor in said housing with an annular space formed between said housing and said inner conductor, said conductor having a plug for forming an upper conductor section and a lower coinductor section, a releasable connection between said conductor and said pipe string, said pipe string having means for establishing fluid communication between said first production zone and said annular space, said housing and said conductor having means establishing fluid communication between said annular space and the interior of said upper conductor section, said pipe string having means for establishing fluid communication between said second production zone and the lower conductor section, said housing and said conductor having an additional means establishing fluid communication between the interior of said lower conductor section and the area above said packers and exterior of said housing, said conductor constituting the only obstruction in the bore of said pipe string, and means for releasing said connection between said conductor and said pipe string whereby said conductor may be removed to provide the full open bore of the pipe string for well operations without removing the entire pipe string from the well casing.

2,717,042
IMPACT OPERATED VALVE
Harry C. Grant, Jr., and Floyd Bancroft Parsons, Ridgewood, N. J., assignors to Specialties Development Corporation, Belleville, N. J., a corporation of New Jersey
Application March 31, 1950, Serial No. 153,064
9 Claims. (Cl. 169—11)

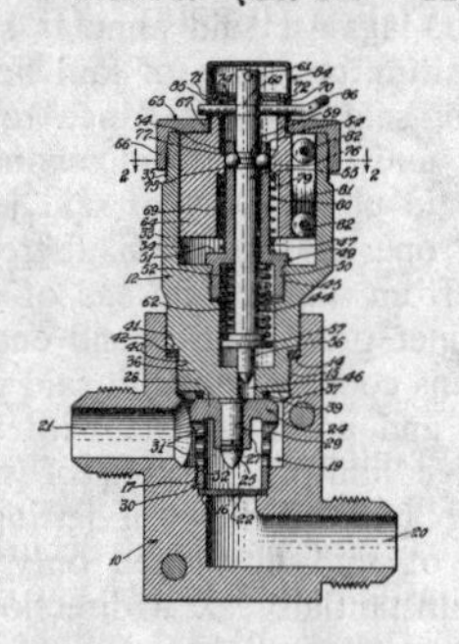

1. An automatically impact actuated valve comprising a housing having an inlet and an outlet, a disc-like closure positioned in said housing for normally preventing the flow of fluid from said inlet to said outlet, means for rupturing said closure including a cartridge chamber adapted for mounting a cartridge therein, said chamber having one end facing one side of said closure and a spring actuated cartridge firing pin facing the other end of said chamber, detent means in said housing for normally restraining said firing pin, and impact responsive means in said housing for rendering said detent means ineffective including a movable mass, means for guiding said mass for movement in a path parallel to the longitudinal axis of said pin and a spring in said housing engaging said mass for opposing movement of said mass, said mass being formed with means for retaining said detent means in position to restrain said pin and with an inclined surface adjacent said detent retaining means for permitting said detent means to move out of pin restraining position upon movement of said mass.

2,717,043
CONTRACTABLE JET-DRIVEN HELICOPTER ROTOR
Vittorio Isacco, Paris, France
Application May 16, 1952, Serial No. 288,136
15 Claims. (Cl. 170—135.4)

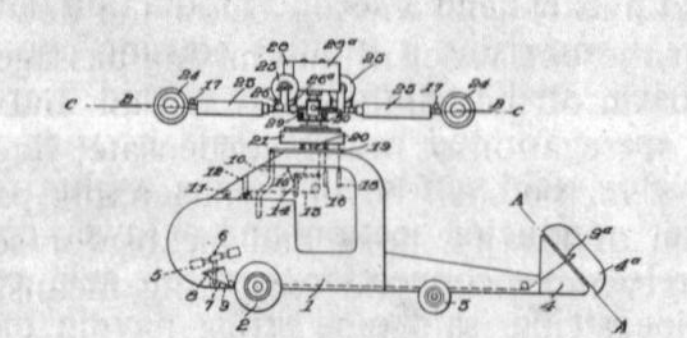

1. A helicopter, comprising a support, an upright shaft on said support, a hub rotatably mounted on said shaft, a plurality of telescopic blades each composed of a plurality of telescoping blade elements adapted to be telescoped to storage position and extended to operative position, means journalling the root elements of said blades on said hub, engines on the tip elements of said blades, fuel feeding means carried by said hub, reel means rotatably mounted at the root ends of said blades, flexible control and fuel lines wound on said reel means and extending longitudinally through said blades to said engines, passage means between the inner ends of said fuel lines and the rotatable mountings of said reel means, passage means between the fuel feeding means and said rotatable mountings of said reel means, and means engaging said control lines for actuating said control lines.

2,717,044
PROPELLER
Hidetsugu Kubota, Niigata-ken, Japan
Application March 13, 1950, Serial No. 149,400
Claims priority, application Japan April 18, 1949
1 Claim. (Cl. 170—165)

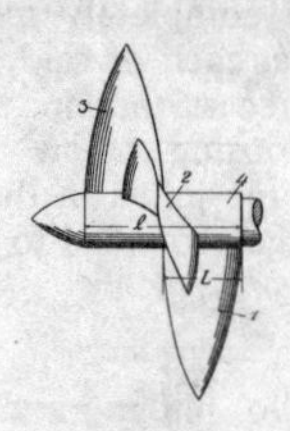

A propeller of the class described, comprising: an elongated forwardly and rearwardly extending boss member; a plurality of propeller blades disposed along said boss member longitudinally thereof at equal distances therealong and at equal angular displacements from each other, the pitch of each blade intermediate the foremost blade and the rearmost blade increasing by a constant increment from the pitch of the adjacent blade immediately forward thereof, the pitch of the rearmost blade being substantially equal to the pitch of the adjacent blade immediately forward thereof, whereby the propelling power produced by each of the blades is substantially constant, said blades having equal root lengths measured longitudinally of the boss member, and the length of the boss member being substantially equal to twice the said root length of each of the blades.

2,717,045
HEATING AND COOLING SYSTEM FOR VEHICLE PASSENGER AND BATTERY COMPARTMENTS
Friedrich K. H. Nallinger, Stuttgart, Germany, assignor to Daimler-Benz Aktiengesellschaft, Stuttgart-Unterturkheim, Germany
Application November 9, 1950, Serial No. 194,848
In Germany September 28, 1949
Public Law 619, August 23, 1954
Patent expires September 28, 1969
3 Claims. (Cl. 180—1)

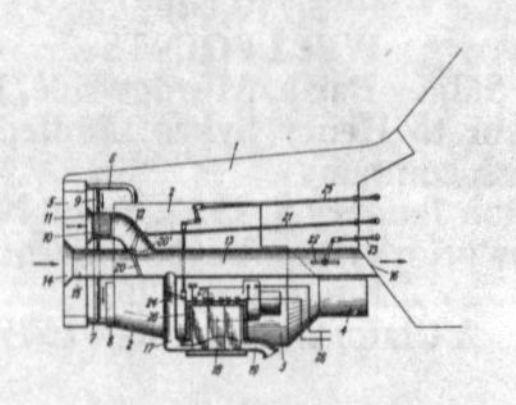

2. In a motor vehicle, a cooling system for the vehicle engine including a radiator and admission and discharge lines communicating with said radiator for the circulation of a coolant therethrough, a battery, a heat insulating housing surrounding said battery and provided with an air discharge port, a main air line leading to the passenger compartment of the vehicle and having a fresh air intake mouth, a branch line connecting said main air line to said housing for the purpose of air-conditioning the latter, means including a heat exchanger operatively connected with said cooling system for heating air flowing through said main air line, means in said main air line for selectively connecting and disconnecting said last-mentioned means for heating air with said main air line, means in said branch line for optionally closing the latter, and means in said main air line for optionally interrupting the flow of air from said main air line to said passenger compartment.

2,717,046
ACOUSTIC CABINET
Wilhelmus Adrianus Jacobus Liebert, Hilversum, Netherlands
Application August 2, 1952, Serial No. 302,308
Claims priority, application Netherlands August 7, 1951
1 Claim. (Cl. 181—31)

A cabinet for a radio, phonograph, television receiver and like instrument having a loudspeaker, comprising, in combination, a cabinet body, means defining a separate box-shaped chamber in the rear of said cabinet provided with a lower opening and an upper opening, a horizontal speaker supporting soundboard positioned in said chamber, a speaker mounted in said soundboard and positioned to emit sound waves upwardly and downwardly from both faces of said plate through said lower opening and said upper opening, a cover hinged to said cabinet positioned to receive the upwardly directed sound waves from said speaker and to reflect them forwardly from said cabinet in a horizontal direction, spaced apart columns supporting said cabinet, a reflecting screen mounted between said columns in a slanting position to receive the downwardly directed sound waves from said speaker and to reflect them forwardly in a horizontal direction, said cover and said screen being substantially equally spaced from said supporting soundboard, whereby said soundboard bisects the angle defined between said cover and said soundboard to provide an acoustic dipole.

2,717,047
WIDE-BAND LOUDSPEAKER
Gerhard Buchmann, Eutingen, Baden, Germany, assignor to International Standard Electric Corporation, New York, N. Y., a corporation of Delaware
Application December 3, 1952, Serial No. 323,774
Claims priority, application Germany December 7, 1951
5 Claims. (Cl. 181—32)

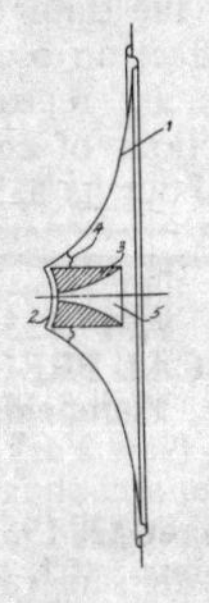

1. A loudspeaker comprising a vibratory diaphragm having a reduced neck portion which is closed by an insert, said diaphragm and insert providing a frequency response which substantially drops off above a predetermined frequency range, a displacer member, means resiliently attaching said displacer member at its periphery to the said vibratory diaphragm adjacent to said neck portion but held in spaced relation to said insert, the mass of said displacer and the resiliency of said attaching means being proportioned so that below said frequency range said displacer and diaphragm vibrate effectively as a unit, while above said frequency range the displacer remains relatively quiet, and a single moving coil attached to said neck portion to operate said diaphragm and displacer.

2,717,048
MUFFLER WITH VIBRATION DAMPING SHELL
Floyd E. Deremer, Detroit, Mich., assignor to Oldberg Manufacturing Company, Grand Haven, Mich., a corporation of Michigan
Application August 28, 1951, Serial No. 243,969
7 Claims. (Cl. 181—61)

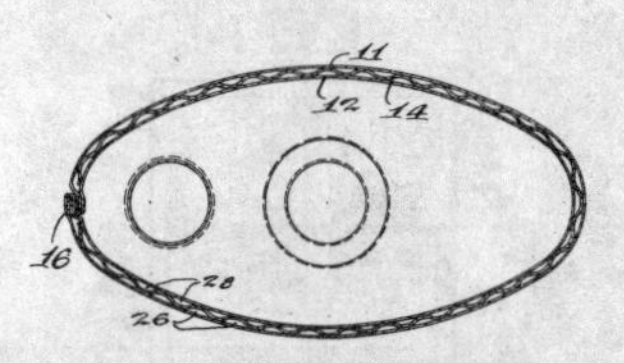

5. Apparatus for attenuating sound in a moving gas stream including, in combination, a tubular shell having a multi-ply wall structure; end heads secured to the shell to form therewith a tubular chamber; gas inlet and outlet ducts formed respectively in said end heads; sound attenuating means formed in said chamber; an intermediate ply of said wall structure being of undulated shape to dispose adjacent plies in spaced relation, the space formed thereby being isolated from the sound attenuating means formed within the shell to provide an air cushion to reduce shell noise.

2,717,049
DEVAPORIZING MUFFLER
John A. Langford, Anaheim, Calif., assignor to The Fluor Corporation, Ltd., Los Angeles, Calif., a corporation of California
Application May 5, 1952, Serial No. 286,134
11 Claims. (Cl. 183—2)

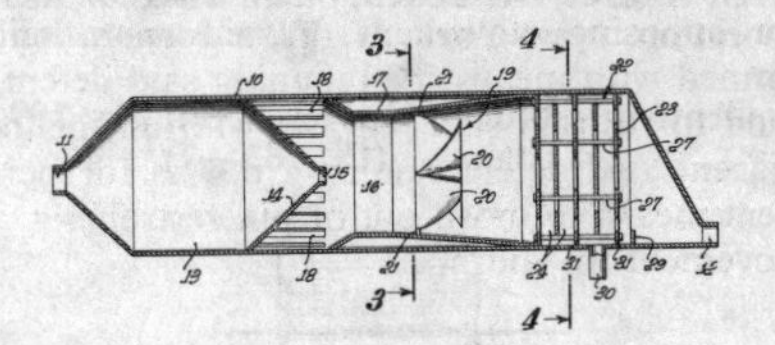

1. An engine muffler comprising an elongated shell having at its opposite ends a gas inlet and an outlet, said shell containing a mixing passage defining a vapor condensing means and the wall of the shell containing air inlets through which inflow of atmospheric air is induced by virtue of the gas flow within the shell for admixture with the gas and to cool the gas below the dew point of water vapor therein, vanes in said mixing passage causing swirling motion of the admixed gas and air and outward centrifugal segregation of water condensate, liquid separating means in the shell beyond said mixing passage in the direction of the gas flow, and means for separately withdrawing water from said separating means.

2,717,050
ROOM DEHUMIDIFIER
James G. Ames, Aurora, Ill.
Application November 28, 1951, Serial No. 258,647
21 Claims. (Cl. 183—4.1)

19. In a room dehumidifier having a casing which contains a pervious bed of solid hygroscopic material in combination with means to circulate a stream of air from a room through said bed and heater means for heating said stream of air before it reaches said bed; automatic means for causing said dehumidifier to pass through a continuous series of dehumidifying and regenerating cycles, comprising: a compartment into which said stream of air passes after passing through said bed; a return valve in said compartment to control the flow of the stream of air into a room; an exhaust valve in the compartment to control the flow of the stream of

air to a stack; means for opening said valves alternately; temperature responsive means in said compartment causing said exhaust valve to open at a predetermined low temperature; means to start said heater simultaneously with the opening of said exhaust valve to start a regenerating cycle; temperature responsive means in said compartment to stop said heater at a predetermined high temperature in said regenerating cycle; temperature responsive means to close said exhaust valve at a predetermined temperature between said low and said high to start a dehumidfying cycle; and holding means to permit said exhaust valve to close only during a period of descending temperature in said regenerating cycle.

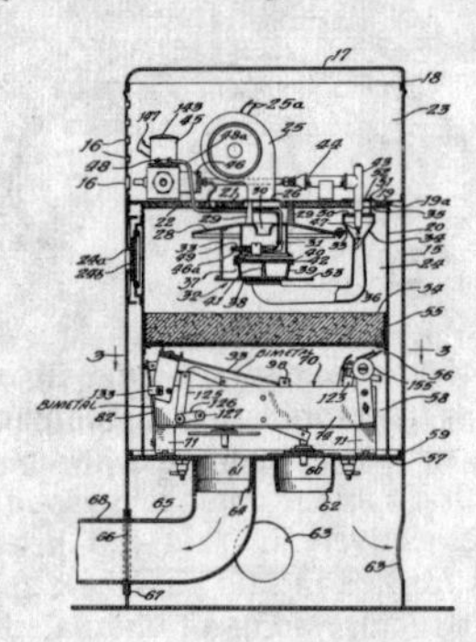

2,717,051
APPARATUS FOR REMOVING SUSPENDED MATERIALS FROM GAS STREAMS

Stanley G. Andres, Metuchen, N. J., assignor to Research Corporation, New York, N. Y., a corporation of New York
Application February 1, 1952, Serial No. 269,524
4 Claims. (Cl. 183—7)

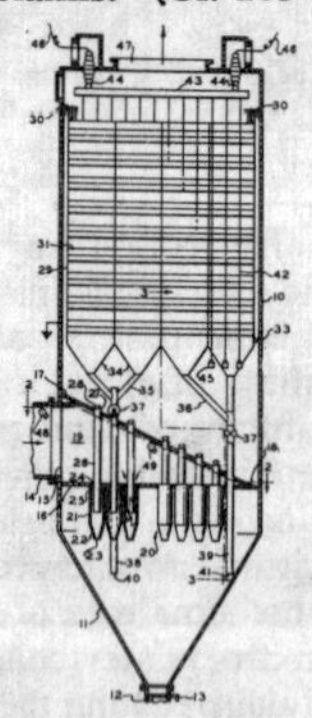

1. An apparatus for separating suspended particulate material from a gas stream comprising, a vertically extending casing, single particulate material receiving hopper at the lower end of said casing, a horizontal header plate extending across the casing above the hopper, an inclined header plate in said casing cooperating with said horizontal header to define a dirty gas inlet into said casing, mechanical gas cleaning means in said casing, said mechanical gas cleaning means having gas inlet means communicating with the dirty gas inlet and gas outlet means discharging into said casing above said inclined header means, particulate material discharge means from said mechanical gas cleaning means into the single particulate material receiving hopper, a complementary discharge and collecting electrode system positioned in said casing vertically above the inclined header plate and the stream of gas discharged from said mechanical gas cleaning means, means including conduits conveying material collected by the collecting electrodes to the particulate material receiving hopper below the mechanical gas cleaning means whereby material collected by both the mechanical gas cleaning means and on the collecting electrodes is discharged to and collected in the same material receiving hopper; clean gas outlet means above the complementary discharge and collecting electrode system, and means in said casing preventing gas flow from said particulate material receiving hopper to the complementary discharge and collecting electrode system whereby the gas stream to be cleaned must pass serially through the mechanical gas cleaning means and between the complementary discharge and collecting electrodes.

2,717,052
RAPPING MECHANISM FOR PRECIPITATOR

Joseph Francis Valvo, Wellsville, N. Y., assignor, by mesne assignments, to Apra Precipitator Corporation, New York, N. Y., a corporation of Delaware
Application November 6, 1953, Serial No. 390,654
3 Claims. (Cl. 183—7)

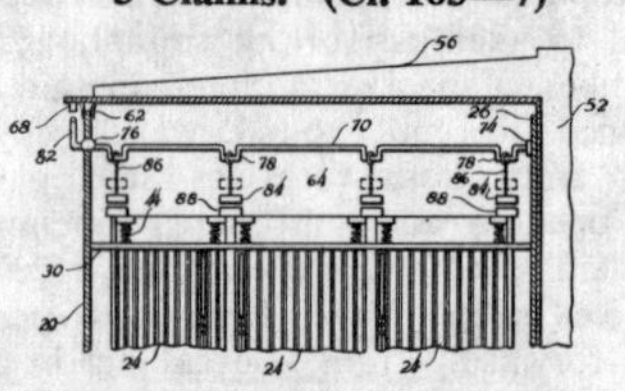

1. An electrostatic precipitator for the removal of dust particles from a gas stream comprising; a pair of concentric cylindrical shells forming an annular space therebetween; radial support beams extending between cylindrical shells dividing the annular space into a plurality of wedge shaped compartments; an assembly of grounded collecting elements suspended from each pair of adjoining support beams; a radially disposed crank arm rotatably supported from the concentric cylindrical shells in each space above the collecting elements; weights depending from each crank arm and arranged to rest upon said collecting element sections when the crank arm is disposed downwardly; a rotor post constructed and arranged to freely rotate within the innermost cylindrical shell; a wedge or sector shaped hood member fixed to said rotor post and extending radially past the peripheral edge of said stator shell; and means fixed to said hood member outwardly of the stator shell to cooperate with the crank arm and effect an oscillatory motion thereof whereby said weights are repeatedly raised from and dropped upon the sections of collecting elements when said hood is rotated about its axis.

2,717,053
ELECTRICAL PRECIPITATOR

Harry A. Wintermute, Plainfield, N. J., assignor to Research Corporation, New York, N. Y., a corporation of New York
Application September 23, 1954, Serial No. 457,957
4 Claims. (Cl. 183—7)

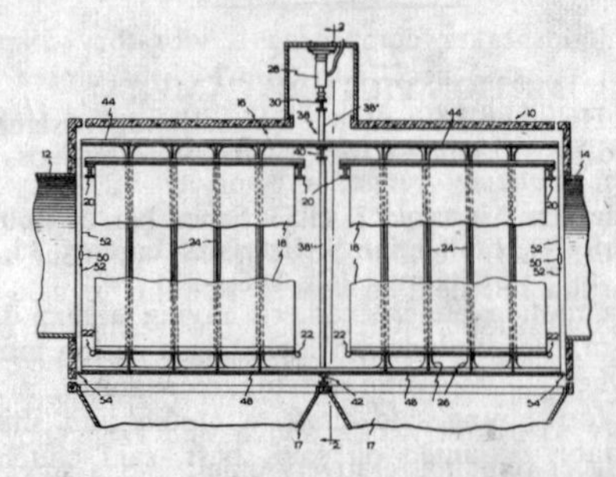

1. A casing having a gas inlet and a gas outlet at opposite ends thereof and a gas treating chamber there-

between, a horizontal beam insulatedly supported centrally of said casing and in the upper portion thereof, a vertical beam secured to the medial point of said horizontal beam and extending downwardly through the gas treating chamber, a pair of vertically spaced support frame members supported at the medial points thereof by said vertical beam in the upper and lower portion of the gas treating chamber, a plurality of flexible high tension discharge electrodes supported between said spaced frame members, and a rapping device positioned to impart vibrations to said vertical beam.

2,717,054
APPARATUS FOR SEPARATING SUSPENDED PARTICLES FROM GASES
Alfred Arnold Petersen, Byram, Conn., assignor to Prat Daniel Corporation, South Norwalk, Conn., a corporation of Connecticut
Application May 19, 1953, Serial No. 356,027
6 Claims. (Cl. 183—81)

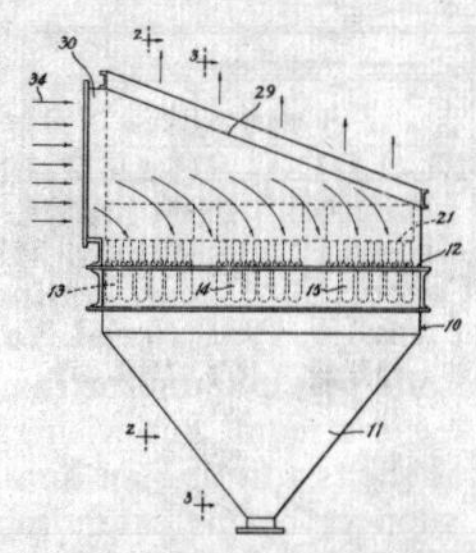

1. Apparatus for separating suspended particles from gases which comprises a dust bin having a top tube sheet, centrifugal separators mounted in said tube sheet in groups, said groups being arranged in spaced rows to form spaces between said rows, the groups being spaced in said rows to form transverse spaces in said rows; plates spaced vertically above said top tube sheet to form plenum spaces, one plate for each group of tubes and having longitudinal and transverse edges spaced to provide longitudinal and transverse passages downwardly to the longitudinal and transverse spaces, respectively, between said groups of tubes, each centrifugal separator having an off-take pipe extending through the plenum space and through the plate of its group and each separator having a rotatory inlet about its off-take pipe from its plenum space into said tube, and a pair of partition walls, one on each side of and joined to, each row of plates and extending upwardly to enclose an exhaust space for said off-take pipes above said plates, parts of said partition walls converging upwardly from the longitudinal edges of said plates to a throat, alternating parts of said partitions converging downwardly from said throat to a union in the transverse passages between said plates and transverse parts extending from the throat to the transverse edges of said plates to complete the enclosure.

2,717,055
MOTION PICTURE CAMERA
Wilfred Heiniger, Yverdon, Switzerland, assignor to Paillard S. A., Sainte-Croix, Switzerland, a Swiss company
Application January 22, 1953, Serial No. 332,600
Claims priority, application Switzerland January 30, 1952
5 Claims. (Cl. 185—39)

1. In a motion picture camera having a film driving mechanism, the combination including a spring motor, a pinion wheel in said film driving mechanism, a shaft, a crank for turning said shaft, a clutch unit shiftably and rotatably mounted on said shaft, said clutch unit including an integral gear wheel engaging said pinion wheel, said gear wheel having a recess, means for winding said motor, said means being mounted on said shaft and including a roller cam angularly secured to said shaft, said roller cam being positioned in said recess, and friction means for driving said gear wheel integral with said

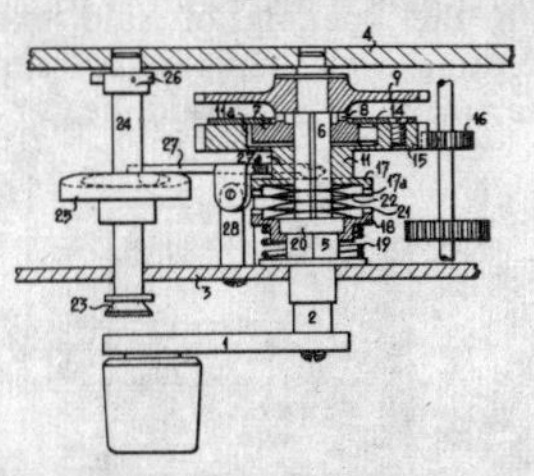

clutch unit, whereby said pinion wheel engaging said gear wheel is driven upon the stopping of the winding of the motor, said friction means being positioned in said recess between and in contact with said roller and said gear wheel.

2,717,056
ELEVATOR DISPATCHING SYSTEMS
Danilo Santini, Tenafly, and John Suozzo, Paramus, N. J., assignors to Westinghouse Electric Corporation, East Pittsburgh, Pa., a corporation of Pennsylvania
Application November 30, 1951, Serial No. 259,066
24 Claims. (Cl. 187—29)

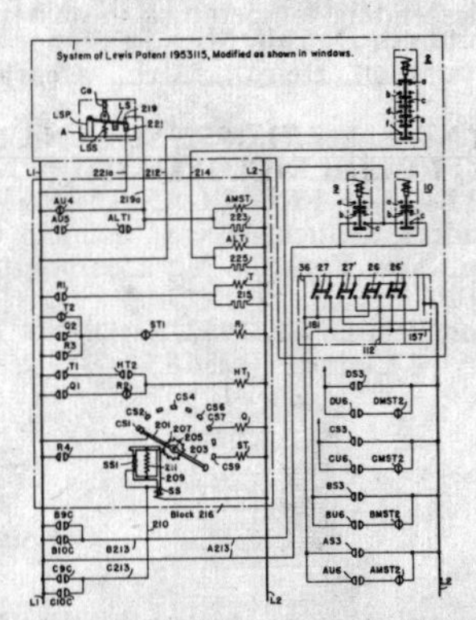

22. In an elevator system, a structure having a plurality of floors, a plurality of elevator cars, means mounting the elevator cars for movement relative to the structure, control means operable for moving each of the elevator cars and stopping each of the elevator cars at the floors for which elevator service is required, and demand means responsive to the time rate of the service demand for all of the elevator cars in service, in combination with means for successively dispatching the elevators from one of the floors, means controlled by said demand means for modifying the dispatch interval between successive ones of the elevator cars dispatched from the dispatching floor, and means responsive to the load of each of the elevator cars at the dispatching floor for expediting the dispatch of the elevator car from the dispatching floor, said last-named means being ineffective following a stopping operation of each of the elevator cars at the dispatching floor for a time sufficient to permit unloading of each of the elevator cars.

2,717,057
SPADE FOR ANCHORING TANK
Willard G. Bankes, Jr., Nescopeck, Pa., assignor to the United States of America as represented by the Secretary of the Army
Application August 17, 1954, Serial No. 450,564
8 Claims. (Cl. 188—5)

1. A spade for arresting retrograde movement of a vehicle comprising a body portion having ground surface

engaging means on one end thereof and an attaching lug formed on its other end, a supporting bracket secured to said vehicle, a horizontally disposed pivot element journalled for rotation in said bracket and fixed in said attaching lug for rotation therewith, a torsion bar secured

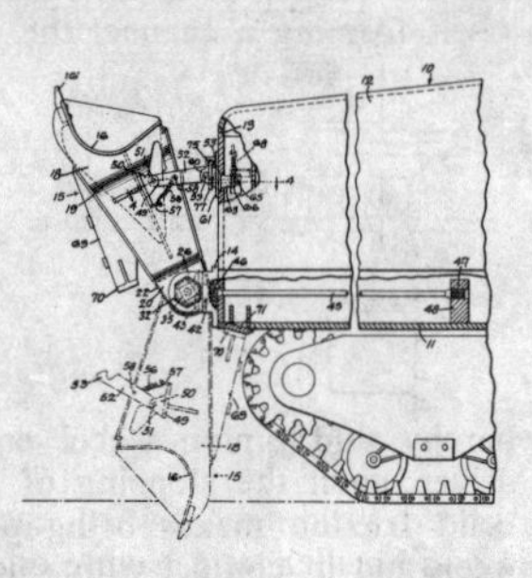

at one end to said vehicle and having its opposite end positioned adjacent said horizontally disposed pivot element, and means connecting said torsion bar and pivot element for transferring the dead load of said spade to said torsion bar to assist manual raising of said spade to travelling position.

2,717,058
SHOCK ABSORBER CONTROL VALVE
George A. Brundrett, Detroit, Mich., assignor to General Motors Corporation, Detroit, Mich., a corporation of Delaware
Application November 20, 1952, Serial No. 321,657
7 Claims. (Cl. 188—88)

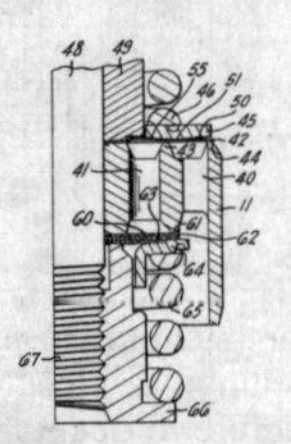

1. A control valve in an hydraulic shock absorber, including in combination, piston means having passage means therein for flow of fluid in one direction between opposite ends of the said piston means, a flexible valve disc on said piston closing said passage means, a rigid valve retainer disc positioned on said valve member and having the valve engaging face thereof formed to provide for flexure of said valve disc relative to said piston and said retainer to open thereby said passage means, pressure of fluid against said valve member first flexing the same to a position against said retainer to provide a first restrictive variably increasing flow of fluid through said passage means to a predetermined maximum flow and thereafter bodily lifting said valve member and said retainer to provide a second restrictive variably in increasing flow of fluid through said passage means greater than said first flow, and resilient means acting on said retainer urging said valve member to close said passage means.

2,717,059
WHEEL STRUCTURE
George Albert Lyon, Detroit, Mich.
Application September 11, 1951, Serial No. 246,131
7 Claims. (Cl. 188—264)

5. A wheel structure comprising a tire rim and a load sustaining body portion, the body portion being formed as a sheet metal stamping having marginal portions secured to the tire rim, said marginal portions having generally axially directed extensions thereon, and said extensions being substantially arrowhead-shaped and being

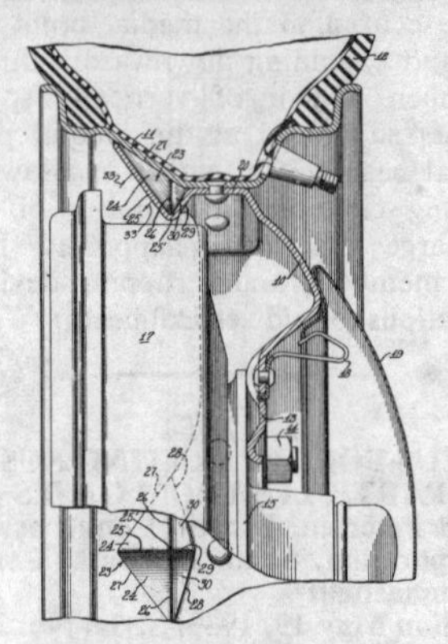

longitudinally creased and turned into divergent vane areas.

2,717,060
ARCHED BUILDING STRUCTURES AND ELEMENTS FOR SAME
Tappan Collins, Grosse Pointe Woods, Mich., assignor to National Steel Corporation, a corporation of Delaware
Application April 6, 1951, Serial No. 219,711
8 Claims. (Cl. 189—1)

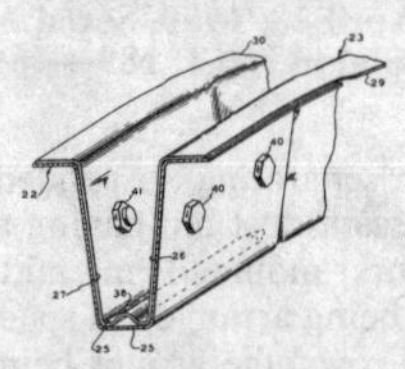

1. An elongated arched rib member of sheet metal having in cross section a single channel shape including an inner web and a pair of side web portions extending outwardly from the two lateral edges of the inner web portion, the side web portions being equally spaced apart by the inner web throughout a major portion of the length of the member with the inner web of said major portion being flat and the inner web in at least one end portion of the member including a rib of inner web metal displaced outwardly a distance such that the side web portions of said one end portion are more closely spaced apart than the side web portions in said major portion by a distance equal to twice the thickness of the sheet metal, each side web portion in said one end portion being parallel with the respective side web portion in said major portion whereby the said one end portion will telescope inside of an end portion of another arched rib member that is the same cross sectional size as said major portion with the telescoping side web portions of the two members in snugly fitting parallel relationship with each other.

2,717,061
PREFABRICATED DOOR FRAMES AND DOOR JAMBS
Robert Katz, Brooklyn, N. Y.
Application December 12, 1949, Serial No. 132,455
1 Claim. (Cl. 189—46)

An adjustable door jamb, comprising in combination, a pair of stiles, a vertically extending stop bead on each stile, a top rail adjustably connected to said stiles for movement vertically thereof in either direction, a sill affixed to the lower ends of said stiles, a fixed rail secured to said stiles intermediate the sill and the top rail to constitute the top of a door frame, said adjustable top rail being slotted at its end portions to accommodate said stop beads for guiding its vertical movement of said stiles and a pair of set screws in said rail positioned for engage-

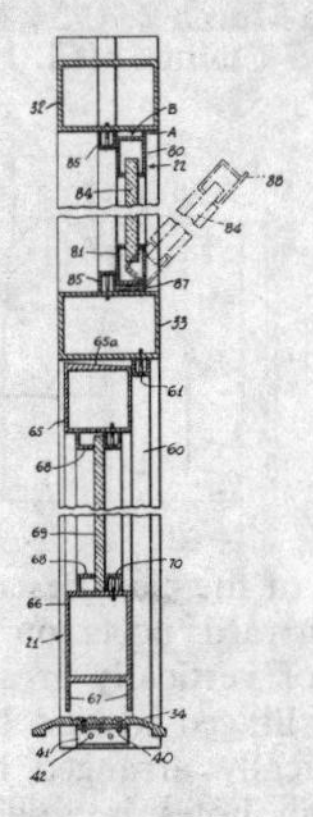

ment with said stop beads to fix said rail in place relative to said stiles.

2,717,062
FIRE DOOR
Leon F. Dusing, Kenmore, N. Y., and Felix H. Saino, Memphis, Tenn.; said Dusing assignor to Dusing & Hunt, Inc., Buffalo, N. Y., and said Saino assignor to F. H. Saino Manufacturing Company, Memphis, Tenn.
Application April 28, 1950, Serial No. 158,896
8 Claims. (Cl. 189—46)

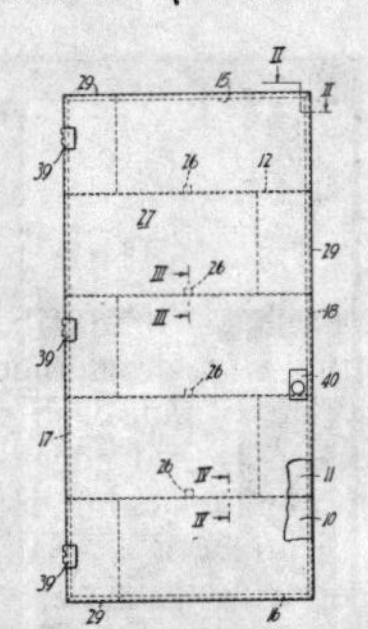

1. In a fire door, a group of generally rectangular relatively rigid slabs of insulating material disposed edge to edge to form a vertical series of horizontally extending rows of slabs to form a door body, complementary tongue and groove formations in the upper and lower abutting edges of adjacent slabs, framing channels at the four edges of the door body with their flanges directed inwardly, marginal recesses in said slabs extending about the front and rear faces of the door body to receive said flanges to form a flush door core construction, said framchannels being welded to each other at their ends at the corners of the door body, and front and rear facing sheets at the front and rear surfaces of the door and having marginal flanges lying against the webs of the framing channels, the facing edges of the flanges of said facing sheets being spaced to form a medial groove extending about the door edges and the front and rear facing sheet flanges being jointly series spot welded to said framing channel webs, and a filler material in said medial groove.

2,717,063
DOOR FRAME TRIM ASSEMBLY
Joseph Sylvan, Berkley, Mich.
Application September 12, 1952, Serial No. 309,246
3 Claims. (Cl. 189—46)

3. A trim assembly for a door frame comprising two cooperating linearly extending strips adapted to be secured to the header and each of the two jambs of the frame, one strip being a generally L-shaped strip having one leg adapted to overlie the outer face of the frame and another leg adapted to overlie the jamb face of the frame, securing means adapted to extend through the last mentioned leg and into the door frame, said leg which is adapted to overlie the outer face of the frame being folded over upon itself forming a channel the side walls of

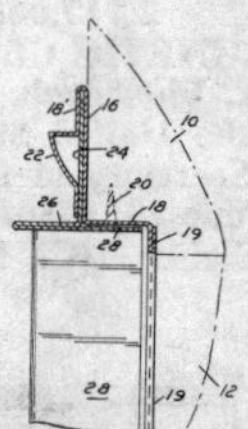

which are tensioned yieldingly toward each other, the other strip being an angular strip having one leg adapted to be received within said channel and held grippingly between the side walls thereof at adjusted positions therein and the other leg being adapted to overlie and cover that leg of the L strip which is secured to the door frame and overlie and cover said securing means.

2,717,064
DOOR LATCHING APPARATUS
Lester R. Hock, Inglewood, Calif.
Application October 6, 1952, Serial No. 313,275
7 Claims. (Cl. 189—46)

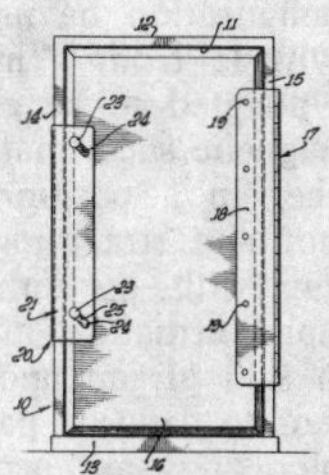

1. In apparatus for latching closure members: a structure having an opening; a closure member pivotally mounted on said structure to open and close said opening; a latch member slidable on said closure member and having a groove therein, said latch member being bodily translatable in upward and downward directions on said closure member; said structure having a rib receivable in said groove; and a pin and inclined slot connection between said latch member and closure member, said slot being inclined in a direction to bring and maintain said groove over said rib when said latch member is moved bodily downward along said closure member to hold said closure member in closed position over said structure opening, bodily movement of said latch member upwardly along said closure member withdrawing said groove from said rib to allow said closure member to be shifted to open position.

2,717,065
HANDLE AND METHOD OF CONSTRUCTION THEREOF
Erdick H. Nelson, Naperville, Ill.
Application July 22, 1952, Serial No. 300,289
14 Claims. (Cl. 190—57)

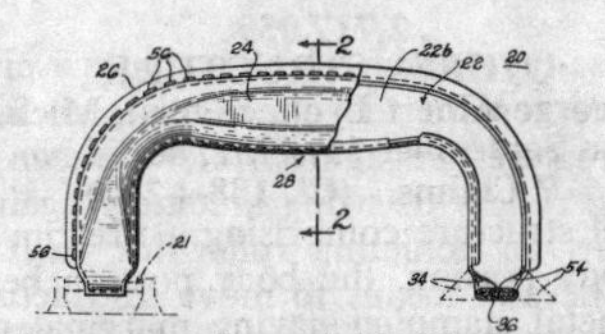

1. A handle comprising a folded hollow shell having a pair of opposed edges, a covering positioned on the outer

surface of said shell and including a seamed portion engaged by the edges, and means joining the covering at a location remote from the seamed portion to apply tensional forces to the covering to force the edges toward the seamed portion.

2,717,066
ELECTROMAGNETIC CLUTCH OPERATOR
Franklin S. Malick, Pittsburgh, Pa., assignor, by mesne assignments, to the United States of America as represented by the Secretary of the Navy
Application June 10, 1950, Serial No. 167,393
2 Claims. (Cl. 192—84)

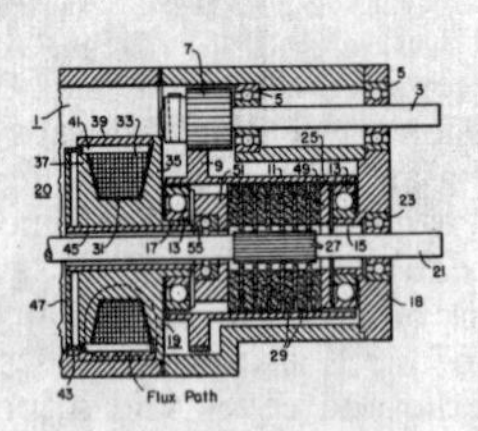

1. In an electromagnetic clutch operator adapted to cause the engagement of a pair of relatively movable clutch members subject to wear, the combination of an operating member for biasing said clutch members in one direction to cause engagement thereof, a member of circular configuration having a circumferential recess therein and made from a magnetic material, a coil in said recess adapted to be energized in varying degree, a ring of magnetic sheet material disposed about said circular member in a position engaging one peripheral edge thereof and straddling said recess to define a radial airgap with the other peripheral edge thereof, a ring-shaped armature of magnetic material fitted partially into said airgap and in spaced relation with the sides thereof to define a pair of radial airgaps, support means supporting said armature member for linear movement transversely of said airgap, and means connecting said armature member with said operating member.

2,717,067
CLUTCH ENGAGING MECHANISM
Charles A. Ramsel, Peoria, Ill., assignor to Caterpillar Tractor Co., Peoria, Ill., a corporation of California
Application July 3, 1953, Serial No. 365,878
7 Claims. (Cl. 192—93)

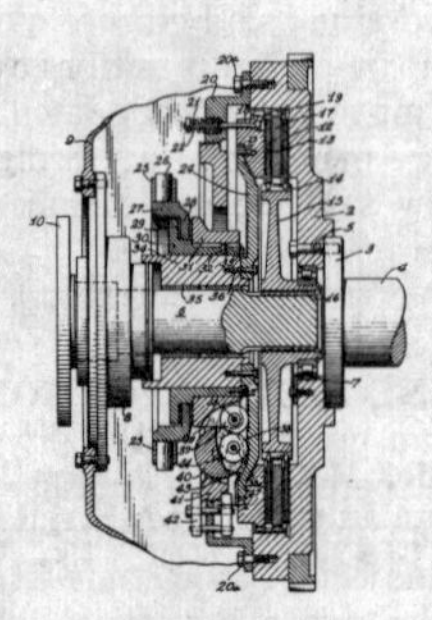

1. In a disc-type clutch which includes a pressure member moveable toward the clutch discs to impart clutch engaging pressure, engaging rollers, means supporting the engaging rollers to swing arcuately against the pressure member, an actuating roller for each of said engaging rollers, and means to move said actuating rollers in contact with the engaging rollers to swing the engaging rollers through their arcuate courses.

2,717,068
CORN HANDLING MEANS
Arthur E. Paschal, Vinton, Iowa
Application March 2, 1954, Serial No. 413,584
5 Claims. (Cl. 193—2)

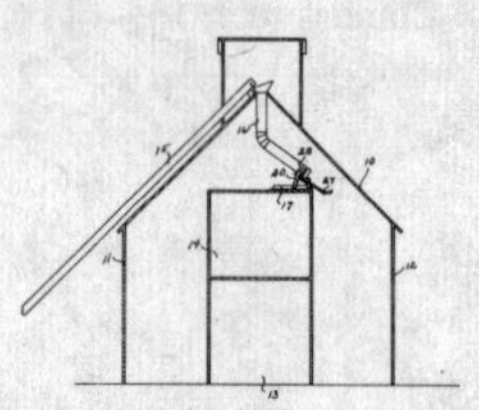

1. In a device of the class described, a base frame, two spaced apart forward posts on said base frame each having a series of vertically arranged holes, two spaced apart rear posts hinged to said base frame each having a series of vertically arranged holes, a rod selectively extending through holes in said first two posts, a rod selectively extending through the holes of said two second posts, a plate secured to said two rods, and a plurality of spaced apart grate rods on said plate.

2,717,069
CHECK INSURING MACHINE
William F. Driscoll, Philadelphia, Pa.
Application January 20, 1950, Serial No. 139,710
4 Claims. (Cl. 194—9)

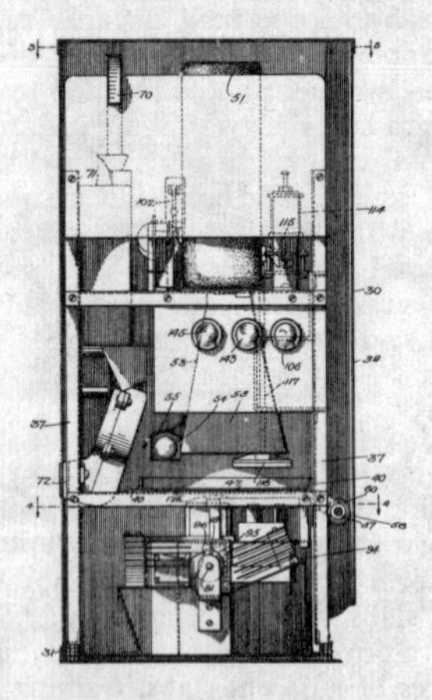

1. In a check insuring machine, a rate wheel, a succession of rate switches selectively closed by the rate wheel, a ratchet accumulator, a plurality of accumulator solenoids operatively connected to the ratchet accumulator and having different throws, a plurality of coin operated switches operatively connected respectively in circuit each with a different solenoid, a plurality of accumulator switches successively closed by the accumulator, a succession of parallel circuit branches each including one of the rate switches and the corresponding accumulator switch in series, a switch solenoid in series with all of the parallel circuit branches and energized by current flowing through any one of the parallel circuit branches, a slidable check supporting tray, a support for the tray, a camera in position to photograph a check on the tray operated by the switch solenoid, a printing numbering device having numerals in position to contact a check on the tray, a press plate reciprocating from a position remote from the check to a position pressing the check against the printing numbering device, a press plate operating solenoid in operative connection with the press plate and means under the control of the switch solenoid for operating the press plate solenoid.

2,717,070
LINE-SPACING MECHANISM
Charles B. Letterman, Hartford, Conn., assignor to Underwood Corporation, New York, N. Y., a corporation of Delaware
Application December 29, 1953, Serial No. 400,871
4 Claims. (Cl. 197—114)

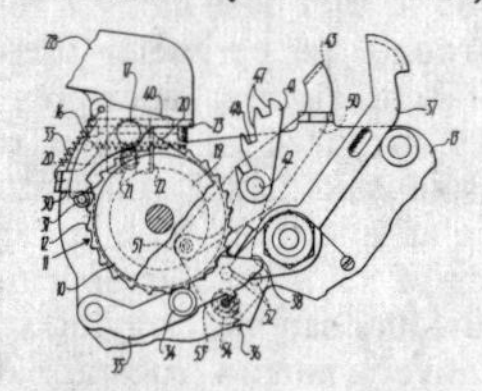

1. A line-spacing mechanism for a typewriter or the like, comprising in combination, an element having a succession of line-spacing teeth and being advanceable in a direction of succession of such teeth, a pivotal pawl actuatable to advance said element in said direction by driving engagement with the teeth thereof and to ratchet in a reverse direction idly over such teeth, means comprising a pivotal support for said pawl, actuatable and restorable respectively to actuate and return said pawl, a member pivotally settable to different angular positions, a pivotal support for said member, and abutment means oppositely on said pawl and said member engageable in the different settings given to the latter to limit the actuation of said pawl to give different integers of tooth advances to said element, said abutment means being arranged so that in all movement limits reached by said pawl there exists an appreciable distance between the said two pivotal supports, said member and said pawl having in each of the movement limits reached by said pawl coacting contacting faces at two different locations intermediate said pivotal supports to provide final lever-interlocks therebetween which will hold the pawl positively in toothed mesh with said element to suppress overthrow of the latter, said contacting faces arranged to allow reverse movement of said pawl over said teeth at the restoration of said actuatable means.

2,717,071
ROLLER SUPPORT FOR POTATO DIGGERS
Curtiss L. Cook, Syracuse, N. Y., assignor, by mesne assignments, to Deere & Company, Moline, Ill., a corporation of Illinois
Application July 19, 1952, Serial No. 299,854
18 Claims. (Cl. 198—189)

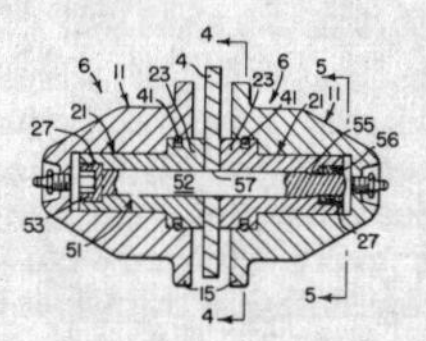

1. A roller support for the conveyor chain of a potato digger or the like having a frame, comprising a fixed spindle having a cylindrical bearing portion and an adjacent base portion concentric with the bearing portion, there being an annular groove in the peripheral portion of said base portion, a conveyor-supporting roller journaled on said bearing portion and having an inner bore dimensioned to receive at least a portion of said base portion, said bore having an annular groove positioned to overlie the groove in said spindle base portion, a spring ring member disposed loosely in said base groove, the latter being at least as deep in a radial direction as the diameter of the stock of which said spring ring member is formed and the circumferential length of said spring ring member being such that when one end of said roller is forced over onto the base portion of said spindle the spring ring member is contracted into the bottom portions of said base groove so as to permit the open end of said roller to be moved axially of said spindle until the annular groove in said roller comes into registry with said base groove, whereby the spring ring member expands into the groove in said roller, said latter groove and said spring ring member being dimensioned so that said spring member, when seated in said annular groove, has sufficient residual resilience to be held in said roller groove but has portions lying radially inwardly of the groove in said roller whereby said spring member serves to hold said roller on said spindle, and means for fixing said spindle to said frame.

2,717,072
STOCK DRAWING APPARATUS
Joseph W. Andrews, Birmingham, and Otto R. Schuler, Detroit, Mich., assignors to Calumet & Hecla, Inc., a corporation of Michigan
Application February 12, 1951, Serial No. 210,602
11 Claims. (Cl. 205—3)

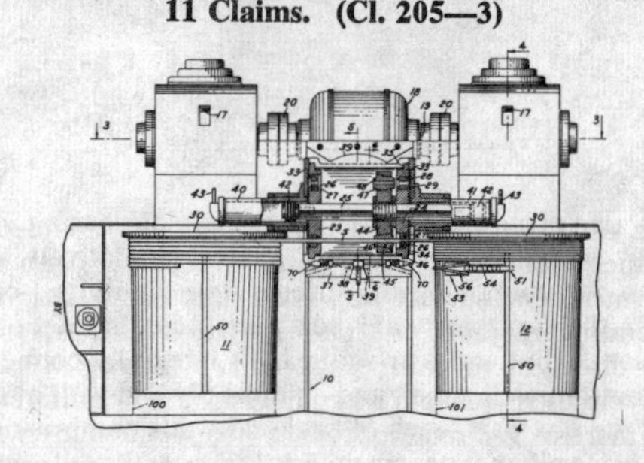

11. Die mechanism for successively drawing stock in opposite directions comprising a plurality of draw dies of different sizes facing in opposite directions, carrier means for said dies operative to move said dies across the fixed path of a length of stock in a drawing operation to successively register oppositely facing dies with said fixed path, gripper means supported for movement along said fixed path and alternately engageable with the opposite ends of the stock, and means for moving said gripper means in opposite directions along said fixed path to push point the stock through a selected die.

2,717,073
TRANSPARENT WALL DISPLAY PACKAGE
Lewis Douglas Young, Providence, R. I., assignor to Douglas Young, Inc., a corporation of Rhode Island
Application April 30, 1953, Serial No. 352,186
1 Claim. (Cl. 206—45.34)

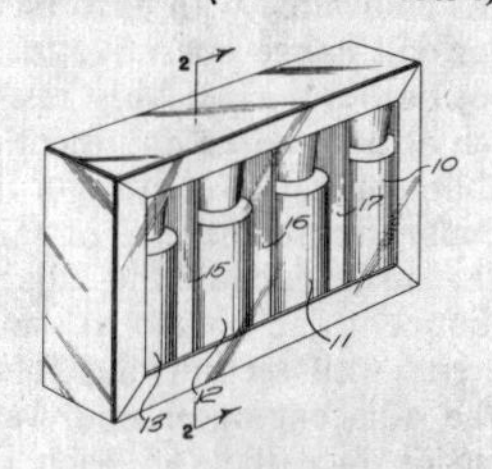

A package for a group of a plurality of articles positioned side by side comprising a transparent casing having telescoping sections each having side and end walls with a top wall on one and a bottom wall on the other, a frame of a single strip of material within said casing extending along and engaging the side and end walls of said casing and held in position thereby with its ends unconnected and in adjacency, a plurality of articles within said frame, a partition within said frame made of a single strip of material folded in zig-zag shape to provide wall sections located between said articles with the connecting portions between the wall sections extending along the side walls of said frame in engagement therewith, said frame having peripheral flanges at opposite sides thereof which meet at the corners in mitered abutting relation and extend into overlapping engagement with said partitions and are held in place by engagement with said casing.

2,717,074
HINGED COVER CARTONS
Marshall I. Williamson, New Haven, and Herman A. Carruth, Northford, Conn., assignors, by mesne assignments, to National Folding Box Company, Incorporated, New Haven, Conn., a corporation of New York
Application September 20, 1951, Serial No. 247,472
7 Claims. (Cl. 206—58)

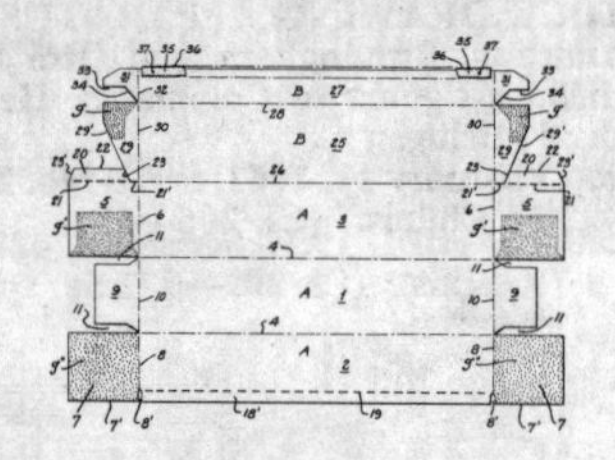

5. A one-piece paperboard carton for dispensing sheet material wound as a sheet roll having a relatively rigid box part for containing the sheet roll and a hinged cover part for the box part, said box part including a bottom wall panel, front and rear wall panels integrally connected to the bottom wall panel, and a plural ply end wall at each end of the box part, each of said end walls comprising an inner end wall section, an intermediate end wall section and an outer end wall section secured together to provide a relatively rigid end wall, each of said inner end wall forming sections being integrally connected to the adjacent end of the bottom panel, each of said intermediate end wall forming sections being integrally connected to the adjacent end of the rear wall panel and having a heighth substantially equal to the heighth of said rear wall panel and a width substantially equal to the width of said bottom wall panel, each of said outer end wall forming sections being integrally connected to the adjacent end of the front wall panel, said cover forming part comprising a top wall panel hinged to the upper edge of said rear wall panel, front and end flanges depending from said top wall panel designed to telescope over the front and end panels of the box part when the cover part is in closed position, sheet tearing means at the lower edge of said cover front flange by means of which withdrawn sectional lengths of the rolled sheet may be severed, and a pair of cooperating abutment elements associated with each box end wall and the adjacent cover end flange operative to lock the cover part in closed position and releasably resist open movement thereof, each of said paired abutment elements comprising a fixed abutment tab presenting a fixed abutment edge integrally connected to the end of said front flange and secured to the inside face of the adjacent cover end flange, and an abutment flap hinged to the upper edge of the intermediate end wall section and foldable over the upper edge of the adjacent outer end wall section and overlying the outside face thereof, each of said hinged abutment flaps presenting an abutment edge designed to releasably interlock with the fixed abutment edge of the cover securing tab when the cover part is in closed position.

2,717,075
SAMPLE HOLDER AND DISPLAY DEVICE
Martin B. Steinthal, New York, N. Y.
Application January 26, 1953, Serial No. 333,180
3 Claims. (Cl. 206—82)

1. A sample holder and display device comprising a relatively rigid substantially rectangular flat base terminating at one edge in a relatively short portion folded over said base in superimposed relation, said relatively short portion terminating in a relatively shorter portion folded under and disposed between said relatively short portion and said base, a display panel superimposed on said base and substantially coextensive therewith, said panel being secured to said base along a relatively narrow border entirely around the perimeter thereof, oppositely disposed pairs of semi-circular apertures in said panel within said border arranged in parallel rows, a row of circular apertures in said panel adjacent each alternate row of semi-circular apertures with a circular aperture opposite each pair of semi-circular apertures, a plurality of removable individual sample carrying strips, each strip comprising a relatively narrow substantially rectangular central portion, a relatively short reduced tapered tab extending from one end of said central portion and providing shoulders at the line of juncture therewith, a rela-

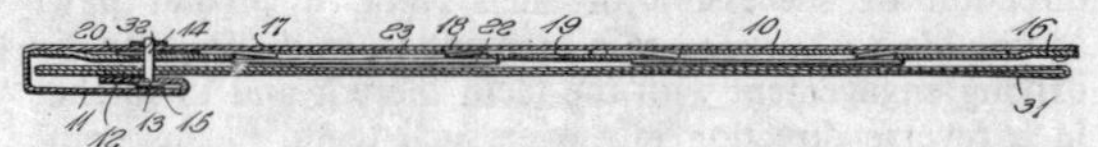

tively long reduced tapered tab extending from the opposite end of said central portion and providing shoulders at the line of juncture therewith, a pressure sensitive adhesive on said central portion for securing an individual sample thereto and indicia on said relatively long tab for identifying the sample, said relatively long tab being received in one of said semi-circular apertures with said indicia visible through the associated circular aperture and said relatively short tab being received in the opposite semi-circular apertures whereby said samples are removably supported in rows on said panel, a large sample substantially coextensive with said panel and having an edge disposed between said relatively shorter portion and said panel and fastening means extending through said relatively shorter portion, said large sample, said panel and said base whereby said large sample is secured to said holder in overlying relationship to said individual samples but permitting turning of said large sample to expose said individual samples.

2,717,076
POTATO HARVESTING MACHINE
Charles R. Leighton and Noel D. Leighton, Limestone, Maine
Application August 21, 1952, Serial No. 305,628
3 Claims. (Cl. 209—138)

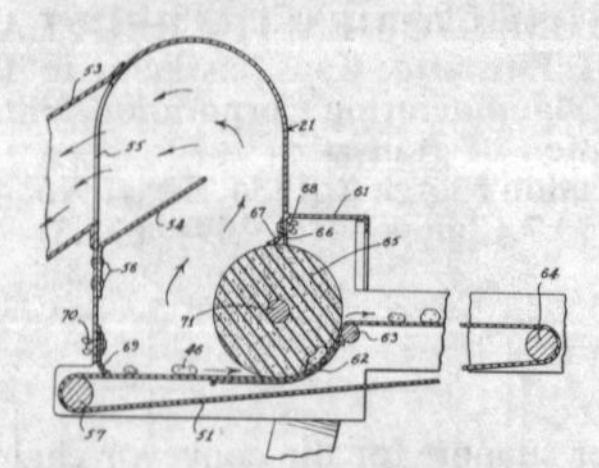

1. A potato separator for potato harvesting machines comprising a housing, a nose portion having a bottom potato inlet opening and extending outwardly from the housing of the separator, power driven air suction means connected to the separator to draw the potatoes through the bottom of the nose portion, said housing having a discharge opening therein, a discharge belt conveyor extending across the bottom of the housing and through the discharge opening, a rigid upwardly curved support for the discharge belt conveyor over which the belt conveyor passes and to elevate a portion thereof, outwardly of the discharge opening, and a compressible roller in the discharge opening and conforming to the rigid belt supporting plate so as to provide an air closure means

for the opening and adapted to be compressed by the potatoes traversing the discharge belt and over the rigid supporting plate of the belt, and a flexible flap on the housing extending into the discharge opening and riding over the compressible roller whereby to seal off the top of the discharge opening and to prevent air leakage over the top of the roller.

2,717,077
MACHINE FOR SEPARATING STONES, POTATOES, AND VINES
Charles R. Leighton and Noel D. Leighton, Limestone, Maine
Application August 2, 1951, Serial No. 239,958
6 Claims. (Cl. 209—139)

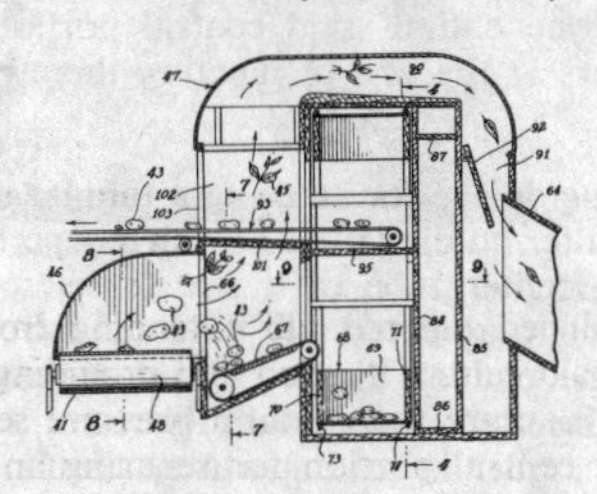

1. A potato separating device to separate the potatoes from stones, comprising a separator casing, an inlet conveyor operable along one side of the casing, suction head means extending over the conveyor, suction fan means connected to the casing to draw air and the potatoes from the inlet conveyor, said casing having a lower section for receiving the potatoes lifted from the inlet conveyor and passageway means extending from the lower section through the casing to the suction fan means, said casing having an upper section and baffle means separating the upper section from the lower section, inclined conveyor means for elevating the potatoes received from the inlet conveyor in the lower section of the casing, a compartmented rotatable drum adjacent to the discharge end of said inclined conveyor for receiving potatoes discharged therefrom and raising the potatoes to the upper section, and discharge conveyor means within the upper section for receiving the potatoes from said drum and discharging the same from the separating device.

2,717,078
AUTOMATIC DENSITY REGULATOR
Cass B. Levi, Pittsburg, Kans., assignor to The McNally-Pittsburg Manufacturing Corporation, Pittsburg, Kans., a corporation of Kansas
Application March 8, 1954, Serial No. 414,789
7 Claims. (Cl. 209—172.5)

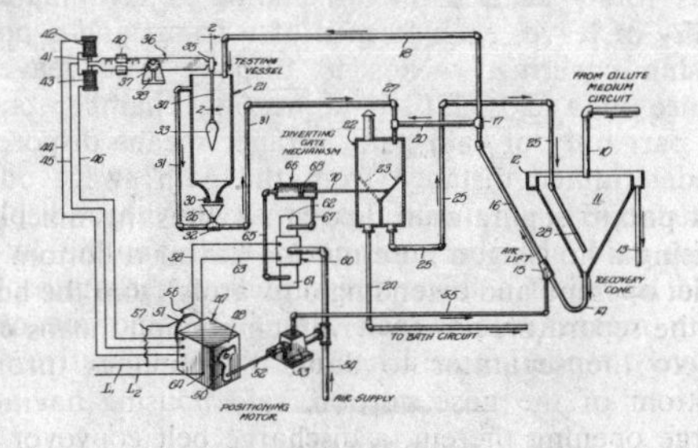

1. In apparatus for automatically regulating the density of a liquid suspension comprising ultra-fine solids in water, the combination including a testing vessel, a balance beam, an hydrometer suspended from one end of the balance beam and having location in the testing vessel, inductance coils associated with the other end of the balance beam in a manner whereby the impedance of the coils is balanced only when the beam is disposed in a horizontal neutral position, means continuously delivering a quantity of said liquid suspension to the testing vessel to submerge the hydrometer therein and effect movement of the balance beam as the density of the liquid suspension may vary, and other means responsive to the balanced and unbalanced conditions of said inductance coils for diverting the liquid suspension delivered thereto to a bath circuit when the same has the correct density or a higher density, and for diverting the liquid suspension to a return circuit for thickening when the density of the same is below the correct density.

2,717,079
BATH REGULATOR FOR DENSE MEDIA SEPARATION SYSTEMS
Cass B. Levi, Pittsburg, Kans., assignor to The McNally Pittsburg Manufacturing Corporation, Pittsburg, Kans., a corporation of Kansas
Application April 1, 1954, Serial No. 420,336
7 Claims. (Cl. 209—172.5)

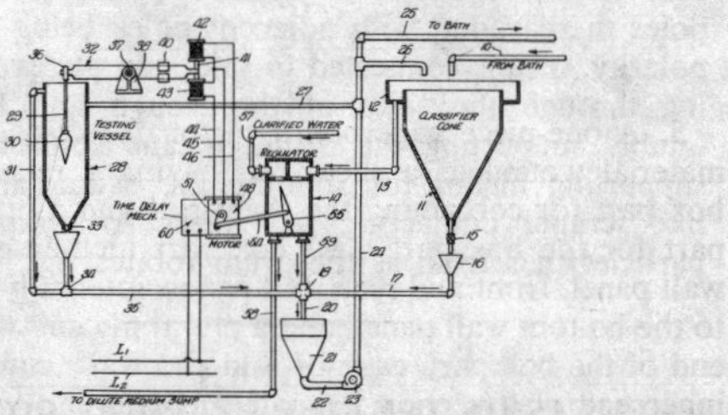

1. In apparatus for regulating the density of a bath medium comprising ultra-fine solids in water, the combination including a testing vessel having a fixed outlet orifice and an overflow, a balance beam, an hydrometer suspended from one end of the balance beam and having location in the testing vessel, means continuously supplying bath medium to the testing vessel to submerge the hydrometer and effect movement of the balance beam as the density of the bath medium may vary, a pair of inductance coils associated with the other end of the balance beam in a manner causing a balanced condition as regards the impedance of the coils when the beam is disposed in a horizontal neutral position and an unbalanced condition when the beam is above or below said horizontal position, a regulator to which is separately supplied a dilute medium and clarified water, a positionable diverting gate within the regulator for controlling the density of the bath medium by controlling flow of the dilute medium and the clarified water, whereby the liquids may be delivered to the bath medium or diverted from the system, said gate having a first operative position for increasing the density of the bath medium by diverting both liquids from the system, said gate having a second operative position for decreasing the density of the bath medium by delivering both liquids to the bath medium, and said gate in a neutral position returning the dilute medium to the bath medium but diverting the clarified water out of the system, and an electric positioning motor responsive to the balanced and unbalanced conditions of said inductance coils for actuating the diverting gate.

2,717,080
MAGNETIC SEPARATOR
Axel Anderson, Rockford, Ill., assignor to Sundstrand Magnetic Products Co., a corporation of Illinois
Application November 26, 1951, Serial No. 258,223
3 Claims. (Cl. 210—1.5)

1. A magnetic separator for separating magnetic material from a liquid comprising a trough having a bottom and spaced sides, a housing disposed above said trough

and having sides depending downwardly adjacent the sides of said trough, a drum disposed above the bottom of the trough, said drum including a pair of spaced circular side plates and an outer exposed cylindrical sleeve consisting of uninterrupted permanently magnetizable material carried by the side plates and extending between the peripheries of said side plates, said sleeve being substantially completely permanently magnetized

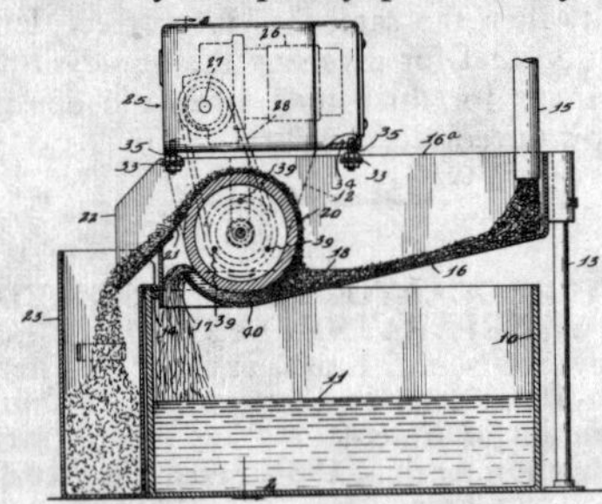

over its entire extent to form a plurality of magnetic poles therearound with adjacent poles being of opposite polarity, means connected to said side plates and extending through the sides of the trough and housing rotatably supporting the drum, means connected to said supporting means for rotating the drum, and a blade-like scraper contacting the sleeve to scrape magnetic particles therefrom as the drum rotates.

2,717,081
EMULSION TREATER PRESSURE CONTROL
Samuel A. Wilson, Tulsa, Okla., assignor to Maloney-Crawford Tank and Manufacturing Company, Tulsa, Okla., a corporation of Delaware
Application February 27, 1952, Serial No. 273,620
2 Claims. (Cl. 210—52.5)

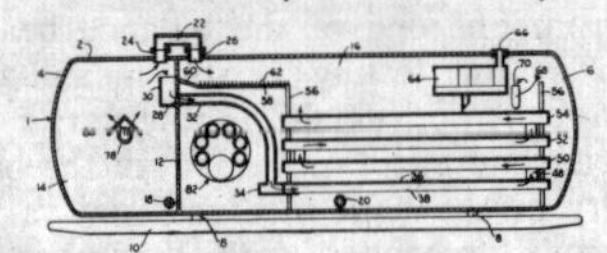

1. An oil-water emulsion processing system, comprising a treater shell, a partition in the shell dividing the shell into a knockout chamber and a separating chamber, an emulsion inlet in the shell for directing the emulsion into the knockout chamber under pressure, whereby the free water is separated from the emulsion and gravitates to the lower portion of the knockout chamber, a single set of emulsion outlets in the partition providing discharge of the emulsion into the separating chamber, inverted trays in the separating chamber communicating with said emulsion outlets to guide the emulsion through the separating chamber in a tortuous path, whereby the emulsion is further separated, an outlet in the upper portion of the separating chamber for discharge of the separated oil therefrom, outlets in the lower portion of the shell to remove the separated water from the knockout and separating chambers, a pressurized water conduit communicating with said last mentioned outlets, means co-operating with said water conduit to control the flow of the separated water therethrough in accordance with the water level in the treater shell, an oil discharge line leading from the emulsion outlet, a spring loaded diaphragm valve interposed in said oil discharge line, conduit means providing imposition of the pressure of said water line on one side of the diaphragm of said valve, whereby said pressure tends to retain said valve in a closed position, and means imposing the pressure of said oil discharge line on the opposite side of the diaphragm of said valve, whereby the pressure in the treater must exceed the pressure of said water conduit to open said valve.

2,717,082
FILTER ELEMENT WITH END SEALING MEANS
Kelly F. McCann, Tulsa, Okla., assignor to Warner Lewis Company, Tulsa, Okla., a corporation of Delaware
Application April 5, 1952, Serial No. 280,783
2 Claims. (Cl. 210—169)

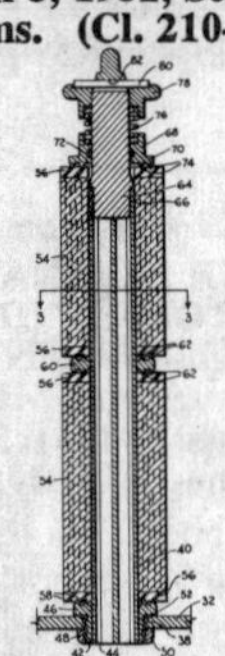

1. A locking device for securing a tubular shaped filter element to a filter holding plate having an aperture therein, comprising a center tube of a size to telescopically receive the element, an externally threaded tubular member secured to one end of the tube, said member being of a size to be inserted in the plate aperture, a nut for securing said member in said aperture, annular gaskets on the opposite ends of the element, a plurality of concentric circumferential flanges on said member to engage one of said gaskets and seal the respective end of the element to the filter holding plate, a stud secured in the opposite end of the tube, a sealing ring slidable on the stud, a plurality of concentric circumferential flanges on the sealing ring to engage the remaining gasket and seal the respective end of the element to the stud, a helical spring surrounding the stud, and means carried by the stud for retaining one end of the helical spring in contact with the sealing ring.

2,717,083
FILTRATION APPARATUS
Edward Walter Wolfe Keene, Kingston-upon-Hull, England
Application February 4, 1952, Serial No. 269,768
4 Claims. (Cl. 210—199)

1. A rotary suction filter including in combination a plurality of leaves each formed of a frame and a porous enclosing covering, means to displace said leaves in sequence in a closed path, a heating chamber enclosing a part only of said path, scraper means disposed at a predetermined distance from the path swept out by the surfaces of the said leaves as they are displaced through said chamber, means to maintain the interior of said leaves at a lower pressure than the atmosphere, heating means within each filter leaf, and means for selectively rendering said heating means operative.

2,717,084
BICYCLE STAND
Ubbo Wilhelm Groenendal, Zurich, Switzerland
Application December 27, 1950, Serial No. 202,871
Claims priority, application Switzerland December 28, 1949
3 Claims. (Cl. 211—20)

1. A bicycle stand comprising a body having top and bottom faces, said body having a major axis and a minor

axis, and an elongated aperture extending along said major axis, through the body from top to bottom, the end walls of said aperture being downwardly converging, and the two side walls of said aperture having two ribs, respectively, on said minor axis, extending perpendicular to said major axis toward and terminating approximately

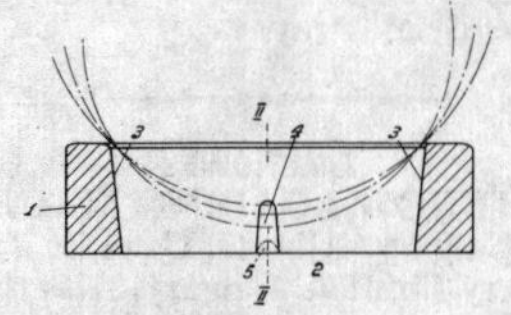

at the mid-height of the body, said ribs having rounded faces tapering from each other from the bottom toward the mid-height of the body, whereby a bicycle wheel received in said aperture is supported at the ends of the aperture and between the faces of said ribs, the relative taper of said ribs providing for the accommodation of wheels of different diameters.

2,717,085
SELF-LEVELING, STORING AND DISPENSING APPARATUS
William H. Waddington, New York, N. Y., assignor to American Machine and Foundry Company, a corporation of New Jersey
Application October 20, 1950, Serial No. 191,105
3 Claims. (Cl. 211—74)

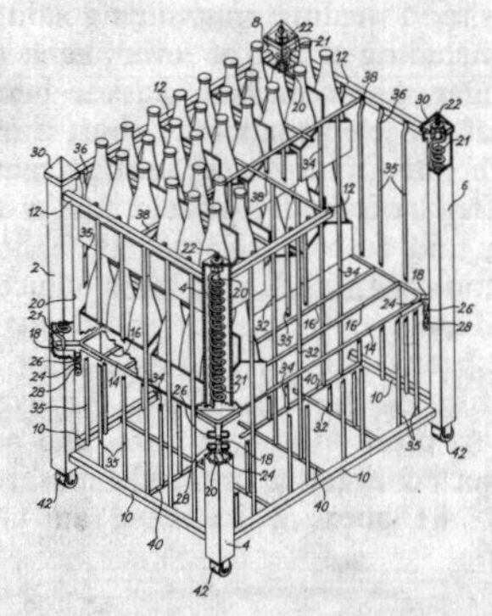

2. A self-leveling, storing and dispensing apparatus comprising a rectangular base, three or more hollow upright posts arranged in non-aligned relationship at corners of the base to which they are connected and extending upwardly therefrom and having vertical slits formed therein, bracing means maintaining said posts in fixed relationship with respect to one another, a carrier platform having arms extending through the slits formed in each of said posts, a calibrated tension spring mounted inside of each of said posts and connected at the lower end to said arms and at the upper end to the top of said posts to counterbalance material supported on said carrier, members attached to said carrier and extending vertically in front of the slitted portion of each of said slits, and sets of vertically spaced rollers connected to said members and engaging with the outside slitted portion of said posts to maintain said carrier platform horizontal.

2,717,086
CONTROL SYSTEM FOR TRAVELING CONVEYORS
George L. Bush, Flushing, N. Y., assignor to The Teleregister Corporation, Stamford, Conn., a corporation of Delaware
Application July 14, 1954, Serial No. 443,314
14 Claims. (Cl. 214—11)

1. In a traveling conveyor system in which a conveyor has a plurality of sections respectively for carrying articles from a loading position for selective distribution to a plurality of unloading positions, means for producing destination signals respectively for preselecting the various unloading positions where the articles are to be ejected, ejector devices at said unloading positions, means for producing timing pulses in synchronism with the travel of

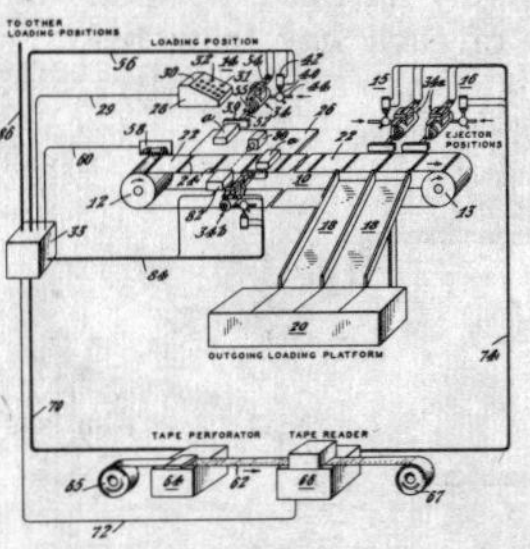

the conveyor sections, signal storage means, means controlled by said timing pulses for transmitting said destination signals successively to said signal storage means, means for successively reading the stored destination signals and means controlled thereby for selectively actuating said ejector devices to unload said articles at the preselected unloading positions.

2,717,087
MECHANIZED STORAGE APPARATUS
Harold Auger, Boyne Hill, Maidenhead, England
Application August 16, 1949, Serial No. 110,642
Claims priority, application Great Britain August 23, 1948
17 Claims. (Cl. 214—16.1)

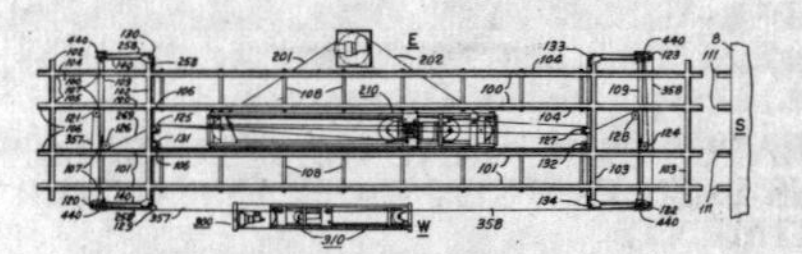

1. In storage apparatus, the combination of a load-carrying platform adapted to move along a first predetermined path, a cross-carriage adapted to move along a second predetermined path and to carry said platform along said second path, a collector adapted to drive said platform along said first path and being itself reciprocable on a course substantially parallel with said first path, means for reciprocating said collector, a collector coupler mounted and fixed upon said collector, and means including actuator means connected with and adapted to operate said coupler to drive said cross-carriage along said second path, said coupler being selectively engageable with and disengageable from said platform by said actuator means as said cross-carriage moves along said second path.

2,717,088
MULTIPLE CAR PARKING APPARATUS AND INSTALLATION THEREFOR
Herbert F. Morley, Chevy Chase, Md.
Application March 31, 1953, Serial No. 345,840
8 Claims. (Cl. 214—16.1)

1. A storage structure comprising an entrance driveway extending from end to end of a storage area; a skeleton framework extending along each side of said driveway and defining at each side thereof a longitudinal series of transversely extending storage bays of a length slightly greater than the length of the articles to be stored, each bay being of a width slightly greater than the width of the article to be stored and said framework extending below the driveway a distance slightly more than twice the height of the article to be stored and extending above the driveway at the end remote from the driveway section a distance substantially less than the extension below said driveway;

a movable storage structure for receiving articles to be stored in spaced superposed relation located in each bay and including a pair of generally transversely extending vertically spaced support structures pivotally connected at their ends remote from said driveway to said skeleton framework and adapted to normally occupy a downwardly inclined position to store objects in that portion of each bay lying below said driveway

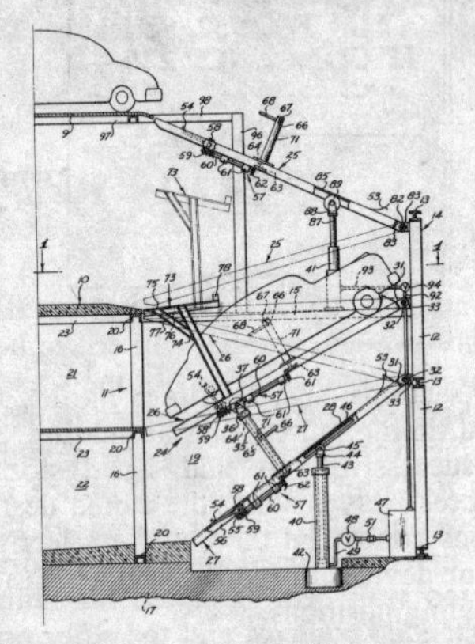

level and be selectively elevated to position their respective driveway ends to receive objects from and deliver objects onto said driveway; and hoist means adapted to engage and raise said movable storage structure to selectively present the driveway ends of said support structures alternately in position to receive objects from said driveway and discharge objects onto said driveway.

2,717,089
APPARATUS FOR MANUALLY BULK LOADING CANS INTO FREIGHT CARS AND OTHER COMPARTMENTS

Harold C. Hebert, Tampa, Fla., assignor to American Can Company, New York, N. Y., a corporation of New Jersey
Application December 23, 1953, Serial No. 400,015
8 Claims. (Cl. 214—83.26)

1. An apparatus for manually bulk loading sheet metal cans into freight cars and other restricted places for shipment or storage, comprising in combination, a portable rack of substantially the same width as a freight car for arrangement in the car transversely thereof, said rack being movable manually from one end of the car toward the other, a substantially flat shallow tray inclined from end-to-end mounted on said rack at substantially waist height of a manual packing operator for receiving cans rolling on their sides and for supporting them in a continuous line, said tray extending transversely of the car and being unobstructed above and below so that an operator standing behind the tray can readily lift the cans from the tray and stack them in front of the tray in transverse rows in the end of the car unobstructed and without extensive walking, and means for conveying the cans in a substantially continuous procession to the high end of said inclined tray to keep said tray filled with cans to be stacked in the car, said can conveying means comprising a plurality of rigid telescoping can runway sections extending longitudinally of the car from a side doorway thereof and connecting with an end of said tray, whereby to permit the latter and said rack to be manually moved as aforesaid longitudinally of the car without interrupting the continuous delivery of cans to said tray.

2,717,090
OVERTHROW LOADING MACHINE

John H. Malo, Calgary, Alberta, Canada, assignor to Barber Machinery Limited, Calgary, Alberta, Canada
Application December 15, 1950, Serial No. 200,892
3 Claims. (Cl. 214—131)

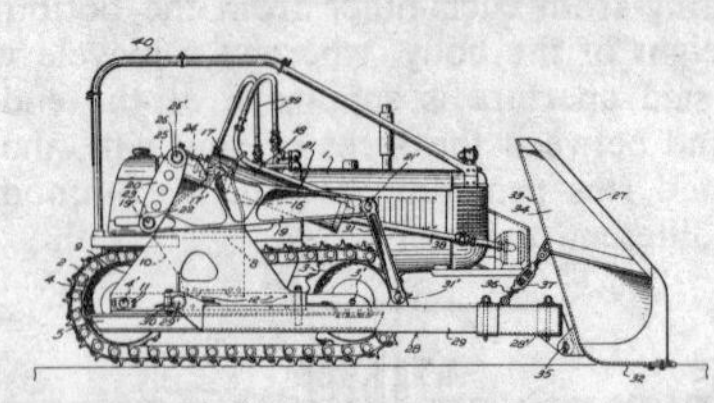

1. In combination with a tractor having a pair of track frames and forward and rearward crawler-supporting sprockets mounted on said frames, an overthrow loading device comprising a supporting frame mounted on each of said track frames, a bucket-carrying yoke having side arms pivotally mounted about a fixed axis on said supporting frames, and means for swinging said yoke about its fixed axis including a pair of levers each of generally scalene triangular shape, each said lever being pivotally connected adjacent the apex of its short and long sides to a respective one of said supporting frames about a fixed axis located rearwardly of said first fixed axis and directly above said rearward sprockets, a link having a pivotal connection with each said lever adjacent the apex of its intermediate and long sides and a pivotal connection with a respective one of said side arms of the yoke located forwardly of the axis of said forward sprockets in the lowermost position of said yoke, and power means pivotally connected about a movable axis to each said lever adjacent the apex of its short and intermediate sides.

2,717,091
POWER SHOVEL AND IMPLEMENT ATTACHMENT

Roy F. Smith, Silt, Colo.
Application March 30, 1953, Serial No. 345,545
1 Claim. (Cl. 214—132)

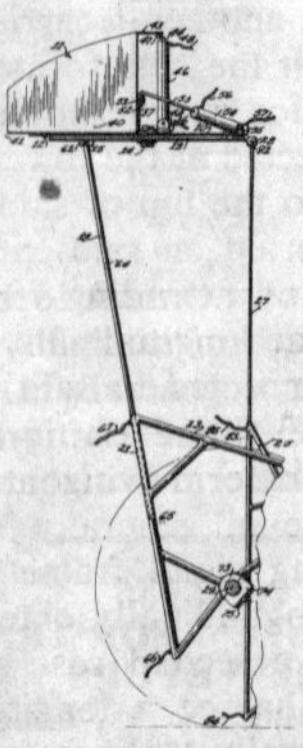

In an attachment for tractors and other vehicles, the combination which comprises spaced vertically disposed frames having fluid pressure actuated telescoping elements in base members thereof, a platform, means pivotally connecting pairs of the frames to the platform at longitudinally spaced points, a turn table journaled on the platform, a hydraulic jack positioned to actuate the turn table, a scoop pivotally mounted on the turn table, and a hydraulic jack for actuating the scoop to a dumping position, and means for pivoting the frames to raise and lower the same.

2,717,092
DEVICE FOR PICKING A LOOSE WIRE FROM ABOUT A SUPPORT
Stanley J. Gartner and Henry J. Zwald, Emporium, Pa., assignors to Sylvania Electric Products Inc., a corporation of Massachusetts
Application March 2, 1948, Serial No. 12,646
4 Claims. (Cl. 214—658)

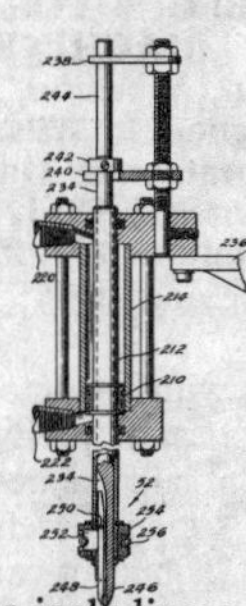

1. A hydraulic picker including a cylinder, a movable piston within said cylinder, a hollow piston rod connected to said piston, and another rod within, connected by, and movable with, said hollow piston rod, a fixed jaw and a pivoted jaw carried by said other rod, means for restricting the movement of said other rod to a lesser extent than that of said hollow piston rod and means responsive to relative motion between said piston rod and said other rod for swinging said pivoted jaw.

2,717,093
SHIPPING CASE OR THE LIKE
Steven Etienne Mautner, Port Jervis, N. Y., assignor to Skydyne, Inc., Port Jervis, N. Y., a corporation of New York
Application October 28, 1950, Serial No. 192,707
5 Claims. (Cl. 217—56)

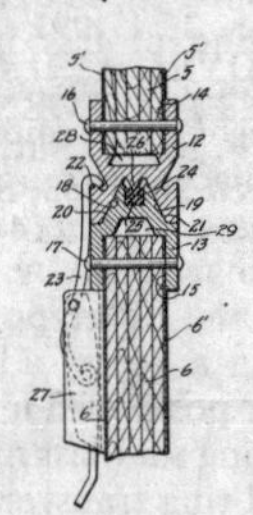

1. In a case, two case members to be joined, one case member comprising a longitudinally extending metal frame member having a generally U-shaped longitudinally extending channel for receiving the edge of a case panel, said frame member having a longitudinally extending locking tongue of generally V-shape, a second longitudinally extending metal frame member having a generally U-shaped longitudinally extending channel for receiving the edge of a second case panel, said second metal frame member having a longitudinally extending locking groove of generally V-shape to receive said locking tongue to form a secure metal-to-metal tight wedging joint between said frame members, said first frame member having said V-shaped locking tongue having a longitudinally extending seal groove for a yielding gasket and formed at the convergence of the walls of said locking tongue and locking groove, said second frame member having said V-shaped locking groove having a longitudinally extending generally V-shaped seal tongue to project into said seal groove and engage a yielding gasket therein.

2,717,094
PETROLEUM CONTAINERS
George Arlington Moore, New York, N. Y.
Application May 17, 1952, Serial No. 288,412
17 Claims. (Cl. 220—4)

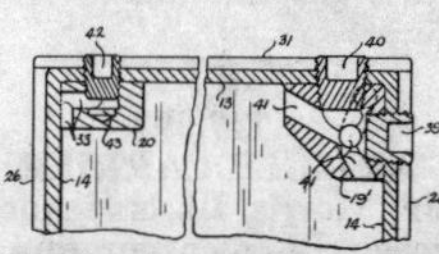

1. In a container having top and side walls meeting at corner portions, which comprises an orifice body member being secured to interior portions of the walls of the container and disposed in corner portions of said wall, said member having an orifice in alignment with an orifice through said top wall, and another orifice in alignment with an orifice through said side wall of the container adjacent the first mentioned orifice, said orifice aligned through the top wall being adapted for filling said container with intended contents and the other orifice aligned through the side wall being adapted for dispensing said contents of the container.

2,717,095
DRAINAGE APPARATUS FOR MOVABLE ROOFS
Myron W. Gable, Houston, Tex., assignor to Shell Development Company, San Francisco, Calif., a corporation of Delaware
Application July 18, 1949, Serial No. 105,413
9 Claims. (Cl. 220—26)

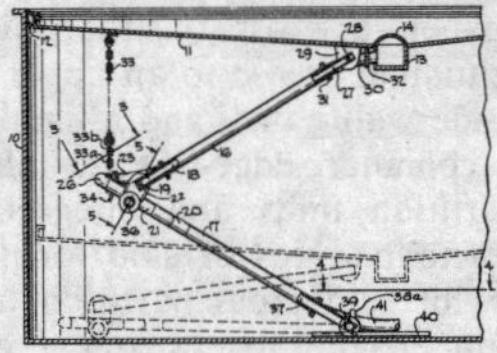

1. Drainage apparatus for a movable roof of a tank comprising: a drain opening in the roof; a discharge pipe extending into and terminating within the tank below the roof; and an articulated down drain connected to establish flow communication between said drain opening and said discharge pipe and having a plurality of rigid pipe sections connected by one or more compound joints in the tank, each joint comprising: a framework carried by and extending transversely of the first conjoined pipe section and beyond the end thereof; a pair of rigid branch pipes fixed to the end of the second conjoined pipe section in flow communication with said second pipe section and extending longitudinally, one on each side of the axis of the second pipe section; pivotal connections having a common axis between said branch pipes and said framework for transmitting mechanical stress between said pipe sections; a transverse pipe beyond said pivotal connection interconnecting said branch pipes; and a short section of flexible hose connected to the end of the first pipe section and extending thence beyond said hinge axis to said transverse pipe for establishing communication between said pipe sections independently of said pivotal connection.

2,717,096
MAGNET GAS TANK CAP
Minnis W. Henderson, Bluefield, W. Va.
Application January 23, 1953, Serial No. 332,832
1 Claim. (Cl. 220—40)

A cap for closing filling pipes of gasoline tanks comprising a circular body having a flat top surface, an annular flange extending at right angles to the margin of the body, a circular magnet secured to the outer surface of said flange with portions of said magnet extending beyond

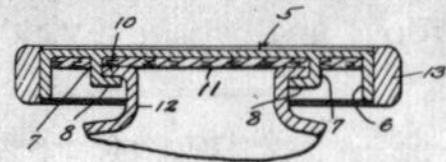

the flat top surface of the body and the free end of the flange for contact with the metal supporting surface, and means on said body for effecting attachment of the cap on a filling pipe of a gasoline tank.

2,717,097
ARTICLE CARRIER
Edwin L. Arneson, Morris, Ill., assignor to Morris Paper Mills, Chicago, Ill., a corporation of Illinois
Application August 10, 1951, Serial No. 241,231
3 Claims. (Cl. 220—113)

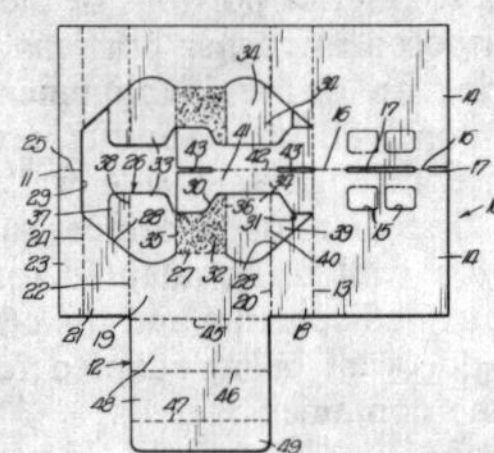

1. A blank of flexible paperboard for use in constructing an article carrier, said blank being subdivided into substantially similar sections along a medial longitudinal line, each of said sections provided with creases and cuts defining, in the successive order named, a side wall panel, an end wall panel integrally hinged to said side wall panel by a transverse crease, a longitudinal partition panel integrally hinged by a transverse crease to said end wall panel, a cross partition unit integrally hinged by a longitudinal crease to an upper edge defining margins of said side wall panel, and a handle reinforcing element lying in coplanar, edge-to-edge relation to each of said cross partition units and integrally hinged by transverse creases to margins of said longitudinal partition panels, parts of said cross partition units and said handle reinforcing element occupying a common area of said blank lying between said side and end wall panels of said respective sections, said element being separated from said units by cuts defining adjacent edges of the element and units.

2,717,098
ARTICLE CARRIER
Edwin L. Arneson, Morris, Ill., assignor to Morris Paper Mills, Chicago, Ill., a corporation of Illinois
Application August 22, 1951, Serial No. 243,058
5 Claims. (Cl. 220—113)

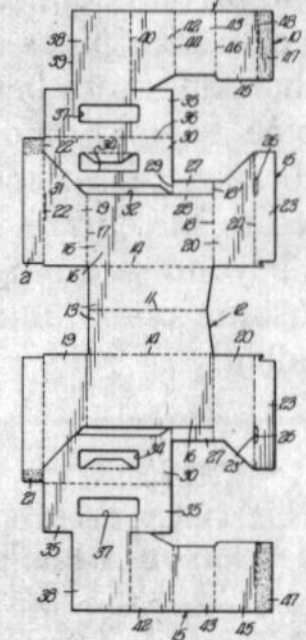

1. A flexible paperboard article carrier blank comprising a side wall forming panel having end wall panels hinged by transverse creases to opposed end margins thereof, a handle panel separated by a cut from said side wall forming panel and integrally hinged by means including transverse creases to margins of said respective end wall panels, and a handle and partition defining unit hinged by a longitudinal crease to a margin of said handle panel remote from said side wall forming panel, said last named unit including a handle panel immediately adjoining said longitudinal crease, a longitudinal partition forming panel integral with said last named handle panel, and a cross partition structure including a series of at least three panels integrally hinged by transverse creases to a margin of said longitudinal partition panel and to one another.

2,717,099
TRANSFER BLOCK FOR RING SETTING MACHINES
George J. Rundblad, Wheaton, and Eremeldo Cairelli, Chicago, Ill., assignors to Wilson-Jones Company, Chicago, Ill., a corporation of Massachusetts
Application March 7, 1951, Serial No. 214,414
12 Claims. (Cl. 221—293)

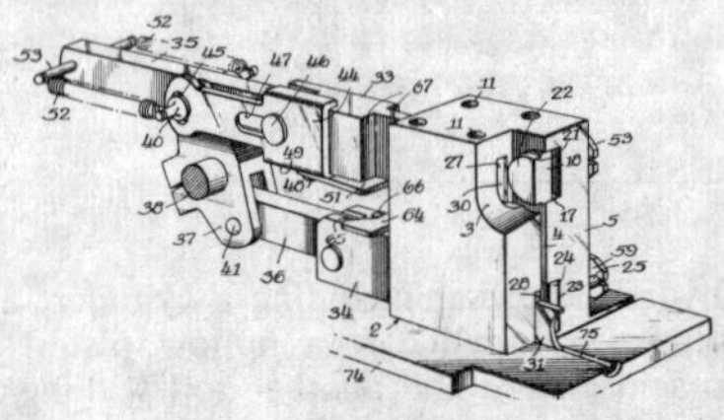

1. A transfer block for ring setting machines, comprising a block having a vertical recess medial of its front edge, the upper front edge of said block being cut away on one side of said vertical recess, a transverse recess in said block aligned horizontally with said cut away portion and extending from the side of said vertical recess opposite said cut away portion, a plunger mounted in said transverse recess, means for moving said plunger transversely of said block within predetermined limits, said plunger having a projection on one side thereof movable into said cut away portion when said plunger is moved to the limit of its movement in one direction, said projection having a ring mounted thereon when said projection is positioned in said cut away portion, said transverse recess being of less height than the overall height of the ring mounted on said projection, means for retracting said plunger into said transverse recess, the edge of said block adjacent said transverse recess engaging said ring and stripping it from said projection upon retraction of said plunger thereby causing it to fall into said vertical recess, the lower portion of the front edge of said block having a recess adjacent said vertical recess, another plunger positioned in said last mentioned recess, means for moving said last mentioned plunger transversely to move said ring transversely of said block from said vertical recess, and a pusher blade aligned with said ring after it has been moved transversely from said vertical recess, said pusher blade being reciprocable in said block to push said ring forwardly out of said block.

2,717,100
GAS FLOW CONTROL UNIT
Arthur E. Engelder, Morenci, Ariz.
Application September 29, 1951, Serial No. 248,929
11 Claims. (Cl. 222—5)

1. In a gas flow control unit, a container adapted to carry a conventional disposable pressurized gas cartridge having a penetrable sealing element, a solid body formed with a first laterally projecting abutment and with a second oppositely directed laterally projecting abutment, said first abutment being adjustably seated in said con-

tainer for relative rotation therein, a needle carried by said first abutment and so positioned and arranged that rotation of said body relative to said container advances said needle into said cartridge to pierce said sealing element and release the contained gas from said cartridge, a movable gas flow-controlling valve in said body, a manually rotatable cap seated on said second abutment for relative rotation thereon to displace said valve to open and close it, an apertured connector member provided on said cap and adapted to connect to a gas delivery tube, said body and said cap forming between them

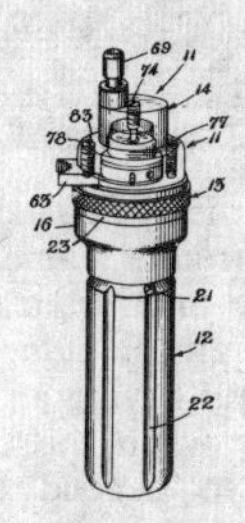

a chamber open to said connector member, a first passage formed through said body and connecting said chamber and said cartridge through said valve, a second passage formed through said body and connecting said chamber to the exterior of said second abutment, said cap normally closing said second passage, and a port formed in said cap and positioned as to be aligned with said second passage upon rotation of said cap to a predetermined position effecting closing of said valve, whereby gas passing through said chamber from said connector member is caused to flow to the atmosphere when said cap is in said predetermined position.

2,717,101
PASTE-TYPE DENTIFRICE DISPENSING TOOTHBRUSH
Ambrose B. Van Handel, North Hollywood, Calif.
Application May 4, 1949, Serial No. 91,284
4 Claims. (Cl. 222—80)

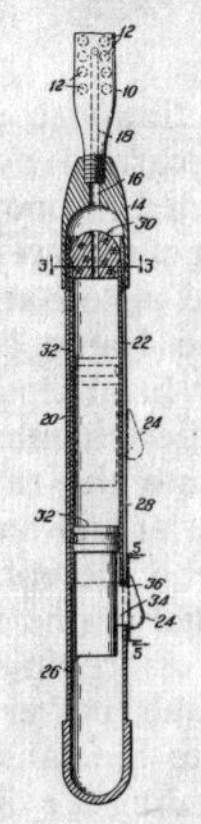

1. In a dispensing device, the combination of a hollow handle having a dispensing opening at one end and closed at its opposite end, a dentifrice-containing cartridge having a dispensing opening at one end and open at its opposite end, the walls of said cartridge being of readily severable material, said cartridge being disposed within said handle with its dispensing opening adjacent the dispensing opening of said handle, a follower closely fitting within said cartridge, a manually operable elongated piston operably connected to and within said handle in a position to engage said follower, said piston adapted to enter the open end of said cartridge to confine the wall of said cartridge between said piston and surrounding handle, and a web including a knife edge traversing the wall of said cartridge and extending from said piston to a point exteriorly of said hollow handle, said knife edge being adapted to cut through the wall of said cartridge as the knife edge is advanced.

2,717,102
FLUID HANDLING DISPENSER
Halcolm D. Rives, Los Angeles, Calif., assignor to Swingspout Measure Company, Los Angeles, Calif., a corporation of California
Application December 11, 1950, Serial No. 200,172
1 Claim. (Cl. 222—89)

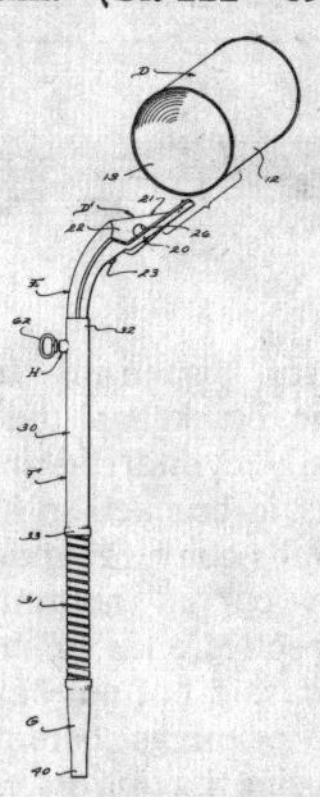

In a fluid handling dispenser for use with a can having a cylindrical side wall and an end held by a bead; a head having a can guide of an elongate trough shape and an elongate can piercing cutter, said can guide and said cutter being located adjacent each other so that the cylindrical side wall of a can may be gripped between them when the end wall of said can is punctured by said cutter, a rigid arcuate tubular neck extending from said cutter, said neck being joined to one end of a straight and rigid duct having a length substantially greater than the combined length of said head and said neck, a manually operable shut-off valve provided in said duct adjacent the juncture of said duct and said neck, a section of flexible tubing joined at one end to the other end of said duct, and an elongated nozzle joined to the other end of said flexible tubing whereby said nozzle may be disposed in various positions out of alignment with said duct.

2,717,103
DISPENSERS FOR COLLAPSIBLE TUBES
Glenn M. Hill, Kansas City, Kans.
Application February 16, 1954, Serial No. 410,617
5 Claims. (Cl. 222—100)

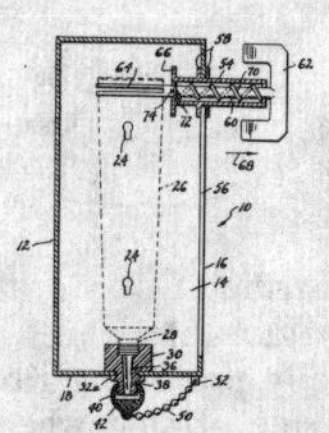

1. In a dispenser for collapsible tubes, the combination of a vertically elongated housing adapted to receive therein an inverted collapsible tube, said housing including a bottom wall and a side wall provided with a vertically extending slot, a combined tube supporting and dispensing member carried by said bottom wall, a substantially tubular guide projecting laterally from said housing and movable upwardly and downwardly in said slot, a shaft rotatably mounted in said guide and having a slotted

inner end portion adapted to operatively engage a collapsible tube for collapsing the same, and a handle provided at the outer end of said shaft.

2,717,104
FERTILIZER SPREADING MACHINES
Lloyd G. Hoppes, Hazelton, Kans., assignor, by mesne assignments, to Lester Wilkinson, Wichita, Kans., as trustee
Application November 1, 1951, Serial No. 254,229
5 Claims. (Cl. 222—177)

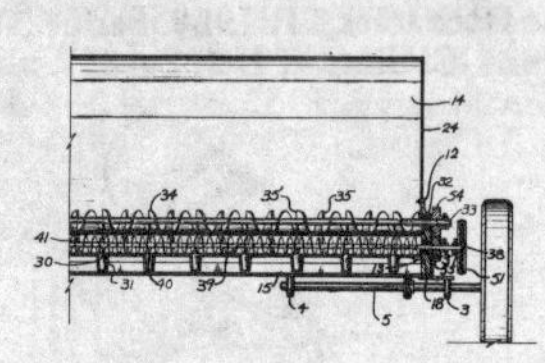

5. Apparatus for dispensing fertilizer comprising a wheel borne frame, brackets at the respective ends of said frame, substantially cylindrical inturned flanges on the lower portion of said brackets, a hopper for the fertilizer having inclined walls converging downwardly terminating in a downwardly curved bottom, means securing the hopper at its respective ends to said brackets above said cylindrical flanges, said bottom having a plurality of longitudinally spaced openings therein, a substantially cylindrical-shaped housing having its respective ends secured to the inturned cylindrical flanges on said brackets below said hopper, said housing having free ends turned upwardly at an angle and a longitudinal opening extending over the openings in the bottom of said hopper, said free ends embracing the lower portions of the inclined sides of said hopper to strengthen the same, said housing having a plurality of longitudinally spaced openings in the bottom thereof in staggered relation to the openings in the bottom of said hopper, auger means in the bottom of said hopper and in said housing, and means operatively connected to both of said auger means and said wheels whereby movement of said apparatus over the ground will cause the auger means to rotate causing the fertilizer to move through said openings in the hopper and housing simultaneously.

2,717,105
SEED DISTRIBUTOR FOR DRILLS
Otto Weitz, Butzbach, Hessen, Germany, assignor to Firma A. J. Tröster, Butzbach, Hessen, Germany
Application October 15, 1949, Serial No. 121,615
Claims priority, application Germany November 12, 1948
3 Claims. (Cl. 222—283)

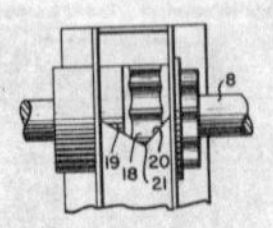

1. In a seed distributor for drills, a rotary shaft, a distributing axially movable wheel upon said shaft, said wheel having radially projecting ribs, an isolating sleeve located upon one side of said wheel; a casing enclosing said wheel, a closing ring connected with said casing and enclosing said wheel, said sleeve having a sloping face, said ring and said face forming a trough-like adjustable chamber receiving and feeding the seeds, a wedge, means pivotally supporting said wedge adjacent said ribs to adjust the distance of said wedge from said ribs, said wedge having an outer edge having a V-shaped recess formed therein, whereby the seeds flow from said wheel and drop over said edge in a continuous stream.

2,717,106
PASTRY ICING MACHINE
Oscar Hammer, Pittsburgh, Pa.
Application December 19, 1952, Serial No. 326,998
3 Claims. (Cl. 222—318)

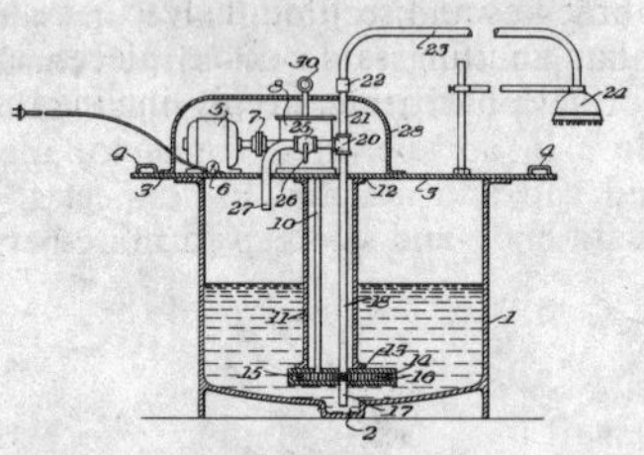

1. A machine for icing baked goods and the like which comprises a portable frame adapted to be moved from one container to another, a scaled tubular housing depending from said frame, a pump mounted on the lower end of said housing and having an inlet adjacent the bottom of the container when the frame is positioned to permit the housing to depend therein, motor means carried by said frame and connected through the tubular housing to operate said pump, a pump outlet including a pipe extending up through the housing to the frame, and a discharge nozzle connected to said pump outlet pipe to dispense the icing from the container.

2,717,107
GUN FOR VISCOUS MATERIAL
George J. Moletz, Jr., Jamaica, and Bennie Pollack, Brooklyn, N. Y.
Application October 29, 1954, Serial No. 465,652
7 Claims. (Cl. 222—334)

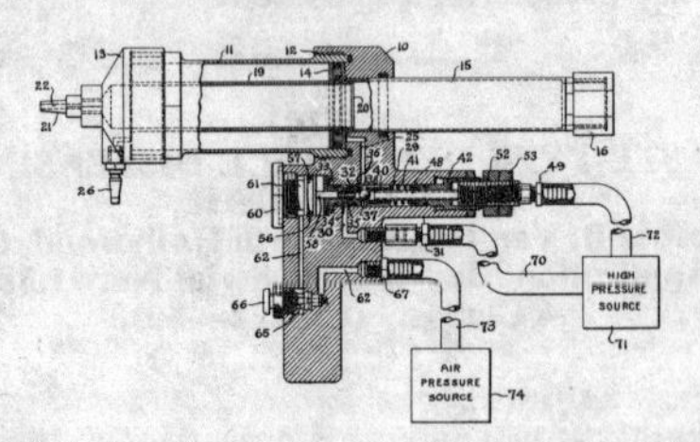

1. A gun for projecting viscous material comprising a frame; power means carried by the frame including a power cylinder connected at one end to the frame, a ring piston movable within the power cylinder, a tubular piston rod carried by the piston at one end of the rod and projecting through the frame and the other end being closed, and a cylinder head closing the other end of the power cylinder; a tubular chamber carried by the cylinder head and projecting into the ring piston and piston rod for receiving the material and having a discharge opening at the cylinder head end thereof, valve means carried by the frame and connected to one end of the power cylinder to feed and exhaust pressure fluid therein including spring means to close the same, power means connected to the valve means to operate the same against the pressure of the spring means including a cylinder and a piston within the cylinder, and an air pressure valve connected with one end of the cylinder of the valve power means to operate the valve power means and its valve means.

2,717,108
GARMENT HANGER
John E. Schaerer, Clifton, N. J.
Application August 24, 1953, Serial No. 375,891
1 Claim. (Cl. 223—91)

A garment hanger comprising a crossmember, an upstanding hook arranged intermediate the ends of said crossmember and secured thereto, a pair of vertically disposed spaced parallel brackets secured to said crossmember adjacent the ends thereof and each bracket including a pair of side pieces, an end piece, and a bottom piece, the lower edges of said pieces being coplanar, said pieces defining an elongated inwardly extending vertical slot in each bracket, a bar triangular in cross section extending between said brackets and secured thereto, one of the surfaces of said bar abutting said bottom pieces, another of the surfaces of said bar being steeply inclined to a horizontal plane and having its lower edge spaced from the opposed side pieces and defining a space therebetween, the upper longitudinal edge of said bar being rounded, and a cylindrical rod having its ends slidably positioned in said slots and movable downwardly in said spaces for coaction with said inclined surface to maintain a garment on the bar, the diameter of said rod being greater than the distance between the lower edge of said inclined surface and either of the opposed side pieces.

2,717,109
PAINT CAN HOLDER
Bernard A. Walsh, San Diego, Calif.
Application October 20, 1952, Serial No. 315,783
3 Claims. (Cl. 224—5)

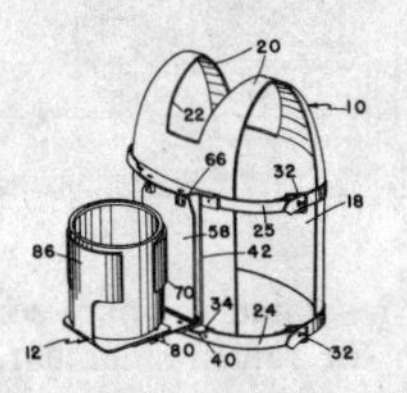

2. Apparatus for painters, comprising a vest, an inflexible panel mounted on the front of said vest, a paint can holder assembly removably secured to said panel, said assembly including a shelf releasably mounted on said panel, and an angle plate member releasably secured to said shelf, said shelf and said member having paint can engaging elements secured thereto, said elements being opposed arcuate clamp plates, said clamp plates being slightly resilient to conform with the curvature of slightly differently sized paint cans.

2,717,110
AUTOMOBILE GARMENT HANGERS
Roger S. Funk, Warrington, Pa.
Application March 30, 1954, Serial No. 419,680
5 Claims. (Cl. 224—42.45)

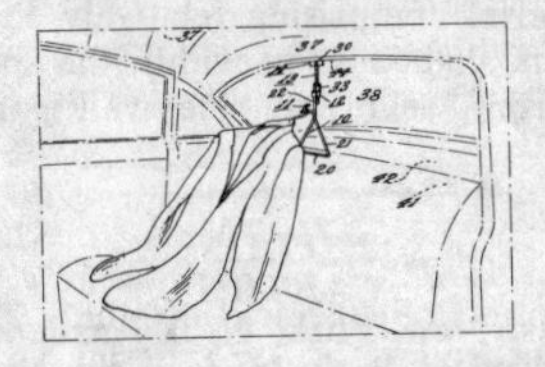

1. A garment hanger for automobiles for disposition in the rear of an automobile betwen the deck behind the rear seat and the frame mounting the rear window thereof, comprising a base element of triangular profile in one plane and having a horizontal base course for stable engagement on the deck behind a seat, an upper oscillatable element generally parallel to the said one plane for engagement with part of the roof adjacent to the upper edge of such rear window of such automobile, means for temporarily locking the respective base and upper element in such engagements, and a hook element mounted on the base element of the hanger to project forwardly over and relative to such seat.

2,717,111
SIPHON TUBE BASKET
William D. Gilardi, Dos Palos, Calif.
Application May 19, 1952, Serial No. 288,680
3 Claims. (Cl. 224—45)

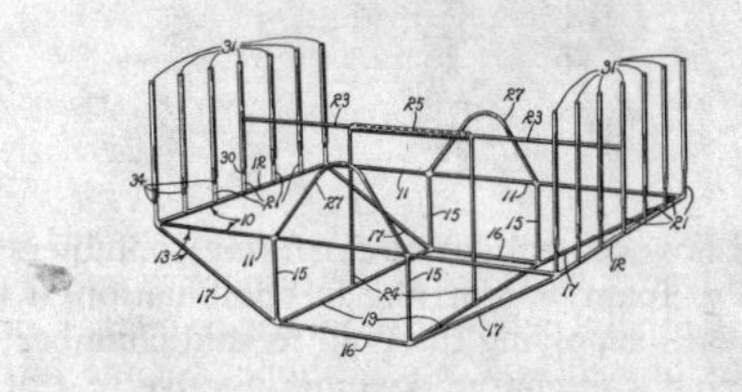

3. A siphon tube basket comprising a substantial rectangular open frame having elongated side members and transverse end members rigidly interconnected and located in a common plane, a pair of legs rigidly mounted on each side member and extended therefrom in normal relation to the plane of the rectangular frame, said legs being mounted on the side members in positions dividing the lengths of their respective side members into three substantially equal parts, base struts adapted to rest on a supporting surface and to support the frame in substantially parallel relation thereto rigidly conneected between the extended ends of the legs of each side member of the frame in substantially parallel relation to its legs' respective side member, an oblique strut rigidly interconnecting the extended end of each leg and the adjacent end of its respective side member, corresponding oblique struts of the side members being adapted concurrently to rest on a supporting surface and to support the rectangular frame in inclined position relative thereto, lower cross members interconnecting the extended ends of corresponding legs of the side members in substantially parallel relation to the end members, each lower cross member lying in a plane common to the pair of oblique struts connected to its respective legs and the transverse end member to which said struts are connected as well as in a plane angularly related thereto common to the base struts and the opposite cross member, a plurality of elongated substantially straight fingers mounted in corresponding equally spaced relation on each of the end members of the frame and extended therefrom in substantially parallel relation normal to the plane of the frame, there being an odd number of fingers on each end member so as to provide a central finger midway between the side members, an arcuate side guard rigidly mounted on each side member having opposite ends connected to its respective side member at the juncture of the legs therewith and a central portion in spaced relation to said side member oppositely from the legs and lying in a plane common to its respective side member and legs, a medial division rod rigidly mounted on each of the lower cross members midway between the legs interconnected thereby and extended therefrom in substantially parallel relation to the legs having extended ends located in spaced relation to the plane of the frame and opposite from the lower cross members relative thereto, and an elongated rod interconnecting the central finger of each end member and the extended ends of the division rods.

2,717,112
FLUID FLOW PRECISE SHUT-OFF VALVE
John D. Ralston, Indianapolis, Ind.
Application October 17, 1952, Serial No. 315,308
8 Claims. (Cl. 226—116)

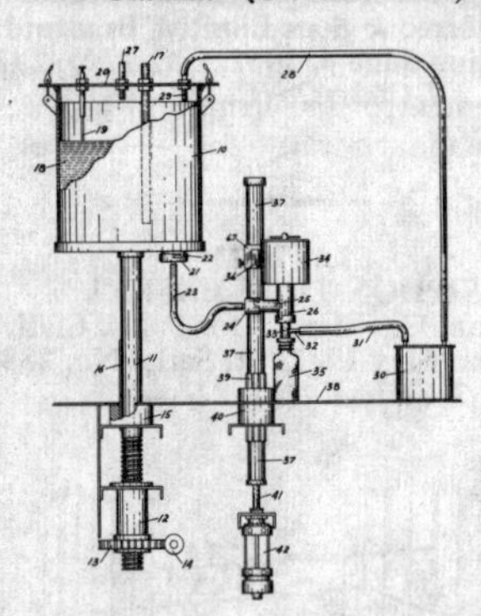

5. A fluid flow control device for filling containers with fluids likely to foam, comprising in combination, a valve chamber; means supplying the fluid to said chamber; said chamber having a discharge opening; a valve seat at said opening; a shiftable valve core seatable on said seat withholding flow of fluid through said opening; means for actuating the core; a fluid discharging tube extending from said chamber opening below said seat; a housing carried in spaced relation about said tube, open at its lower end and closed about the tube at its upper end; a sealing member carried by the housing against which a container may sealably connect to have the housing open into the container and also have said tube discharge into the container; said housing having a discharge opening externally of said container; and means applying a partial vacuum at said housing discharge opening.

2,717,113
ATTACHMENT FOR SAUSAGE STUFFING MACHINES
Robert H. Clark, Denver, Colo.
Application January 24, 1952, Serial No. 267,984
1 Claim. (Cl. 226—125)

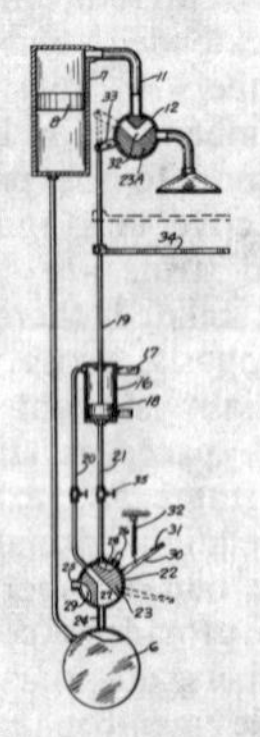

A ground meat paste dispensing device of the type having an elongated vertical cylinder open at the top and closed at the bottom, a floating piston in the cylinder, a removable cover attached to the top of the cylinder in sealing relation thereto, the chamber above the piston forming a receptacle for the material to be dispensed, the chamber below the piston forming a compressed air receiving chamber, a materials discharge pipe in communication with the upper chamber adjacent the upper end thereof, a discharge valve in said pipe, a source of air under pressure, a pipe communicating the air pressure source with the chamber beneath the floating piston to thereby place the contents on the upper chamber under sufficient pressure to cause it to flow through the discharge pipe when the discharge valve therein is open; in combination with the above, means for opening and closing the discharge valve by power derived from the compressed air comprising, a power cylinder positioned below the discharge valve, a piston slidable in said cylinder, a piston rod operatively interconnecting the power piston with the discharge valve to move the latter to open position when the power piston approaches its upper limit of travel, means comprising a manually controllable valve of the slide valve type and cooperating conduits, for alternately communicating opposite sides of the power piston with the said air supply and with the atmosphere, to move the said piston and the discharge valve at will to either of their extreme positions, spring means operatively associated with the slide valve forming means to normally hold it in discharge valve closed position, and a table carried by the piston rod at a point below the outlet end of the discharge pipe for holding a vessel in position to receive extruded material during the time that the discharge valve is open and to lower the table and hold it in its lowermost position during the times that the discharge valve is closed.

2,717,114
THERMAL CONTAINER
Charles L. Parham, Jr., Dearborn, Mich.
Application April 27, 1950, Serial No. 158,469
9 Claims. (Cl. 229—14)

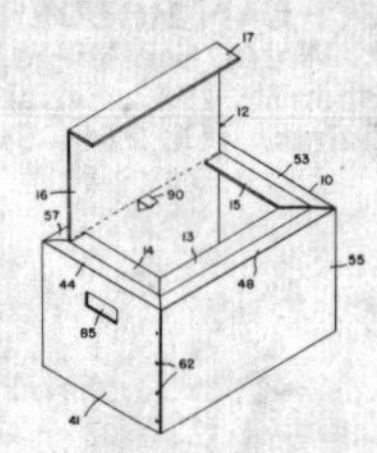

7. A thermally insulated shipping container adapted to be knocked down for return shipment or storage in flat condition comprising: a one-piece outer, open topped box having a bottom, front, rear and two side wall portions, ledge portions extending inwardly from the top edges of said front, rear and side wall portions, liner portions extending downwardly from the inner edges of said ledge portions; an inner box structure comprising a first element having a bottom wall portion, a rear wall portion extending upwardly from said bottom wall portion, a cover portion extending forwardly from the top of said rear wall portion, and a flap portion at the front edge of said cover portion, and a second element having connected front and side wall portions, the rear edges of the side wall portions of said second element and the side edges of the rear wall portion of said first element having releasably engaging interlock means thereon; and insulation material interposed between the bottom wall portions of said outer and inner box and between the front, rear and side wall portions of said outer and inner box, said insulation material comprising relatively strong self-supporting blankets shaped to fit within the spaces between the bottom, front, rear and side wall portions of the boxes.

2,717,115
CARTON
Oscar L. Vines, New York, N. Y., assignor to Alford Cartons, Ridgefield Park, N. J., a corporation of New Jersey
Application December 27, 1950, Serial No. 202,914
2 Claims. (Cl. 229—28)

1. A partitioned carton comprising a top panel, a bottom panel and two side panels, the side panels being pro-

vided with a plurality of pairs of spaced and substantially congruent and oppositely disposed trough-shaped cut lines disposed longitudinally thereof and having their free ends terminating at the top panel, each pair of cut lines being connected substantially centrally thereof by a transverse cut line extending across the top panel so as to define a pair of fold sections adapted to be folded downwardly and divergently inwardly into the interior of the erected carton, the depth of each fold section being less than the height of the side panels, the fold sections being folded as aforesaid about transverse fold lines joining the respective ends of each pair of longitudinal cut lines, said transverse fold lines being provided by score lines extending divergently inwardly from opposite ends of each

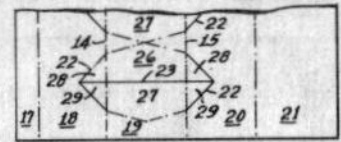

longitudinal cut line toward the other cut line of each pair thereof, the central portions of the resulting pairs of transverse fold lines being spaced further apart than the extremities thereof whereby folding of each section downwardly and inwardly into the erected carton causes the central portion of the top panel adjacent each score line to be arched and causes each fold section to assume a substantially concave form, the arching of the top panel causing the side walls to be drawn toward one another so as to firmly engage the lateral ends of each fold section, and the trough-shaped cut lines in the side panels permitting the fold sections to swing freely about the transverse fold lines and thus form partitions adapted to accommodate articles of substantially variant size.

2,717,116
CARTON
Oscar L. Vines, New York, N. Y., assignor to Alford Cartons, Ridgefield Park, N. J., a corporation of New Jersey
Application April 11, 1952, Serial No. 281,791
4 Claims. (Cl. 229—28)

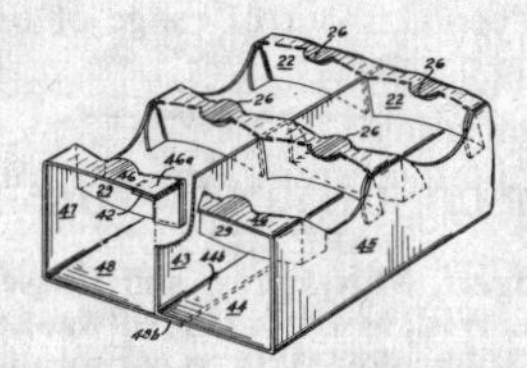

4. A partitioned carton composed essentially of a top panel, a bottom panel, and at least two side panels, the top panel being provided with a plurality of pairs of fold sections each pair of which is adapted to define a compartment within the boundaries of said panels, each pair of fold sections being defined by a pair of transversely disposed fold lines extending across the top panel between adjacent side panels and by a transverse cut line positioned intermediate the transverse fold lines and connected terminally with the terminal portions of the transverse fold lines, each of said top panel transverse fold lines comprising a central cut portion communicating with distal scored portions, said distal scored portions of the transverse fold lines for each pair of fold sections extending divergently inwardly from each side panel, each fold section being folded downwardly and divergently inwardly into the interior of the erected carton about said transverse fold lines with resulting divergent bowing of the fold sections of each pair thereof and upward arching of the top panel between proximate transverse fold lines of proximate pairs thereof, the central cut portion of each transverse fold line having a contour such that the center portion of the top panel defined thereby projects longitudinally over a portion of each compartment defined by each pair of bowed fold sections.

2,717,117
PROCESS AND APPARATUS FOR PIECING THE WARP ENDS OF A WARP BEAM TO THE WARP ENDS OF ANOTHER WARP BEAM
William Felton, Bradford, England, assignor of one-half to David Crabtree & Sons Limited, Bradford, England
Application June 6, 1952, Serial No. 292,037
7 Claims. (Cl. 28—49)

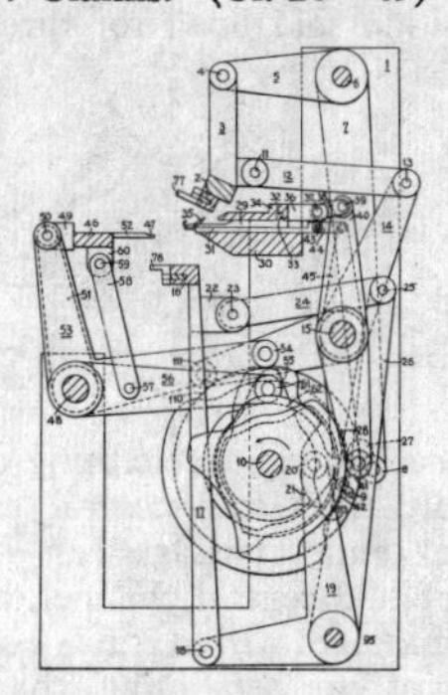

1. A process of piecing the warp ends of a warp beam to the warp ends of another warp beam, consisting in simultaneously forming on each of a plurality of the warp ends of the one warp beam an untightened loop and an untightened loop extending through the first named loop, then threading a different warp end of the other warp beam through each of the first named loops and thereupon tightening each of the first named warp ends so as to draw each of the first named loops through the second named loop and also draw the threaded warp end through the second named loop into a loop form and finally to tighten the remaining loop, all for producing weaver's knots which connect warp ends of the said warp beam to the warp ends of the said other warp beam.

2,717,118
TURBO-COMPRESSOR
Hellmuth Walter, Upper Montclair, N. J., assignor to Worthington Corporation, a corporation of Delaware
Application March 7, 1952, Serial No. 275,301
10 Claims. (Cl. 230—116)

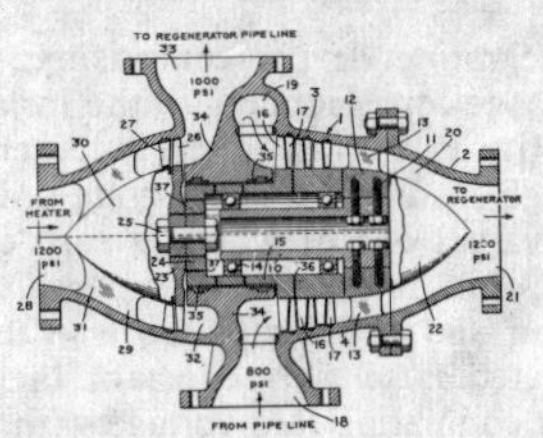

1. In a turbo-compressor, a casing, a turbine in said casing, an axial compressor in said casing, an annular supporting ring in said casing, a quill shaft attached to said ring and extending axially in the casing, a hollow hub shaft rotatably mounted on said quill shaft and having one end closed, said turbine including a rotor, said turbine rotor attached to the closed end of said hub shaft, said compressor including rotary blades, said blades attached to said hollow hub shaft for rotation therewith.

2,717,119
CENTRIFUGAL SEPARATOR
Leo D. Jones, Philadelphia, Pa., assignor to The Sharples Corporation, a corporation of Delaware
Application November 3, 1951, Serial No. 254,683
9 Claims. (Cl. 233—21)

5. In a centrifuge, the combination of a rotatable centrifugal bowl having a separating chamber, an accelerating-decelerating device in said chamber, said accelerating-

decelerating device having a central open channel extending in axial direction therethrough and having liquid-flow connection about and with said central channel to provide communication inwardly from said accelerating-decelerating device toward the axis of rotation of said bowl, means for mounting said accelerating-decelerating device in said bowl, means for feeding liquid to said bowl, means within said bowl for directing the feed of

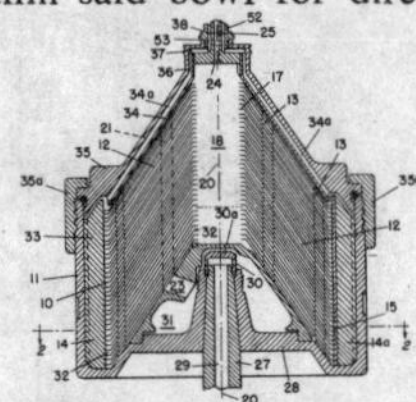

liquid to said accelerating-decelerating device at a region spaced outwardly from said central channel, a central light component outlet passage for said bowl leading from one end of said central channel, and a heavy component outlet passage within said bowl leading from the outer part of the separating chamber toward the axis of rotation of said bowl and terminating adjacent said axis.

2,717,120
INTEGRATING DEVICE FOR INDICATING GROUND POSITION OF AIRCRAFT
John C. Bellamy, Chicago, Ill., assignor to Cook Electric Company, Chicago, Ill., a corporation of Illinois
Application July 12, 1950, Serial No. 173,443
11 Claims. (Cl. 235—61)

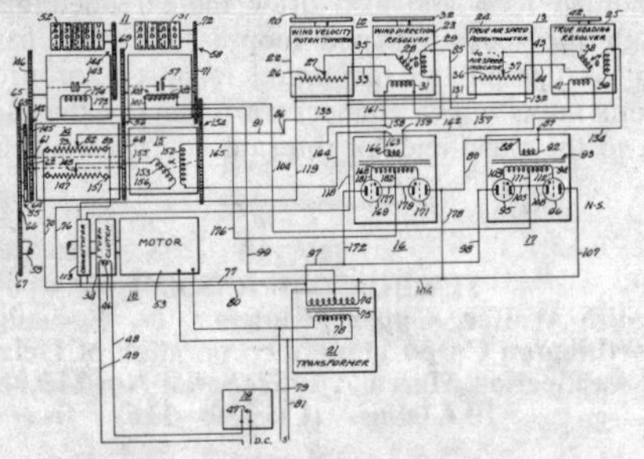

3. An integrating device comprising, means for obtaining a voltage corresponding to the instantaneous value of the quantity to be integrated, a potentiometer to be driven through a voltage cycle varying uniformly between two certain values one of which is zero during each one of successive time intervals, said intervals being substantially constant and sufficiently short so that the quantity to be integrated is sensibly constant therein, means for continuously comparing said voltage with said potentiometer voltage varying from one of said certain values to the other during each of said intervals, a cumulative indicator, and means controlled by said comparing means for driving said indicator in each of said intervals an amount proportional to the magnitude of said quantity during said interval.

2,717,121
KEYBOARD
Hans P. Luhn, Armonk, N. Y., assignor to International Business Machines Corporation, New York, N. Y., a corporation of New York
Original application January 19, 1949, Serial No. 71,734. Divided and this application October 27, 1950, Serial No. 192,581
15 Claims. (Cl. 235—145)

1. In a keyboard structure, a plurality of denominational banks of keys, a key-latching bail associated with each key bank, a reset key carried by each of said key-latching bails for manually resetting said bails individually, a pair of reset bars in common association with said bails, and means associated with each of said reset

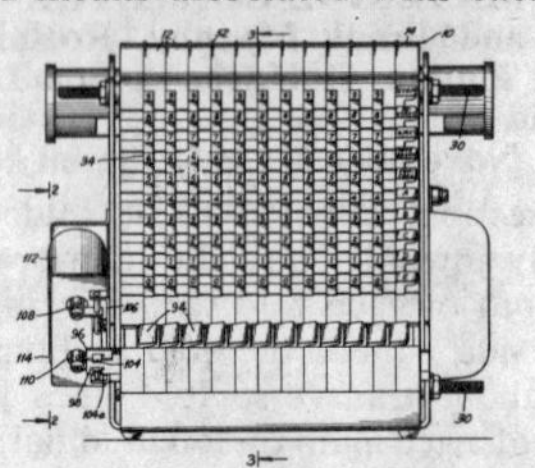

keys for selectively connecting any of said bails in operative relation with either of said reset bars.

ERRATUM

For Class 236—21 see:
Patent No. 2,717,381

2,717,122
REGAIN CONTROL METHOD AND APPARATUS
Eugene C. Gwaltney, Jr., Atlanta, Ga.
Application July 26, 1951, Serial No. 238,739
9 Claims. (Cl. 236—44)

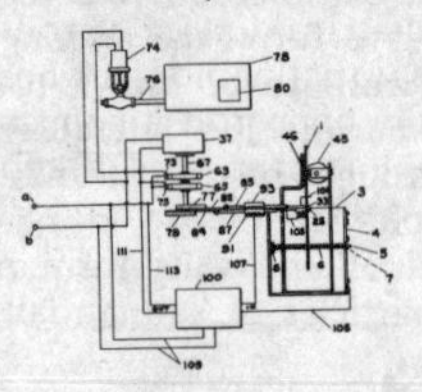

1. Apparatus for measuring and controlling the moisture content of textile materials, having in combination a balance beam carrying a sample of textile material, a pointer on the beam, a scale traversed by the pointer, contacts spaced apart along the scale, a humidifier starting and stopping in response to engagement of any one of the contacts by the pointer, and means withdrawing such contacts out of range of the pointer when engaged by the pointer.

2,717,123
LOW POWER CONDITION RESPONSIVE CONTROL APPARATUS
Adolph J. Hilgert, Milwaukee, and Russell B. Matthews, Wauwatosa, Wis., assignors to Milwaukee Gas Specialty Company, Milwaukee, Wis., a corporation of Wisconsin
Application May 22, 1952, Serial No. 289,242
16 Claims. (Cl. 236—75)

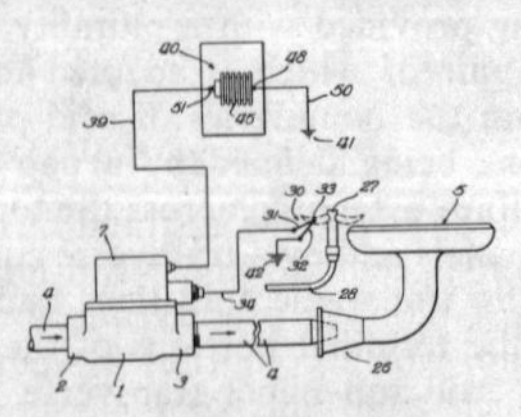

14. Control apparatus comprising, in combination, a low resistance electric circuit, condition responsive means having in said circuit enclosed low resistance contacts at which the resistance of said circuit is varied, a main valve for directly controlling a main flow of fluid, a single thermocouple for energizing said circuit, and an electromagnetic operator under the direct control of said condition responsive means, said electromagnetic operator being energized by electric energy from said thermocouple and acting electrically and directly to operate said valve at values of said resistance below a given value and deenergized to release said valve at resistance values above a given value.

2,717,124

ROLLER MILL WITH PNEUMATIC DISCHARGE-MATERIAL CONVEYOR

Fred Fielden and Frank Murphy, Rochdale, England, assignors to Thomas Robinson & Son Limited, Rochdale, England

Application November 21, 1951, Serial No. 257,455

3 Claims. (Cl. 241—60)

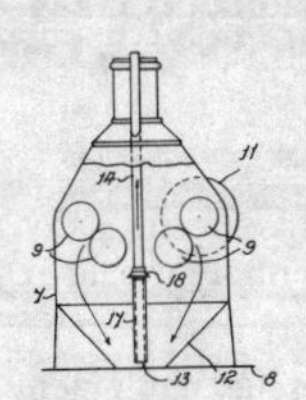

1. A rolling mill comprising a casing having an opening at the top for the stock to enter, pairs of rollers mounted for rotation inside said casing, one roller of each pair being positively driven, a hopper inside said casing and located below said pairs of rollers to receive stock falling from the rollers, and an upwardly directed suction pipe intermediate said pairs of rollers with its lower end terminating in a suction nozzle whose inlet is in communication with the hopper at a point near to but above the bottom of said hopper, said suction pipe having a bend therein to avoid fouling the said stock opening as it passes through the top of the casing.

2,717,125

APPARATUS FOR ADVANCING STRANDS

Vincent A. Rayburn, Baltimore, Md., assignor to Western Electric Company, Incorporated, New York, N. Y., a corporation of New York

Application July 25, 1951, Serial No. 238,546

5 Claims. (Cl. 242—25)

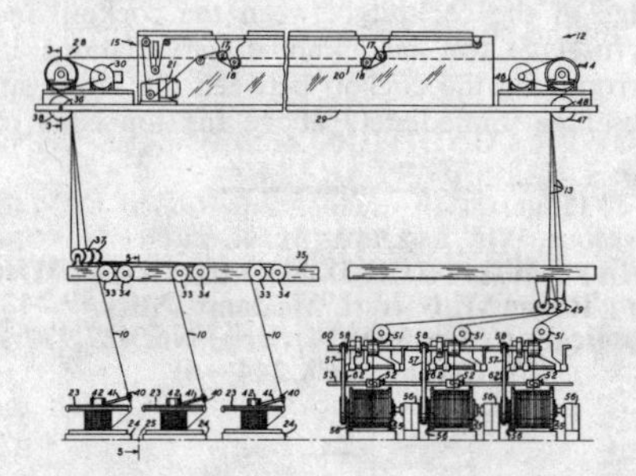

1. Apparatus for advancing a strand of magnetic material at a constant speed and under constant tension from a supply thereof along a predetermined elongated path and to a takeup, which comprises a feeding capstan for withdrawing such a strand from its supply, means for rotating the feeding capstan at a constant peripheral speed, magnetic means provided on the feeding capstan and so designed as to hold the strand magnetically against the feeding capstan with such force that the strand is advanced by said capstan at a linear speed equal to the peripheral speed of the capstan, a tensioning capstan engaged by the strand at the end of said path, and means for rotating the tensioning capstan at such a speed that it exerts a constant force on the strand to create a predetermined tension in the strand between the feeding capstan and the tensioning capstan the magnitude of which normally is insufficient to cause slippage of the strand over the feeding capstan.

2,717,126

CONTROL MECHANISM FOR DRIVES OF BOBBIN WINDING MACHINES

Walter Warren Egee, Yeadon, Pa., assignor to Fletcher Works Incorporated, Philadelphia, Pa., a corporation of Pennsylvania

Application February 12, 1953, Serial No. 336,469

6 Claims. (Cl. 242—39)

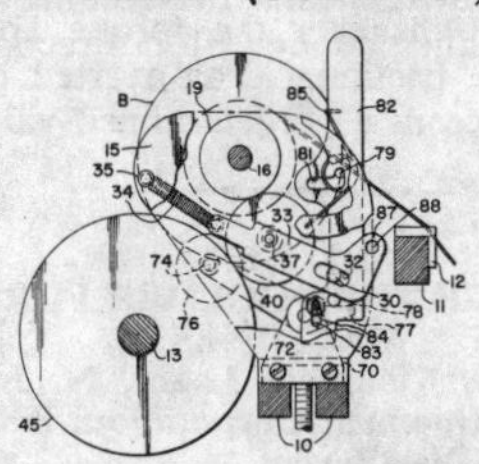

1. In a bobbin winding machine having a frame and means carried by said frame for rotatably supporting a bobbin to be wound, a driving member for driving engagement with a portion of said supporting means and having a peripheral drive surface, a control arm having a pulley for engagement with said peripheral surface, a resilient member for urging said control arm in a predetermined direction towards driving position, a control lever having a slidable pivotal mounting at one end, a latching pin carried by said frame, said lever having an L-shaped slot for engagement with said latching pin, said control arm and said control lever having portions for engagement upon movement of said control lever in a predetermined direction for controlling the positioning of said control arm, and a resilient member for urging said control lever in a direction for effecting movement of said control arm away from driving position upon release from said latching pin.

2,717,127

APPARATUS FOR ADVANCING STRANDS

Vincent A. Rayburn, Baltimore, Md., assignor to Western Electric Company, Incorporated, New York, N. Y., a corporation of New York

Application July 8, 1954, Serial No. 442,015

12 Claims. (Cl. 242—45)

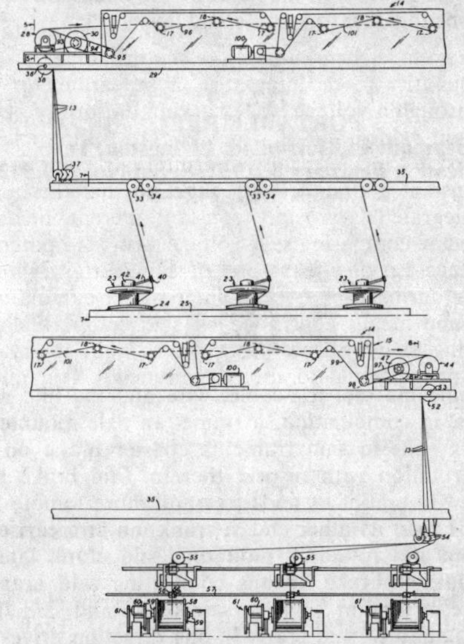

1. Apparatus for advancing a strand from a supply thereof along an elongated tortuous path to a take-up while maintaining a predetermined constant linear speed and a substantially constant predetermined tension in any given portion of the path, which comprises a strand-

metering and hold-back means for withdrawing the strand from its supply and letting it out at a predetermined constant linear speed, and a strand-tensioning capstan for pulling the strand against the drag created by the strand-metering and hold-back means with a force tending to advance the strand faster than said predetermined linear speed, said captsan being designed to yield rotationally whenever the torque applied thereto resulting from the frictional drag exerted by the strand on the capstan exceeds a predetermined value.

2,717,128
YARN PACKAGE TUBE HOLDER FOR WINDING MACHINES
Edward J. Heizer, Mountain Lakes, N. J., assignor to Specialties Development Corporation, Belleville, N. J., a corporation of New Jersey
Application June 24, 1952, Serial No. 295,253
5 Claims. (Cl. 242—72)

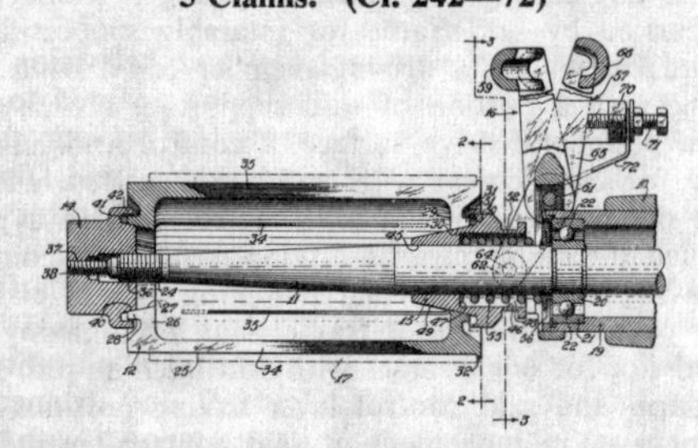

1. In a yarn package winding machine, the combination of a spindle shaft having a free end, a support for the other end of said shaft, a stationary handle on said support adjacent said shaft, means for operating a yarn package tube holder including a member slidably mounted on said shaft, a spring for urging said member in a direction to operate the holder to secure a tube thereon, and manually operable means including a lever pivotally mounted on said support and connected at one end to said member for moving said member in a direction in opposition to said spring to operate the holder to release the tube thereon, said lever having a handle at the other end thereof closely adjacent said first handle to facilitate grasping said handles in the palm of the hand and squeezing the same together to operate said lever.

2,717,129
PORTABLE REEL
Seymour F. McDonald, Livingston, Tex.
Application February 8, 1952, Serial No. 270,542
2 Claims. (Cl. 242—99)

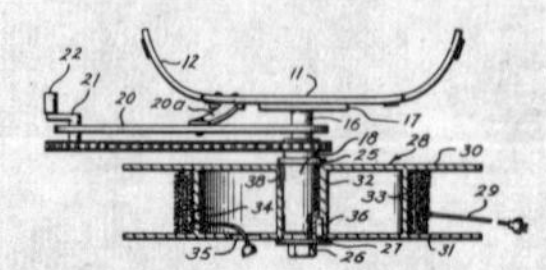

1. A portable reel for cable, wire and the like which comprises, in combination, a frame, an axle mounted by one of its ends to said frame, a coaster brake on said axle for rotation with respect thereto, said brake being disposed with one of its ends free and more remote from said frame than its other end, a crank, an arm carried by said frame and rotatably mounting said crank laterally of said coaster brake, means connecting said crank to said brake at a point between said brake and said frame so that rotation of said crank in one direction drives said brake and in the other direction actuates the brake to cause a braking action, a drum removably mounted on said coaster brake, said drum having an axial portion freely slidable onto said brake from either end of the axial portion, means preventing relative rotation between the drum and brake, and removable fastening means connecting the brake and drum to maintain the drum on said brake.

2,717,130
FLEXIBLE BOAT COAMING
Francis M. Johnson, Dayton, Ohio
Application March 3, 1953, Serial No. 340,171
7 Claims. (Cl. 244—2)
(Granted under Title 35, U. S. Code (1952), sec. 266)

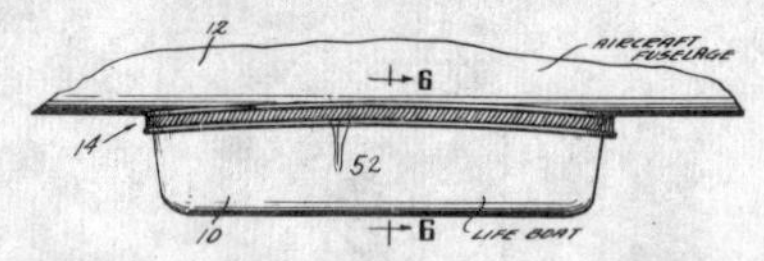

3. Airplane-sea rescue equipment comprising an airplane having a fuselage, a lifeboat "bombed up" to the bottom of the fuselage in closely spaced relation thereto, said lifeboat having a cockpit, a flexible vertically yieldable boat coaming surrounding the top of the cockpit on the inner surface thereof for closing the space between the top of the cockpit and the bottom of the fuselage comprising, a channel shaped receiver fixed to the boat around the top of the cockpit below the top edge thereof, an adjustable flexible elongated closure member vertically slidable in said channel, extending substantially throughout the length thereof, projecting upwardly beyond the top edge of the cockpit toward the said fuselage, spring means confined between the base and side walls of the said channel shaped receiver and closure member urging said closure member vertically upward toward said fuselage, means between the said receiver and said closure member limiting vertical upward movement of said closure member throughout its length toward said fuselage, and a yieldable elongated cushion member fixed on the top of said closure member, throughout its length, in contact with the bottom surface of the fuselage above the top edge of the cockpit, said flexible closure member and said cushion member forming a substantially continuous yieldable closure around the upper edge of the cockpit between the cockpit and the airplane fuselage for accommodating variations in the space surrounding the cockpit between the lower surface of the fuselage immediately above the top edge of the cockpit.

2,717,131
AIRCRAFT WITH FIXED AND ROTARY WINGS
Roger M. Barrett, McComb, Miss.
Application May 6, 1954, Serial No. 427,977
7 Claims. (Cl. 244—6)

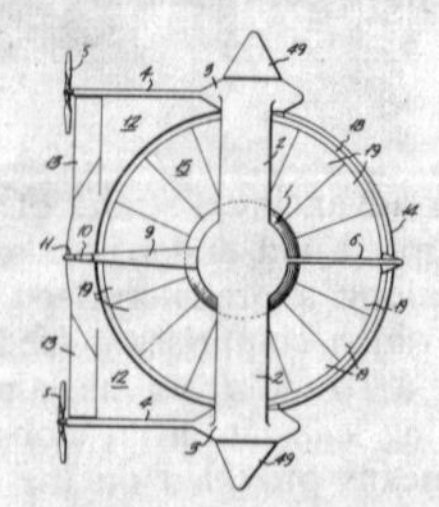

1. Convertible aircraft comprising, a cab, a disk having a plurality of tiltable vanes mounted upon axes radial to the disk and movable to positions at an angle to the horizontal plane of the disk and to form substantially continuous upper and lower surfaces for the disk, said disk being mounted upon and for rotation about said cab, a rigid wing connected to and extending transversely to either side of said cab, means carried by said wing to support the periphery of said disk, means to tilt said vanes, means to rotate said disk and means to propel said aircraft forwardly.

2,717,132
AUTOMATIC APPROACH SYSTEM
James M. Cooper, Schenectady, N. Y., assignor to General Electric Company, a corporation of New York
Application January 7, 1953, Serial No. 329,959
13 Claims. (Cl. 244—77)

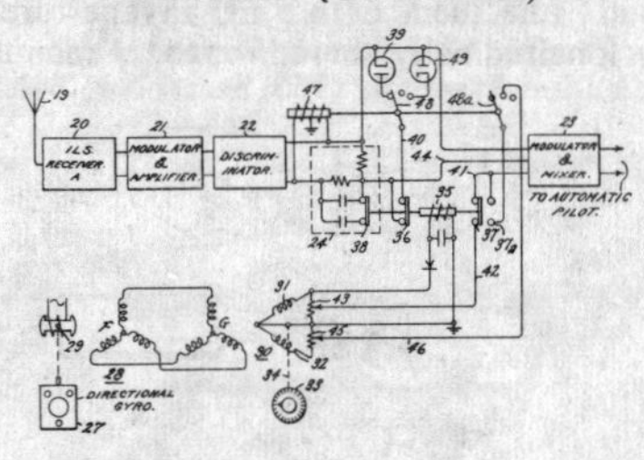

1. In a radio beam controlled automatic approach and landing system for aircraft having an automatic pilot, radio guidance means for detecting the presence of said radio beam and energizing the autopilot to turn the aircraft toward the beam, reference approach means, precalibrated by the pilot, for automatically disabling said radio guidance means upon the aircraft being positioned at a predetermined approach angle toward the beam; and simultaneously therewith automatically energizing the autopilot to maintain this predetermined aircraft approach angle toward the beam, and automatic control means energized by the radio guidance means responsively to the aircraft's reaching the radio beam to energize the autopilot in opposition to said reference approach means for aligning the aircraft with the beam direction, the reference approach means being disengaged upon the aircrafts being diverted from the predetermined approach angle to thereafter enable the radio guidance means to regain control of the aircraft heading.

2,717,133
PARACHUTE PACK
James Gregory, Guildford, England
Application June 29, 1953, Serial No. 364,751
2 Claims. (Cl. 244—148)

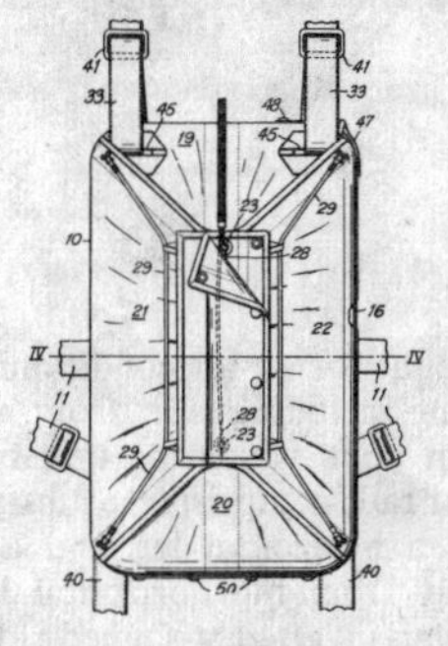

1. A parachute pack comprising a flexible container of substantially rectangular shape, including a plurality of flaps normally connected together to form one wall of said container, an opposite wall adapted for contact with the body of the wearer, a flexible partition located between said walls, a frame secured to said container and extending around said first-mentioned wall substantially in the plane of said flaps, longitudinal rows of loops upon said partition adapted for retaining the rigging lines of a parachute canopy stowed in said container between said first-mentioned wall and said partition, said container being adapted for longitudinal passage of tensile members of a parachute harness between said partition and said opposite wall in alignment with said rows of retaining loops, and means for resisting bulging of said opposite wall due to presence of the parachute canopy stowed in said container.

2,717,134
STAND WITH DAMPING DEVICE FOR PHOTOGRAPHIC OR TELEVISION CAMERAS
Robert Ferber, Courbevoie, France, assignor to Usines Gallus Societe Anonyme, Courbevoie, France, a corporation of France
Application January 6, 1951, Serial No. 204,782
Claims priority, application France March 21, 1950
10 Claims. (Cl. 248—11)

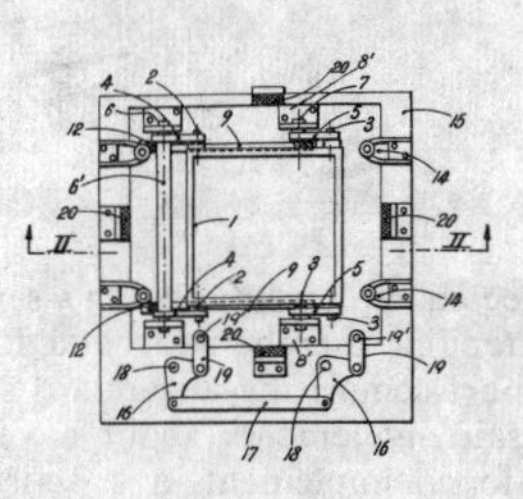

1. A support for photographic and television apparatus and the like, said support being adapted to rest on a platform subject to vibrations to prevent transmission of the vibrations to the apparatus, said support comprising a frame for receiving the apparatus and adapted to be rigidly secured to the apparatus, a base plate adapted to be rigidly secured to the platform, and means for suspending said frame upon said base plate, said suspending means comprising a plurality of supports secured to said base plate, and connecting means flexibly interconnecting said supports with said frame, said connecting means including horizontally pivoted lever arms and rigid supports carrying said lever arms and flexible jointing means positioned to receive the movements of said lever arms, one of said lever arms and said flexible jointing means being connected to said frame, a mechanical connection between said lever arms to compel simultaneous oscillating movement of said arms, and means for compelling the frame to resist rotational movements upon oscillating movement of said arms, said last-named means being pivotally mounted upon said base plate.

2,717,135
RESILIENT SUPPORT FOR LAUNDRY APPARATUS
Peyton W. Douglas, Syracuse, N. Y., assignor to Easy Washing Machine Corporation, Syracuse, N. Y., a corporation of Delaware
Application December 29, 1948, Serial No. 67,897
8 Claims. (Cl. 248—20)

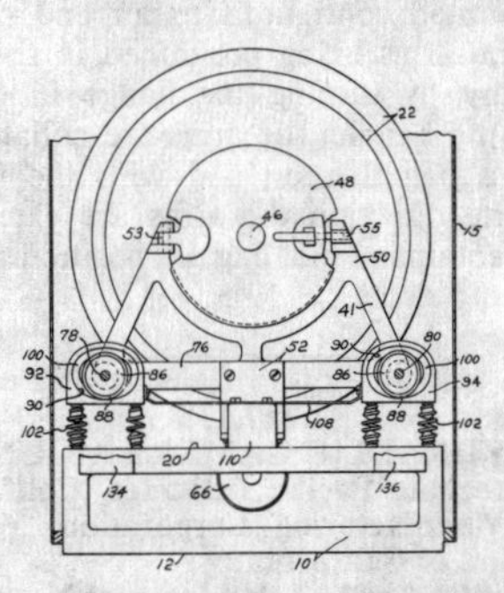

1. In a support for a washing machine having an extracting drum rotatable about an axis out of the vertical and subject to unbalanced loads, a journal supporting carriage for the revolving extractor drum, a stationary frame, a plurality of resiliently supported means comprising arcuate tracks lying in planes substantially transverse to the axis of drum rotation, and rollers in engagement therewith carried by said carriage and frame.

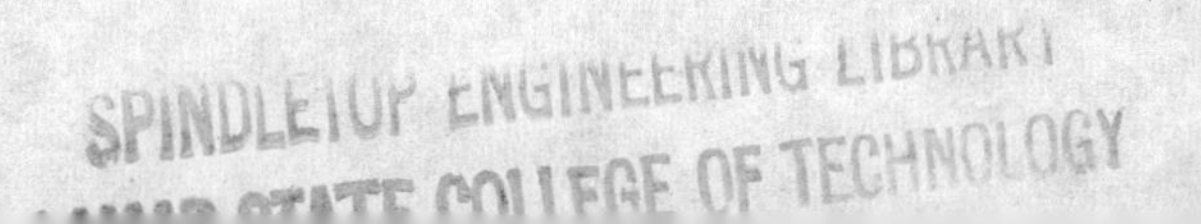

2,717,136
HOSE HOLDER
Clarett B. Greeson, Rockford, Ill., assignor to J. I. Case Company, Racine, Wis., a corporation of Wisconsin
Application September 29, 1949, Serial No. 118,644
6 Claims. (Cl. 248—75)

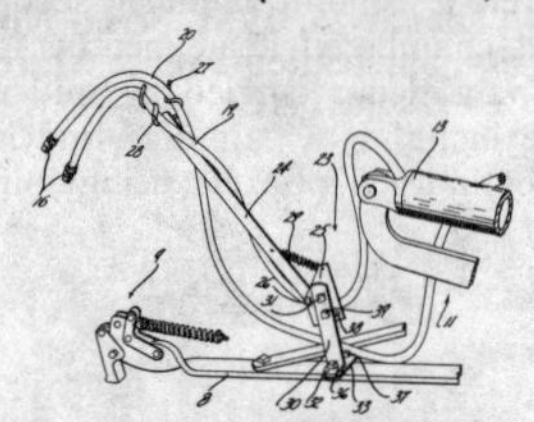

1. On connected implements having a separable hydraulic conduit extending therebetween and fastened thereto, a conduit support comprising a standard swivelly secured with one of said implements about a vertical pivot on the last mentioned implement, a support lever secured with said standard and having a fulcrum for up-and-down movement for supporting said conduit, gripping means for holding said conduit as to longitudinal movement therethrough said gripping means being secured to said support lever, resilient means for continuously urging said support lever and said conduit upwardly about said fulcrum for promptly lifting said condit to an elevated position upon separation thereof from the other implement, said resilient means being anchored between said standard and said support lever, and means for increasing the horizontal moment of force exerted by said resilient means upon said support arm for swinging said lever upwardly.

2,717,137
COTTON PICKING MACHINE
David H. Grogan, Montezuma, Ga.
Application May 19, 1950, Serial No. 162,996
1 Claim. (Cl. 248—99)

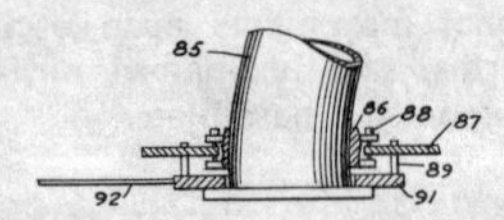

A bag holder for a cotton picking machine having discharging ducts extended from blowers comprising a band having circumferentially spaced bearings thereon, arms pivotally mounted in the bearings of the band, said arms having pins extended downwardly therefrom, a flange including a handle and said flange having radially disposed spaced apart slots therein, said flange being rotatably mounted contiguous to an end of said band, the pins of the arms being positioned in the slots of the flange whereby by turning the flange in one direction by means of the handle, the arms are collapsed to facilitate placing a bag thereon and with the flange turned in the opposite direction the arms are extended to grip the bag for retaining the bag in position for receiving cotton from the ducts.

2,717,138
HYDRAULIC CAMERA MOUNT
James W. Sheehan, Pacific Palisades, Calif., assignor to Arcturus Manufacturing Corporation, Venice, Calif., a corporation of California
Application April 22, 1954, Serial No. 424,879
3 Claims. (Cl. 248—183)

1. In an arrangement of the character described for controlling the movement of a camera support which is rotatably supported on a supporting base, a stationary circular bearing member extending upwardly from said base, a hydraulic cylinder, said bearing member passing centrally through said cylinder, a vane affixed to said bearing member, said vane having opposite ends thereof conforming respectively with the circular bearing member and the circular inner wall of the cylinder, and a second vane member mounted on said cylinder and movable therewith, said second vane member having an apertured portion extending transversely therethrough, said second vane member having an apertured portion extending longitudinally therethrough, a manually oper-

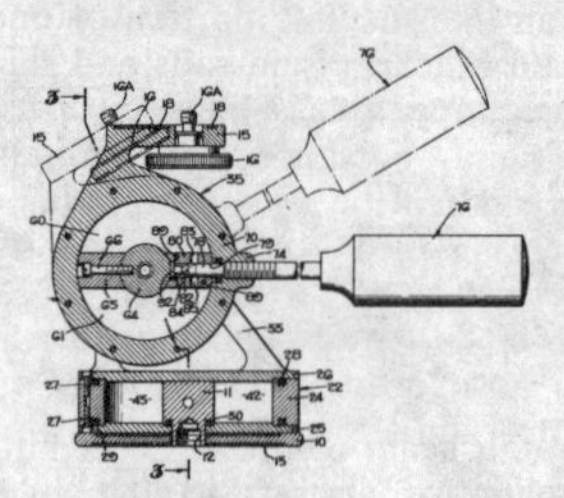

able valve member adjustably mounted on said cylinder and extending into said centrally apertured portion of said second vane member for controlling the flow of fluid through said vane member, said second vane member including a movable brake element which is engageable by said valve element and movable therewith into engagement with said bearing member.

2,717,139
MOUNTING MEANS FOR ROOM AIR CONDITIONER
Bernard W. Jewell, Wichita, Kans., assignor to The O. A. Sutton Corporation, Inc., Wichita, Kans., a corporation of Kansas
Application December 11, 1951, Serial No. 261,000
1 Claim. (Cl. 248—208)

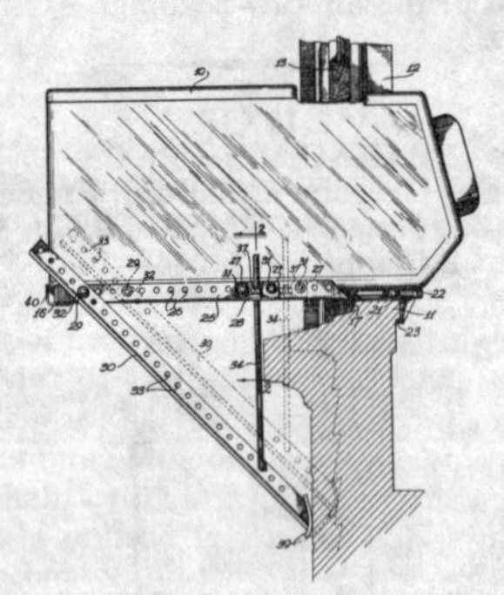

A window mount for an air conditioning unit comprising a metal frame including side, front and rear rails defining a platform upon which the unit may rest, each of said side and rear rails comprising a substantially L-shaped member having a horizontal leg and a vertical leg extending upwardly from the horizontal leg, each of said side rails having an inverted V-shaped ridge formed therein whereby an air conditioning unit may be supported upon said ridges and will be confined upon said frame by the vertical legs of said side rails and said rear rail, said front rail comprising a wide, horizontal portion adapted to rest upon a window ledge and a downwardly directed flange adapted to engage the outside of the window ledge, adjustable clamping means mounted on said front rail comprising a pair of elongated, threaded rods each extending into one of the grooves provided by the ridges of said side rails, a pair of threaded blocks fixed to said front rail in alignment with said ridges and each cooperating with one of said elongated, threaded rods whereby said rods may be adjusted longitudinally with respect to said side rails, and hook means pivotally mounted on the outer end of said threaded rods and adapted for engagement with the inside of the window ledge whereby the frame may be clamped to a window ledge in outwardly projecting relation to the window opening, a pair of braces each pivotally connected along a horizontal axis to an outer portion of one of said side rails and projecting diagonally downward and inwardly therefrom to engage the outer surface of the wall beneath the window, a pair of brackets each secured to an intermediate portion of one of said side rails, a pair of vertical tie rods each pivotally connected to and connecting an intermediate portion of one of said braces to one of said brackets, said rails and said brackets comprising cooperating, releasable securing means whereby said brackets may be shifted longitudinally of said rails, said braces and said tie rods comprising cooperating, releasable securing means whereby said tie rods may be shifted to various relative positions with respect to said braces, each of said brackets comprising a horizontal portion having an aperture therethrough, and each of said tie rods comprising a threaded portion projecting through the aperture of its associated bracket, and a nut threadedly engaging said threaded portion and engaging the upper surface of said horizontal portion, said braces being independent of each other whereby said nuts may be independently adjusted to vary the angular relationship of one brace with respect to its associated side rail without affecting the other brace.

2,717,140
CLAMP FASTENERS
Alexander B. Hulse, Jr., Dallas, Tex.
Application July 23, 1949, Serial No. 106,421
7 Claims. (Cl. 248—223)

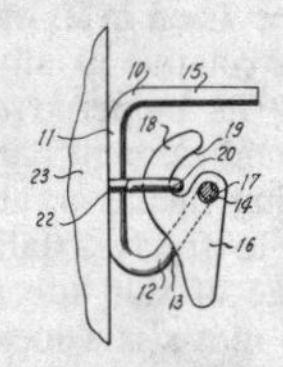

2. A clamp fastener including, a frame member having a base portion and resilient arms extending outwardly from said base portion, a lever member pivotally mounted on said arms at a point spaced from the base of the frame, said lever member being disposed to swing on said pivotal mounting in a plane normal to the base portion of the frame member, a hook portion formed on the lever arm spaced from the pivotal mounting and having a spiral wedge surface converging inwardly toward and facing said pivotal mounting and a latching portion at its inner end concentric with said pivotal mounting, and a latch member having an opening therein adapted to receive the hook portion of the lever arm and engage the wedge surface thereon whereby the lever arm may be swung to draw the latch member toward the pivotal mounting of the lever arm, the concentric latching portion of the wedge surface of the hook of the lever arm being disposed to engage the latch member for preventing undesired pivotal movement of the lever member after said latch member has been drawn by the wedge surface toward the pivotal mounting of the lever arm.

2,717,141
ADJUSTABLE SUPPORTS PROVIDING UNIVERSAL MOVEMENT
Harry F. Livingston, New York, N. Y.
Application May 17, 1951, Serial No. 226,821
10 Claims. (Cl. 248—278)

10. An angularly adjustable support providing for controlled movement with axes angularly related with respect to each other which comprises, a barrel member formed with screw threaded portions adjacent the ends thereof, closing members, formed with openings through the bases thereof, mounted in screw threaded engagement with said screw threaded portions of said barrel, a spring extending throughout the length of said barrel and engaging the inner surfaces of said closing ends, coacting means between said barrel and one of said ends for limiting the relative rotation thereof, a chambered head member overlying one of said ends on the remote side thereof from said barrel, a conduit member secured within an opening in said chambered head member and extending away therefrom, a sleeve carried by said conduit member, said sleeve including a pair of cylindrical pockets at its ends, said cylindrical pockets being joined by a web

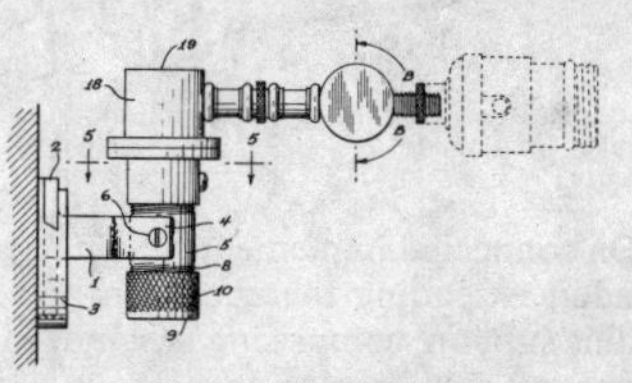

portion of substantially less circumferential extent than the circumference of said sleeve, said conduit being engaged with said web portion, a collar overlying said sleeve, the center portion of said collar being recessed circumferentially for a substantial portion of the circumference thereof to embrace said conduit, spring members in said cylindrical pockets and screw threaded cap members engaged with the ends of said collar, the interiors of said cap members engaging said springs to press the same against the base of said pockets.

2,717,142
SCALES FOR THE WEIGHING OF LIQUIDS
Cecil Walter Murray, Holly Bank, Repton, England, assignor to George Fletcher & Co. Limited, Derby, England, a British company
Application June 21, 1951, Serial No. 232,698
Claims priority, application Great Britain August 2, 1950
28 Claims. (Cl. 249—3)

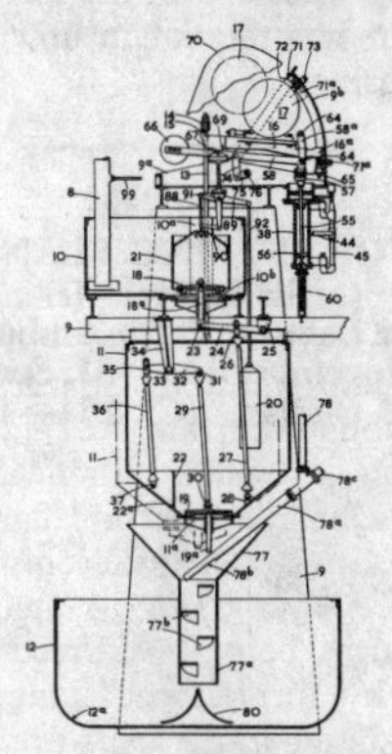

1. In a liquid weighing scale of the type described, the combination of a stationary supply tank having an outlet opening provided with a valve seat, a lift valve adapted to be lifted and lowered bodily from and to said seat to open and close said outlet opening, said lift valve serving as a delivery valve to a weigh tank, a counterweighted weigh beam, said weigh tank being suspended from the beam and disposed below the supply tank, said weigh tank having a discharge opening provided with a valve seat, a lift valve adapted to be lifted and lowered bodily from and to said seat to open and close said discharge opening, said lift valve serving as a discharge valve from the weigh tank, a fixed fulcrum above the weigh tank, a lever pivoted on said fulcrum and to said delivery valve and a weigh tank connected link pivoted to the lever at a point nearer the fulcrum than the distance between the fulcrum and the connection with the delivery valve to move the said valve bodily relative to its seat a greater distance than the travel of the weigh tank.

2,717,143
WEIGHING AND PACKAGING MACHINE
George L. McCargar, Chicago, Ill.
Application December 13, 1952, Serial No. 325,871
5 Claims. (Cl. 249—19)

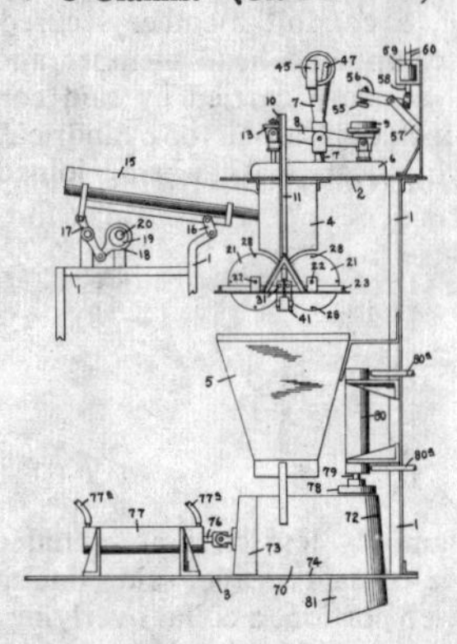

1. A weighing machine comprising, a weighing scale, a supporting bracket hung on said weighing scale, two weighing wheels rotatably mounted on said bracket with their axis in substantially the same horizontal plane and with their peripheries closely adjacent each other, each of said weighing wheels having a plurality of outwardly opening compartments, means for continuously feeding material into upwardly opening adjacent compartments of both of said weighing wheels, said weighing wheels and connected parts lowering by gravity upon accumulation in said compartments thereof of a predetermined weight of material, means for holding said weighing wheels against rotation when in elevated position, means for releasing said weighing wheels for rotation upon said lowering thereof, said holding means becoming reactive to hold said weighing wheels with the next in rotation of the adjacent compartments thereof in upwardly opening, material receiving position.

2,717,144
DEVICE FOR SETTING PLANKING AND SHEATHING
Walter Labuza, Perth Amboy, N. J.
Application November 13, 1951, Serial No. 255,956
3 Claims. (Cl. 254—16)

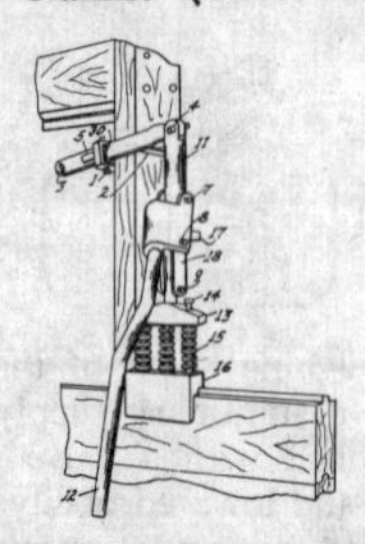

2. A board jack comprising a supporting member having a socket extending inwardly from one end thereof; means for anchoring the supporting member at a position displaced transversely from the location of the supporting member, said means including an arm pivotally mounted to the supporting member, a first clamping element fixedly secured to the arm and a second clamping element slidably mounted on the arm, said second element having limited pivotal movement on the arm; a connecting rod slidably mounted in the socket in the supporting member; a board engaging member connected to the rod; spring means interposed between the rod and the board engaging member permitting relative movement therebetween; and means connecting the supporting member and the connecting rod whereby movement of the rod within the supporting member in a certain direction will transmit a pressure on the board-engaging member through the spring means and cause the anchoring means to be secured at the displaced position.

2,717,145
STRETCHER JACK
James M. Andrew, Findlay, Ohio
Application October 12, 1953, Serial No. 385,533
6 Claims. (Cl. 254—98)

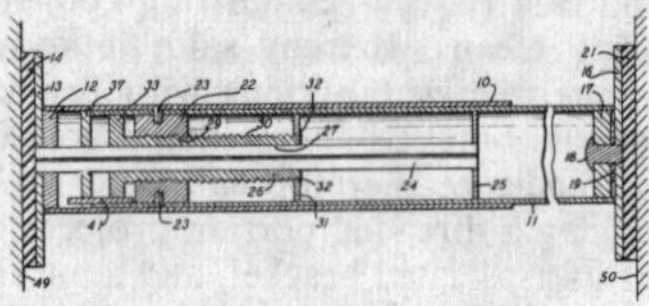

1. A stretcher jack comprising a pair of telescoped tubes adapted to be freely axially slidable with respect to each other and adapted to be forced in opposite directions by relative rotation of the tubes, a guide rod fixedly mounted against rotation and having a non-circular shape in cross section, means associated with one of said tubes for forcing said tubes in opposite directions axially by the relative rotation of said tubes and comprising an element having a through non-circular bore through which said rod extends and at all times interlocks so that said element is fixed against rotation, locking means for preventing the free relative axially slidable movement of said tubes when in its operative position and for permitting the free relative axially slidable movement of said tubes when in its inoperative position, and coupling means between said element and said locking means for moving said locking means to its operative position when said tubes tend to freely axially slide with respect to each other, said one of said tubes adapted to engage said locking means to move said locking means to its inoperative position when said tubes are relatively rotated.

2,717,146
HEAVY DUTY FLEXIBLE DRILL PIPE
John A. Zublin, Los Angeles, Calif.
Application April 9, 1953, Serial No. 347,665
6 Claims. (Cl. 255—28)

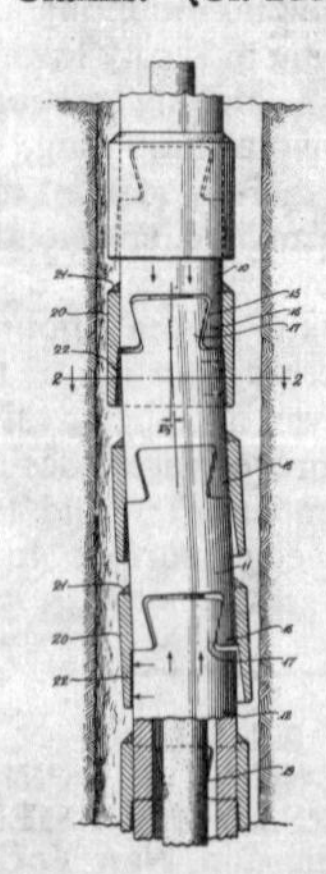

1. A heavy duty flexible drill pipe comprising an elongated tubular member subdivided into a plurality of sections of rigid pipe of substantial wall thickness in end to end relationship with a plurality of teeth and complementary recesses on the opposite ends of the intermediate sections, the teeth of one section being positioned in the recesses and loosely interlocking the teeth of the adjacent sections with appreciable clearance to form a loose joint permitting limited relative angular movement in

any direction between the sections, a tubular member coaxially secured to each of said sections and projecting beyond the teeth at one end thereof, the extended portion of said tubular member overlapping and being spaced from an end portion of the next adjacent section to be contacted by and limit the relative angular movement of said next adjacent section.

2,717,147
MIXING MACHINES FOR MAKING CONCRETE OR THE LIKE

Erik Valdemar Fejmert and Bernhard Valdemar Fejmert, Nykoping, Sweden; said Bernhard Valdemar Fejmert assignor to said Erik Valdemar Fejmert
Application August 19, 1952, Serial No. 305,243
Claims priority, application Sweden August 31, 1951
4 Claims. (Cl. 259—178)

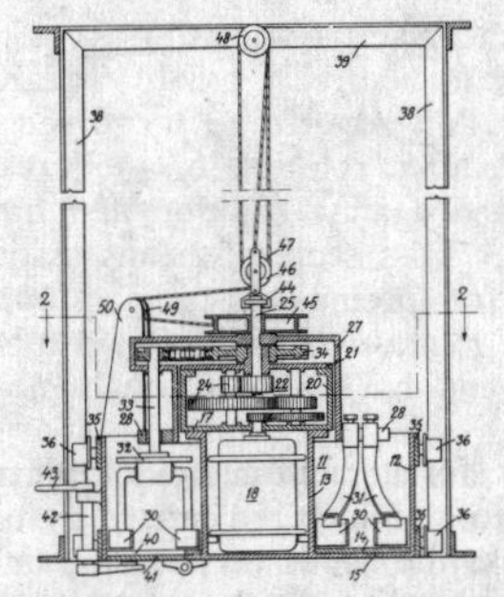

1. A mixing machine for concrete or the like comprising an annular tank having an open top and having an inner annular wall, an outer wall and a base plate closing the bottom of the tank, mixing devices in the tank, a prime mover, gearing driven by said prime mover, a housing completely enclosing said prime mover and isolating it from the contents of the tank and comprising a portion of said base plate, said inner wall and an upper wall mounted on said inner wall, a casing completely enclosing said gearing and comprising a side wall and a top wall, the upper wall of said housing forming the bottom of said casing, and means connecting said gearing to said mixing devices for operating said mixing devices.

2,717,148
AIR CLEANER AND HUMIDIFIER

Michael Frank Hall, Huntington Park, Calif.
Application November 19, 1953, Serial No. 393,190
2 Claims. (Cl. 261—16)

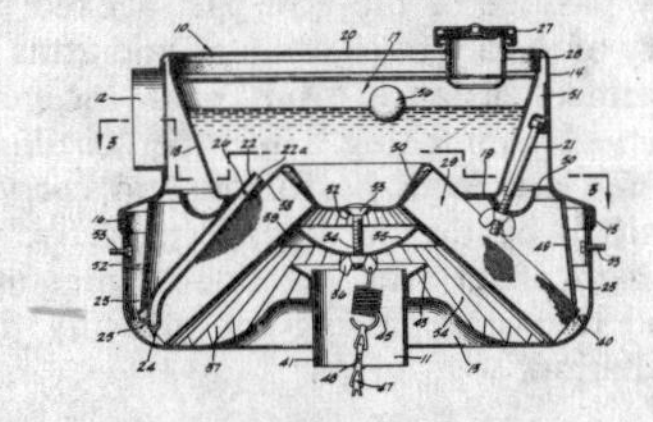

1. In a humidifier and air cleaner for the air intake of an internal combustion engine, the combination of: a cover and a pan cooperating to define an enclosure, a water reservoir formed within said enclosure, a conical filter element positioned within the enclosure and below the water reservoir, a metering tube adapted to deliver water from the reservoir to form a pool in the lower portion of the pan and to regulate the depth of the pool, the lower portion of the filter element projecting into said pool so that the filter element may be saturated by capillary action, means on the pan forming a central air inlet into the space within the filter element, whereby the air is caused to pass through the interstices of the filter element, a skirt on the cover extending into the interior of the pan adjacent the filter element, the cover having supplementary air inlet openings communicating with the space between the skirt and the pan, and means on the cover forming an outlet for delivery of moist air.

2,717,149
FLUID FEED DEVICE

David E. Anderson, Cleveland Heights, Ohio, assignor to Thompson Products, Inc., Cleveland, Ohio, a corporation of Ohio
Application October 9, 1951, Serial No. 250,481
7 Claims. (Cl. 261—18)

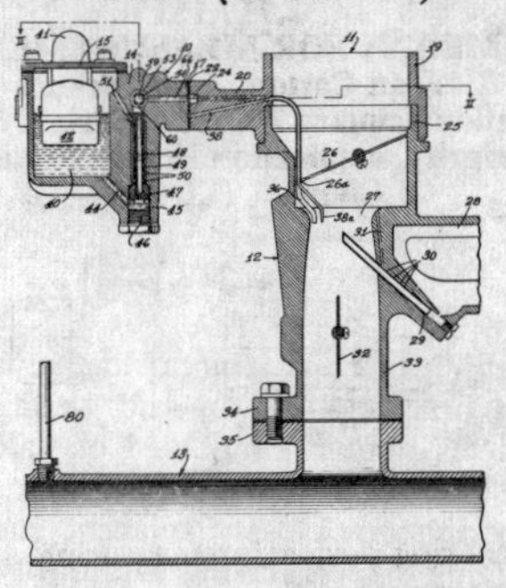

4. In a device for metering fuel to the fuel intake of an internal combustion engine, a body having a horizontal bore extending inwardly from one side thereof, a fuel supply chamber in communication with an inner portion of said bore and a fuel outlet passage in communication with a portion of said bore spaced from said inner portion thereof, means defining a valve seat intermediate said spaced portions inside the bore, a valve member movable against said seat, removable means for closing said bore at said one side of said body, spring means between said removable means and said valve member for urging said valve member against said seat, said removable means bein adjustable to adjust the applied pressure of said spring means, a cap on the opposite side of said body, a diaphragm between said cap and said body, means for subjecting said diaphragm to engine intake vacuum, and means between said diaphragm and said valve member for actuating said valve member against the action of said spring means and thereby controlling the fuel flow in response to engine intake vacuum.

2,717,150
CHOKE VALVE ASSEMBLY

Kenneth C. Agar, Ann Arbor, Mich., assignor to Clinton Machine Company, Clinton, Mich., a corporation of Michigan
Application July 6, 1951, Serial No. 235,406
4 Claims. (Cl. 261—64)

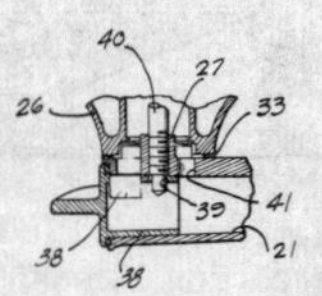

3. In a fuel and air mixing device of the class described, a body having a passage formed therethrough for flow of fluid therethrough, said body having adjacent one end an opening formed therethrough to communicate with the passage therein; an air cleaner bowl mounted on said body adjacent one end thereof and having an opening formed therein in communication with the open-

ing in said body; a choke valve rotatably mounted in the end of said body and serving as a closure for said end, said choke valve having a circumferential slot formed therein; a threaded stem for securing said bowl on said body and threaded at one end through said body and extending at said end into said slot for preventing axial movement of said choke valve, said choke valve having a cutaway portion movable into and out of registration with said opening in said body for controlling communication of the passage through said body with the interior of said air cleaner.

2,717,151
MEANS FOR MEASUREMENT
Jean Carton, Paris, France
Application January 10, 1950, Serial No. 137,877
Claims priority, application France January 18, 1949
8 Claims. (Cl. 265—7)

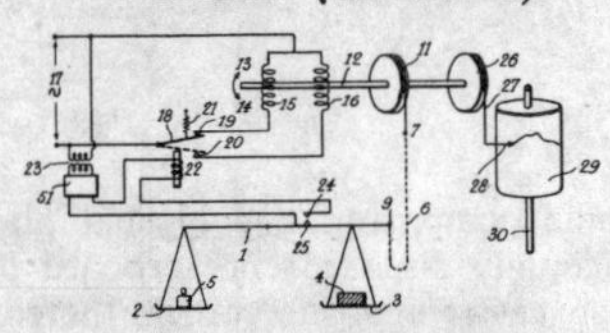

1. A measuring device for measuring a physical magnitude by comparison with a reference physical magnitude, the said device comprising a movable member arranged to hunt about and pass through a position of equilibrium indicative of a condition of balance between the two magnitudes, reversible movable means operatively connected with the movable member and arranged to exert a moving force upon the movable member for moving the same in one or the other direction relative to said position of equilibrium, and control means including contact means controlled by the position of the movable member and controlling the direction of movement of said reversible means, the said contact means being actuated for reversal of the reversible means by the movable member moving into a position of unbalance on one or the other side of said position of equilibrium, the said reversible means being arranged to be continually moving during a measuring operation thereby causing the movable member continually to swing through a range including said position of equilibrium, the median position of the movable member being indicative of a value of the magnitude to be measured.

2,717,152
REVERSED SPRING SUSPENSION FOR VEHICLE BODIES
Roscoe B. Hopkins, Lewiston, Idaho, assignor of one-half to Art L. Ketchum, Lewiston, Idaho
Application January 21, 1953, Serial No. 332,161
1 Claim. (Cl. 267—20)

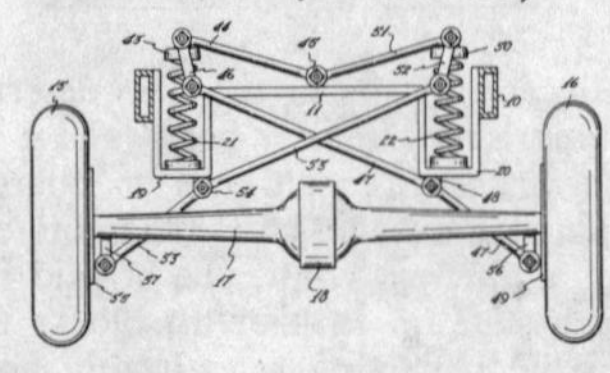

A wheel suspension to suspend a vehicle frame between two road wheels on their wheel mounts comprising a cross frame piece having a spring seat at each end, coiled springs on said seats, spring caps on said springs, links pivoted on the caps and pivoted to the frame to hold said springs upright in place, two oppositely disposed frame supporting levers, each having one end suspended by a link from one of the caps and the other end suspended by a link from the wheel mount most remote from that cap, and a frame supporting pivot on each lever adjacent to the wheel mount supported end thereof.

2,717,153
TIMING MECHANISM FOR AUTOMATIC FOLDERS
Harry D. Abell and Homer E. Abell, St. Albans, and Norman R. Heald, Chester Depot, Vt., assignors, by direct and mesne assignments, to Fisher & Christen, Washington, D. C., a firm
Original application April 27, 1949, Serial No. 89,904, now Patent No. 2,709,585, dated May 31, 1955. Divided and this application September 25, 1952, Serial No. 311,499
2 Claims. (Cl. 270—81)

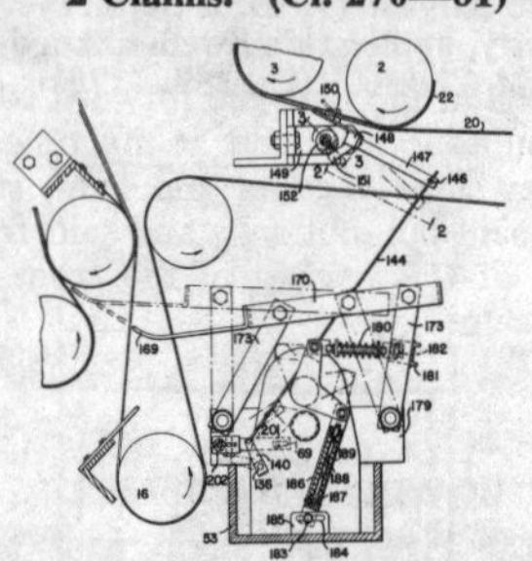

1. A timer for an automatic folding machine wherein flat goods to be folded are carried by tapes, and a folding blade tucks the goods between a pair of folding rolls comprising a plurality of rotatably mounted control arms having a common axis of rotation, a coaxial positively driven rotary positive drive member for intermittently driving said control arms, a coaxial rotary friction drive member for intermittently driving said control arms, means for driving the friction drive member optionally at one-half or two-thirds the speed of the positive drive member and a self locking escapement at a fixed point of the periphery of the arc of rotation of said control arms for releasing said arms one at a time to be driven by said driving means, said arms having means engageable by said escapement, means for actuating said escapement to release an arm when the front edge of the article reaches a fixed point, said last means including a feeler for said flat goods positioned between the tapes ahead of the blade, means for maintaining an arm in driving engagement with the friction member while the article is passing said feeler and in driving engagement with the positive driving member thereafter, and a mechanical trip for starting the folder actuating means, said trip being activated by passage of the arm past a fixed point along the periphery of the arc of rotation of said arms and spaced a predetermined distance from said escapement, and said escapement comprising a member oscillated by the passage of the article to be folded and carrying a pair of wings shifted by the oscillation respectively into and out of the path of an arm, the line of pressure between the arm and the wing passing substantially through the axis of oscillation.

2,717,154
SHEET FEEDER
James R. Wood, Cleveland, Ohio, assignor to Harris-Seybold Company, Cleveland, Ohio, a corporation of Delaware
Application January 6, 1951, Serial No. 204,756
6 Claims. (Cl. 271—12)

6. In mechanism for feeding sheets from a pile selectively in a stream or sheet by sheet, sheet separating means having a sheet forwarding action operated at constant speed, a feed board, front stops at the forward end of said feed board, means for conveying sheets over said feed board, pull-out rolls cooperating with said conveying

means at the rear end of said feed board, drive means for said conveying means and pull-out rolls adjustable to a constant low speed for stream feeding or to a non-uniform speed for sheet by sheet feeding, said constant low speed corresponding to the speed of operation of said sheet separating means, said non-uniform speed drive operating

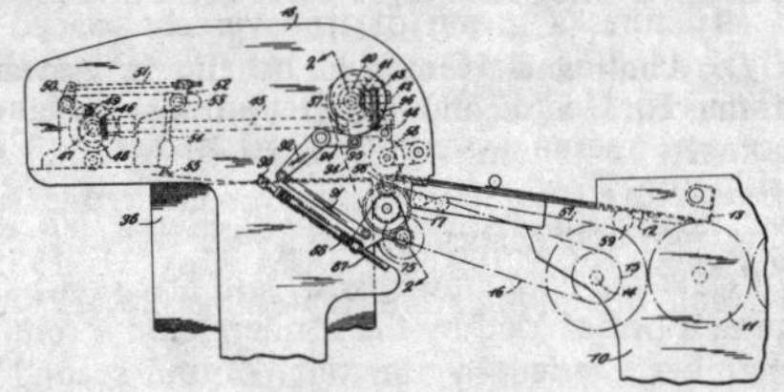

at said low speed at the time the forward edge of a sheet engages said pull-out rolls and at the time the forward edge of another sheet engages said front stops, the distance between said pull-out rolls and said front stops being a multiple of the spacing of the front edges of the sheets in the stream.

2,717,155
BOWLING BALL LIFT
William F. Huck, Forest Hills, N. Y., assignor to The Brunswick-Balke-Collender Company, a corporation of Delaware
Application September 18, 1950, Serial No. 185,407
13 Claims. (Cl. 273—49)

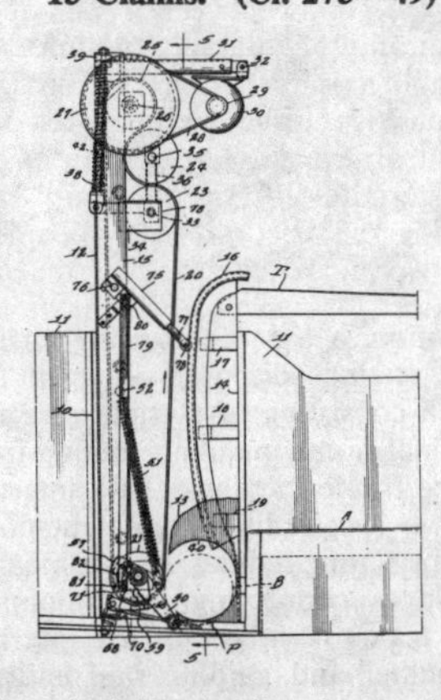

1. A bowling ball lift having, in combination, an upright casing, a pair of substantially parallel and curved ball tracks positioned vertically in said casing, an endless belt having one run positioned along said tracks and operable when driven to roll a ball over the tracks, means for supporting and driving said belt, and means for inserting a ball between said belt and the lower portion of said tracks comprising an actuator having a pair of arms pivotally supported in said casing and normally occupying a lower inoperative position adjacent the bottom of the casing and vertically below said tracks, means for rotating said actuator arms in a direction to move a ball into engagement with said tracks and belt, and a latch device normally holding said actuator arms in their lower position and having a trip arm normally positioned to be engaged by a ball rolled through an opening in the casing and cause operation of said actuator arms and means for resetting the actuator arms after an operation thereof comprising a reset lever pivotally supported in said casing adjacent the upper end of said tracks and with a portion extending into the path of a ball passing over the tracks, a link connected at one end to said lever and at its other end having a lost motion connection with said actuator arms, said lost motion connection permitting the reset lever to swing down by gravity to its operative position after a reset operation and said link operating to reset the actuator arms when the reset lever is actuated by a ball rolling up over the tracks.

2,717,156
EDUCATIONAL GAME APPARATUS
George E. Nelson, Westwood, N. J.
Application June 26, 1952, Serial No. 295,669
1 Claim. (Cl. 273—134)

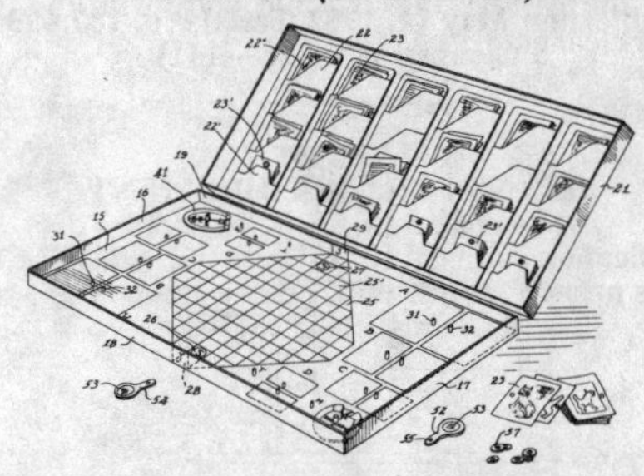

A quiz game comprising a polygonal playing board having a diagonally disposed checkerboard play area, a pair of goal spaces at opposite corners thereof, a pair of card displaying areas oppositely disposed adjacent the remaining corners of said play area, a plurality of pairs of adjacent pins in each of said card displaying areas, detachable arcuate guide members connecting each of said pairs of pins, a set of punched cards on each of said pairs of pins, the cards of each of said sets having on one side indicia of one category to be answered and on the other side answer indicia, a spinning element in each of said card displaying areas having a plurality of card category denoting segments below said spinning elements, and a plurality of distinctive tokens to be moved on said play area toward said goals as question indicia are correctly identified.

2,717,157
EDUCATIONAL GAME EQUIPMENT
Stanley A. Dylewski, Baltimore, Md.
Application June 26, 1952, Serial No. 295,621
3 Claims. (Cl. 273—135)

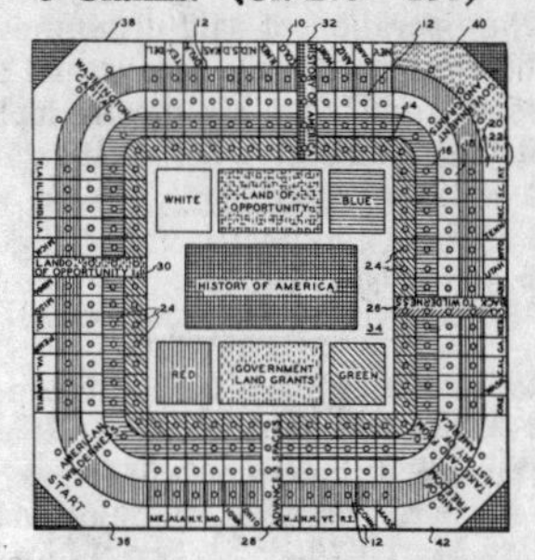

1. In an education game of the class described, player equipment comprising a master board and a plurality of individual player boards, said master board having a plurality of annular rows of blocks thereon, each of said rows being of a different color, one of said rows having the names of a state on each of the blocks therein, and the blocks in the remaining rows having holes therein, the blocks having holes being arranged so that one block of each color is aligned with each block having the name of a state thereon, said names corresponding to states outlined on a map printed on each of said player boards, each of said player boards being provided with parallel rows of state names corresponding to the names on said master board, a plurality of differently colored blocks, each having a hole therein, aligned with each of the names in said parallel rows, and indicating means place-

able into one of the holes on the master board, whereby an individual player board is used by an individual player to record the result of his individual play with said indicating means on said master board.

2,717,158
RANDOM SELECTOR FOR AMUSEMENT DEVICE OR THE LIKE
Joseph O. E. Dieterich, Jamaica, N. Y.
Application May 14, 1952, Serial No. 287,673
5 Claims. (Cl. 273—141)

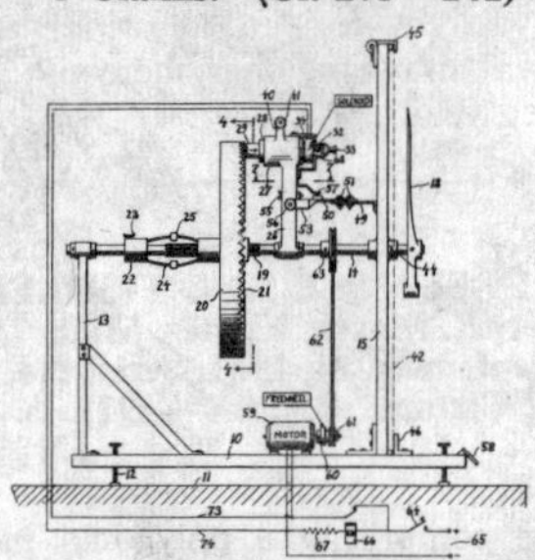

1. A random selector comprising a movable and a stationary indicator member positioned adjacent each other for relative movement through a plurality of operative relative positions separated by inoperative positions, a serrated element, an indexing element adapted for camming engagement with the serrations of said serrated element, one of said elements being positively coupled with said movable member, means for preventing the other of said elements from following the movement of said one of said elements, said serrations by their engagement with said indexing element defining respective ones of said operative positions of said members, means yieldingly urging said elements into camming engagement with each other, electromagnetic decoupling means operatively connected with one of said elements and energizable to withdraw the latter from the other of said elements, drive means for displacing said movable indicator member with respect to said stationary indicator member, circuit means coupled with said drive means for energizing said electromagnetic decoupling means during at least part of the operation of said drive means, and coupling means between said drive means and said movable indicator member, said coupling means including a free wheeling mechanism enabling continued relative displacement of said indicator members, against the camming action of said elements, by inertia following de-activation of said means and of said decoupling.

2,717,159
EXTENSIBLE SLED
Clifford A. Thomas, Cumberland, Md., assignor of one-half to Richard Diamond, Cumberland, Md.
Application March 20, 1953, Serial No. 343,733
2 Claims. (Cl. 280—12)

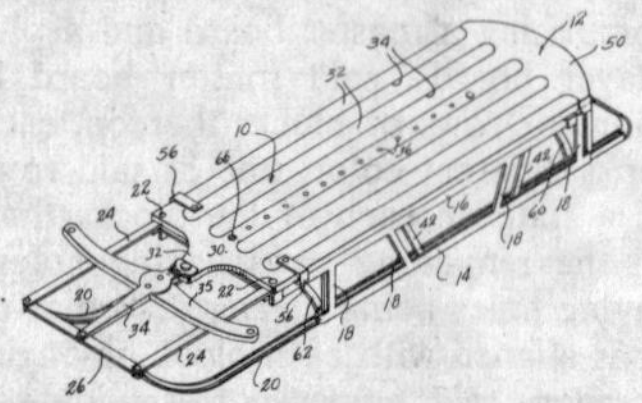

1. An extensible sled including spaced parallel runners each comprising a pair of vertically spaced channels having their open sides in opposed relation, struts carried by the channels and extending therebetween to define with said channels a trussed structure, a second pair of vertically spaced channels having their open sides in opposed relation, struts carried by the second mentioned channels and extending therebetween to define with said second pair of channels a second trussed structure capable of telescoping into the first mentioned trussed structure with the channels of the second mentioned trussed structure slidably fitting into the channels of the first mentioned trussed structure, a group of transversely spaced longitudinally extending slats carried by the first mentioned trussed structure, a second group of transversely spaced longitudinally extending slats carried by the second trussed structure, the slats of the second group of slats extending into the spaces between the slats of the first group of slats and cooperating with said first group of slats in forming a bed carried by the runners, and a reinforcing cross member carried by the slats of the second group of slats and extending transversely across the sled beneath the first group of slats for guiding said slats and cooperating with them in supporting weight imposed on the bed.

2,717,160
FISHING SHELTER
John B. Schmidt, Minneapolis, and Harold J. Hanson, St. Paul, Minn., assignors to Farmgard Products Company, Minneapolis, Minn., a partnership
Application February 3, 1954, Serial No. 407,862
2 Claims. (Cl. 280—20)

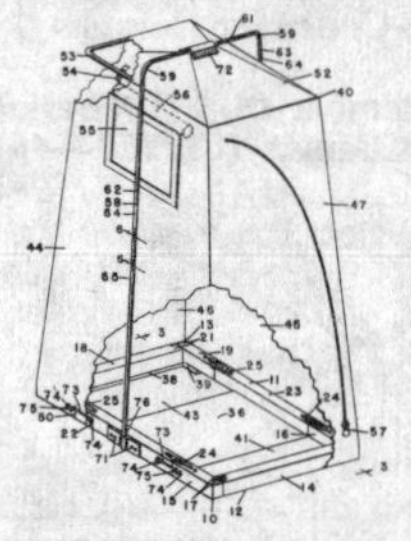

2. In combination, a base comprising an intermediate section and end sections disposed adjacent thereto, said end sections each consisting of a transverse frame member and longitudinal frame members issuing from the ends of said transverse frame members, said intermediate section including two longitudinal frame members lying in continuation of the longitudinal frame members of the end sections, hinges connecting the adjoining ends of the longitudinal frame members of the intermediate and end sections together and guiding said end sections for movement from positions in which the longitudinal frame members of said intermediate and end sections lie in continuation of one another to positions in which the longitudinal frame members of said end sections become disposed at right angles to the longitudinal frame members of the intermediate section, U-shaped runners having spaced parallel legs disposed beneath the longitudinal frame members of said end sections, fasteners extending through said legs and the longitudinal frame members of said end sections for securing the runners thereto and inwardly of the ends of said frame members, said runners upon disposition of the end sections in the plane of the intermediate section lying substantially in the plane of the intermediate section and upon disposition at right angles thereto extending below the intermediate section and in ground engaging position, and a flexible enclosure carried by and extending upwardly from said base.

2,717,161
COLLAPSIBLE WHEELBARROW OR HAND TRUCK
James V. Orr, Pullman, Wash.
Application April 28, 1954, Serial No. 426,248
3 Claims. (Cl. 280—36)

1. In a collapsible wheelbarrow, the combination which comprises a platform the sides of which are di-

verged from the forward end to the rear end, a wheel positioned below the platform and rotatably mounted on the forward end thereof, stationary tubes having longitudinally disposed slots with transversely positioned end sections in inner surfaces thereof mounted on the sides of the platform, telescoping tubes slidably mount-

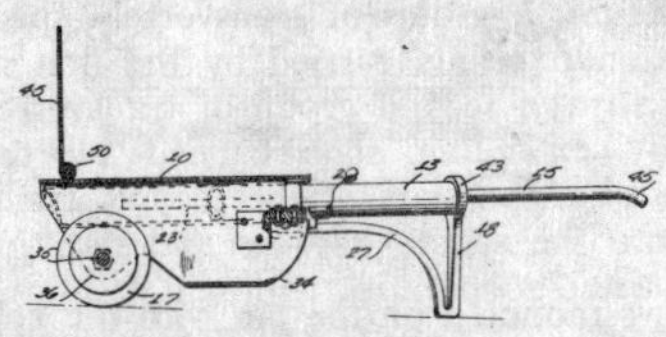

ed in the stationary tubes, rods providing handles slidably mounted in the telescoping tubes, legs carried by the telescoping tubes and having arms with projections on extended ends thereof positioned with the projections in the slots of the stationary tubes, and keys positioned to coact with collars on the rods forming the handles for locking the handles in extended positions.

2,717,162
SAFETY GUARD FOR AUTOMOBILE PASSENGERS
Albert F. Walters, Kansas City, Mo.
Application August 19, 1952, Serial No. 305,181
5 Claims. (Cl. 280—150)

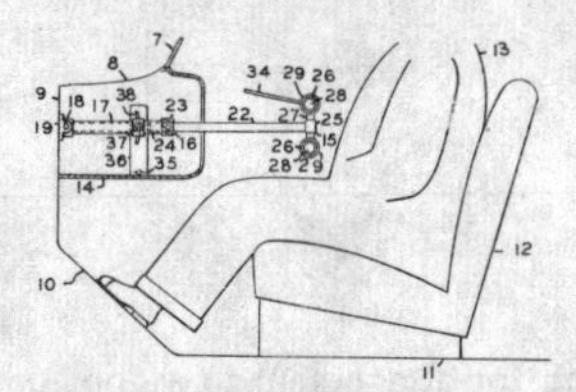

1. A safety guard for passengers in a motor car having at least a windshield and a firewall spaced in front of a seat and a transverse guard in front of a passenger having a longitudinal post connected thereto and to the firewall comprising; a tube member, a telescoping second tube member, the first mentioned said tube member being slotted on one end thereof and slidable over the second mentioned telescoping tube member, taper threads, said taper threads being on the end of said first mentioned tube member having the slots therein, a nut, said nut being taper threaded to fit said first mentioned threads, and said nut clamping said first mentioned tube to said second mentioned tube against movement to form the post between the firewall and the transverse guard.

2,717,163
CONNECTION FACILITATING VEHICLE DRAFT MEANS
Paul H. Martin, Kutztown, Pa.
Application September 18, 1953, Serial No. 381,069
1 Claim. (Cl. 280—477)

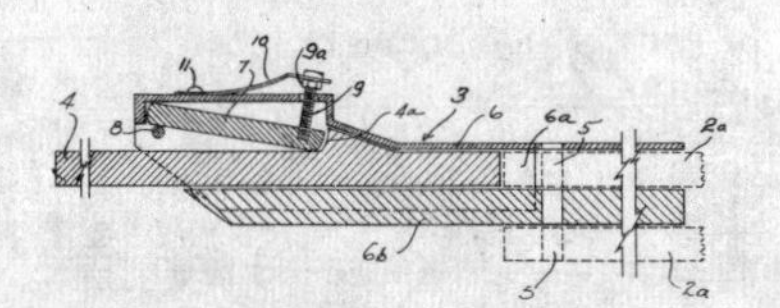

A tractor hitch comprising a drawn bar of rectangular cross-section attachable to a towed vehicle, said drawn bar having an upstanding integral pawl, a substantially tubular casing of rectangular cross-section closely surrounding said drawn bar and adapted to be longitudinally telescoped therewith, and having a coupling hole, a tractor having a draw bar in the form of tongues having holes adapted to register with said coupling hole, said casing having a rear, upwardly projecting portion forming an interior space, a pivot pin fastened in said casing and extending laterally across said space, said pivot pin serving as a stop member engageable with said pawl to prevent complete withdrawal of said casing from said drawn bar, a latching lever having the side of one end resting upon said pivot pin and having a pin extending upwardly on the other end for allowing selective lifting of said other end above the crest of said pawl to permit free sliding movement of said casing relative to said draw bar, or, upon lowering of said pawl to effect latching engagement with said pawl whereby forward drawing movement of said draw bar will draw together therewith said drawn bar and the towed vehicle.

2,717,164
COMBINATION BUMPER AND TRAILER HITCH
Evert E. Meyer, Kiester, Minn.
Application September 14, 1951, Serial No. 246,635
2 Claims. (Cl. 280—491)

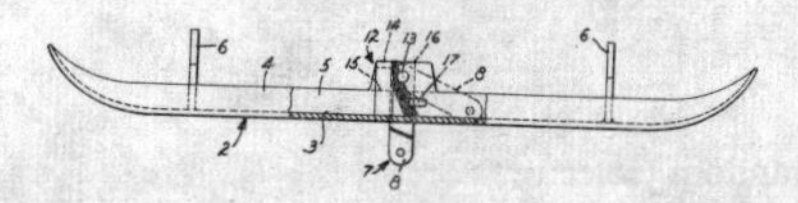

1. In a combination bumper and trailer hitch, a bumper bar of angular cross-section having an elongated upright wall portion and a lower horizontally disposed forwardly extending flange, a wall member secured to the underside of said bumper flange intermediately of the bumper ends and spaced downwardly therefrom to provide an elongated guide opening, an elongated coupling member supported in said guide opening, means pivoting the forward end of said coupling member to the forward end portions of the forwardly extending spaced apart wall portions of said bumper flange and wall member adjacent to one end of said elongated opening, the over-all length of said guide opening being relatively greater than the over-all length of the coupling member, whereby when said coupling member is not in use, it may be swung into and completely concealed within said opening and within the confines of the bumper, and the over-all length of the coupling member and its pivotal connection with the bumper flange being such that when the coupling member is swung rearwardly into load-engaging position against said abutment, said coupling member will be disposed at substantially right angular relation to the longitudinal axis of the bumper with its rear end disposed well rearwardly of the bumper, thereby to facilitate coupling a trailer or implement thereto.

2,717,165
MEMORANDUM PAD
George J. Shapiro, New York, N. Y.
Application October 23, 1952, Serial No. 316,346
2 Claims. (Cl. 281—16)

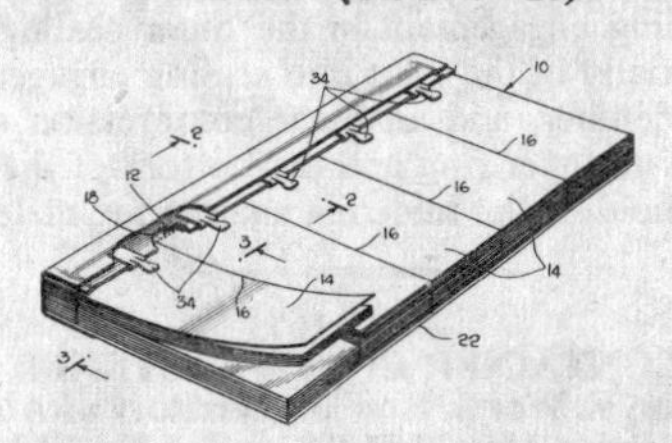

1. A memorandum pad comprising in combination: a package of superimposed sheets, each sheet having a full end portion and having a front portion divided into a series of sections, each of said sections being individually detachable, a pair of reinforcing sheets, said package of sheets being placed between said pair of reinforcing sheets, the upper reinforcing sheet having sub-

stantially the same length and width as said full end portion and covering the full end portion of the uppermost sheet, a strip of material of substantially the same length as said upper reinforcing sheet, said strip being wider than said upper reinforcing sheet and being attached to the upper surface of the latter, said strip of material projecting from said upper reinforcing sheet toward the front of the pad so as to be capable of carrying signals, and a binder embracing said strip, the end portion of the package of sheets and the reinforcing sheets and being connected thereto.

2,717,166
SELF ALIGNING, FLUID HANDLING SEALED SWIVEL CONNECTION
Robert R. Hedden, Fullerton, Calif., assignor to Chiksan Company, Brea, Calif., a corporation of California
Application April 24, 1950, Serial No. 157,706
4 Claims. (Cl. 285—10)

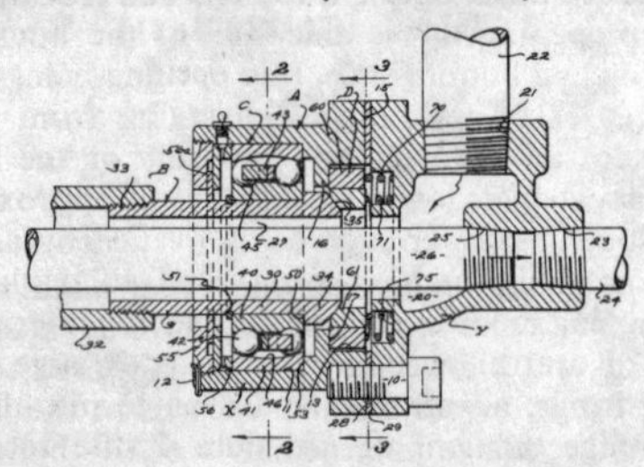

1. A self aligning fluid handling swivel joint of the character described including, an elongate rigid tubular female member with a socket opening entering it from one end and a bore inward of the opening with a flat bottom the bore having a flat radially inwardly projecting annular shoulder therein spaced from and opposing the bottom of the bore, an elongate rigid tubular male member extending into the opening to terminate therein, antifriction bearing means in the female member and rotatably and pivotally carrying the male member and including an outer ring releasably engaged in the socket opening, an inner ring releasably engaged on the male member, and axially spaced annular rows of balls between the rings, the inner ring having axially spaced grooves carrying the balls and the outer ring having a spherically curved concave race carrying the balls, the center of said concave race being coincidental with the central longitudinal axis of the joint, and axially and radially shiftable sealing means in said bore and sealing between the members, the sealing means including, an outer sealing ring in said bore, said outer ring being smaller in diameter than the base and having sliding sealing engagement with said shoulder and shiftable axially in the bore and on the shoulder, an outer compression spring between said outer ring and the bottom of the bore and yieldingly holding the ring against the shoulder, an inner sealing ring having sliding sealing engagement in the outer sealing ring and shiftable axially of the joint into sealing engagement with the male member, and an inner compression spring between the said inner ring and the bottom of the bore and yieldingly holding the said ring against the male member.

2,717,167
CONTAINER COVER FASTENER
William E. Worth, Kodiak, Territory of Alaska
Application May 29, 1953, Serial No. 358,305
2 Claims. (Cl. 292—30)

1. A fastener for container covers having a handle, said fastener including a pair of bellcranks pivotally mounted in the cover, an operating bar slidable on the handle, means operatively connecting said bar to one end portion of the bellcranks, and detents on the other ends of the bellcranks engageable with a container for locking the cover thereon, said means including a pin suspended from the bar, said pin being slidable through the cover and extending between the bellcranks, a pair of opposed,

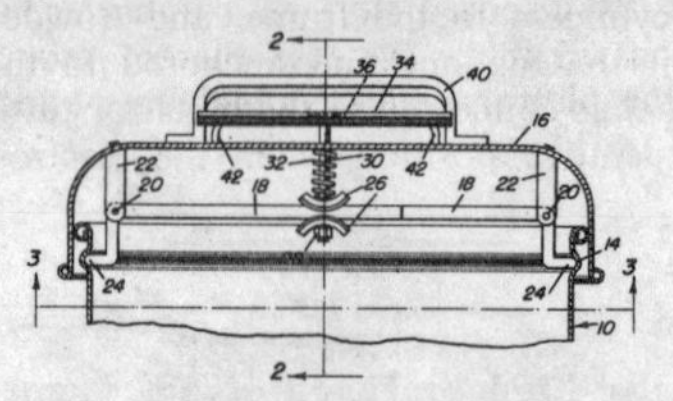

arcuate jaws mounted on the pin with the bellcranks engaged therebetween, and a coil spring on the pin engaged with one of the jaws for yieldingly urging said bellcranks toward locking position.

2,717,168
COMBINATION DOOR HANDLE AND LATCH
Cecil R. Patterson, Rivera, Calif.
Application September 6, 1951, Serial No. 245,310
1 Claim. (Cl. 292—70)

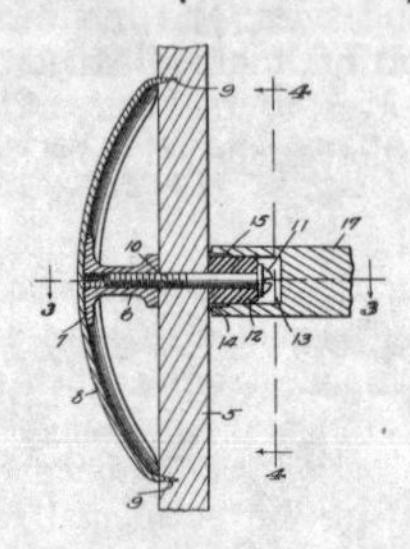

A door latch for attachment to a door opposite to one edge of a stationary support having a keeper opening formed with a straight wall against which the door abuts, comprising a body adapted to serve as a handle secured to one side of a door, a tubular internally threaded shank extending laterally from the central portion of said body, said door having an opening registering with the internally threaded shank, a bolt having a head on one of its ends, extending through the door and secured in said threaded shank, a tapered plug constructed of yieldable material fitted on the bolt, engaging the other side of said door adjacent to said opening, the head of the bolt compressing and expanding said plug as said bolt is threaded into the internally threaded shank, and said plug adapted to move into said keeper opening frictionally contacting the straight wall thereof, securing the door closed.

2,717,169
VEHICLE DOOR LOCK
Franz Lindbloom, Chicago, Ill.
Application March 5, 1954, Serial No. 414,348
1 Claim. (Cl. 292—144)

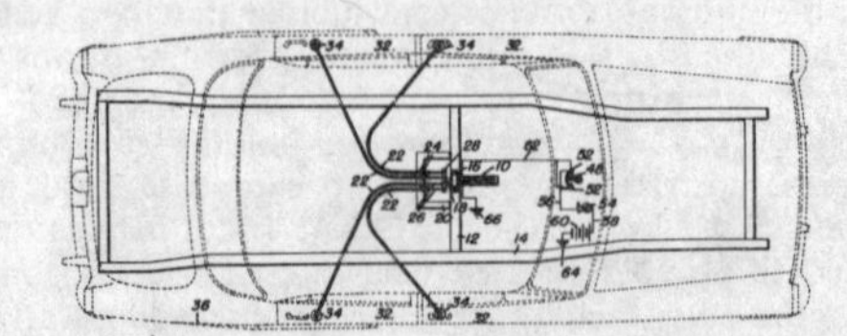

Locking means for the doors of vehicle bodies comprising cylinders mounted in the body adjacent the doors, spring retracted bolts slidable in the cylinders and engageable with the doors for securing said doors in closed position, a substantially U-shaped frame mounted on the vehicle, an electro-magnet including an armature mounted on the vehicle adjacent the frame, switch controlled means for electrically connecting said electro-magnet to a source of electric current, a bar mounted on the armature of the electro-magnet, flexible conduits extending from the cylinders to the frame, and Bowden wires in the conduits having one end anchored to the bar and their other ends connected to the bolts for projecting said bolts to operative position when the electro-magnet is energized.

2,717,170
SAFETY SEAL
Isabel C. Percival and Joseph E. Tierney, Ottawa, Ontario, Canada
Application October 3, 1952, Serial No. 312,984
6 Claims. (Cl. 292—325)

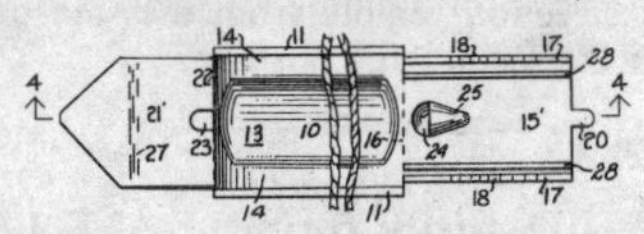

4. A one piece safety seal comprising a dished base portion having upstanding sides, a central rib, channels between the rib and sides; and intermediate leaf at one end of the base portion, toothed flanges upstanding along the longitudinal edges of the leaf, a tongue projecting from the free end thereof spaced from said edges, a transverse scoring at the junction of the base portion and leaf; a third section at the other end of the base portion having a second scoring at the junction of said section and base portion, said section provided with an orifice housing the full length of the tongue therein when the leaf is folded onto the base portion, leaving the tongue projecting therethrough when the section is subsequently folded onto the leaf and means to prevent accidental raising of the section when the leaf and section are folded over onto the base portion.

2,717,171
CONTAINER-CARRYING DEVICE
Adolph J. Gottstein, Woodbridge, N. J.
Application January 22, 1952, Serial No. 267,634
2 Claims. (Cl. 294—16)

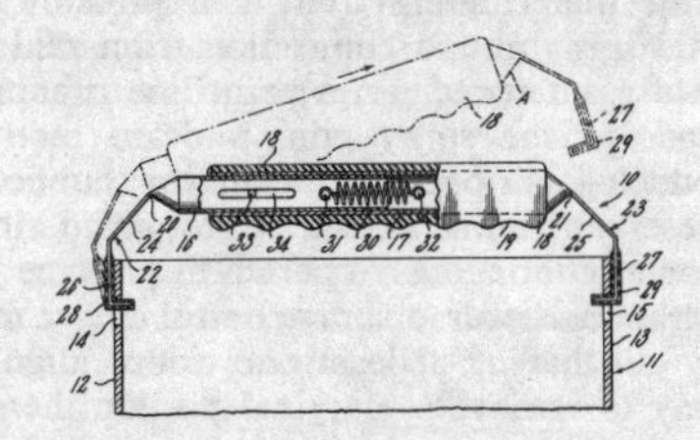

1. A lift attachment securable to a case having upstanding walls formed at locations spaced from the upper wall edges with opposed finger-receiving slots comprising a pair of tubular members telescopically mounted one within the other for movement relative to each other, the outermost one of said tubular members serving as a handle for said attachment, gripping means dependingly supported from said extensible members and operatively engaging said case for holding said tubular members in straddling relation to said case, said gripping means each including a planar frictional abutment adapted to provide an area bearing and friction contact against the outer face of the adjacent upstanding wall and an inwardly-extending hook receivable within the adjacent finger-receiving slot, the length of said planar frictional abutments being selected in relation to the location of said finger-receiving slots to maintain said tubular members spaced above the upper edges of said upstanding walls, and resilient means operatively connected to said extensible members and biasing said gripping means into operative engagement with said case.

2,717,172
WELL FISHING TOOL
Charles J. Boyd, Santa Fe, N. Mex.
Application July 17, 1953, Serial No. 368,749
2 Claims. (Cl. 294—86)

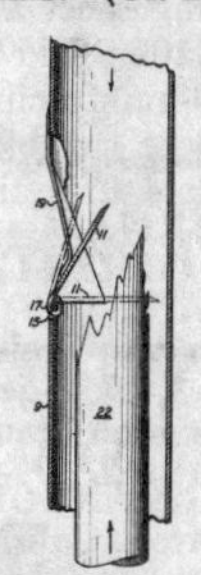

1. A well rod extracting device comprising an elongated tubular housing having a top end and a bottom end, at least one opening in the side wall of the housing between the top and bottom ends, said opening being formed by lateral edges of the housing diverging from a point on the surface downwardly and inwardly of the housing to encompass at their lower ends an arc of approximately 120°, said opening lower edge being formed by an upper edge portion of the housing lying in a radial plane and ending between the lower ends of the opening lateral edges, an elongated central portion of said upper edge portion of the housing being recessed and a hinge pin affixed to the upper edge portions of the sides of the recess and bridging said recess and having its upper surface on a plane substantially in the plane of said upper edge portion of the housing, an elongated gripping dog having a length slightly less than the diameter of the housing and having a base end and a gripping end, said base end having curved edges with the curvature being proximate the curvature of said opening lower edge, said curved edges terminating in a centrally located looped portion, said looped portion surrounding said hinge pin to thereby hinge said gripping dog; spring means affixed to the housing and coercing the gripping end of said gripping dog inwardly and downwardly, and means for attaching a tractor device to the upper end of said housing, and the interior of said bottom end being unobstructed.

2,717,173
VENTILATING STRUCTURE FOR MOTOR VEHICLE BODIES
Karl Rabe, Stuttgart, Germany, assignor to Dr. Ing. H. C. F. Porsche K.-G., Stuttgart-Zuffenhausen, Germany
Application October 22, 1952, Serial No. 316,115
Claims priority, application Germany November 20, 1951
8 Claims. (Cl. 296—28)

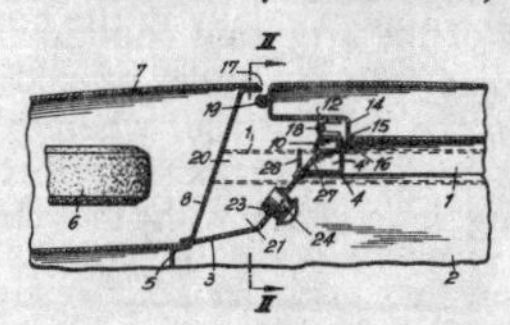

1. A ventilating arrangement for motor vehicles of the type including a body having a door at the side, means located adjacent the forward portion of the door inside the outer skin of the body forming an air collecting chamber which is shut off from the interior of the vehicle the outer surface of said door forming substantially a continuation of the exterior surface of the body of the vehicle, said exterior surface being interrupted at the forward edge of the door by a door chink through which the air-collecting chamber communicates with the air outside the vehicle, said door chink being of sufficient width when the door is closed to permit the flow of air therethrough into said air-collecting

chamber, said chink being located in a pressure region of the flow of air around the side of the vehicle when in motion and adapted to conduct air under pressure into said chamber, means for conducting air from said chamber to the interior of the vehicle to effect ventilation of the vehicle, and means for controlling the flow of air in said conducting means.

2,717,174
POST-CARD OR OTHER CARD WITH A FRAGRANT PASTILLE
Michel Casanovas, Frejus, France
Application August 1, 1951, Serial No. 239,809
Claims priority, application France August 2, 1950
1 Claim. (Cl. 299—24)

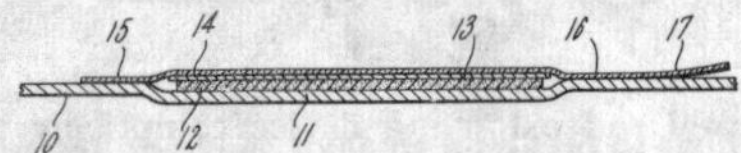

A post-card made of sheet material and provided with a relatively shallow recess in its surface to thereby define an open receptacle therein, a concentrated perfume pastille substantially conformed to the shape of and having a thickness approximately corresponding to the depth of said recess, and protective flap means flatly overlying said receptacle and said pastille and adhesively secured to said surface to retain said pastille in position within said recess and in substantially airtight condition, said flap means including at one end thereof a pull tab, the sheet material of said post-card being made of relatively heavy stock with respect to the thickness of said flap means.

2,717,175
DISH WASHING DEVICE
Katherine B. Anderson, Vancouver, British Columbia, Canada
Application February 11, 1953, Serial No. 336,427
Claims priority, application Canada August 25, 1952
5 Claims. (Cl. 299—83)

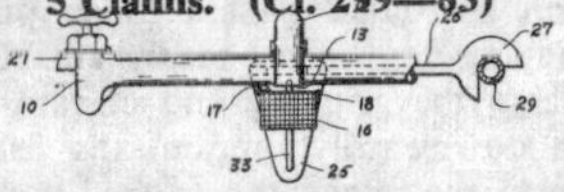

1. A device for assisting in the washing of articles, such as dishes or the like, under a water tap, comprising a vertical mounting plate to be suspended from a tap or a pipe leading thereto, a vertical rod supported by and spaced outwardly from the plate, an arm slidably and swingably mounted at one end on the rod, means for positioning the arm on the rod without interfering with the swinging action thereof, and a container for holding a cleansing agent and through which water may run connected to the opposite end of the arm, said container being moved into and out of line with the outlet of said tap when the arm is swung on the vertical rod, whereby articles may be washed when held in water passing through the container after being in contact with the cleansing agent therein.

2,717,176
DETERGENT DISPENSER
Leonard Osrow and Harold Osrow, Queens Village, N. Y.
Application September 27, 1954, Serial No. 458,509
9 Claims. (Cl. 299—84)

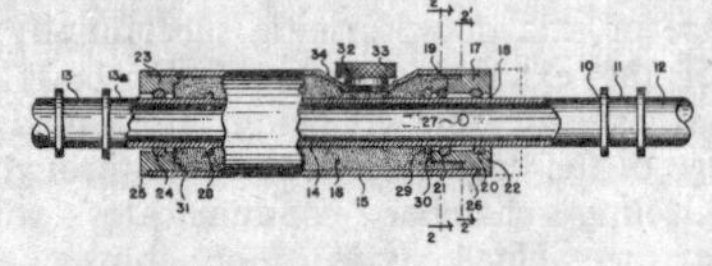

9. A dispenser including a tube adapted for connection to a fluid pipeline, in combination with a chamber encompassing a portion of the tube, said chamber being enclosed by an outer shell and the exterior surface of the tube including fluid tight sealing means between said tube and said shell, said means being in movable relationship relative to said tube and affixed to said shell in further combination with a perforation through said tube located to communicate with the said chamber when the sealing means is in one position, the movement of the sealing means longitudinally to a second position preventing such communication, access means for filling said chamber, one of said end walls including one sealing means and the other end wall including two sealing means, said two sealing means being spaced to fall on each side of the perforation when in the first said position and both of said sealing means are arranged to lie beyond the perforation in a direction remote from the said one sealing means when in the second said position.

2,717,177
SPRAY DEVICE
Lawrence B. Goda, Sr., Los Angeles, Calif.
Application September 4, 1951, Serial No. 244,943
7 Claims. (Cl. 299—86)

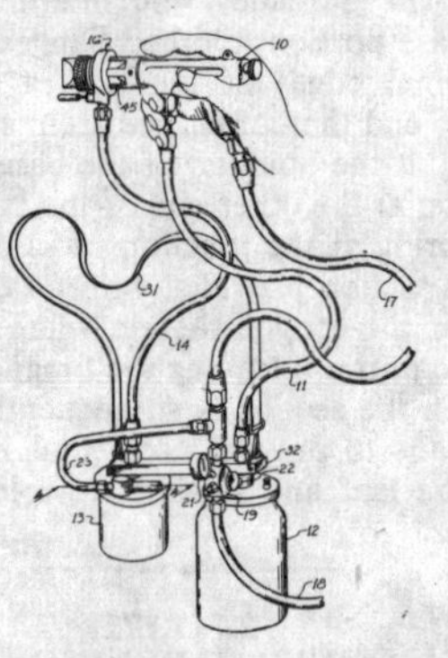

1. A spray device of the character described comprising a spray gun, a paint container, air pressure control means adapted to receive air from a suitable pressure source, an air line running from said pressure control means to said spray gun, a connection from said control means to said paint container, a paint line running from said container to the spray gun, a drain reservoir, a vacuum-producing air-operated aspirator connected to said drain reservoir, a line extending from said air line to said aspirator to continuously operate to produce a vacuum in said drain reservoir, a spray control device mounted to the spray gun having at least one orifice aligned with the said spray to remove a marginal portion thereof and reduce its cross-sectional area, and a suction drain line extending from said control device to the drain reservoir.

2,717,178
SPRAYERS
Nelson F. Cornelius, Anoka, Minn.
Application August 12, 1953, Serial No. 373,745
6 Claims. (Cl. 299—97)

1. In combination, a receptacle for a liquid to be sprayed, a cap attached to said receptacle, a nozzle carried by said cap, means forming a liquid conducting passageway between said receptacle and said nozzle, pump means communicating with said passageway and alternately drawing liquid from said receptacle through said passageway and forcing it through said nozzle, a shut-off valve at said nozzle including a valve head engageable with a valve seat, a valve stem issuing from said valve head, a cylinder communicating at one end with said passageway, a piston in said cylinder connected to said valve stem and operating said valve head, a spring held from movement at one end relative to said cap and

engaging said piston at its other end and urging said valve head into closing position, means forming a by-pass around said piston and sealing means engageable with

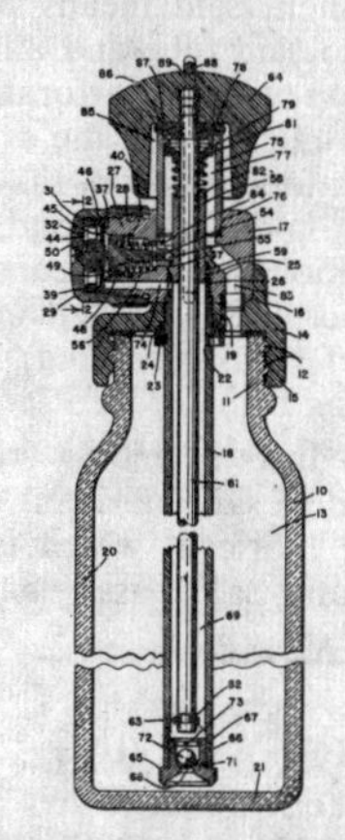

said piston to close said by-pass when the piston reaches a position in which passage through the orifice in the nozzle is effected.

2,717,179
COVERS FOR VEHICLE WHEELS

George R. Pipes, Mayfield Heights, and Howard J. Findley, Cleveland, Ohio, assignors to Eaton Manufacturing Company, Cleveland, Ohio, a corporation of Ohio
Application June 4, 1952, Serial No. 291,722
12 Claims. (Cl. 301—37)

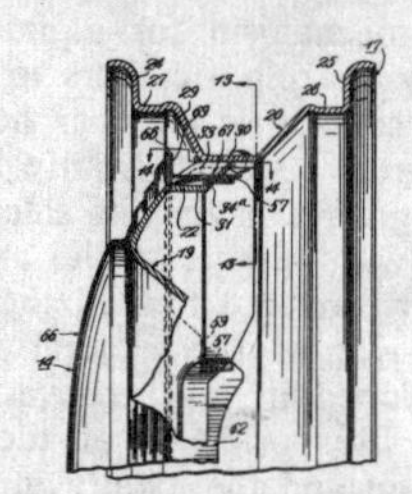

1. A wheel cover comprising, substantially circular sheet metal cover means, and an annular series of retaining members in the form of circumferentially spaced and substantially circumferentially extending and edgewise aligned sheet metal webs connected with said cover means and projecting generally axially rearwardly therefrom, said webs having bowed intermediate portions such that said retaining members are resiliently flexible for compression and expansion substantially edgewise thereof and substantially in the direction of the circumference of said series.

2,717,180
SLUG-TRAP

Barton S. Snow, Chicago, Ill., assignor to T. W. Snow Construction Company, Inc., a corporation of Illinois
Application August 29, 1950, Serial No. 182,034
6 Claims. (Cl. 302—48)

1. A slug-trap, for use in a pipe line transporting dry sand by gas pressure, comprising: a body having an elongated chamber provided with a closed bottom, an inlet port for connection to a generally horizontal inlet pipe line, and spaced above said bottom, a wall spaced from said inlet port and extending across the path of said material, and an outlet port for connection to the other side of the pipe line, said ports being offset from each other with the outlet port being intermediate said inlet port and the wall, so that masses of sand entering through the inlet port travel under force of gas pressure through the chamber past said outlet port and impinge upon sand stacked up against the wall and bottom of

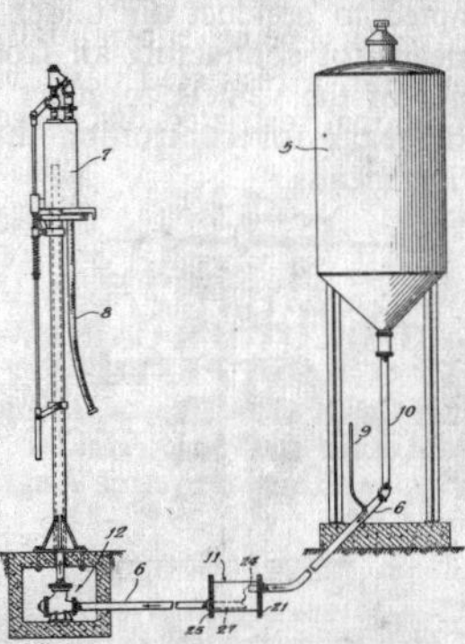

said chamber to break up slugs of sand and to entrain the granular sand in said gas before entering said outlet port.

2,717,181
SAND HANDLING EQUIPMENT

Barton S. Snow, Batavia, Ill., assignor to T. W. Snow Construction Company, Inc., a corporation of Illinois
Application December 13, 1954, Serial No. 474,876
7 Claims. (Cl. 302—59)

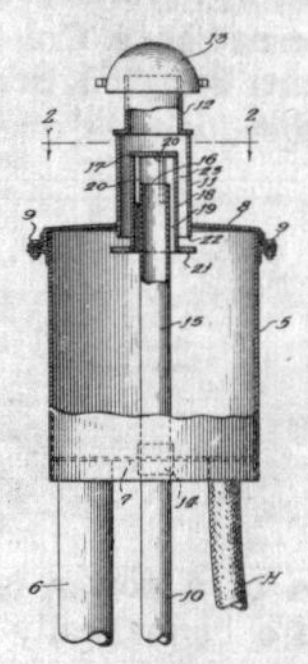

1. In a granular material storage tank, means for admitting granular material to the tank, comprising: an upwardly open pipe within the tank for discharging granular material carried in an air stream; a sleeve telescoped loosely about the pipe and having a closed end above the open end of the pipe so that incoming material may impinge on the closed end and then fall into the tank through the space between the sleeve and pipe; an air duct for conducting air out of the tank and having a lower edge about the sleeve; and a lip on the sleeve adapted to be raised into contact with the lower edge of the duct when the sleeve is raised by incoming material and excessive air pressure, temporarily closing the air duct and preventing the air from blowing material out of the tank.

2,717,182
SHAFT-POSITIONING MECHANISM FOR TURBINE-DRIVEN PUMPS

Robert H. Goddard, deceased, late of Annapolis, Md., by Esther C. Goddard, executrix, Worcester, Mass., assignor of one-half to The Daniel and Florence Guggenheim Foundation, New York, N. Y., a corporation of New York
Application September 28, 1948, Serial No. 51,542, now Patent No. 2,616,373, dated November 4, 1952, which is a division of application Serial No. 598,755, June 11, 1945, now Patent No. 2,450,950, dated October 12, 1948. Divided and this application February 23, 1952, Serial No. 273,131
4 Claims. (Cl. 308—9)

1. Apparatus for restoring and maintaning a predetermined axial position of a rotated shaft comprising a rotor

fixed on a rotated shaft, a separate and first pressure chamber at the upper side of said rotor, and a separate second pressure chamber at the lower side of said rotor, each of said chambers being substantially enclosed at one side by said rotor, a restricted passage between said chambers, an additional restricted discharge port for said

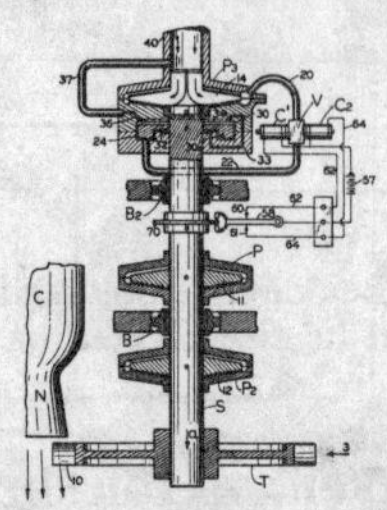

first and upper chamber, a pipe to supply liquid under pressure to the second and lower chamber, a valve in said pipe, and automatic means to move said valve to vary the liquid supply on an axial displacement of said shaft in either direction, whereby the initial axial position of said shaft is restored and maintained.

2,717,183
POWER TAKE-OFF SHAFT AND BEARING FOR HIGH SPEED MOTORS
Guilbert Francis Soucy, Cohoes, N. Y.
Application September 24, 1953, Serial No. 382,140
3 Claims. (Cl. 308—77)

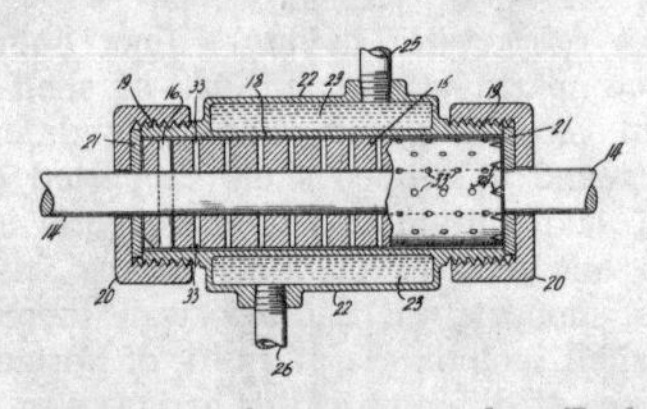

1. The combination of a power take-off shaft for a high speed motor and a hermetically sealed bearing for said shaft, which shaft has a substantially enlarged cylindrical bearing portion provided with peripheral edge notches; said bearing comprising a bearing-sleeve for said enlarged bearing portion; means encircling said shaft and secured to the ends of said bear-sleeve for encasing and sealing said enlarged cylindrical bearing portion; fluid cooling means encircling said bearing-sleeve; and means containing mercury communicating with said bearing-sleeve and said enlarged bearing portion for supplying mercury thereto, as a lubricant, under pressure, and to expel air therefrom; whereby centrifugal action of said shaft, because of said edge notches on said enlarged bearing portion, will force said mercury back from the ends of said bearing and away from said end sealing means.

2,717,184
CROWN AND TRAVELING BLOCK LUBRICATION SYSTEM
Jack A. Amerman, Houston, Tex., assignor to Emsco Manufacturing Company, Los Angeles, Calif., a corporation of California
Application June 1, 1953, Serial No. 358,699
9 Claims. (Cl. 308—187)

1. In means for lubricating a plurality of bearings disposed on a shaft having a longitudinal opening open at at least one end, there being lubricant passages extending in said shaft from each of said bearings to said opening: a unitary lubricant distributing member insertable in said opening through said open end thereof and being of such length within said opening that it passes across the inner ends of passages in said shaft, said member having ports therein spaced therealong so that they will communicate with the inner ends of said lubricant passages when the distributing member is inserted as a unit

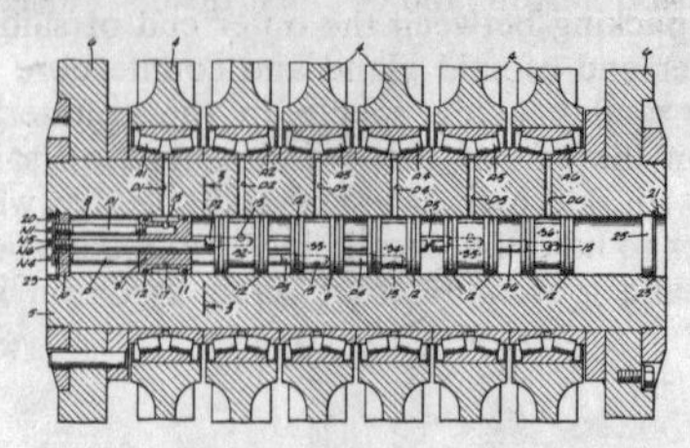

in said opening, and ducts extending from the respective ports to the front end of said lubricant distributing member so as to receive lubricant which is to be delivered through said ports and said passages to said bearings.

2,717,185
BEARING SEAL
Ralph E. Baumheckel, Connersville, Ind., assignor to Link-Belt Company, a corporation of Illinois
Application May 12, 1952, Serial No. 287,321
4 Claims. (Cl. 308—187.2)

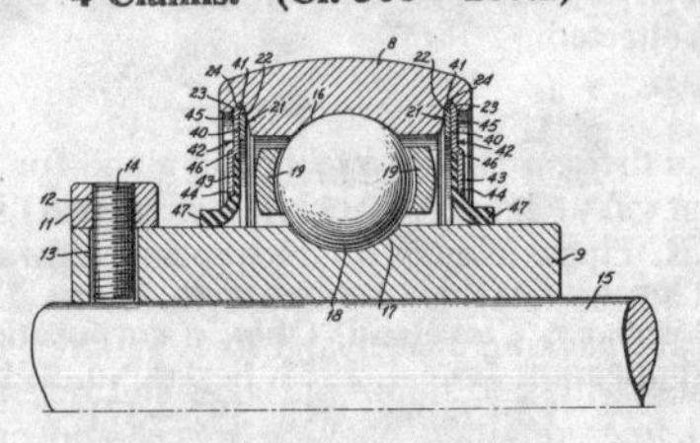

3. In a sealing mechanism for a pair of concentric members spaced by bearing members to permit relative rotation, the outer concentric member having a radially inwardly opening groove spaced axially from said bearing members, said groove having its side walls arranged at opposite angles with respect to the axis of said concentric members to converge toward the bottom of the groove, a first seal ring having a radially arranged marginal portion positioned in said groove for sealing engagement between the periphery of the inner face of said marginal portion and the axially inner side wall of said groove, the remainder of said inner face being in spaced relationship with the angularly related axially inner side wall, and a second seal ring having its marginal portion wedged between the periphery of the outer face of the marginal portion of said first seal ring and the axially outer side wall of said groove to clamp the two rings in sealing engagement with the side walls of said groove out of contact with its bottom.

2,717,186
METAL-TO-METAL STABILIZED PUMP LINER WITH INDEPENDENT PACKING ADJUSTMENT
George E. Campbell, Chattanooga, Tenn., assignor to The Wheland Company, Chattanooga, Tenn., a corporation of Tennessee
Application March 29, 1954, Serial No. 419,468
7 Claims. (Cl. 309—3)

1. Metal-to-metal stabilized pump liner with independent packing adjustment, comprising a pump cylinder having a bore with a stop shoulder and a counterbore enlargement at the outer end of the cylinder, a liner seated in said bore and having a shoulder in engagement with said stop shoulder, a packing spacer engaged in said counterbore about the outer end portion of the liner, packing at the inner end of said spacer between the liner, packing spacer and cylinder, a liner retainer within the packing spacer in engagement at its inner end with the outer end

of the liner, a cylinder head in direct metallic pressure applying connection with said retainer, a gland in the outer end of the counterbore about the outer end of said retainer, packing between the outer end of said spacer and the inner end of said gland and counterbore sealing said retainer and packing spacer in the counterbore, screw studs on the end of the cylinder, nuts on said studs engaged with said cylinder head and effective with said stop shoulder to hold the liner in metal-to-metal stabilized condition, and a second set of screw studs on the cylinder

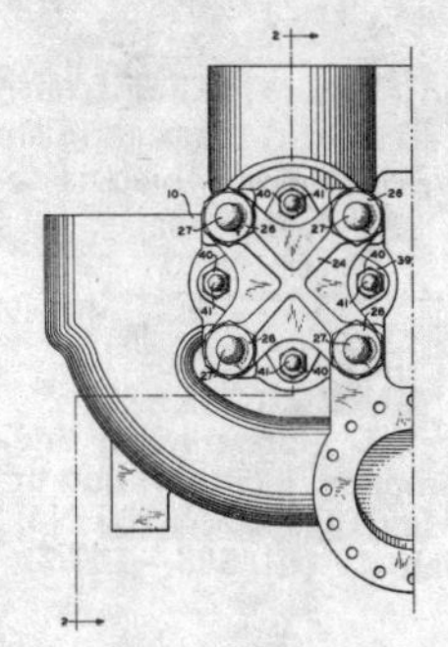

and nuts thereon engaged with said gland for effecting adjustment of said liner, packing spacer and liner retainer packings independently of and apart from any pressure applied to the liner, said pump cylinder having fluid passages, said packing spacer and liner retainer having ports to register with said pump passages and registering means on said liner retainer, packing spacer and cylinder head for aligning said parts in proper relation to register said ports in the liner retainer and packing spacer with the passages in the pump cylinder.

2,717,187
LAMINATED TABLE TOP WITH EDGING
Erving B. Morgan, Bruno S. Jurewicz, and Alexander J. Plachecki, Grand Rapids, Mich., assignors to American Seating Company, Grand Rapids, Mich., a corporation of New Jersey
Application August 13, 1953, Serial No. 374,006
3 Claims. (Cl. 311—106)

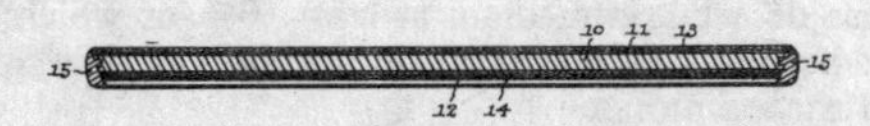

1. A top for a table or the like, comprising: a flat core; a finishing sheet bonded to the upper surface of the core to form a plied-up panel having its peripheral edge provided with a bevel extending inwardly from top to bottom; and finishing strips provided with complementary bevels bonded to the peripheral edge of said plied-up panel to form a composite top; said composite top having the upper peripheral corner edge thereof bevelled outwardly-downwardly from the upper line of junction between the finishing sheet and the finishing strips.

2,717,188
MOUNTING FOR CHASSIS AND THE LIKE
Ferdinand Kuss, Philadelphia, Pa.
Application August 6, 1954, Serial No. 448,313
6 Claims. (Cl. 312—323)

1. In a mounting including a chassis, slides on either side of the chassis, rails extending longitudinally adjoining the slides and means on the slides guided on the rails, the combination which comprises a pivot head extending longitudinally at the top of one of the slides, supports on the sides of the chassis extending up over the slides and at the one slide engaging over the pivot head in pivotal position on a longitudinal axis and latch means between

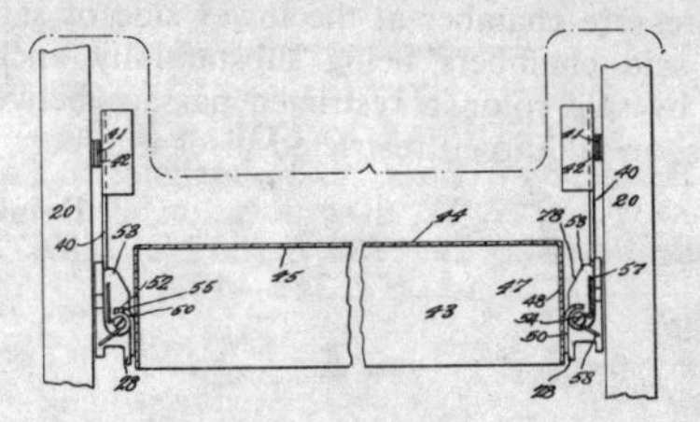

the chassis and the one slide holding the chassis in a position tilted up on its side pivoted on the pivot head.

2,717,189
REFRIGERATOR SHELVES
Walter D. Teague, Annandale, N. J., and Seymour D. Wassyng, Brooklyn, N. Y., assignors to Servel, Inc., New York, N. Y., a corporation of Delaware
Application January 31, 1952, Serial No. 269,228
3 Claims. (Cl. 312—351)

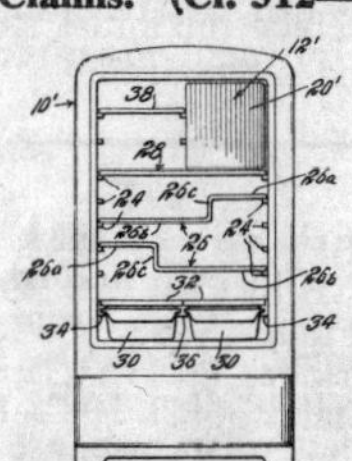

1. In a refrigerator having a liner bounding a food storage compartment, a plurality of shelf supports arranged in vertical spaced relation on different walls of said liner and a stepped shelf supported on said shelf supports, said plurality of shelf supports including supports on one wall of the liner in the same horizontal plane as supports on another wall thereof, and said stepped shelf including a plurality of horizontal supporting surfaces of different widths arranged in different horizontal planes and connected by vertical connecting means into a unitary shelf structure of substantially the same width as the width of said compartment, the height of said vertical connecting means being substantially the same as the vertical spacing of said plurality of shelf supports, and said stepped shelf being reversible end-for-end and top-for-bottom relative to said shelf supports and to said compartment from certain shelf supports located in two certain horizontal planes to other shelf supports located in the said two horizontal planes.

2,717,190
CATHODE-RAY TUBE ANALYZER AND RESTORER
Carl A. Shoup, Detroit, Mich., assignor to Thomas A. Demetry, Detroit, Mich.
Application October 13, 1954, Serial No. 462,020
12 Claims. (Cl. 316—2)

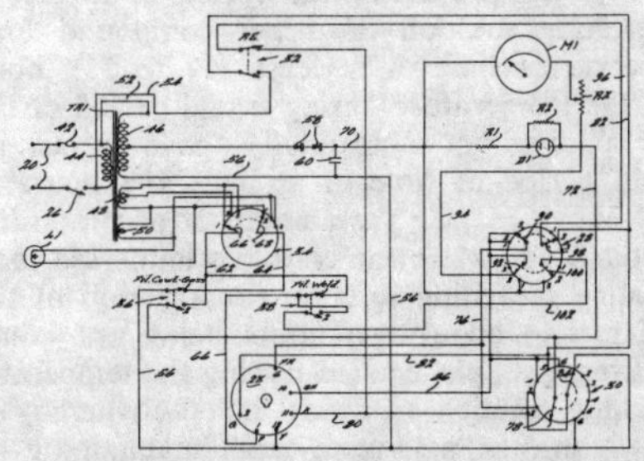

1. The method of renewing the condition of a cathode-ray tube having a cathode and a grid which com-

prises applying an alternating voltage having an amplitude greater than 250 volts directly between the cathode and grid.

2,717,191
SINUSOIDAL RECORDER
Frank A. Hester, New York, N. Y., assignor to Faximile, Inc., New York, N. Y., a corporation of Delaware
Application May 20, 1952, Serial No. 288,877
7 Claims. (Cl. 346—101)

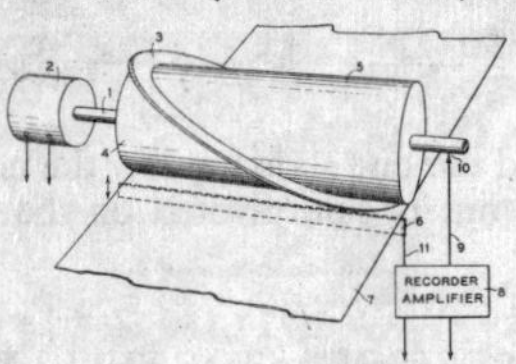

1. A recorder for electrically marking a recording medium comprising a rotatable member, a substantially elliptical electrode mounted on said member, a linear electrode disposed to contact the elliptical electrode during rotation thereof with the recording medium therebetween, and means for supplying a recording current to said electrodes.

2,717,192
RECORDING APPARATUS
Ralph R. Chappell, Belfort, Md.
Application June 14, 1949, Serial No. 98,912
15 Claims. (Cl. 346—139)

11. Recording apparatus comprising a ratchet movable in a plurality of directions in response to variable functions, a pawl engageable with and drivable by said ratchet on movement thereof in one of said directions, means yieldably opposing drive of said pawl in said one direction for holding said ratchet and pawl in driving engagement

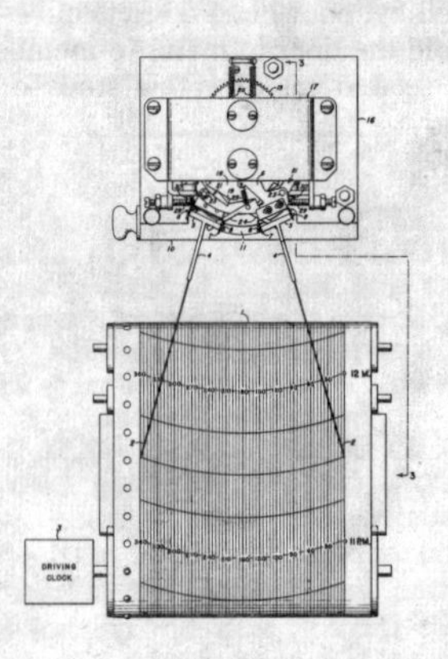

on movement of said ratchet in another of said directions, recording means drivably connectable to said ratchet through said pawl, and means for limiting movement of said recording means under drive of said ratchet in either of said directions.

CHEMICAL

2,717,193
BLEACHING PROCESS FOR COTTON OF LOW GRADE FOR COLOR
Simon A. Simon, Longmeadow, and Harvey Clayton Ruhf, Springfield, Mass., assignors to Chicopee Manufacturing Corporation, a corporation of Massachusetts
Application November 19, 1952, Serial No. 321,378
14 Claims. (Cl. 8—111)

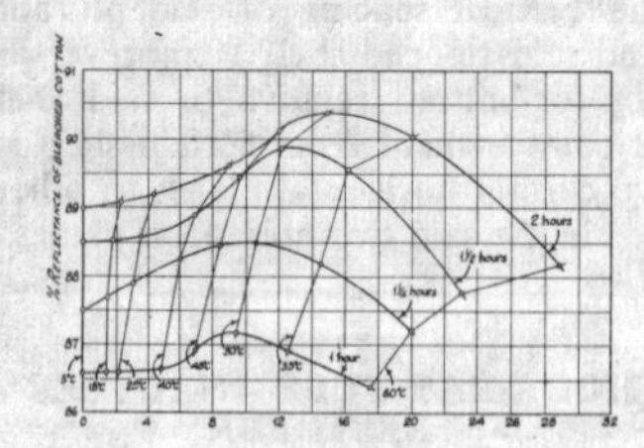

1. The method of bleaching cotton of low grade for color to a per cent reflectance of at least 88.5 Rd comprising the steps of: immersing the cotton in an inactive alkaline bleaching solution of a per compound for a minimum period of time of at least 1¼ to 1½ hours, maintaining the temperature during a said period of time of at least 1¼ hours not substantially above 50° C., and during a said period of time of at least 1½ hours not substantially above 55° C.; the amount of decomposition of the per compound while the cotton is immersed in the inactive alkaline bleaching solution for a period of time of not longer than about two hours being not greater than about twenty-six per cent; elevating the temperature of the bleaching solution to about 85° C., whereby the bleaching solution is activated; and maintaining the bleaching solution in an activated state until the per compound is substantially completely exhausted.

2,717,194
BETA-PROPIOLACTONE MODIFICATION OF WOOL
William Gordon Rose and Harold P. Lundgren, Berkeley, Calif., assignors to the United States of America as represented by the Secretary of Agriculture
No Drawing. Application October 21, 1954, Serial No. 463,837
4 Claims. (Cl. 8—112)
(Granted under Title 35, U. S. Code (1952), sec. 266)

1. A process for chemically modifying wool comprising reactiong wool with beta-propiolactone in the presence of a liquid medium at least 10% of which is a high-molecular weight aliphatic alcohol containing at least 8 carbon atoms.

2,717,195
METHOD FOR TREATING A FIBROUS MATERIAL
Bruce Armstrong, Saginaw, Mich., assignor to Jackson and Church Company, Saginaw, Mich., a corporation of Michigan
Application August 2, 1954, Serial No. 447,371
18 Claims. (Cl. 8—156)

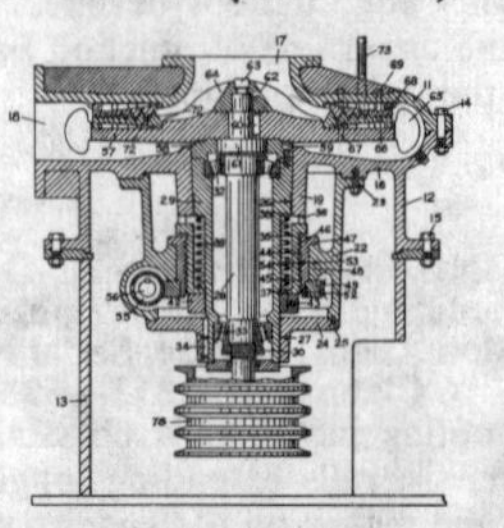

1. The process for treating a bulk, unoriented fibrous substance with a treating agent, which includes: forwarding a mass of bulk, unoriented fibers through a fiber-

separating and fiber-treating zone; subjecting the fibers in the zone to a mechanical fiber-separating action of progressively increasing intensity to separate substantially the fibers from one another, introducing a fiber-treating agent into a region within the zone intermediate the regions of least and of greatest fiber-separating action to contact substantially the surfaces of the separated fibers with the treating agent; and withdrawing fibers having the treating agent substantially evenly dispersed therein from the fiber-separating and fiber-treating zone.

2,717,196
SHEET WRAPPING MATERIAL CONTAINING NITRO-PHENOL COMPOUND
Aaron Wachter and Robert J. Moore, Oakland, Calif., assignors to Shell Development Company, San Francisco, Calif., a corporation of Delaware
No Drawing. Application February 25, 1950, Serial No. 146,386
10 Claims. (Cl. 21—2.5)

1. A corrosion-inhibiting wrapping material comprising a substantially solid inactive sheet wrapping material having associated therewith in the solid state a compound of the group consisting of nitrophenols and base metal salts thereof, said nitrophenol having a molecular weight of less than about 350 and containing no groups other than nitro, hydroxy and hydrocarbyl groups.

2,717,197
COMPLEX FLUORIDE SALTS OF TITANIUM
Gerald Taylor Brown, St. Helens, England, assignor to Peter Spence and Sons Limited, Widnes, Lancashire, England, a British company
No Drawing. Application January 18, 1954, Serial No. 404,779
Claims priority, application Great Britain January 26, 1953
2 Claims. (Cl. 23—88)

1. A process for preparing potassium fluotitanate, K_2TiF_6, which comprises adding potassium fluoride to an aqueous solution of hydrofluoric acid, adding titanium tetrachloride to the resultant solution, and thereafter crystallising potassium fluotitanate from the solution.

2,717,198
AMMONIA-PHOSPHORUS PENTOXIDE REACTION PRODUCTS AND METHOD OF PRODUCING SAME
Otha C. Jones, Campbell, Calif., and Peter G. Arvan, Anniston, Ala., assignors to Monsanto Chemical Company, St. Louis, Mo., a corporation of Delaware
Application October 25, 1951, Serial No. 253,112
8 Claims. (Cl. 23—106)

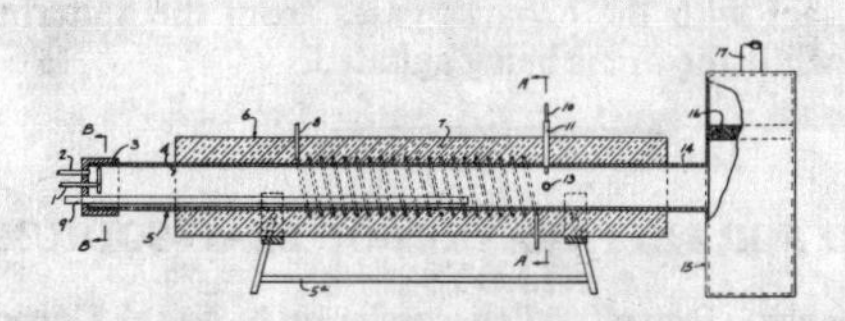

1. The method of producing valuable products, which comprises supplying dry air and elemental phosphorus to a combustion chamber where it is ignited to form phosphorus pentoxide vapor, introducing substantially anhydrous gaseous ammonia into said vapor to effect a reaction with said phosphorus pentoxide vapor, cooling the resulting reaction product in a period of from 3 to 5 seconds to a temperature below about 200° C., and then subjecting the resulting product, while in contact with an atmosphere consisting of 100% by volume of ammonia gas, to a temperature within the range of about 20° C. to about 300° C., said initial reactants being employed in the proportions yielding a product having an ammonia to phosphorus pentoxide molar ratio of about 2.2/1 to about 3.25/1 and said initial reaction being carried out at a temperature substantially in the range of about 240° C. to about 725° C.

2,717,199
HYDROGEN CHLORIDE RECOVERY
David H. Campbell, Baton Rouge, La., assignor to Ethyl Corporation, New York, N. Y., a corporation of Delaware
No Drawing. Application November 8, 1954, Serial No. 467,633
5 Claims. (Cl. 23—154)

1. A process of recovering concentrated hydrogen chloride gas from an anhydrous mixture with other gases comprising contacting said mixture with a tetraalkyl ammonium halide selected from the group consisting of tetramethyl ammonium chloride and tetramethyl ammonium bromide, forming thereby a solid addition product of the hydrogen chloride and tetramethyl ammonium halide, said addition product containing not more than one mole of hydrogen chloride to one mole of tetramethyl ammonium halide, and then heating said addition product and decomposing at least a portion of the addition product into concentrated anhydrous hydrogen chloride and tetramethyl ammonium halide.

2,717,200
MANUFACTURE OF HYDRAZINE
William E. Hanford, Short Hills, N. J., assignor to The M. W. Kellogg Company, Jersey City, N. J., a corporation of Delaware
Application April 4, 1950, Serial No. 153,984
25 Claims. (Cl. 23—190)

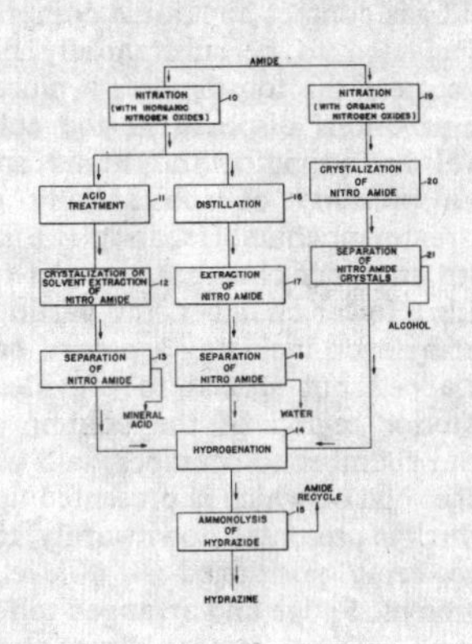

23. A process for making hydrazine which includes reacting a hydrazide with ammonia, and thereafter separating the resultant hydrazine from the reaction mixture.

2,717,201
PRODUCTION OF HYDRAZINE
Herbert J. Passino, Englewood, N. J., assignor to The M. W. Kellogg Company, Jersey City, N. J., a corporation of Delaware
No Drawing. Original application October 23, 1950, Serial No. 191,735, now Patent No. 2,675,301, dated April 13, 1954. Divided and this application May 20, 1952, Serial No. 288,997
17 Claims. (Cl. 23—190)

1. A method for producing hydrazine which comprises: contacting urea with a particulate carbonyl-forming metal in a reaction zone, said carbonyl-forming metal being

continuously maintained in an amount equivalent to at least 2 per cent by weight of the quantity of urea present in said reaction zone, at a temperature between about 40° C. and below the melting point of urea to convert at least a substantial quantity of urea present to hydrazine; and recovering hydrazine as a product of the process.

2,717,202
COUNTERFLOW LIQUID-GAS CONTACT APPARATUS
Alton E. Bailey, Louisville, Ky., assignor, by mesne assignments, to National Cylinder Gas Company, Chicago, Ill., a corporation of Delaware
Application September 7, 1949, Serial No. 114,335
12 Claims. (Cl. 23—283)

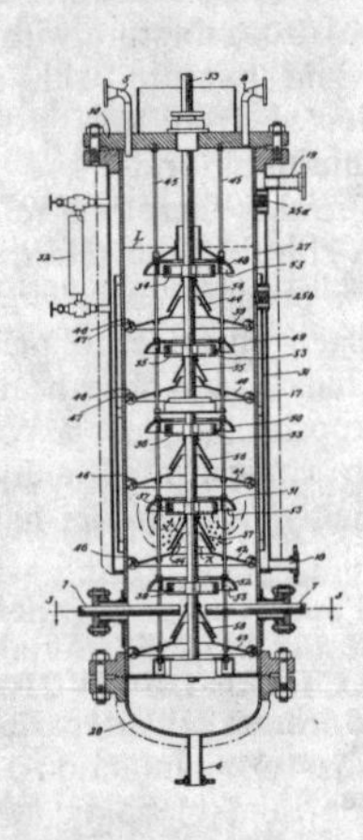

1. In a liquid-gas contact apparatus comprising an upright column adapted to be substantially filled with a liquid to a level near the top thereof, a plurality of vertically spaced partitions disposed in the column below the liquid level and dividing the liquid space of the column into a plurality of substantially independent superimposed treatment chambers, an inlet for liquid communicating with an upper chamber, an inlet for gas communicating with a lower chamber, the partitions between adjacent chambers each including a pair of partition wall elements overlapped but spaced to provide a gas passage in an interior region of the column for upward flow of gas from chamber to chamber, said wall elements having free edges one of which is presented upwardly and the other of which is presented downwardly, the upwardly presented edge being positioned at a level above the downwardly presented edge and arranged to form a liquid trap, and a liquid delivery passage disposed radially outwardly from said interior gas passage for downward flow of liquid from chamber to chamber.

2,717,203
METHOD OF PRODUCING APERTURE IN HOLLOW ARTICLE
Jan Anton Willem van Laar, Eindhoven, Netherlands, assignor to Hartford National Bank and Trust Company, Hartford, Conn., as trustee
No Drawing. Application May 29, 1953, Serial No. 358,541
Claims priority, application Netherlands June 25, 1952
8 Claims. (Cl. 41—42)

1. A method of making hollow articles furnished with an aperture, comprising, filling the cavity of the article with a liquid, placing the article in a liquid etching bath until the liquid flows from said cavity into the etching bath after an aperture has been formed by etching.

2,717,204
BLASTING INITIATOR COMPOSITION
George Adelbert Noddin, Clarksboro, and Charles Philip Spaeth, Woodbury, N. J., assignors to E. I. du Pont de Nemours & Company, Wilmington, Del., a corporation of Delaware
Application May 2, 1952, Serial No. 285,601
3 Claims. (Cl. 52—2)

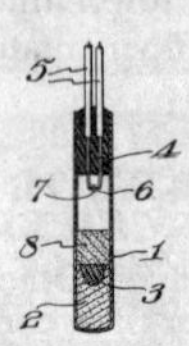

1. A burning charge for use in delay electric blasting caps which comprises amorphous boron and red lead in the respective proportions of 0.5–3.0% and 99.5–97.0%.

2,717,205
PROCESS OF TREATING LOW GRADE ORES
Enoch F. Edwards, Nashwauk, Minn., assignor to BeVant Mining & Refining Corporation, Duluth, Minn., a corporation of Minnesota
Application July 12, 1950, Serial No. 173,305
3 Claims. (Cl. 75—5)

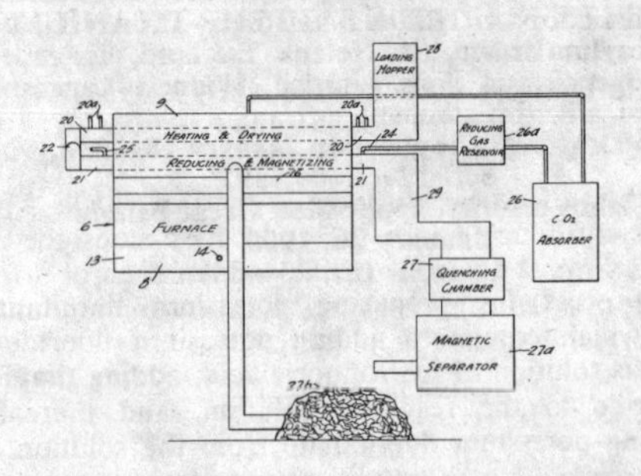

1. The process of treating iron ore which comprises sintering pretreated concentrated ore in a furnace and producing hot gases thereby, using said hot gases for preheating and drying raw ore to be treated out of contact with said gases, conveying the preheated ore to a reducing ore treating chamber, obtaining reducing gases from the expended hot exhaust gases by removing the CO_2 from said gases, agitating the preheated ore in the reducing chamber and passing the reducing gas from which the CO_2 has been substantially removed through the agitated ore, and heating the ore in the reducing chamber by and out of contact with the exhaust gases from the sintering operation, while the ore is being agitated.

2,717,206
METHOD FOR PREPARATION OF LEAD-SODIUM ALLOYS
Hymin Shapiro, Detroit, Mich., assignor to Ethyl Corporation, New York, N. Y., a corporation of Delaware
No Drawing. Application November 27, 1951, Serial No. 258,529
3 Claims. (Cl. 75—167)

1. A method of converting the lead of finely divided predominantly lead residues from a tetraalkyllead process to sodium lead alloy comprising mixing sodium hydride with said finely divided lead in the proportion of about 11.6 parts by weight to 100 parts by weight of lead, then heating to a temperature of from about 250° C. to 400° C. for a time sufficient to convert the lead to sodium lead alloy.

2,717,207
ANIMAL FEEDS
Artemy A. Horvath, Santa Fe, N. Mex.
No Drawing. Application February 2, 1954, Serial No. 407,803
7 Claims. (Cl. 99—2)

1. The process comprising incorporating in animal feed alfalfa residue having substantially no content soluble in aliphatic alcohol lower than C_4.

2,717,208
METHOD OF ACCELERATING THE GROWTH OF CHICKS
Charles M. Ely, Springdale, Ohio, assignor to National Distillers Products Corp., a corporation of Virginia
No Drawing. Application April 9, 1952, Serial No. 281,483
2 Claims. (Cl. 99—4)

1. Method of accelerating the growth of chicks which comprises introducing into the solid food consumed by such chicks from .05–.25% of the amide of a fatty acid of from 8–18 carbon atoms based on the amount of solid foods, air dry basis, supplied.

2,717,209
METHOD OF ACCELERATING THE GROWTH OF CHICKS
Charles M. Ely, Springdale, and Stuart Schott, Cincinnati, Ohio, assignors to National Distillers Products Corp., a corporation of Virginia
No Drawing. Application April 9, 1952, Serial No. 281,486
2 Claims. (Cl. 99—4)

1. Method of accelerating the growth of chicks which comprises introducing into the solid food consumed by such chicks from .05–.50% of the sodium soap of a fatty acid of from 8–18 carbon atoms based on the amount of solid foods, air dry basis, supplied.

2,717,210
METHOD OF OBTAINING CONCENTRATES OF CAROTENE
Marinus Cornelis de Witte, Zwyndrecht, Netherlands
No Drawing. Application August 30, 1952, Serial No. 307,397
4 Claims. (Cl. 99—11)

1. In a method of obtaining carotene concentrates from edible oils by dissolving the edible oil in a hydrocarbon solvent and crystallizing the solution by cooling, the improvement which comprises extracting substantially all of the carotene content by mixing the oil and a hydrocarbon solvent having volatility permitting separation at a temperature subjecting the carotene to not exceeding 70° C., cooling the oil and solvent mixture rapidly at a rate of 5° C. per minute until a temperature of about 10° C. is reached, then cooling slowly a a rate of 0.5–1.5° C. per minute until —25° C. is reached, filtering off the solids and removing the solvent from the filtrate.

2,717,211
BABY FOOD
Hoy A. Cranston, Chicago, Ill., assignor to Carl S. Miner, Chicago, Ill.
Application November 3, 1952, Serial No. 318,421
15 Claims. (Cl. 99—54)

1. A baby food component comprising a mixture of whole milk solids produced by heating and evaporating whole milk, and concentrated, pathogen-free mammary gland phosphatase.

2,717,212
METHOD AND APPARATUS FOR CONTINUOUS CHEESE MANUFACTURE
Bernard T. Hensgen and Johan C. Vanden Bosch, Chicago, Ill., Albrecht M. Lederer, New York, and Peirce M. Wood, Mamaroneck, N. Y., assignors to Swift & Company, Chicago, Ill., a corporation of Illinois
Application April 20, 1949, Serial No. 88,676
7 Claims. (Cl. 99—116)

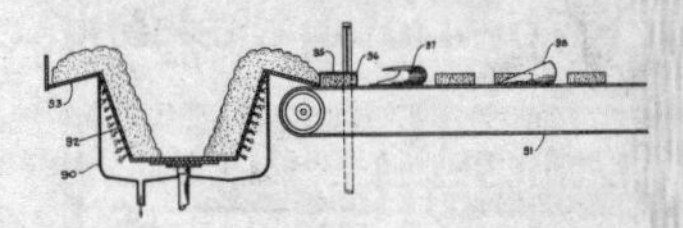

6. The method of operating a centrifuge to separate the whey from a whey and milk curd mixture contained therein and for cheddaring the curd including the steps of rotating the whey and curd mixture at high speed in the centrifuge to separate the majority of the whey from the curd, removing said separated whey from contact with the curd, with the curd remaining on the walls of the centrifuge, reducing the speed of rotation of the separated curd in the centrifuge and continuing its rotation for a period of time sufficient to cheddar the curd.

2,717,213
PROCESS OF PREPARING A PAPER COATING COMPOSITION
Gilbert Stevens, Minneapolis, Minn., assignor to Minnesota and Ontario Paper Company, Minneapolis, Minn.
Application February 10, 1951, Serial No. 210,418
5 Claims. (Cl. 106—214)

1. A continuous process of preparing paper coating composition which has on completion a solid content in excess of 50% comprising blending native starch, mineral pigment, and water; continuously flowing the blended composition through a dispersion zone, a heating zone, introducing into the flowing material an enzyme, then flowing the composition in a confined layer and violently agitating the composition passing through a second heating zone to convert the starch therein, and then inactivating the enzyme.

2,717,214
COATING COMPOSITIONS AND METHODS OF PREPARING SAME
Ralph Marotta, Malden, and Carl R. Martinson, Everett, Mass., assignors to Monsanto Chemical Company, St. Louis, Mo., a corporation of Delaware
No Drawing. Application September 30, 1950, Serial No. 187,818
17 Claims. (Cl. 106—228)

3. A process of preparing flatted and semi-gloss drying oil resin varnishes which comprises grinding a mixture of a silica aerogel and a drying oil resin varnish containing an organic solvent and from about 15 to 35% by weight of solids in a pebble mill in the presence of from about 4 to 20%, on the weight of the aerogel, of a non-ionic surface active condensation product of 7 to 15 mols of ethylene oxide and 1 mol of a tertiary alkyl mercaptan containing from 8 to 16 carbon atoms, said condensation product being soluble in said varnish in amounts of at least 0.05% by weight, until the aerogel is flocculated in said varnish and then adding at least one additional increment of a drying oil resin varnish containing the same ingredients as the initial varnish to said mixture and grinding the mixture in a pebble mill until the aerogel in said varnish remains in a flocculated condition when the varnish is allowed to stand, the amount of varnish thus added being sufficient to provide a composition which contains from about 20 to 50% by weight of solids, the amount of aerogel employed being sufficient to provide from about 5 to 15% of aerogel, based on the total varnish solids.

2,717,215
METHOD FOR DRYING CORDAGE
Clarence F. Faulkner, Athens, Ga., assignor to Puritan Cordage Mills, Inc., Louisville, Ky., a corporation of Kentucky
Application July 8, 1952, Serial No. 297,703
3 Claims. (Cl. 117—7)

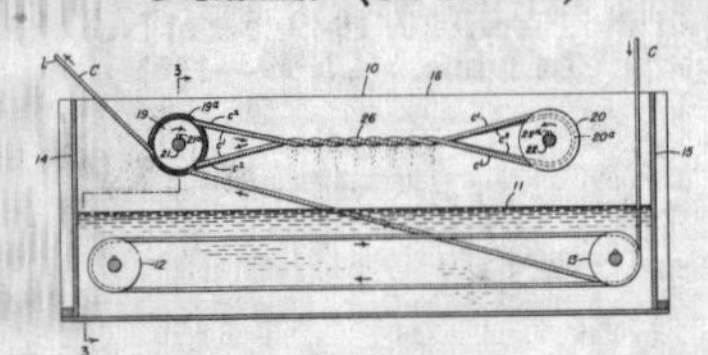

1. The method of removing excess treating liquid from cordage, that has been passed through a treating bath, and of drying said cordage, said method comprising arranging a continuous length of said treated cordage, while in its wet state, in at least two flights with intermediate portions of said flights wound one about the other so as to bring said wound portions in intimate frictional contact, maintaining said flights taut, and drawing said length of cordage longitudinally to cause the same to move progressively in its length and said flight portions in opposite directions.

2,717,216
FLAME-RETARDANT INSULATED CONDUCTORS, METHOD OF MAKING SAME, AND COMPOSITIONS USED TO PREPARE THE SAME
Nicholas F. Arone, Upper Darby, Pa., assignor to General Electric Company, a corporation of New York
Application July 1, 1954, Serial No. 440,784
8 Claims. (Cl. 117—75)

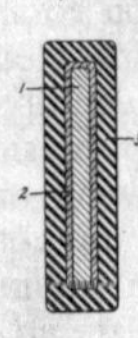

6. A composition of matter comprising, by weight, (1) 100 parts of a complex epoxide resin having a melting point below 76° C. and comprising a polyether derivative of a polyhydric organic compound containing epoxy groups, (2) from 80 to 120 parts of a vinyl chloride resin selected from the class consisting of polyvinyl chloride and copolymers of vinyl chloride and vinyl acetate, (3) from 8 to 50 parts di-(2-ethylbutyl) phthalate, and (4) from 3 to 10 parts antimony trioxide.

2,717,217
PROCESS OF PREPARING COATED FABRICS
David J. Sullivan, Bridgeport, Conn., assignor to E. I. du Pont de Nemours and Company, Wilmington, Del., a corporation of Delaware
No Drawing. Application January 12, 1954, Serial No. 403,650
6 Claims. (Cl. 117—76)

3. Process of preparing dense bubble-free coatings on a fabric substrate which comprises base coating at least one side of a fabric substrate with a solution of a synthetic rubber in a volatile organic liquid to close the interstices of said fabric, passing the coated fabric through a heat zone to evaporate the organic liquid, separately preparing an aqueous slurry of dry particulate compounding ingredients, separately emulsifying oily compounding ingredients, blending said slurry and said emulsion with a synthetic rubber latex to form a coating composition having a viscosity too low for doctor knife application and containing occluded air, the synthetic rubber in said solution and said latex being selected from the group consisting of neoprene, copolymer of butadiene and acrylonitrile, and copolymer of butadiene and styrene, simultaneously concentrating and deaerating said composition by subjecting a thin wet film thereof on a moving surface to heated moving air, continuing the concentration and deaeration until said composition has a vicosity of 6,000 to 16,000 centipoises and is substantially free of occluded air, applying the concentrated and deaerated composition to the coated side of said coated fabric by means of a doctoring device, passing the coated fabric through a heat zone to evaporate the volatile portion of said coating composition and produce a dense bubble-free coating and further heating the coated fabric to cure the coating.

2,717,218
CHEMICAL NICKEL PLATING METHODS AND APPARATUS
Paul Talmey, Barrington, and William J. Crehan, Hinsdale, Ill., assignors to General American Transportation Corporation, Chicago, Ill., a corporation of New York
Application July 19, 1952, Serial No. 299,784
39 Claims. (Cl. 117—97)

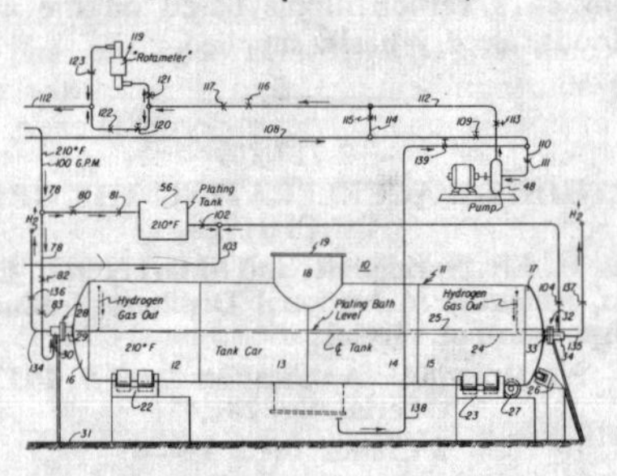

9. The method of chemically plating with nickel the interior of a hollow container formed essentially of an element selected from the group consisting of copper, silver, gold, aluminum, iron, cobalt, nickel, palladium and platinum; which method comprises providing an aqueous chemical nickel plating solution of the nickel cation-hypophosphite anion type having substantially a predetermined composition and characterized by a high plating rate at a temperature within a given range disposed near the boiling point thereof, rotating said container throughout a given time interval about a substantially horizontal axis, maintaining during said rotation and throughout said time interval said container at least partially filled with said solution, and circulating during said rotation and throughout said time interval said solution from the exterior into said container and therethrough and back to the exterior, wherein said solution when introduced into said container has a temperature within said given range, and wherein the rate of circulation of said solution through said container is sufficiently high to maintain the temperature of said fill within said given range and to prevent substantial departure of the composition of said fill from said predetermined composition.

2,717,219
ASBESTOS FIBER ELECTRICAL INSULATING MEMBER IMPREGNATED WITH METHYL HYDROGEN POLYSILOXANE
James G. Ford, Sharon, and Clinton L. Denault, Sharpsville, Pa., assignors to Westinghouse Electric Corporation, East Pittsburgh, Pa., a corporation of Pennsylvania
No Drawing. Application March 29, 1952, Serial No. 279,458
4 Claims. (Cl. 117—126)

1. An electrically insulating member comprising a body of asbestos fibers and an inorganic binder therefor, and at least 2%, based on the weight of said body, of an impregnant applied to said body, the impregnant com-

prising solely a linear polysiloxane having the recurring group

$$\left[\begin{array}{c} CH_3 \\ | \\ -Si-O \\ | \\ H \end{array} \right]$$

and end-blocking groups comprising $(CH_3)_3$ Si, the body with the applied polysiloxane impregnant having been heated at a temperature of from about 140° C. to 300° C. for at least several hours.

2,717,220
HEAT RESISTANT FABRIC COATED WITH A FUSED COMPOSITION COMPRISING POLYTETRAFLUOROETHYLENE AND CRYOLITE, AND METHOD OF PRODUCING SAME
Robert E. Fay, Jr., Highland Mills, N. Y., assignor to E. I. du Pont de Nemours and Company, Wilmington, Del., a corporation of Delaware
No Drawing. Application June 4, 1953, Serial No. 359,640
6 Claims. (Cl. 117—126)

1. A heat resistant fabric coated with a fused composition comprising polytetrafluoroethylene and cryolite, said cryolite representing 5% to 40% of the combined weight of cryolite and polytetrafluoroethylene.

2,717,221
METAL WORKING METHOD
Robert M. Christner, Beaver Falls, Pa.
No Drawing. Application January 12, 1950, Serial No. 138,274
2 Claims. (Cl. 117—127)

1. In the process of drawing stainless steel the steps which comprise coating the steel with a slurry formed of water and the reaction product of ferrous sulphate and lime in water in substantially stoichiometric amounts, drying the slurry on the stainless steel to form a coating thereover and during which drying the iron compound present is oxidized by contact with air.

2,717,222

WITHDRAWN

2,717,223
PROCESS FOR PRODUCING DUCTILE MAGNETIC COBALT-IRON ALLOY MEMBERS
Martin H. Binstock, Pittsburgh, Pa., and Hans A. Steinherz, Beverly, Mass., assignors to Westinghouse Electric Corporation, East Pittsburgh, Pa., a corporation of Pennsylvania
Application February 13, 1952, Serial No. 271,348
3 Claims. (Cl. 148—122)

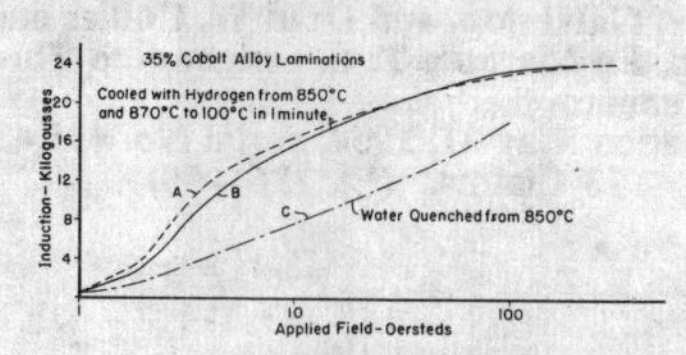

1. The process for treating laminations of a magnetic material composed of from 20% to 50% cobalt, up to 3% vanadium, up to 1% manganese, up to 2.5% chromium, up to 0.5% of arsenic, up to 0.5% silicon, less than 0.1% of carbon, other impurities not exceeding 0.5%, and the balance being iron, comprising solely the steps of heating the laminations in a substantially non-carburizing, non-oxidizing atmosphere to a temperature of from 800° C. to 925° C. for at least several minutes and thereafter cooling the laminations in a substantially non-oxidizing atmosphere from this temperature to approximately 100° C. at a rate of from 5° C. to 60° C. per second, whereby the laminations have a high magnetic permeability of over 20,000 gausses at 100 oersteds, a ductility sufficient to enable laminations 0.025 inch thick to be bent at least 90° around a ¼ inch radius without failing and possess low electrical losses in an alternating current field.

2,717,224
METHOD OF BONDING PAPER AND COMPOSITION THEREFOR
Albert L. McConnell, Morton, and Robert W. Medeiros, Swarthmore, Pa., assignors to Scott Paper Company, Chester, Pa., a corporation of Pennsylvania
Application July 20, 1954, Serial No. 444,673
19 Claims. (Cl. 154—117)

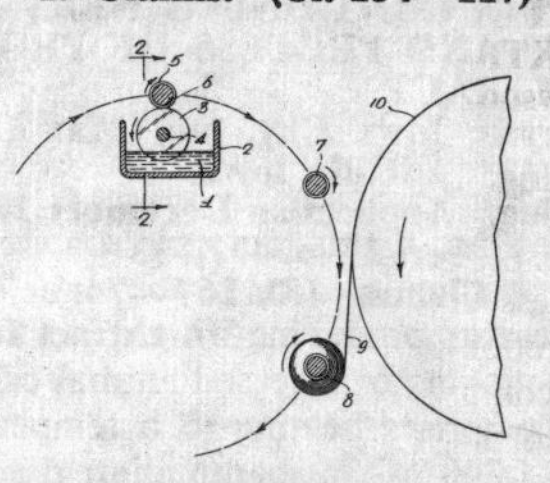

1. The method of bonding paper temporarily to another surface, at least one of which members is absorbent, which comprises bringing together the paper and the surface to which the paper is to be adhered temporarily with hydrogenated rosin therebetween.

2,717,225
SINTERED REFRACTORY MASS
Albert Etheridge Williams, London, England
Application August 25, 1949, Serial No. 112,215
Claims priority, application Great Britain August 30, 1948
6 Claims. (Cl. 154—128)

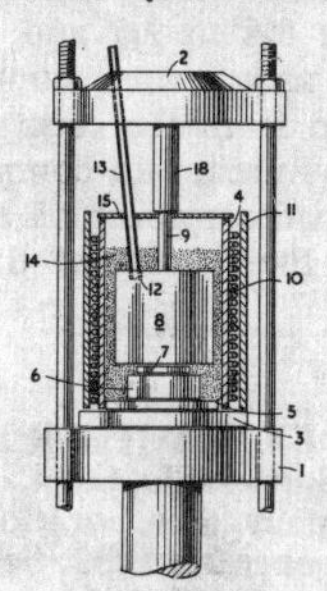

1. A method of joining pre-sintered shaped bodies of refractory oxide together while retaining their configurations which comprises maintaining the bodies in contact with each other by application of a pressure of about ½ to 1 ton per square inch of contacting surface while they are at a sintering temperature in the range 1400° C. to 2000° C.

2,717,226
PROCESS FOR THE MANUFACTURE OF A RAT EXTERMINATING COMPOUND CONTAINING DESICCATED, STORABLE SALMONELLA BACTERIA IN THEIR VIRULENT CONDITION
Aage Thorsen Skovsted, Valby, near Copenhagen, Denmark
No Drawing. Application April 3, 1951, Serial No. 219,132
Claims priority, application Denmark April 11, 1950
3 Claims. (Cl. 167—46)

1. A process for the manufacture of a rat exterminating compound containing desiccated, storable Salmonella

bacteria in their virulent condition, which comprises placing unfrozen Salmonella bacteria in an ampule in a zone containing a chemical drying agent to provide a desiccating atmosphere, exhausting the atmosphere from said ampule to reduce the pressure inside said ampule to a final value of about ½ mm. of mercury, the reduction in pressure being effected in step-wise manner by reducing the pressure inside the ampule to about 100 mm. of mercury the first day and then reducing the pressure further for a series of successive days to desiccate said bacteria, said pressure being reduced on each of said successive days to a value of about ½ the value on the preceding day, and sealing said ampule under vacuum, whereby to maintain said reduced pressure in the ampule during storage.

2,717,227

COMPOSITION CONTAINING NERVE TISSUE EXTRACT AND PROCESS OF PRODUCING SUCH EXTRACT

Helen L. Dawson, Iowa City, and Max D. Wheatley, Hills, Iowa

No Drawing. Application December 22, 1954, Serial No. 477,133

6 Claims. (Cl. 167—74)

1. The process of producing an extract for curing or healing malfunctioning or diseased animal cells, tissues or organs which comprises heating to a temperature of the order of 100 to 300° C. a mammalian tissue consisting essentially of nerve tissue in the presence of an oxidizing agent to oxidize said tissue and extracting the oxidized material with a water soluble solvent compatible with the human body.

2,717,228

HAIR WAVING COMPOSITION

Alfred E. Brown, Takoma Park, Md., assignor to The Gillette Company, a corporation of Delaware

No Drawing. Application April 23, 1951, Serial No. 222,527

12 Claims. (Cl. 167—87.1)

1. An aqueous hair waving composition having a pH within the range 8.6 to 9.5 and comprising a thiol hair-reducing agent and a urea compound capable of swelling hair selected from the class consisting of urea, methyl urea and ethyl urea, said composition being from 0.35 to 0.8 molar in the respect of the thiol and from 0.5 to 4.0 molar in the respect of the urea compound.

2,717,229

SOLVENT EXTRACTION PROCESS

Robert A. Findlay, Bartlesville, Okla., assignor to Phillips Petroleum Company, a corporation of Delaware

Application September 22, 1952, Serial No. 310,861

6 Claims. (Cl. 196—14.15)

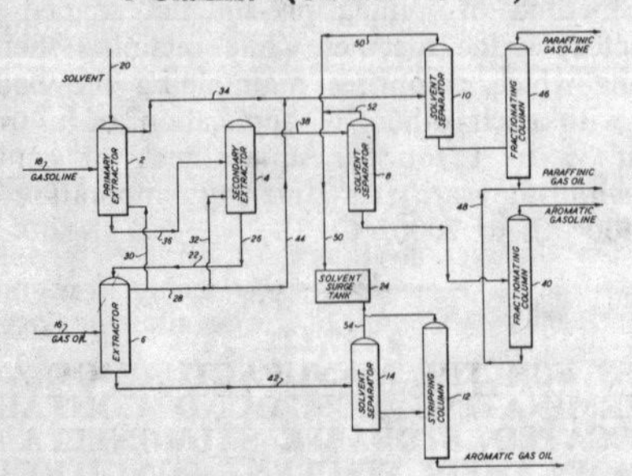

1. In a process which comprises concomitantly solvent extracting a gasoline stream and a gas oil stream for separating aromatics therefrom the steps which comprise extracting gasoline in a first extraction zone with a selective solvent and in the presence of gas oil raffinate as a stripping agent to produce a primary extract containing an aromatic gasoline component and a raffinate containing a paraffinic gasoline component, separately stripping said primary extract in a secondary extraction zone with gas oil raffinate to remove the aromatic gasoline component from said extract, extracting a gas oil fraction with the extract from said stripping operation, withdrawing as a single stream from said gas oil extraction a gas oil raffinate for use in the two stripping operations, any excess being bypassed to the raffiniate stream from the gasoline extraction step, flashing each gasoline extraction raffinate and gas oil extract to remove solvent therefrom and subsequently distilling each stream to recover paraffinic and aromatic components.

2,717,230

CATALYTIC REFORMING OF HYDROCARBON CHARGE STOCKS HIGH IN NITROGEN COMPOUNDS

Maurice J. Murray, Naperville, and Vladimir Haensel and Henry W. Grote, Hinsdale, Ill., assignors to Universal Oil Products Company, Chicago, Ill., a corporation of Delaware

Application June 19, 1951, Serial No. 232,394

7 Claims. (Cl. 196—50)

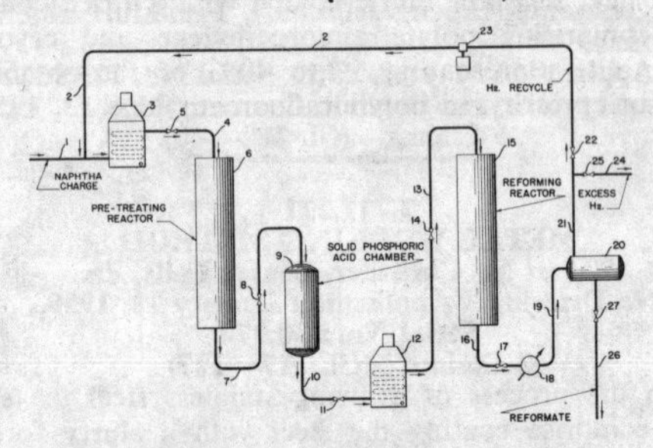

5. In the conversion of a gasoline fraction in a reforming process involving hydrocracking in the presence of a catalyst comprising platinum and alumina, said gasoline fraction containing nitrogen compounds having a depressing effect on the hydrocracking reaction and a poisoning effect on said catalyst, the method which comprises contacting the gasoline fraction in the presence of hydrogen with a hydrogenation-dehydrogenation catalyst to crack said nitrogen compounds and form ammonia therefrom, contacting the resultant products with a solid phosphoric acid catalyst which has been calcined at a temperature of from about 600° C. to about 1000° C. to separate the ammonia from the gasoline fraction, and thereafter reforming the gasoline fraction in the presence of said catalyst comprising platinum and alumina.

2,717,231

PROCESS FOR TREATING AROMATIC-DIOLEFIN MIXTURES

Irvin H. Lutz, Galveston, and Oran W. Collier and David J. Bellman, La Marque, Tex., assignors to The American Oil Company

Application May 21, 1954, Serial No. 431,432

13 Claims. (Cl. 196—50)

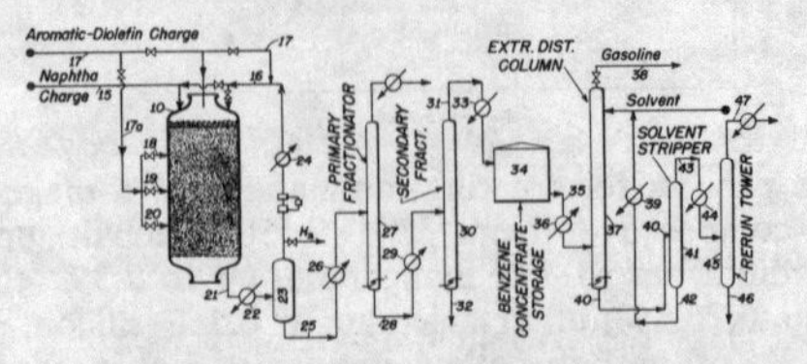

1. A method which comprises introducing a vaporized petroleum naphtha charge into a space above the catalyst bed in a hydroforming reactor; introducing as a separate stream to said space between about 5% and about 30% by volume, based upon total hydrocarbon charge,

of a liquid hydrocarbon fraction resulting from the pyrolysis of a petroleum fraction and boiling chiefly in the gasoline boiling range which contains a volume percentage of diolefins substantially in excess of that present in a cracked naphtha fraction and a volume of aromatics substantially in excess of said diolefin content, said hydrocarbon fraction being introduced at a temperature below the polymerization temperature of said diolefins; introducing into admixture with said naphtha and said hydrocarbon fraction in said space a substantial amount of uncombined hydrogen; and contacting the admixture of naphtha petroleum fraction, normally liquid hydrocarbon fraction and uncombined hydrogen with hydroforming catalyst under hydroforming conditions and recovering an aromatic hydrocarbon, initially present in said hydrocarbon fraction, from the hydroformer product.

2,717,232
DEHYDRATION AND FRACTIONATION OF CRUDE PYRIDINE
Julius Geller, Bad Homburg vor der Hohe, and Heinrich Ratte, Frankfurt am Main, Germany, assignors to Rutgerswerke-Aktiengesellschaft, Frankfurt am Main, Germany
Application May 12, 1951, Serial No. 225,952
Claims priority, application Germany May 13, 1950
6 Claims. (Cl. 202—40)

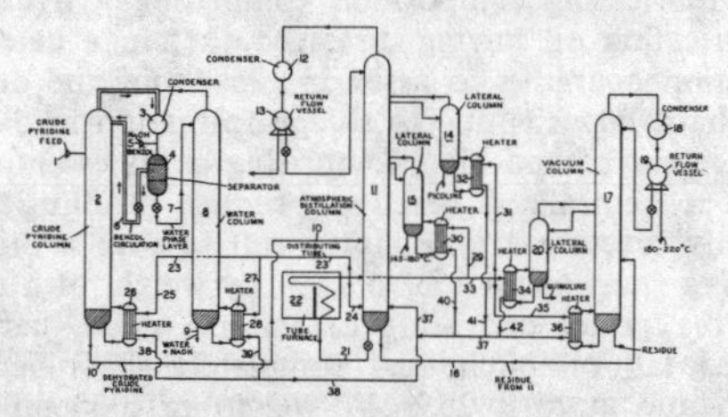

1. A process for continuous dehydration and fractionation of crude pyridine, comprising in combination the following steps: (*a*) subjecting crude pyridine to fractional distillation in a first fractionating column in order to obtain a distillation residue consisting of crude pyridine substantially freed from water and an overhead distillate consisting substantially of a constant boiling mixture of pyridine and water; (*b*) condensing and cooling said overhead distillate, mixing it in a separating vessel with caustic soda lye and benzene simultaneously, and allowing it to separate into a first, predominantly benzene- and pyradine- containing layer and a second, predominantly alkali- and water-containing layer; (*c*) supplying liquid from said first layer to the upper tray of said first column; (*d*) supplying liquid from said second layer to the upper tray of a second fractionating column in order to obtain as a distillation residue water and an overhead distillate consisting substantially of a constant boiling mixture of pyridine and water; (*e*) condensing and introducing said overhead distillate of said second column into said separating vessel and subjecting the resulting dehydrated crude pyridine to fractional distillation.

2,717,233
SEPARATION OF GAMMA BENZENE HEXACHLORIDE BY DISTILLATION
Percy W. Trotter, Baton Rouge, La., assignor to Ethyl Corporation, New York, N. Y., a corporation of Delaware
No Drawing. Application October 30, 1952, Serial No. 317,825
2 Claims. (Cl. 202—57)

2. A method for treating benzene hexachloride isomer mixture containing the gamma and delta isomers comprising distilling the isomer mixture in the presence of a polar solvent selected from the group consisting of triethylene glycol and glycerine until a major portion of the gamma isomer is distilled from the mixture and interrupting said distillation while a major portion of the delta isomer still remains undistilled, and recovering the gamma isomer fraction separate from the delta isomer fraction.

2,717,234
METHOD OF PREPARING K_2UF_6 FOR FUSED BATH ELECTROLYSIS
Rudolph Nagy, Bloomfield, and John W. Marden, East Orange, N. J., assignors to the United States of America as represented by the United States Atomic Energy Commission
No Drawing. Application March 6, 1943, Serial No. 478,271
11 Claims. (Cl. 204—10)

1. The method of producing di-potassium uranous fluoride (K_2UF_6) which comprises heating uranous oxide with potassium acid fluoride at an elevated temperature.

8. A fusible electrolyte comprising as its essential ingredient K_2UF_6 and containing a material selected from the group consisting of alkali metal and alkaline earth metal halides.

2,717,235
METHOD OF PREPARING SULFUR HEXAFLUORIDE
Maurice Prober, Schenectady, N. Y., assignor to General Electric Company, a corporation of New York
Application November 23, 1951, Serial No. 257,693
5 Claims. (Cl. 204—59)

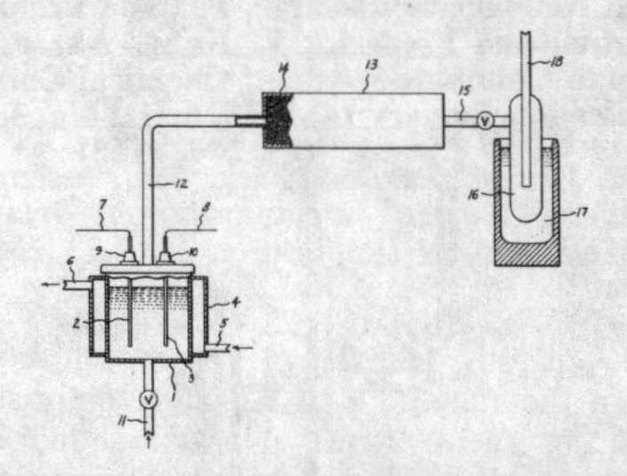

1. The method of preparing sulfur hexafluoride in an electrolytic cell with an insoluble anode, which method comprises bringing an inorganic covalent sulfur compound selected from the class consisting of hydrogen sulfide, carbon disulfide and sulfur monochloride into contact with an electrolyte consisting essentially of substantially anhydrous hydrogen fluoride and a conductivity-promoting solute maintained in said electrolytic cell while maintaining a cell voltage insufficient to generate free fluorine, carrying away gaseous reaction products consisting of sulfur hexafluoride and lower fluorides of sulfur, purifying said products and separating sulfur hexafluoride therefrom.

2,717,236
ELECTROLYTIC PREPARATION OF A DIHYDROSTREPTOMYCIN SULPHATE
Morris A. Dolliver, Stelton, and Serge Semenoff, New Brunswick, N. J., assignors, by mesne assignments, to Olin Mathieson Chemical Corporation, a corporation of Virginia
No Drawing. Application December 9, 1949, Serial No. 132,219
1 Claim. (Cl. 204—75)

The process of converting a streptomycin hydrochloride into a dihydrostreptomycin sulfate, which essentially comprises charging an electrolytic cell, having anode and cathode compartments separated by a semi-permeable diaphragm, with a nonalkaline, electric current-conducting aqueous solution of said streptomycin hydrochloride as the catholyte and an aqueous solution of a strong inorganic acid as the anolyte, the acid being substantially non-reactive with the anode, and the cathode being substantially non-reactive with the components of the catholyte, passing an electric current between the anode and the cathode in the respective compartments while adding sulfuric acid to the catholyte until said hydrochloride salt is substantially completely converted to a sulfate salt, the electrolysis being effected at a temperature below that at which the streptomycin and the dihydrostreptomycin decomposes in the catholyte and until the catholyte is substantially free of chloride ions, and recovering the sulfate salt substantially free of the hydrochloride salt from the catholyte.

2,717,237
PRODUCTION OF CHLORINE DIOXIDE
Nikolaus Rempel, deceased, late of Leverkusen-Bayerwerk, Germany, by Erika G. Rempel, administratrix, Leverkusen-Bayerwerk, Germany, assignor to Farbenfabriken Bayer Aktiengesellschaft, Leverkusen, Germany
No Drawing. Application June 25, 1952, Serial No. 295,551
4 Claims. (Cl. 204—101)

1. In the process for the production of substantially chlorine-free chlorine dioxide by electrolysis of a chlorite, the improvement which comprises electrolyzing an aqueous solution of a chlorite in the presence of a water-soluble metal sulfate.

2,717,238
PREPARATION OF BENZENE HEXACHLORIDE
Joseph A. Neubauer, Pittsburgh, Pa., and Franklin Strain, Barberton, and Frederick E. Kung, Akron, Ohio, assignors to Columbia-Southern Chemical Corporation
Application January 14, 1953, Serial No. 331,184
27 Claims. (Cl. 204—163)

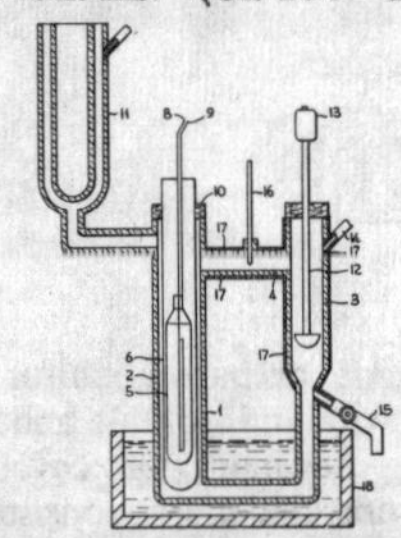

4. A method of preparing benzene hexachloride which comprises irradiating with actinic light a mixture of benzene and an inert polar solvent which at 20° C. has a dielectric constant of at least about 4, introducing chlorine into said mixture at a rate sufficient to maintain in the mixture an appreciable chlorine concentration of from 0.001 per cent up to 1.5 per cent by weight of the unreacted benzene and solvent and maintaining the reaction mixture at a temperature below the freezing point of benzene but above the temperature at which the mixture is solid.

2,717,239
ELECTRICALLY CONDUCTIVE OIL-BASE DRILLING FLUIDS
Paul W. Fischer, Whittier, and John W. Scheffel, Fullerton, Calif., assignors to Union Oil Company of California, Los Angeles, Calif., a corporation of California
No Drawing. Application June 30, 1952, Serial No. 296,496
20 Claims. (Cl. 252—8.5)

1. A conductivity additive composition for oil-base drilling fluids, comprising 100 parts by weight of water, between about 10 and about 40 parts by weight of a mixture of an alkali-metal hydroxide and water-soluble salt of a strong base and a weak acid, and between about 20 and about 80 parts by weight of an alkali-metal salt of carboxymethyl cellulose.

2,717,240
METHOD OF MAKING MAGNESIUM METABORATE SOLUTIONS
George D. Fronmuller, Mamaroneck, N. Y., assignor to Commonwealth Color & Chemical Co., New York, N. Y., a corporation of New York
No Drawing. Application February 1, 1952, Serial No. 269,603
7 Claims. (Cl. 252—8.6)

1. A method which comprises mixing a solution of magnesium chloride with a solution of an alkali metal meta borate in substantially equimolecular proportions at a temperature not substantially higher than 25° C. while maintaining a pH of 6.5–9.0 in the mixed solution whereby a stable solution of magnesium meta borate is formed, the concentration of said solutions being sufficient to form at least a 2% solution of magnesium meta borate which is stable at temperatures lower than 50° C.

2,717,241
TALL OIL PITCH-PHOSPHORUS SULFIDE REACTION PRODUCT AND METALLIC SALTS AS DISPERSANTS FOR LUBRICATING OILS
Edwin O. Hook, New Canaan, and Lindley C. Beegle, Darien, Conn., assignors to American Cyanamid Company, New York, N. Y., a corporation of Maine
No Drawing. Application February 9, 1953, Serial No. 336,006
7 Claims. (Cl. 252—32.7)

4. A lubricating composition comprising a hydrocarbon lubricating oil having dissolved therein a detergent composition operative to assist in preventing the deposition of hard deposits due to sludge formation in the oil, said detergent composition comprising as its essential ingredient a neutralized reaction product resulting from chemically reacting a phosphorus sulfide with a mixture comprising from about 1% to 75% by weight of a member of the group consisting of tall oil pitch, partially saponified tall oil pitch, and completely saponified tall oil pitch and at least 25% by weight of a member of the group consisting of an unsaturated ester wax, a partially saponified unsaturated ester wax and a completely saponified unsaturated ester wax, and neutralizing the reaction product with a suitable reactive metal compound.

2,717,242
POLYOXYALKYLENE LUBRICANT COMPOSITION
Edward G. Foehr, San Rafael, Calif., assignor to California Research Corporation, San Francisco, Calif., a corporation of Delaware
No Drawing. Application December 5, 1951, Serial No. 260,113
6 Claims. (Cl. 252—49.6)

1. A lubricant composition comprising a dialkyl poly-1,2-propylene glycol diether having a molecular weight of from about 400 to 800 and wherein the alkyl end groups contain from 1 to 8 carbon atoms, together with an amount of from about one-fourth to about four-fold, in terms of the volume of said ether, of a member of the group consisting of tetraalkyl silicate and hexaalkoxy disiloxane wherein said alkyl groups contain from 5 to 8 carbon atoms each and have a branched-chain structure.

2,717,243
NON-CAKING ALKYL ARYL SULFONATE DETERGENT COMPOSITIONS
Herman S. Bloch, Chicago, and George L. Hervert, Downers Grove, Ill., assignors to Universal Oil Products Company, Chicago, Ill., a corporation of Delaware
No Drawing. Application June 19, 1951, Serial No. 232,478
18 Claims. (Cl. 252—138)

1. A comminuted detergent composition comprising a free-flowing water-soluble mixture of finely divided solid particles of: (1) an anionic detergent selected from the group consisting of the alkali metal, ammonium, alkyl

ammonium and hydroxy-alkyl ammonium salts of an alkyl aromatic sulfonic acid containing an alkyl group of from about 9 to about 18 carbon atoms; (2) an alkyl benzene sulfonate salt of a metal selected from the group consisting of magnesium, calcium, strontium, barium, aluminum and zinc, the last-mentioned salt having an alkyl group of from 1 to about 18 carbon atoms and being in sufficient amount to prevent caking of said detergent, and (3) a sufficient amount of a sequestering agent to enhance the solubility of the last-mentioned sulfonate salt.

9. The composition of claim 1 further characterized in that said sequestering agent is an alkali metal, molecularly dehydrated complex phosphate salt.

2,717,244
LUMINESCENT MATERIAL
Pieter Zalm and Jan van den Boomgaard, Eindhoven, Netherlands, assignors to Hartford National Bank and Trust Company, Hartford, Conn., as trustee
Application January 14, 1953, Serial No. 331,188
Claims priority, application Netherlands January 15, 1952
3 Claims. (Cl. 252—301.6)

1. A luminescent manganese-activated zinc-phosphatosilicate with a peak emission between 6000 and 6200 Å. and being stable at room temperature having the general formula:

$$aZnO.bSiO_2.cP_2O_5.dMnO$$

wherein

$a{:}b$ lies between 2.2:1 and 1:1,
$d{:}a$ lies between 1:200 and 1:5 and
$c{:}a$ lies between 1:400 and 1:10

2,717,245
COATING COMPOSITION FOR FLOORING, WALLS, AND THE LIKE
Annis G. Asaff, Auburndale, Mass., assignor to Callaghan Hession Corporation, Boston, Mass., a corporation of Massachusetts
No Drawing. Application December 18, 1953, Serial No. 399,181
8 Claims. (Cl. 260—31.2)

1. The method of forming a surface coating mastic comprising adding and mixing an aliphatic alcohol having less than four carbon atoms with a mixture of aggregate filler material and a vehicle therefor comprising firm-forming material selected from the group consisting of chlorinated rubber and polystyrene dissolved in a solvent comprising between about 50 and 75 per cent by weight of an acetate ester of an aliphatic alcohol having less than 6 carbon atoms and between about 50 and 25 per cent by weight of a liquid aromatic hydrocarbon in the proportion of between about 30 and 60 grams of film-forming material to 100 cc. of solvent, said alcohol being added until a two-phase separation of the mixture occurs, one phase being the mastic composition which is plastic and cohesive and of such non-sticky character that it can be spread with a trowel, and separating the two phases to obtain said mastic.

2,717,246
TITANIUM DIOXIDE PIGMENT COATED WITH A HYDROUS OXIDE AND A POLYSILOXANE
Roy H. Kienle, Bound Brook, John W. Eastes, Somerville, and Theodore F. Cooke, Martinsville, N. J., assignors to American Cyanamid Company, New York, N. Y., a corporation of Maine
No Drawing. Application December 10, 1952, Serial No. 325,231
13 Claims. (Cl. 260—37)

1. Titanium dioxide pigment particles coated first with between about 0.1% and 2.5% their weight of a water-insoluble polyvalent metal hydrous oxide and then with between about 0.005% and 15% of their weight of a thermocured polyorganosiloxane, said polyorganosiloxane being the hydrolysis product of at least one silane of the formula A_2SiXY, wherein A is a substituent selected from the group consisting of Cl and H, X is selected from the group consisting of A and Y, and Y is a hydrocarbon radical having fewer than 22 carbon atoms.

2,717,247
INTERPOLYMERS OF STYRENE, ALLYL ACETATE AND ALKYL HALF ESTERS OF MALEIC ACID AND PROCESS FOR PREPARING THE SAME
Leo L. Contois, Jr., Springfield, Mass., assignor to Monsanto Chemical Company, St. Louis, Mo., a corporation of Delaware
No Drawing. Application December 21, 1953, Serial No. 399,618
15 Claims. (Cl. 260—78.5)

9. A low molecular weight interpolymer of styrene, allyl acetate and a member of the group consisting of unsubstituted and halogen-substituted saturated C_1 to C_8 alkyl half esters of maleic acid and mixtures thereof, a 15% solids aqueous ammoniacal solution of the ammonium salt of said interpolymer at normal temperatures having a viscosity of less than 600 centipoises, said interpolymer consisting of 1–2 mols of styrene and 0.075–0.15 mol of allyl acetate per mol of alkyl half ester.

2,717,248
POLYMERIZATION OF VINYL CHLORIDE WITH MONOPERMALONATE CATALYST
William E. Vaughan, Berkeley, and Fred E. Condo, El Cerrito, Calif., assignors to Shell Development Company, San Francisco, Calif., a corporation of Delaware
No Drawing. Application August 2, 1948, Serial No. 42,153
1 Claim. (Cl. 260—92.8)

A process comprising polymerizing vinyl chloride in a mildly alkaline aqueous emulsion in the presence of 0.01% to 0.5% by weight of O,O-tertiary-butyl O-ethyl monopermalonate having the structure:

$$CH_2\begin{cases} C(=O)-O-O-C_4H_7 \\ C(=O)-O-C_2H_5 \end{cases}$$

at a temperature between —10° C. and +50° C.

2,717,249
METHOD OF MAKING PYROPHOSPHORTETRAAMIDES OF SECONDARY ALIPHATIC AMINES
Arthur D. F. Toy, Park Forest, and James R. Costello, Jr., Chicago Heights, Ill., assignors to Victor Chemical Works, a corporation of Illinois
No Drawing. Application July 28, 1950, Serial No. 176,526
9 Claims. (Cl. 260—247.5)

1. The method of making a pyrophosphortetraamide of a secondary aliphatic amine comprising reacting a member of the class consisting of bis(dialkylamido) phosphoryl chlorides, dipiperidino phosphoryl chloride and dimorpholido phosphoryl chloride with water and triethylamine and separating the resulting pyrophosphortetraamide.

2,717,250
FORMYLATION OF TETRAHYDRO-10-FORMYLPTEROIC ACID AND AMINO ACID AMIDES
Martin E. Hultquist, Bound Brook, and Barbara Roth, Middlesex Borough, N. J., assignors to American Cyanamid Company, New York, N. Y., a corporation of Maine
No Drawing. Application April 29, 1950, Serial No. 159,152
6 Claims. (Cl. 260—251.5)

1. A method of preparing a member of the group consisting of tetrahydro-10-formylpteroic acid and amino

acid amides of tetrahydro-10-formylpteroic acid and amino acid amides of tetrahydro-10-formylpteroic acid which comprises treating a compound of the group consisting of tetrahydropteroic acid and amino acid amides of tetrahydropteroic acid with formylating agents of the group consisting of formic acid, lower alkyl formates and glycol formate.

2,717,251
1-METHYL-4-AMINO-N'-PHENYL-N'-THENYL-PIPERIDINES

Arthur Stoll, Arlesheim, near Basel, and Jean-Pierre Bourquin, Basel, Switzerland, assignors to Sandoz A. G., Basel, Switzerland
No Drawing. Application November 26, 1952, Serial No. 322,792
Claims priority, application Switzerland November 30, 1951
4 Claims. (Cl. 260—293.4)

1. A member selected from the group consisting of compounds which correspond to the type formula

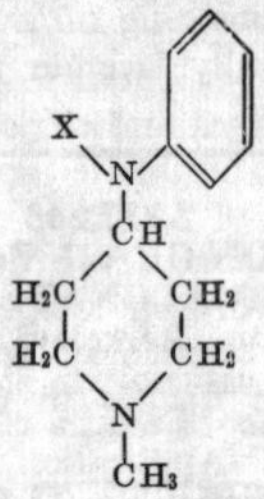

X—N(C₆H₅)—CH< (H₂C—CH₂, H₂C—CH₂) >N—CH₃

and salts thereof with acids, wherein X is a member selected from the group consisting of

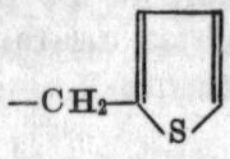

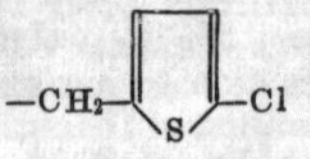

—CH₂—(thienyl, S)   —CH₂—(thienyl, S)—Cl

and

—CH₂—(thienyl, S)—Br

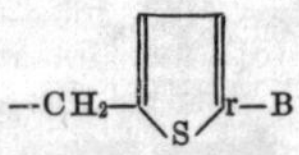

2,717,252
PREPARATION OF METHIONINE HYDANTOIN

David Oliver Holland, Dorking, England, assignor to Beecham Research Laboratories Limited, Betchworth, England, a company of Great Britain
No Drawing. Application December 15, 1952, Serial No. 326,168
Claims priority, application Great Britain December 21, 1951
10 Claims. (Cl. 260—309.5)

1. Process for preparing methionine hydantoin by reacting methionine nitrile with carbon dioxide in the presence of an amine of basicity substantially greater than the nitrile.

2,717,253
PREPARATION OF METHIONINE HYDANTOIN

David Oliver Holland, Dorking, England, assignor to Beecham Research Laboratories Limited, Betchworth, England, a company of Great Britain
No Drawing. Application December 15, 1952, Serial No. 326,169
Claims priority, application Great Britain December 21, 1951
14 Claims. (Cl. 260—309.5)

1. Process for preparing methionine hydantoin by reacting β-methylthiopropaldehyde with an ammonium carbonate and an inorganic cyanide in the presence of a tertiary alkylamine.

2,717,254
NITRO ALKYL DERIVATIVES OF NITROTHIOPHENEMETHANOL

Robert E. Miller, Dayton, Ohio, assignor to Monsanto Chemical Company, St. Louis, Mo., a corporation of Delaware
No Drawing. Application August 5, 1954, Serial No. 448,144
4 Claims. (Cl. 260—332.3)

1. A compound having the formula

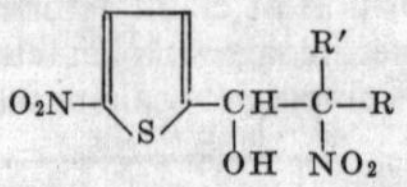

O_2N—(thienyl, S)—CH(OH)—C(R')(NO₂)—R

in which R is selected from the class consisting of hydrogen and alkyl radicals of from 1 to 3 carbon atoms and R' is selected from the class consisting of hydrogen and the methyl radical.

2,717,255
NEW INTERMEDIATES OF THE ANTHRAQUINONE SERIES

Harry Edward Westlake, Jr., Somerville, and William Baptist Hardy, Bound Brook, N. J., assignors to American Cyanamid Company, New York, N. Y., a corporation of Maine
No Drawing. Application June 10, 1952, Serial No. 292,714
3 Claims. (Cl. 260—381)

1. Compounds having the formula:

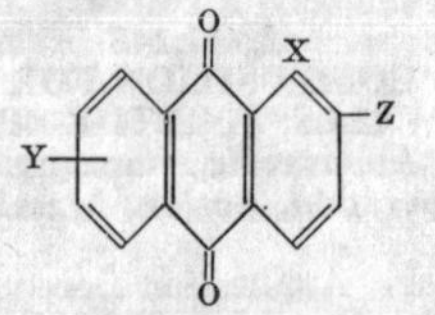

(anthraquinone with Y on one outer ring; X and Z on the other outer ring)

in which X is a halogen, and Y is an α-substituent selected from the class consisting of NO_2 and NH_2 groups, and Z is the same radical as Y.

2,717,256
OIL BLEACHING METHOD

Charles E. McMichael, John W. Godbey, and Victor L. Zehnder, Louisville, Ky., assignors, by mesne assignments, to National Cylinder Gas Company, Chicago, Ill., a corporation of Illinois
Application August 28, 1951, Serial No. 244,016
9 Claims. (Cl. 260—428)

1. In a process for bleaching oil with a bleaching adsorbent, which includes the steps of subjecting the oil to the bleaching action of partially spent bleaching adsorbent, mixing the partially bleached oil thus obtained with fresh bleaching adsorbent, and separating the resulting partially spent adsorbent therefrom for use in bleaching additional oil according to the first bleaching step, the improvement which comprises forming a slurry by mixing a fractional portion of the fully bleached oil with fresh adsorbent and adding this slurry to the partially bleached oil in the second bleaching step.

5. A method for bleaching oil with a bleaching adsorbent comprising mixing fresh adsorbent with a small amount of oil to form a slurry, dispersing said slurry in space and subjecting the dispersed slurry to the action of stripping steam to effect deaeration and dehydration of the adsorbent, mixing the deaerated and dehydrated adsorbent with partially bleached oil to effect a second stage of bleaching thereof, filtering the adsorbent from said oil, and subjecting fresh oil to a first stage bleaching by contact with the bleaching adsorbent separated by the filtration.

2,717,257
REDISTRIBUTION OF ORGANOSILANES
Ben A. Bluestein, Schenectady, N. Y., assignor to General Electric Company, a corporation of New York
No Drawing. Application September 9, 1952, Serial No. 308,719
7 Claims. (Cl. 260—448.2)

1. The process for effecting reaction on a continuous basis between (1) a preformed compound corresponding to the general formula

$$(R)_mSiCl_{(4-m)}$$

and (2) a preformed compound corresponding to the general formula

$$(R')_nSiCl_{(4-n)}$$

where R and R′ are alkyl radicals, *m* is a whole number equal to from 1 to 4, *n* is a whole number equal to from 0 to 3, and *n* may equal 4 where R and R′ are dissimilar alkyl radicals, which process comprises (*a*) simultaneously passing compounds (1) and (2) in the vapor phase through a reaction zone maintained at a temperature of from 250° to 500° C. and containing an inert carrier supporting an alkali-metal halogenoaluminate catalyst for the reaction, the reaction zone being substantially free of unsupported, sublimed catalyst at the reaction temperature and (*b*) continuously withdrawing the reaction product from the reaction zone.

2,717,258
METHODS FOR PREPARING DIMETHYL SILANEDIOL
Simon W. Kantor, Schenectady, N. Y., assignor to General Electric Company, a corporation of New York
No Drawing. Application February 25, 1953, Serial No. 338,873
2 Claims. (Cl. 260—448.2)

1. The process for making dimethyl silanediol which comprises forming a mixture of ingredients completely free of any traces of acid and alkali and comprising a dimethyldialkoxysilane and neutral water, the said water being in an amount in excess of that required for complete hydrolysis of the dimethyl dialkoxysilane to dimethyl silanediol, thereafter heating the reaction mixture, and isolating the dimethyl silanediol.

2,717,259
HYDROCARBON SYNTHESIS EMPLOYING AN IRON CATALYST IN THE PRESENCE OF A HALOGEN CONTAINING REGULATOR
Hubert G. Davis and Thomas P. Wilson, Charleston, W. Va., assignors to Union Carbide and Carbon Corporation, a corporation of New York
Application January 4, 1950, Serial No. 136,813
14 Claims. (Cl. 260—449.6)

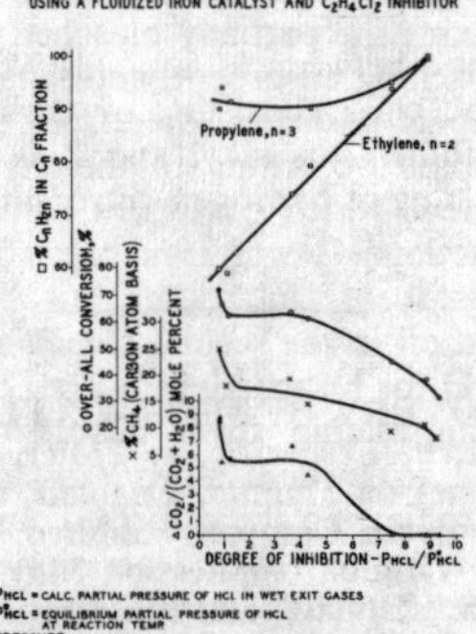

1. In a process for producing liquid and gaseous hydrocarbons wherein an iron-containing hydrocarbon synthesis catalyst is contacted with a gaseous mixture of H_2 and CO at synthesis conditions of temperature and pressure, the improvement comprising contacting said iron-containing catalyst during the synthesis reaction with a mixture comprising H_2 and CO and one member selected from the group consisting of (*a*) chlorine, (*b*) bromine, (*c*) iodine, (*d*) hydrogen chloride, (*e*) hydrogen bromide, (*f*) hydrogen iodide, (*g*) organic chlorides, (*h*) organic bromides, and (*i*) organic iodides.

2,717,260
HYDROCARBON SYNTHESIS
Hubert G. Davis and Thomas P. Wilson, Charleston, and Abraham N. Kurtz, St. Albans, W. Va., assignors to Union Carbide and Carbon Corporation, a corporation of New York
No Drawing. Application January 17, 1952, Serial No. 266,991
4 Claims. (Cl. 260—449.6)

1. In the process for making hydrocarbons containing a substantial amount of olefins having from two to four carbon atoms by the reaction of carbon monoxide and hydrogen in the presence of an iron-base catalyst, the improvements which comprise incorporating an amount of sulfur in the catalyst equal to from 0.02% to 0.5% by weight of the catalyst and subjecting the catalyst during the reaction to intimate contact with regulated amounts of members of the group consisting of (1) chlorine, bromine and iodine, (2) hydrogen chloride, hydrogen bromide and hydrogen iodide and (3) halide-containing compounds capable of releasing the aforesaid halogens or halogen halides in contact with the catalyst.

2,717,261
SYMMETRICAL CYANO-CONTAINING COMPOUNDS AND THEIR PREPARATION
Carl George Krespan, Wilmington, Del., assignor to E. I. du Pont de Nemours and Company, Wilmington, Del., a corporation of Delaware
No Drawing. Application May 6, 1954, Serial No. 428,110
8 Claims. (Cl. 260—465.8)

1. A compound of the formula $R—[C(CN)_2—C(CN)_2]_n—R$ where *n* is an integer from 1 to 2, inclusive, and R is an organic radical containing a tertiary carbon bonded to the central unit and bearing a single cyano group.

6. Process of preparing symmetrical cyano-containing compounds which comprises reacting tetracyanoethylene with an alpha,alpha′-azobisnitrile having bonded to each nitrogen of said azo, —N=N—, group a tertiary carbon atom bearing a single cyano group in the proportion of one mole of said tetracyanoethylene to at least one-half mole of said alpha,alpha′-azobisnitrile.

2,717,262
PROCESS OF PRODUCING CHRYSANTHEMIC ACID ESTERS OF CYCLOPENTENYL KETONIC ALCOHOLS AND PRODUCTS
Robert M. Cole, Bryn Athyn, Pa., assignor to Chemical Elaborations, Inc., Philadelphia, Pa., a corporation of Delaware
No Drawing. Application August 15, 1949, Serial No. 110,459
13 Claims. (Cl. 260—468)

8. The method of producing an insecticide which comprises the steps of chlorinating acetonyl acetone and reacting, under anhydrous conditions, the chlor acetonyl acetone so obtained with a compound selected from the class consisting of mono carboxylic chrysanthemic acid and its salts to produce an ester.

9. A method in accordance with claim 8 and including the further step of reacting under anhydrous conditions, the ester produced by the method of claim 8

with a compound selected from the class consisting of alkene and alkadiene halides and aldehydes in the presence of an alkaline catalyst.

2,717,263
CONDENSATION PRODUCTS OF α-AMINO ACIDS AND PHENOLS
Leonard L. McKinney, Eugene A. Setzkorn, and Eugene H. Uhing, Peoria, Ill., assignors to the United States of America as represented by the Secretary of Agriculture
No Drawing. Application October 3, 1952, Serial No. 313,086
6 Claims. (Cl. 260—471)
(Granted under Title 35, U. S. Code (1952), sec. 266)

1. A compound of the formula:

$$R-\bigcirc(OH)_x-(CH_2-\underset{R_2}{\overset{R_1}{N}}-CH-COOX)_y$$

in which x is an integer selected from the group consisting of 1 and 2, y is an integer selected from the group consisting of 1, 2, and 3, R is a member of the group consisting of H and alkyl, R_1 is a member of the group consisting of H, CH_2CH_2COOX and an organic acyl radical, X is a member of the group consisting of a cation and an alkyl radical, and R_2 is the residual group of an α-amino acid of natural occurrence.

2,717,264
PROCESS FOR DECOMPOSING 1-ALKYLCYCLOHEXYL HYDROPEROXIDES
Frederick F. Rust, Orinda, and Edward R. Bell, Concord, Calif., assignors to Shell Development Company, Emeryville, Calif., a corporation of Delaware
No Drawing. Application July 17, 1951, Serial No. 237,292
4 Claims. (Cl. 260—488)

1. A process for producing a ketol formate which comprises contacting a 1-alkylcyclohexyl hydroperoxide having the general formula

$$\begin{array}{c} R_1 \quad OOH \\ C \\ H_2C \qquad CH_2 \\ R\;\; C \qquad C\;\; R \\ R \quad C \quad R \\ R \qquad R \end{array}$$

wherein R_1 is a lower alkyl radical and the R's are selected from the group consisting of hydrogen and lower alkyl radical with formic acid.

2,717,265
SULFONATION OF MINERAL OIL
Herbert L. Johnson, Media, Pa., assignor to Sun Oil Company, Philadelphia, Pa., a corporation of New Jersey
No Drawing. Application August 11, 1952, Serial No. 303,841
4 Claims. (Cl. 260—504)

1. In a process for preparation by sulfonation, of mahogany sulfonic acids from a mineral oil having at least 1.2 aromatic rings per molecule in the aromatic portion thereof, the improvement which comprises: prior to said sulfonation, contacting said mineral oil in liquid phase with hydrogen under hydrogenation conditions to effect a decrease of at least 0.005 in the refractive index N_d^{20} of the aromatic portion of the oil, and to effect a decrease of not more than 15 percent in the aromatic content of the oil.

2,717,266
PRODUCTION OF PENTADECANEDIOIC ACID
Raymond U. Lemieux, Saskatoon, Saskatchewan, Canada, assignor to National Research Council, Ottawa, Ontario, Canada, a body corporate
No Drawing. Application April 9, 1952, Serial No. 281,455
4 Claims. (Cl. 260—530)

1. In the production of pentadecanedioic acid its esters and salts the method which comprises treating an ester of 15, 16-dihydroxyhexadecanoic acid in acetic acid solution with an oxidizing agent from the group consisting of lead tetraacetate, lead tetraoxide and sodium bismuthate to form an aldehyde and treating the aldehyde with an aqueous alkaline solution of hydrogen peroxide to form pentadecanedioic acid.

2,717,267
PREPARATION OF HIGH DENSITY CALCIUM PANTOTHENATE
John Joseph Garbarini, Dumont, N. J., assignor to American Cyanamid Company, New York, N. Y., a corporation of Maine
No Drawing. Application August 30, 1952, Serial No. 307,380
4 Claims. (Cl. 260—534)

1. A method of increasing the apparent bulk density of calcium pantothenate to at least 0.5 gram per cubic centimeter which comprises the step of agitating calcium pantothenate crystals in a liquid carrier for at least 8 minutes.

2,717,268
PROCESS FOR PREPARING PHENYLDICHLOROACETAMIDOPROPANEDIOLS
Mildred C. Rebstock, Detroit, and Elizabeth L. Pfeiffer, Ypsilanti, Mich., assignors to Parke, Davis & Company, Detroit, Mich., a corporation of Michigan
No Drawing. Application December 16, 1952, Serial No. 326,344
8 Claims. (Cl. 260—562)

1. Process for the production of a 1-phenyl-2-dichloroacetamidopropane 1,3 diol compound of formula,

$$R-C_6H_4-\underset{OH}{CH}-\underset{NH-\overset{O}{\overset{\|}{C}}-CHCl_2}{CH}-CH_2OH$$

which comprises reacting a compound of formula,

$$Cl_2CH-\overset{O}{\overset{\|}{C}}-O-X(OZ)_2$$

with an amino diol compound of formula,

$$R-C_6H_4-\underset{OH}{CH}-\underset{NH_2}{CH}-CH_2OH$$

at a temperature between 20 and 110° C.; where X is a member of the class consisting of trivalent phosphorus and arsenic atoms, Z is alkyl and R is a substituent of the class consisting of hydrogen, nitro, phenyl, $-S-CH_3$, $SOCH_3$, and $-SO_2CH_3$.

2,717,269
RECOVERY OF META-2-XYLIDINE
Stanley F. Birch, Frederick A. Fidler, and Ronald A. Dean, Sunbury-on-Thames, England, assignors to The British Petroleum Company Limited
No Drawing. Original application May 21, 1947, Serial No. 749,598. Divided and this application September 8, 1952, Serial No. 308,508
Claims priority, application Great Britain May 24, 1946
2 Claims. (Cl. 260—582)

1. A process for the recovery of meta-2-xylidine from xylidine mother liquors, said xylidine mother liquors hav-

ing been obtained by treating a crude mixture of xylidines with acetic acid to form the acetate of meta-4-xylidine and a residue, separating the acetate of meta-4-xylidine from the residue, treating the residue with hydrochloric acid to form crude para-xylidine hydrochloride and xylidine residues, separating said para-xylidine hydrochloride by recrystallisation from water obtaining pure para-xylidine hydrochloride and xylidine mother liquors and separating said mother liquors, which comprises regenerating free amines from said mother liquors by treatment with alkali and fractionally distilling said mother liquors in a fractionating column having the equivalent of at least 20 theoretical plates and operated at a reflux ratio of not less than 15.1 and under a pressure not exceeding 50 mm. pressure of mercury to yield a fraction having a final boiling point not exceeding 96° C. at 10.5 mm. pressure of mercury and consisting essentially of meta-2-xylidine.

2,717,270
BASIC POLYGLYCOL ETHERS
Jakob Bindler, Riehen, near Basel, Switzerland, assignor to J. R. Geigy A. G., Basel, Switzerland, a Swiss firm
No Drawing. Application September 3, 1953, Serial No. 378,412
Claims priority, application Switzerland June 23, 1950
2 Claims. (Cl. 260—584)

1. Basic polyglycol ethers having the general formula:

$$C_{18}H_{37}-O-(CH_2-CH_2-O)_n-CH_2-CH_2-N\begin{matrix} R_1 \\ R_2 \end{matrix}$$

wherein n represents an integer from 14–24, and R_1 and R_2 represent lower alkyl radicals selected from the group

2,717,271
PROCESS FOR THE RECOVERY OF POLYGLYCEROLS FROM DISTILLATION RESIDUES
Richard Rowe, Altrincham, England, assignor to Victor Wolf Limited, Manchester, England
No Drawing. Application June 6, 1950, Serial No. 166,537
Claims priority, application Great Britain June 14, 1949
6 Claims. (Cl. 260—616)

4. In a process for obtaining polyglycerols from glycerine pitch containing salts of organic acids by solvent extraction, the employment of dioxane as a solvent.

2,717,272

TREATMENT OF BENZENE HEXACHLORIDE

Percy W. Trotter, Baton Rouge, La., assignor to Ethyl Corporation, New York, N. Y., a corporation of Delaware

No Drawing. Application July 19, 1952, Serial No. 299,926

4 Claims. (Cl. 260—648)

1. A process for recovering the gamma isomer from an essentially solvent-free mixture of benzene hexachloride isomers, including the gamma, alpha and delta isomers, and having at least one part of delta isomer per two parts of gamma isomer comprising heating said mixture to a temperature within the range of from 30° to 100° C., separating the liquid phase formed while the mixture is within said temperature range from the solid phase, said liquid phase containing principally the alpha and gamma isomers and said solid phase containing principally the delta isomer.

ELECTRICAL

2,717,273

BUSHING FOR BATTERIES AND THE LIKE

Walter E. Anderson, Fox Point, Wis., assignor to Globe-Union Inc., Milwaukee, Wis., a corporation of Delaware

Application April 14, 1953, Serial No. 348,751

2 Claims. (Cl. 136—168)

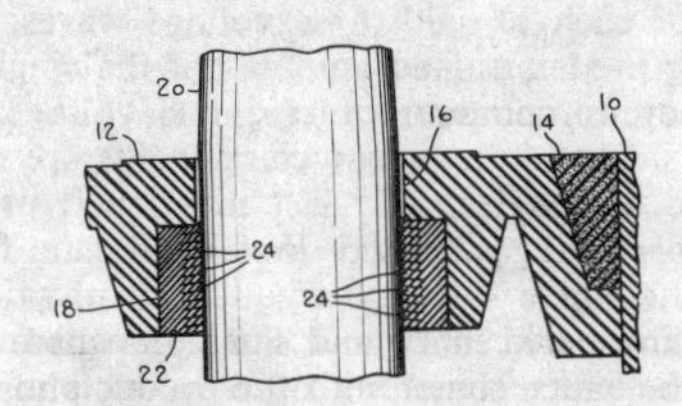

1. A bushing for forming a seal between a lead post and the cover of a storage battery, said cover having an opening for said post and a bushing seat, comprising a hollow rubber cylinder having an outer wall adapted to fit in said seat, and an inner wall having an inner diameter smaller than the outer diameter of said lead post, said inner wall having a multiplicity of spaced slits forming therebetween ring-like flanges, said flanges having flat inner edges providing ninety degree corners between said inner edges and the sides of said flanges, said flanges being in side to side contact and flexed out of their normal unstressed position when said bushing is in position on said post to bring said corners into engagement with said post.

2,717,274
TRANSPOSITION BRACKET
Rogers Case, Orange, N. J.; Ethel Case, executrix of the estate of said Rogers Case, deceased, assignor to Transandean Associates, Inc., Orange, N. J., a corporation of Delaware
Application June 21, 1951, Serial No. 232,737
4 Claims. (Cl. 174—33)

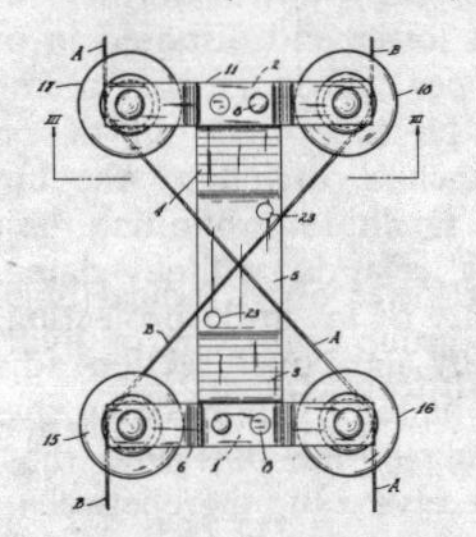

1. A transposition bracket for paired line wires the frame structure of which comprises a central cradle having two arms of approximately equal length downwardly convergent and interconnected at their lower ends and two relatively spaced base members on said arms in a common horizontal plane, two bars secured on opposite faces of each of said base members, said two bars associated with each of the said base members having at both ends thereof offset regions at different levels with respect to the base member, the offset regions of the two said bars being superposed and spaced vertically from each other to provide two outwardly open insulator-mounting clevises extending one chiefly below and the other chiefly above the horizontal plane of the said associated base member, the relative arrangement of the said clevis-

forming bars being reversed at the two said horizontally spaced base members of the bracket to give a diagonal matching of higher and lower clevises, and spool form insulators rotatably mounted in the said four clevises of the bracket for crossing of the two line wires at different levels within the cradle of the bracket.

2,717,275

HOUSINGS FOR ELECTRICAL APPARATUS

Vernon H. Hayden, Pittsburgh, Pa., and Merrill G. Leonard, Nutwood, Ohio, assignors to Westinghouse Electric Corporation, East Pittsburgh, Pa., a corporation of Pennsylvania

Application May 16, 1951, Serial No. 226,606

1 Claim. (Cl. 174—37)

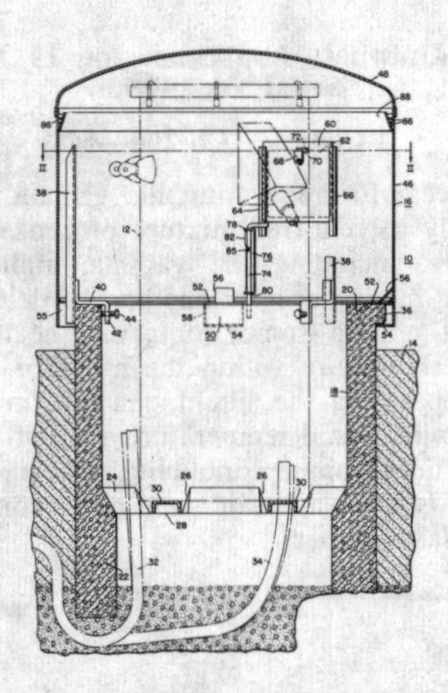

In a transformer housing to be mounted in contact with the ground, the combination comprising, a cylindrical base member disposed on end in contact with the ground, a plurality of outwardly extending spaced lugs carried by the upper end of the base member, means carried inwardly of the base member intermediate the ends thereof for supporting a transformer, a metallic cover section having a lower end disposed in overlapping relation with the upper end of the base member to cover the transformer to be supported out of contact with the ground, guide means carried by the upper end of the base member to facilitate positioning the cover section in said overlapping relation without damaging the transformer, a plurality of inwardly projecting spaced pairs of flanges carried by the cover section adjacent the lower end thereof, the pairs of flanges being spaced corresponding to the spacing of the lugs with the flanges of each pair spaced to receive a lug therebetween, a stop disposed between the flanges of one of the pair of flanges, the upper cover section being disposed to be rotated in a predetermined direction when the cover section is in overlapping relation with the base member and the pairs of flanges are aligned with the corresponding lugs to position each of the lugs between the flanges of a corresponding pair of flanges, the stop associated with said one pair of flanges being disposed to butt against the corresponding lug when the lugs are positioned between the flanges of the corresponding pairs to prevent further rotation of the cover section in said predetermined direction, and a movable stop means carried by the cover section disposed for movement, when the lugs are positioned between the flanges of the corresponding pairs, to a position adjacent one of the lugs to provide a stop for said lug to butt against when the cover member is rotated in a direction reverse to said predetermined direction to thereby prevent movement of the lugs from between the flanges of each corresponding pair of flanges and maintain the upper cover section in an assembled interlocked relation with the base member.

2,717,276

COLOR TELEVISION SYSTEM

Alfred C. Schroeder, Upper Southampton Township, Bucks County, Pa., assignor to Radio Corporation of America, a corporation of Delaware

Application August 11, 1953, Serial No. 373,621

The terminal 15 years of the term of the patent to be granted has been disclaimed

11 Claims. (Cl. 178—5.4)

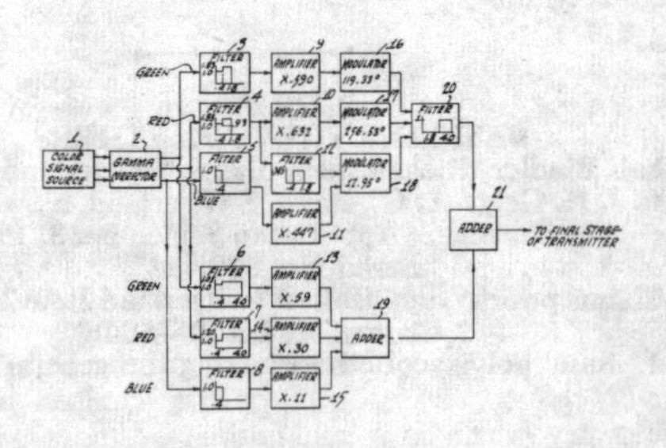

1. A color television system comprising in combination, a source of three gamma corrected voltage waves representative of three primary colors of televised objects respectively, means coupled to said source for modulating each of three phases of a subcarrier by the double sideband frequency components of each of said three voltage waves respectively, means for modulating said three phases of said subcarrier by portions of the single sideband frequency components of two of said voltage waves, said portions of said single sideband components of said two voltage waves being adjusted to compensate for the omission of the single sideband components of the third of said voltage waves in the modulation of said phases, means for producing a signal representative of the luminance of said televised objects, said luminance signal producing means comprising means for combining predetermined amounts of the double sideband frequency components of each of said three voltage waves, means for combining predetermined amounts of the single sideband frequency components of said two voltage waves whose single sideband frequency components are used to modulate said subcarrier, said last mentioned predetermined amounts being adjusted to compensate for the omission of the single sideband frequencies of said third voltage wave from said luminance signal, means for combining said luminance signal with the products of modulation of said subcarrier by said voltage waves to produce a composite video wave, means for detecting said composite wave, means coupled to said detecting means for extracting said luminance signal from said composite wave, means coupled to said detecting means for demodulating said composite wave from which said luminance signal has been extracted, said demodulating means producing three color difference signals respectively, a display device having three input circuits, means for applying one of said color difference signals to a first of said input circuits, means for applying portions of a second and third of said color difference signals to a second of said input circuits, means for applying only the double sideband frequencies of the third of said color difference signals to a third of said inputs, and means for applying said luminance signal to each of said input circuits.

2,717,277

DEVICE FOR SYNCHRONIZING A RECEIVER TO THE TRANSMITTER IN A TIME DIVISION MULTIPLEX SIGNALLING SYSTEM AND CONVERTING TIME POSITION MODULATED PULSE TRAINS INTO AMPLITUDE MODULATED PULSE TRAINS

Lars Bernhard Person and Carl Henric Von Sivers, Stockholm, and Reid Kurt Wadö and Klas Rudolf Wickman, Hagersten, Sweden, assignors to Telefonaktiebolaget L. M. Ericsson, Stockholm, Sweden, a company of Sweden

Application June 9, 1952, Serial No. 292,566

Claims priority, application Sweden June 12, 1951

3 Claims. (Cl. 179—15.6)

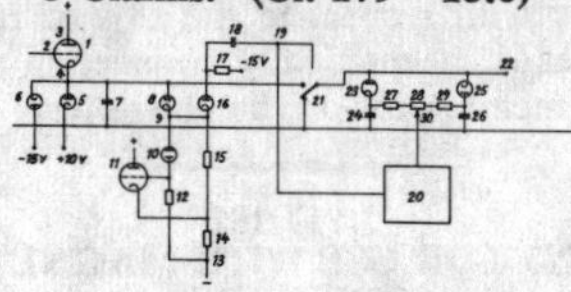

1. A device for synchronizing a receiver to the transmitter in a time division multiplex signalling system, said system being of the type in which each of the pulses in each channel is time-position modulated, and in which synchronizing pulses, having the same repetition frequency as that of the pulses of a single channel, are time-position modulated to a time displacement which is the maximum allowed displacement of the pulses of each single channel, by a synchronizing channel modulation frequency which is a submultiple of the repetition frequency of the synchronizing pulses, and for converting said time-position modulated pulses into amplitude modulated pulses, said device comprising: an input circuit, a condenser connected to said circuit to be charged by the time-position modulated pulses to a fixed, predetermined positive voltage which is equal for all pulses, a constant-current discharge device connected to said condenser to discharge the same by a constant current, during the time between pulses, to a negative voltage equal to the said positive voltage in magnitude during a time corresponding to the maximum allowed time displacement of said pulses, and to a greater but limited voltage of negative sign if the discharge continues for a longer time, a pulse generator arranged to deliver a pulse train whose repetition frequency is equal to the product of the number of channels by the repetition frequency of a single channel, an output terminal, an electronic switch, said pulse generator being connected to control said electronic switch to connect said condenser to said output terminal during the times when said generator is emitting pulses, to provide amplitude modulated pulse trains at said output terminal, the amplitudes of the pulses of said pulse train varying according to the time positions of corresponding time-position modulated pulses, a peak detecting amplitude comparing device connected to said output terminal and arranged to deliver a regulation voltage corresponding to the difference between the maximum positive and negative amplitudes of said train pulses, and means for applying said regulation voltage to said generator to control the repetition frequency and the time position of the train pulses delivered by said generator so that they are equal to the repetition rate and the time position of the time-position modulated pulses arriving at said input circuit.

2,717,278

IMPULSE CIRCUIT FOR AUTOMATIC TELEPHONE SYSTEM

James M. Blackhall, Galion, and Donald S. Baker and Per Olaf Dahlman, Kenton, Ohio, assignors to The North Electric Manufacturing Company, Galion, Ohio, a corporation of Ohio

Application February 26, 1951, Serial No. 212,808

12 Claims. (Cl. 179—16)

1. An impulse correcting arrangement comprising an incoming and an outgoing circuit, a set of relays, sequence control means for controlling said relays to operate in a given sequence to measure a predetermined time period, means for operating one of said relays immediately with receipt of each impulse over said incoming circuit, means controlled responsive to operation of said one relay to

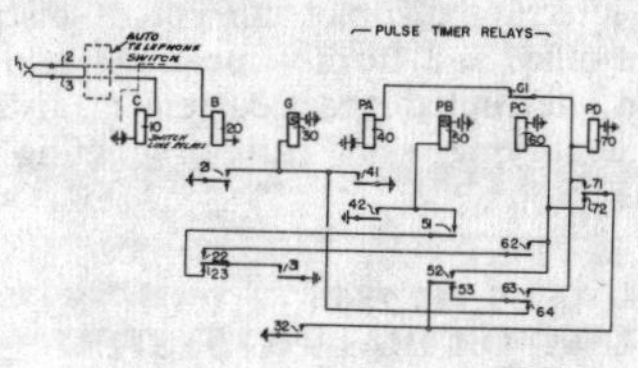

energize said relay set, means operative to connect said one relay for further operation in each operating sequence initiated thereby, and impulsing means operated for a period as measured by said relays alone to transmit an impulse of a corresponding duration over said outgoing circuit.

2,717,279

MULTIPARTY SELECTIVE SIGNALING AND IDENTIFICATION SYSTEM

Richard C. Matlack, Summit, and Frederick W. Metzger, Rutherford, N. J., assignors to Bell Telephone Laboratories, Incorporated, New York, N. Y., a corporation of New York

Application December 28, 1951, Serial No. 263,742

7 Claims. (Cl. 179—17)

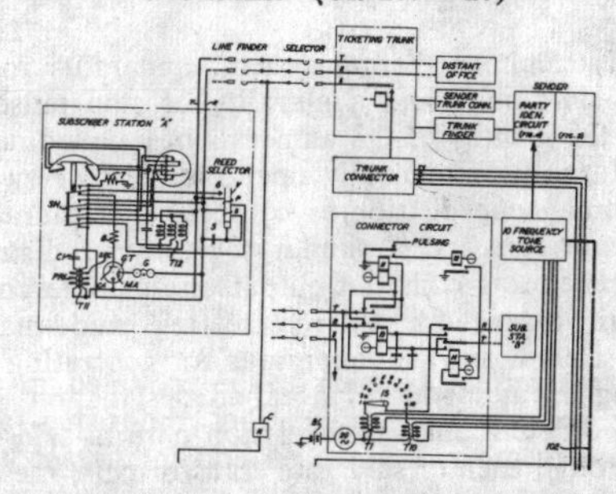

6. In a telephone switching system, a central office comprising multifrequency signal generating means and a battery, a plurality of outlying stations connected with said office by means of metallic lines, certain of said outlying stations being party stations on a common party line, said stations adapted to originate and receive calls, a tuned reed relay at each of said party stations, each of said relays at said party stations on a common party line tuned to a different frequency of said central office generating means, said tuned relays at each said party station operable to complete a ringing circuit thereat in response to a ringing signal from said office comprising energy of the frequency to which said relay is tuned, switching means at said office for transmitting signal current of required frequency over any of said lines to signal the outlying station to be called, calling means at each station for originating a call to said central office over one of said lines, said calling means at each of said party stations including switchhook contacts for connecting the tuned reed relay thereat to the party line thereof upon origination of a call therefrom, means at said office responsive to the origination of a call from any of said party stations for connecting said multifrequency signal generating means to said party line to transmit current of all of said frequencies over said party line to said party stations whereby the reed relay at the call originating party station is caused to vibrate, switching means at said office for disconnecting said multifrequency signal generating means from said party line after a measured time interval and for connecting said battery to said party line, means at said call originating party station including a vibrator contact of

the reed relay thereat responsive to direct current over said party line from said battery for sustaining the vibration of said relay at its tuned frequency and for interrupting said direct current at a distinctive frequency controlled by vibration of said reed relay, means at said call originating party station including said party line for transmitting the interrupted direct current of distinctive frequency to said office and further means at said office responsive to said interrupted direct current of distinctive frequency for identifying said call originating party station.

2,717,280
SIGNALING OR DIALING SYSTEM
Clarence A. Lovell, Summit, N. J., assignor to Bell Telephone Laboratories, Incorporated, New York, N. Y., a corporation of New York
Application March 1, 1954, Serial No. 413,283
3 Claims. (Cl. 179—90)

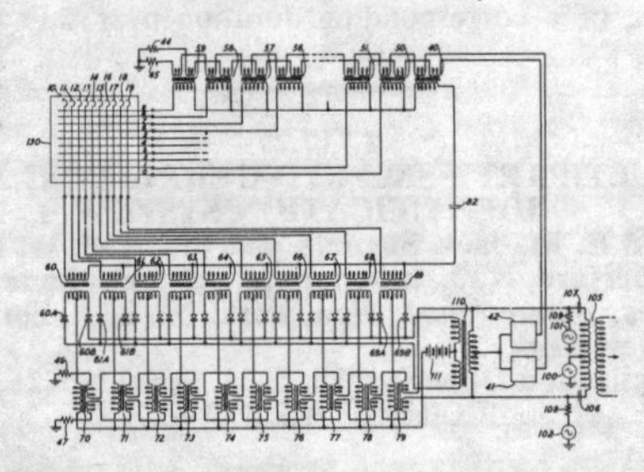

1. In a telephone dialing system, apparatus for generating a start pulse and a plurality of stop pulses representing the identity of the numerals of a called designation, a plurality of manually operable selector switches, an electromagnetic distributor comprising a plurality of gate transformers and a similar plurality of distributor impulse generating coils, an output circuit, interconnecting circuit means for interconnecting said manually operable selector and said apparatus for generating pulses identifying the numerals of a called designation for extending a different pulse transmission path to said output circuit through each of said gate transformers, means for interconnecting said distributor impulse coils with said gate transformers, and means for energizing said distributor impulse coils in succession for successively generating activating pulses for each of said gate transformers whereby a path is successively and operatively established for transmitting said pulses designating the numerals through said gate transformers to said output circuit.

2,717,281
MOTOR DRIVE
Harold W. Bauman, Chicago, Ill., assignor to Ampro Corporation, Chicago, Ill., a corporation of Illinois
Application March 8, 1951, Serial No. 214,601
4 Claims. (Cl. 179—100.2)

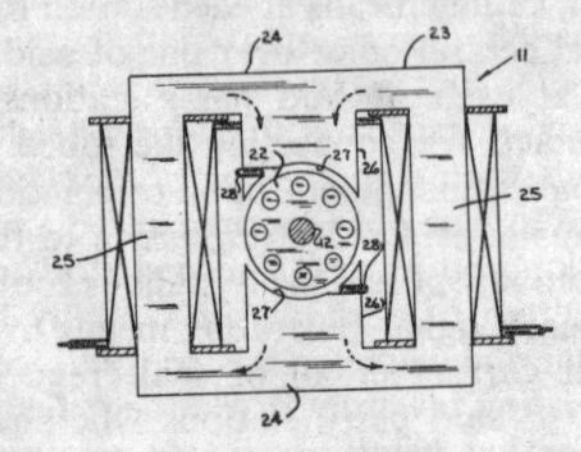

1. In an instrument that includes a pickup device for developing from a recording on a moving record body a corresponding signal output, and that includes magnetic means energizable by stray magnetic fields to produce an error signal output component, and said instrument including a drive member that is rotatable to move a record body relative to said pickup device; a two-pole shaded pole induction motor of the type including a rotor and a highly magnetically permeable stator structure including a pair of pole-forming parts having at their respective ends polar surfaces that are respectively disposed to opposite sides of the periphery of said rotor and radially spaced from its axis, and shading coils mounted on said parts adjacent their polar surfaces, and a core portion of said stator structure forming a continuous magnetic circuit structure extended completely about the periphery of said rotor and about said pole-forming parts and their polar surfaces, said motor being displaced from said magnetic means radially with respect to the axis of rotation of its rotor, and means mechanically connecting said rotor to said rotatable member for driving the latter.

2,717,282
EQUIPMENT FOR USE WITH MAGNETIC TAPE RECORDS
Frank J. Reed, Philadelphia, Pa., assignor to The International Electronics Company, Philadelphia, Pa., a corporation of Pennsylvania
Application July 15, 1953, Serial No. 368,172
16 Claims. (Cl. 179—100.2)

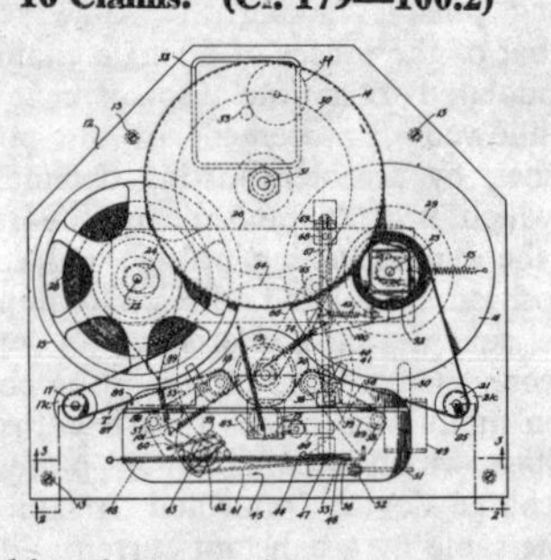

15. A machine for use with elongated magnetic tape wound on reels, comprising means for supporting a pair of said reels with the tape leading from one to the other; tape engaging elements along the tape between said reels including a driving capstan and a magnetic scanning head, said elements confining an inter-reel run of the tape in a scanning feed path with the tape in driving engagement with the capstan and in sliding contact with the head as the tape is fed from one reel to the other, and certain of said elements being respectively located at the two sides of the tape and being relatively movable to release the confinement of said inter-reel run of the tape and to open a threading path between and along said elements; a cover structure extending along said threading path, said structure having therein an elongated tape threading slot through which a tape may be inserted edgewise into said threading path, said slot being sufficiently straight to permit a plane directed lengthwise thereof and to the threading path to extend therethrough, and said machine being configured to have two spaces, adjacent said cover structure and into which said slot opens at its respective ends, adapted to receive at said plane respective pairs of manual fingers grasping the tape between them, whereby to allow a straight length of the tape held between the hands to be inserted edgewise through the slot to the threading path.

2,717,283
SOUND FILM APPARATUS AND THE LIKE
Hans Friess, Karlsruhe-Daxlanden, Germany, assignor to Klangfilm Gesellschaft mit Beschraenkter Haftung, Berlin and Karlsruhe, Germany, a German corporation
Application June 19, 1951, Serial No. 232,251
Claims priority, application Germany August 7, 1950
17 Claims. (Cl. 179—100.3)

1. Sound film apparatus comprising guide means for transporting a film, an optical scanning device, a magnetic

scanning device, means for disposing said optical scanning device so as to scan a film transported by said guide means for reproducing sound from an optical sound track carried thereby on one side thereof, and means for disposing

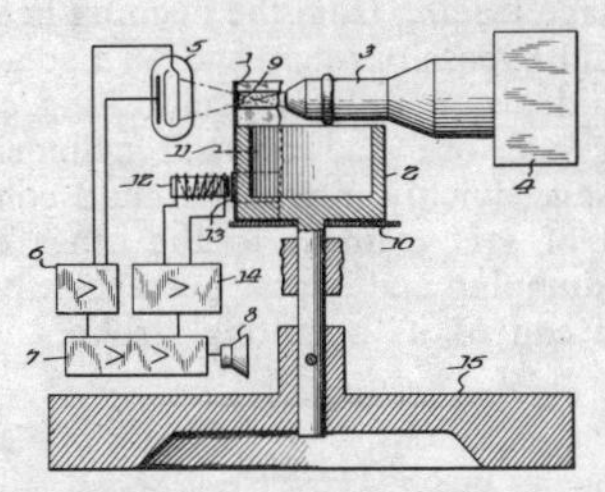

said magnetic scanning device alongside said optical scanning device as viewed in a direction transverse to said film so as to scan a film transported by said guide means for reproducing sound from a magnetic sound track carried thereby on the other side thereof.

2,717,284
LINE-CLEARING APPARATUS FOR A TELEPHONE SYSTEM
John L. Culbertson, Harvey, Ill., assignor to International Telephone and Telegraph Corporation, a corporation of Maryland
Application September 30, 1952, Serial No. 312,276
4 Claims. (Cl. 179—175.2)

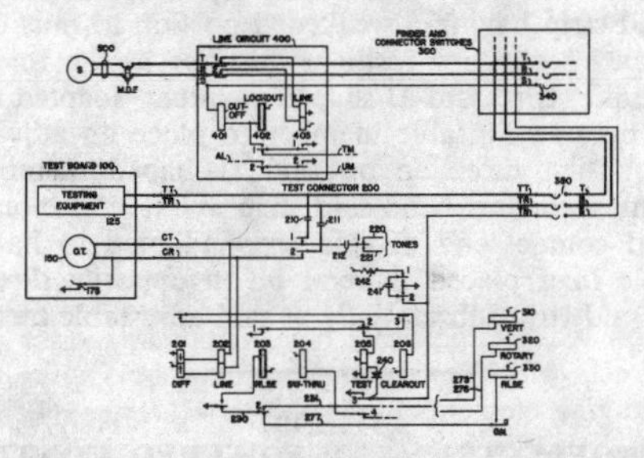

1. In a switching system, lines and switching apparatus common thereto for extending connections between respective calling and called lines, individual sleeve conductors and individual line equipments for respective lines, each line equipment including a normally connected line attachment and a cutoff relay for disconnecting it, a current source having first and second poles, the cutoff relay being normally connected serially in a circuit path extending between the first pole of the current source and the associated sleeve conductor, means in each line equipment for locking the associated line out of normal service, including means for effecitvely transferring the cutoff relay thereof from connection with the first pole into connection with the second pole of the current source, a test board and means controllable therefrom for operating a portion of said switching apparatus to extend a test connection to any desired line, including a connection to its associated sleeve conductor from the said second pole of the current source, said test connection being extended in preparation for testing the selected line with the line attachment thereof disconnected therefrom, means rendering the consequent current flow over the said sleeve connection sufficient to operate the cutoff relay when the selected line is idle, but not when such line is in the said locked-out condition, and means controllable from said test board for substituting a potential on the sleeve connection of such a value and polarity that the resulting current flow operates the cutoff relay of a selected locked-out line, whereby the line attachment thereof is disconnected to permit the desired test of the selected line to be made.

2,717,285
CAPSTANS FOR ADVANCING STRANDS
Vincent A. Rayburn, Baltimore, Md., assignor to Western Electric Company, Incorporated, New York, N. Y., a corporation of New York
Application May 23, 1952, Serial No. 289,459
5 Claims. (Cl. 191—1)

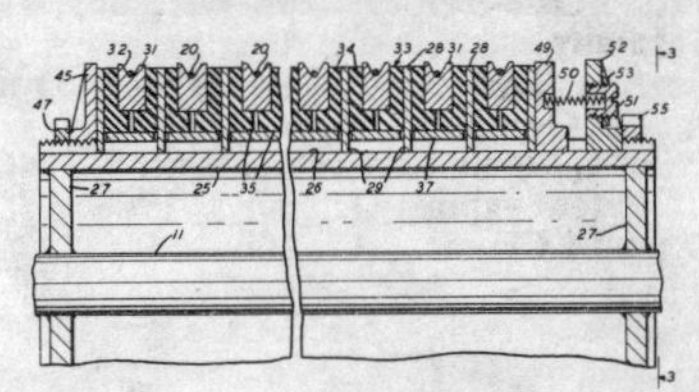

1. A capstan for advancing a plurality of wires comprising a rotatable cylindrical member having a plurality of axial splines therein, a plurality of annular driving discs splined on said cylindrical member for rotation therewith and for axial movement relative thereto, cylindrical bushings mounted on said rotatable cylindrical member between adjacent driving discs, a plurality of metal rings having strand-engaging outer peripheries and flat end faces and being disposed between adjacent driving discs, means comprising a plurality of annular members of insulating material having portions thereof disposed between and engageable with said ring and said driving discs and having portions thereof engageable with the inner periphery of said rings and with said bushings for supporting said rings in concentric relation to said cylindrical member, and means on said cylindrical member for compressing said driving discs, said annular insulating members, and said rings against each other to establish a friction driving connection between said rings and said rotatable cylindrical member.

2,717,286
STABILIZING DEVICE
Max G. Bales, Anderson, Ind., assignor to General Motors Corporation, Detroit, Mich., a corporation of Delaware
Application February 6, 1953, Serial No. 335,562
5 Claims. (Cl. 200—31)

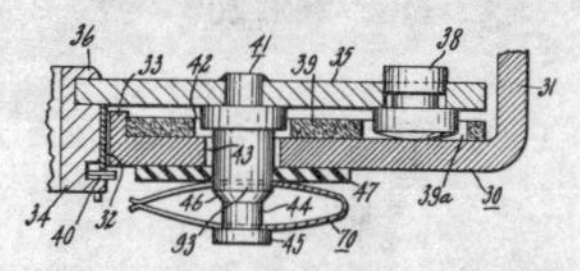

1. In an ignition timer, a friction drag device, comprising in combination, a pair of plates one movable relative to the other, a stud carried by one of the plates and projecting through a slot in the other plate, said stud having a first diameter portion projecting through said other plate and a second reduced diameter portion extending beyond said first portion and terminating in a head portion of larger diameter than said second portion, a substantially U-shaped leaf spring having a first slot in one leg portion thereof of a width not less than the diameter of said stud second portion but not so great as the diameter of said stud first portion with the inner end of the slot forming an opening of a diameter substantially the same as but not less than the diameter of said stud first portion and having a second slot in the second leg portion thereof of a width substantially the same as but not less than the diameter of said stud second portion, said slots extending inwardly of said legs from like edges thereof, said spring being positioned on said stud with said one leg portion engaging said stud first portion with said opening in said one leg portion partially encircling said stud first portion and with said second leg portion partially encircling said stud second portion, said opening in said one leg locking said spring on said stud.

2,717,287
SWITCHING APPARATUS FOR ELECTRIC POWER TRANSMISSIONS
Immanuel Sihler, deceased, late of Hamburg, Germany, by Emilie Luise Sihler, née Orlowsky, executrix, Hamburg, and Joseph Biersack, Berlin-Siemensstadt, Germany, assignors to Siemens-Schuckertwerke Aktiengesellschaft, Berlin-Siemensstadt, Germany, a corporation of Germany
Application June 15, 1951, Serial No. 231,686
In Germany October 1, 1948
Section 1, Public Law 690, August 8, 1946
Patent expires October 1, 1968
14 Claims. (Cl. 200—48)

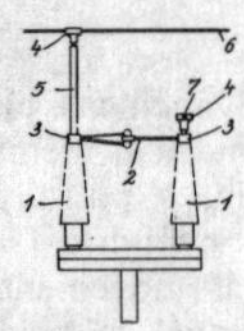

1. Switching apparatus for power-line substations with stratified and crossing conductor systems, comprising a disconnect switch having two insulating support pillars spaced from each other, said pillars having mutually engageable switch contact members respectively and being revolvable about their respective axes for moving said members between switch-closing and opening positions, respective switch terminal members each forming a through-conductor clamp and having a pivot adjacent to said clamp, said terminal members having said pivots joined with said respective pillars so as to be revolvable relative thereto, said terminal members being disposed in different strata corresponding to the conductor strata and forming intermediate supporting means for the respective system conductors.

2,717,288
PNEUMATICALLY CONTROLLED OPERATING DEVICE FOR ELECTRIC CIRCUIT BREAKERS
Milton L. Heintz, Philadelphia, Pa., assignor to General Electric Company, a corporation of New York
Application January 4, 1952, Serial No. 264,978
5 Claims. (Cl. 200—97)

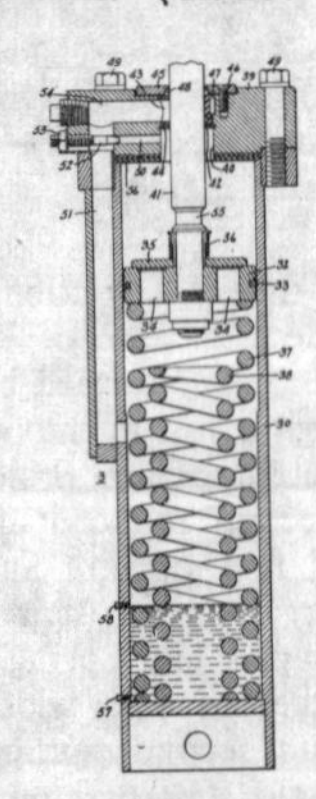

4. In combination, a cylinder containing a compressible working fluid and a piston movable against the fluid and having a piston rod, said piston having an initial position intermediate the ends of said cylinder, an electric circuit breaker operably related with said piston rod, a latch for holding said breaker in the closed position, means for moving said piston from its initial position toward one end of said cylinder to open the breaker in response to tripping of said latch, a closure member mounted at said one end of said cylinder, said closure member having an opening therein through which said piston rod is slidable, a bypass passage having its ends in communication with opposite ends of said cylinder, adjustable valve means in said passage for controlling the flow of fluid from said one end of said cylinder to the other end thereof during an operating stroke of said piston, a dumping passage leading from the opening in said closure member to said bypass passage, and an undercut portion formed on said piston rod for establishing communication between said one end of said cylinder and said dumping passage quickly to transfer fluid compressed in said one end of said cylinder to the other end thereof through said dumping and bypass passages when said piston nears the end of its operating stroke.

2,717,289
WIDE RANGE THERMOSTAT
Edward Bletz, Lexington, Ohio, assignor to Stevens Manufacturing Company, Inc., a corporation of Ohio
Application December 17, 1952, Serial No. 326,503
16 Claims. (Cl. 200—138)

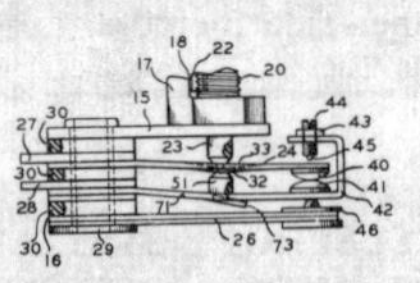

2. A contact carrying blade for a thermostat, comprising, a resilient strip of metal having a mounting end and a contact end, said mounting end adapted to be fastened in a stack of a thermostat for co-operation therewith, said strip having a weakened portion to thus define a U-shaped secondary resilient member at the mid-portion of said strip, said U-shaped member adapted to be abutted by an adjustable member to place an adjustable load in a first direction on said U-shaped member to bend same and thus bend said strip about said mounting end, said contact end of said strip adapted to have an adjustable load placed thereon in an opposite direction to bend said strip substantially at said adjustable member.

2,717,290
THERMOELECTRICALLY POWERED CONTROL DEVICE FOR WATER HEATERS AND THE LIKE
John H. Thornbery, Whitefish Bay, Wis., assignor to Milwaukee Gas Specialty Company, Milwaukee, Wis., a corporation of Wisconsin
Application September 14, 1951, Serial No. 246,576
11 Claims. (Cl. 200—140)

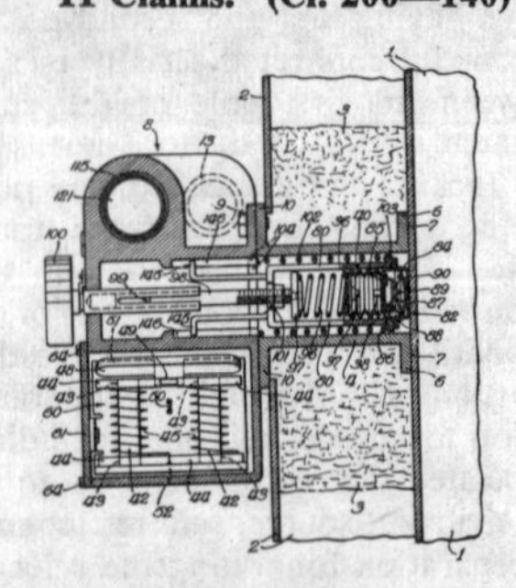

1. A control device for a heater having a wall comprising, in combination, a control body, a bracket carried by said control body for rectilinear movement relative thereto and projecting from said control body, temperature responsive means carried by the outer end of said bracket for movement bodily with said bracket, and a spring interposed between a shoulder on said control body and a shoulder on said bracket and biasing the outer end of said bracket yieldingly to projecting position related to the distance between said wall and said control body so as to position said temperature responsive means

in heat receiving relation with respect to said wall, said temperature responsive means comprising an enclosure expansible and contractible with changes in temperature and contacts encapsulated within said enclosure for relative movement with expansion and contraction of said enclosure.

2,717,291
DOMESTIC APPLIANCE
Millard E. Fry, Dayton, Ohio, assignor to General Motors Corporation, Detroit, Mich., a corporation of Delaware
Application December 4, 1953, Serial No. 396,209
2 Claims. (Cl. 200—140)

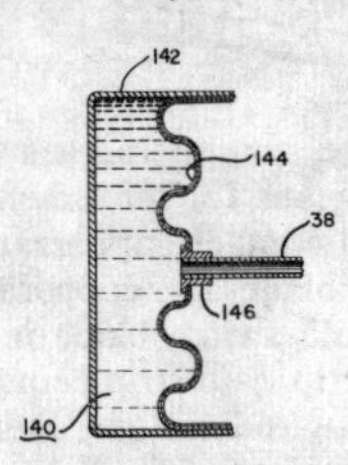

2. A control including a fluid motor, a control device operatively connected to and operated by said fluid motor, a control bulb comprising an enclosure having a bimetal wall portion adapted to deflect inwardly upon increases in its temperature to reduce the interior volume of the enclosure, a small conduit means connecting the interior of said control bulb with the interior of said fluid motor, and a liquid which expands and contracts uniformly at a relatively large rate with changes in temperature completely filling said fluid motor and said control bulb and said conduit means, said control bulb being formed of two members of bimetal material bonded together adjacent their edges and separated within the bonded portions to form the enclosure, said bimetal members being arranged with their high expanding sides facing inwardly, the expansion of said liquid in said bulb and the reduction in volume of the bulb both causing movement of the liquid from the bulb to the fluid motor.

2,717,292
AIR CIRCUIT BREAKER
Russell E. Frink and Paul Olsson, Pittsburgh, Pa., assignors to Westinghouse Electric Corporation, East Pittsburgh, Pa., a corporation of Pennsylvania
Application April 18, 1952, Serial No. 283,038
8 Claims. (Cl. 200—144)

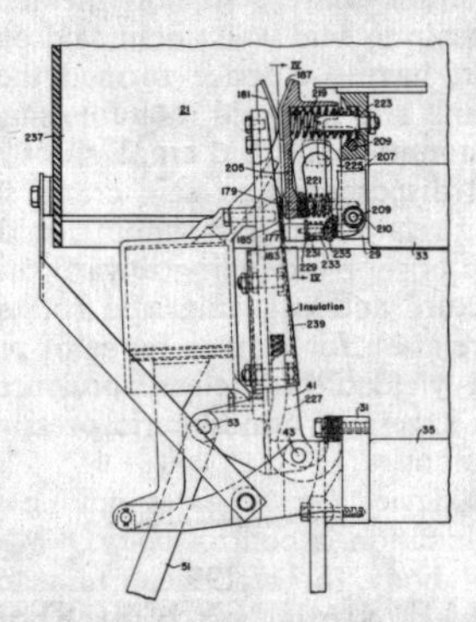

1. In a circuit breaker, stationary contacts comprising a fixed contact member, a pair of spaced movable contact carrying plates disposed one on each side of said fixed contact member and mounted on said fixed contact member for limited movement relative thereto, springs providing pressure contact between said contact carrying plates and said fixed contact member, main stationary contacts on said movable contact carrying plates, a contact member disposed between said contact carrying plates and supported between said movable contact carrying plates for limited movement relative thereto, intermediate and arcing contacts on said contact member, and a movable switch member having movable main, intermediate and arcing contacts thereon for cooperating, respectively, with said stationary main, intermediate and arcing contacts to open and close the circuit.

2,717,293
ELECTRIC CIRCUIT INTERRUPTER
Charles H. Titus, Havertown, and Richard E. Bednarek, Philadelphia, Pa., assignors to General Electric Company, a corporation of New York
Application January 23, 1953, Serial No. 332,914
4 Claims. (Cl. 200—150)

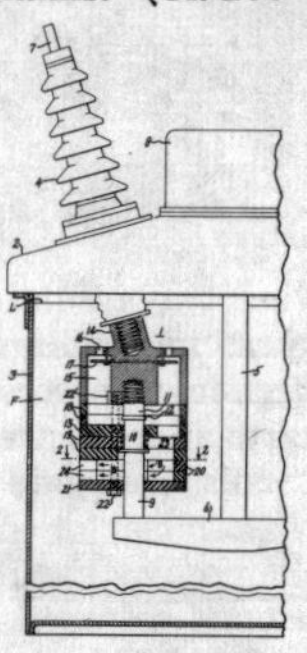

1. A fluid-blast circuit interrupter comprising a pressure confining structure containing an arc-extinguishing fluid; relatively movable contacts within said structure adapted to separate to draw an arc to be extinguished by fluid-blast action; said structure including a baffle stack having internal openings for directing said fluid blast into the path of said arc, said baffle stack containing a resiliently expansible baffle which comprises a bight portion and a pair of arms connected to said bight portion, said arms having end portions which together define an exhaust vent at one side of said baffle stack, said arms being mutually separable to vary the size of said exhaust vent in accordance with arc-generated pressure within said structure, and a pair of parallel spaced-apart rod members fixedly positioned in said baffle stack, said resiliently expansible baffle having spaced-apart apertures each receiving one of said rod members, said apertures being large relative to the cross-section of said rod members and providing sufficient clearance with respect to said rod members to permit said baffle to expand and contract relative to said rod members and within limits determined by said rod members.

2,717,294
ELECTRIC CIRCUIT INTERRUPTER
Conrad J. Balentine, Philadelphia, Pa., assignor to General Electric Company, a corporation of New York
Application February 20, 1953, Serial No. 337,997
2 Claims. (Cl. 200—150)

1. A fluid-blast type of interrupting unit having means for drawing an arc and establishing a pressure therein comprising a composite assembly of stacked insulating baffle plates, an intermediate one of said stacked plates constituting a division baffle fixedly positioned in said assembly and sectionalizing said assembly into a first and second chamber, said first chamber being located adjacent the point at which the arc is initiated, openings in said plates defining a main arc passage extending through said chambers and said division baffle, said openings further defining a first supplementary vented arcing passage for said first chamber disposed adjacent said main arc passage on one side of said division baffle and a second

supplementary vented arcing passage for said second chamber disposed adjacent said main arcing passage on the opposite side of said division baffle, a blast duct communicating between said chambers and adapted to direct a blast of fluid across said arc passage and into one

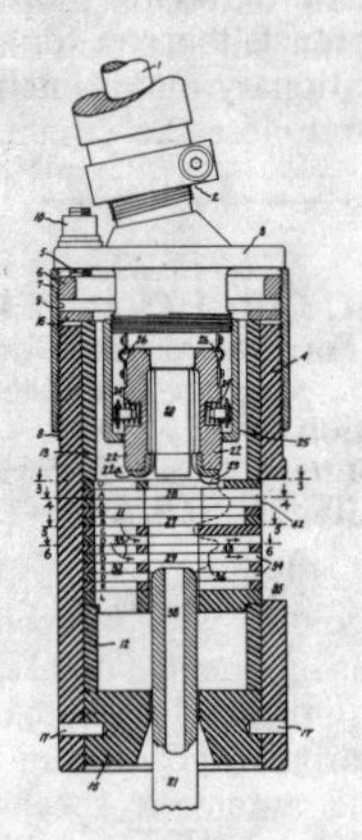

of said supplementary arcing passages, said supplementary arcing passages terminating in vents, the vent from said first supplementary arcing passage being substantially larger than the vents from said second supplementary arcing passage.

2,717,295
ELECTRIC SWITCH
John T. Marvin, Xenia, Ohio, assignor to General Motors Corporation, Detroit, Mich., a corporation of Delaware
Application September 23, 1952, Serial No. 311,082
8 Claims. (Cl. 200—159)

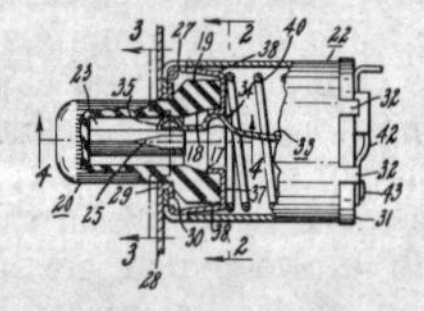

1. In a push-type switch, the combination comprising, a housing; a plunger having a cavity movable into and out of said housing; a movable contact member controlled by said plunger upon movement thereof; a stationary contact member engageable with said movable contact member in certain normal positions of said plunger; means within said cavity of the plunger for preventing said engagement in said certain normal positions; and additional means within the cavity of the plunger operative upon each movement of said plunger for nullifying the effect of the first named means.

2,717,296
ELECTRICAL SWITCH CONTACTS
Robert T. Foley, Lanesboro, and Orin P. McCarty, Pittsfield, Mass., assignors to General Electric Company, a corporation of New York
Application September 14, 1953, Serial No. 379,844
4 Claims. (Cl. 200—166)

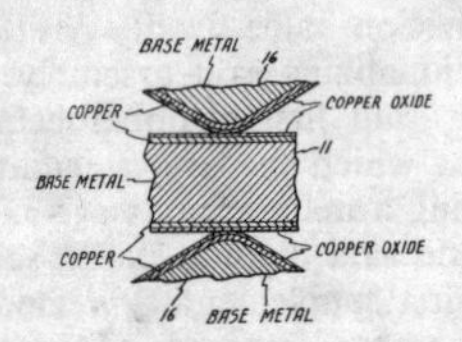

1. An electrical contact member comprising a conducting base, and a layer of copper oxide 0.002–0.008 mil in thickness overlying said base, said layer of copper oxide serving as a contacting surface.

2,717,297
SELF-CLEANING ELECTRIC SWITCH
Neville E. Walker, Portland, Oreg.
Application September 7, 1954, Serial No. 454,449
7 Claims. (Cl. 200—166)

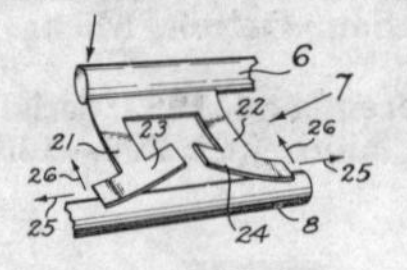

1. A switch device, comprising a switch blade carried by a rotatable coil means for movement therewith, a pair of elongated parallel contact bars arranged one on either side of said switch blade for cooperation and selective contact therewith, said switch blade including a flat thin laterally extending strip, said strip being flexible and being twisted about the longitudinal axis thereof whereby contact pressure against either of the bars simultaneously will untwist and slide the strip along the surface of the contact bar.

2,717,298
SEALED ENCLOSING STRUCTURE FOR ELECTRIC CIRCUIT BREAKERS
Lewis J. Woodward, Philadelphia, and Ralph Baskerville, Drexel Hill, Pa., assignors to General Electric Company, a corporation of New York
Application July 22, 1952, Serial No. 300,182
4 Claims. (Cl. 200—168)

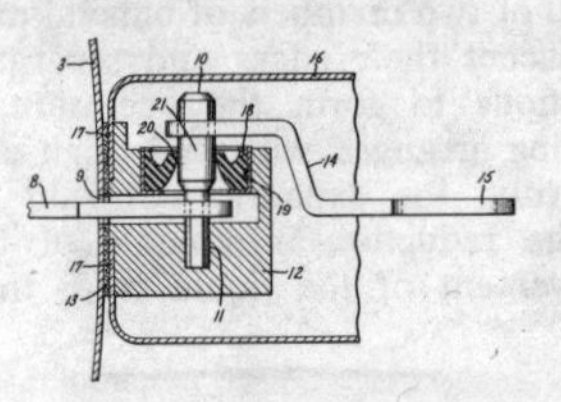

1. An electric circuit breaker comprising an enclosing tank, separable contacts disposed within said tank, a contact-operating crank arm extending through an opening in a wall of said tank and having its inner end operably connected with a movable one of said contacts, a rotatable shaft secured to the outer end of said arm to constitute a crank shaft, a support member secured in sealing relationship to said wall about said opening therein, said support having a space formed therein for receiving said crank arm and said rotatable shaft and forming a journal mounting for said crank shaft, a seal of resilient material disposed about said crank shaft and secured in sealing relation to said support, said seal having a peripheral lip portion extending toward the exterior of said tank structure and engaging said rotatable shaft in pressure tight relation for a given range of pressures, and said seal being yieldable to release pressure established in said tank in excess of atmospheric pressure by a predetermined amount.

2,717,299
TEMPERATURE-DEPENDENT RESISTOR
Werner Jacobi, Grafelfing, near Munich, Germany, assignor to Siemens & Halske, Aktiengesellschaft, Munich, Germany, a corporation of Germany
Application February 3, 1953, Serial No. 334,831
Claims priority, application Germany February 14, 1952
12 Claims. (Cl. 201—63)

1. A temperature-dependent resistor device for controlling high-frequency oscillations comprising a resistance

element, a heater for said element and current conductor means therefor, the inductive resistance of said element and associated current conductor means being at high frequencies small as compared with the ohmic resistance thereof, a casing for enclosing said resistance element and said heater and current conductor means, said casing comprising two potlike members, said current conductor means extending solely within said casing and being in engagement with said potlike members at corresponding areas on the inside thereof, each potlike member carrying a flange extending from the side wall thereof, said flanges being disposed in face-to-face slightly spaced alignment, at least the areas of said potlike members which are engaged by said current conductor means and the faces of said flanges being electrically conductive to form terminals for said current conductor means, and insulating means disposed peripherally of said aligned flangelike extensions forming a vacuumtight seal.

2,717,300
THERMAL-EXPANSION EXTREME-PRESSURE APPARATUS
George Henry Tyne, Nashville, Tenn.
Application December 15, 1953, Serial No. 398,391
8 Claims. (Cl. 219—1)

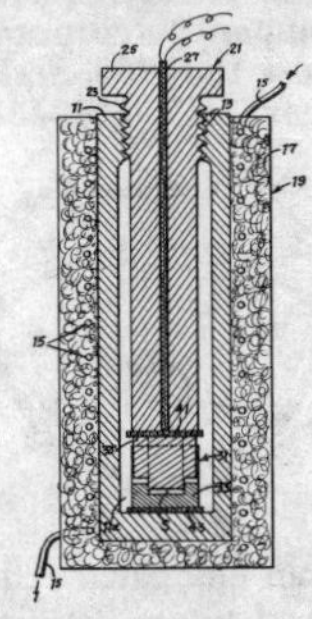

1. Extreme pressure apparatus, comprising: a first body having a relatively high tensile strength, a second body having a relatively high compressive strength, both of said bodies having relatively high positive coefficients of thermal expansion, said bodies being co-extensive for a considerable portion of one of their major dimensions, being rigidly connectable together adjacent one end of their co-extent, and having portions adjacent the other end of said co-extent presenting opposed pressure-exerting surfaces, and means for simultaneously cooling said first body and heating said second body, whereby the thereby induced contraction of said first body and expansion of said second body will cause a high-pressure-producing relative movement between said opposed surfaces.

2,717,301
DEVICE FOR SOLDERING PIN, SOCKET AND THE LIKE TO THE END OF WIRE AND THE LIKE
Henry Johnson Dixon, West Kirby, Joseph Edward Geoffrey Chapman, Workington, and Frederick Arthur Harwood, Maghull, near Liverpool, England, assignors to British Insulated Callender's Cables Limited, London, England, a British company
Application May 18, 1953, Serial No. 355,674
Claims priority, application Great Britain May 20, 1952
10 Claims. (Cl. 219—12)

1. A machine for automatically carrying out a cycle of operations involved in the soldering of a terminal piece to a conductor comprising a support for the conductor, a movable carriage carrying means for gripping the terminal piece, means for actuating the gripping means for the terminal piece upon insertion of a terminal piece, means for moving the carriage to carry the terminal piece towards the conductor into a position for attachment to the conductor, means for heating the terminal piece,

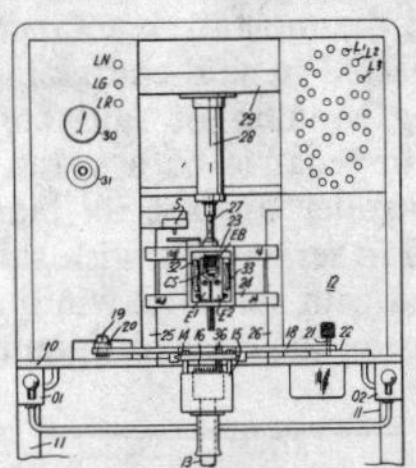

means for blowing cooling air on to the terminal piece while it is held in contact with the conductor, means for automatically releasing the terminal piece after an interval for cooling, means for automatically returning the carriage to its original position on completion of the cooling interval and automatic means for resetting the machine for a further cycle of operations.

2,717,302
HEATERS FOR YARN
Walter Warren Egee, Yeadon, and John W. Bennett, Philadelphia, Pa., assignors to Fletcher Works Incorporated, Philadelphia, Pa., a corporation of Pennsylvania
Application May 31, 1952, Serial No. 290,861
3 Claims. (Cl. 219—19)

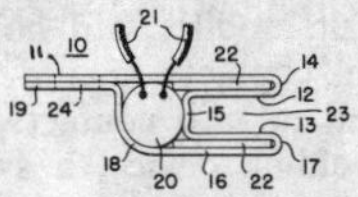

2. A heater for yarn having a body portion formed of a single piece of sheet metal to provide an outer wall portion, a first inner wall connected to said outer wall portion, a second inner wall parallel to and spaced from said first inner wall, a connecting wall between the inner ends of said inner walls, a second outer wall portion connected to said second inner wall and having a part extending towards said first outer wall portion, said inner walls and said connecting wall being in fixed relation and bounding an open sided and open ended heating channel for passage of the yarn therethrough, said first outer wall portion, said connecting wall and said part of said second outer wall portion having a space therebetween, and an electric heating unit in said space for heating said connecting wall and said inner walls.

2,717,303
ELECTRICAL CONTROL CIRCUIT FOR HEATING APPARATUS
Harvey R. Karlen, Chicago, Ill., assignor to Cory Corporation, Chicago, Ill., a corporation of Delaware
Application September 25, 1951, Serial No. 248,223
4 Claims. (Cl. 219—20)

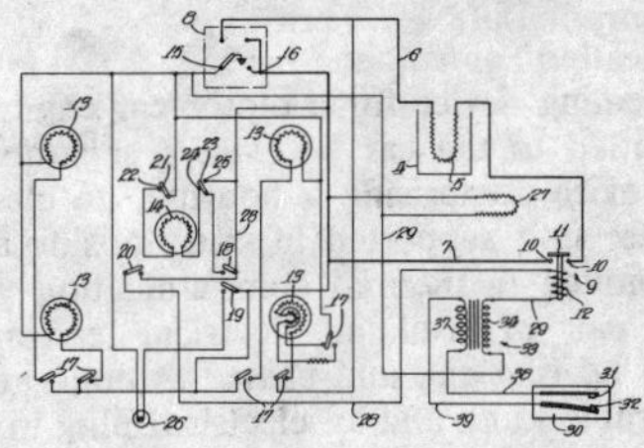

1. In a fluid heating apparatus having a vessel for containing a body of fluid and a heating device for heat-

ing the body of fluid, a control system for governing the heating device comprising: an electrically actuated controller for the heating device biased in one direction and including an actuating winding; a closed energizing circuit for said controller independent of the heating device; a make and break thermostat responsive to the temperature of the body of fluid to be heated; a governing circuit including said thermostat; reactance means having a first winding connected in said energizing circuit in series with the actuating winding of said controller and normally having an impedance of a value preventing actuation of said controller against its bias; and a second winding connected in series only with said thermostat and inductively coupled with said first winding to vary the impedance thereof as an incident to opening and closing of said thermostat.

2,717,304
CIGARETTE LIGHTER
Arns Hoagland, Cincinnati, Ohio
Application August 14, 1952, Serial No. 304,345
1 Claim. (Cl. 219—32)

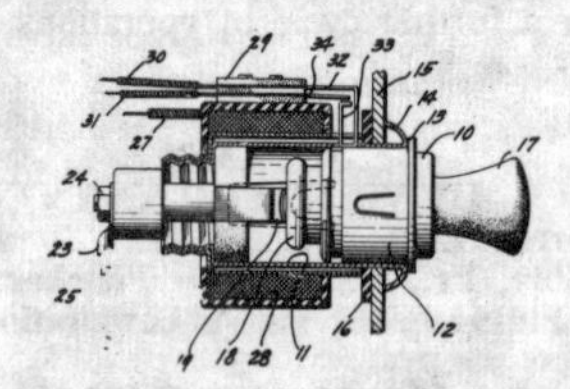

A cigarette lighter, a holding device attached to the dashboard of an automobile, said holder device having an aperture therein, contacts in said aperture, a source of energy connected to said contacts, an electromagnet surrounding said holding device, a switch connected to said electromagnet, a removable igniting unit adapted to fit into said aperture in said holding device, a tubular shoulder on said igniting device, a source of energy connected to said switch, said switch having an arm contacting said tubular shoulder to maintain said electromagnet deenergized and to energize said magnet when said igniting device is removed from the aperture in said holding device.

2,717,305
AUTOMOBILE ENGINE HEATER
James M. Guthrie, Pittsburgh, Pa.
Application January 8, 1953, Serial No. 330,271
7 Claims. (Cl. 219—38)

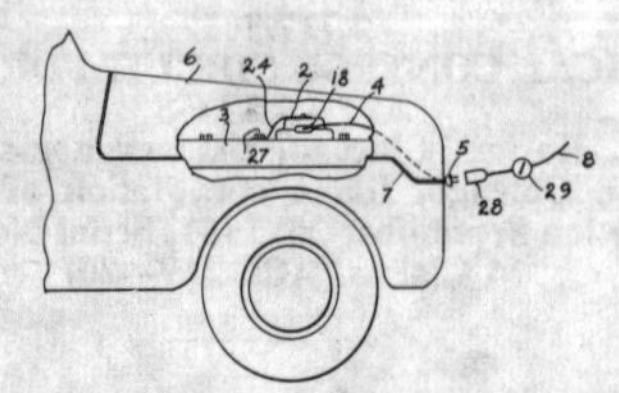

1. A heating device for a motor car engine covered by a hood fitted to the car body in a separable joint, said heating device comprising a housing, an electrical resistance heater unit supported in said housing and arranged to be mounted in thermal communication with the said motor, a pair of insulated electrical leadwires extended from said heating unit and passed through and secured in said separable joint, and an electrical plug-in adapter connected to the outer ends of said leadwires on the outside of said hood and immediately adjacent to said joint.

2,717,306
CANDLE LAMPS
William F. Meara, Oak Park, Ill., assignor to Bloomfield Industries, Inc., Chicago, Ill., a corporation of Illinois
Application February 4, 1953, Serial No. 335,019
4 Claims. (Cl. 240—13)

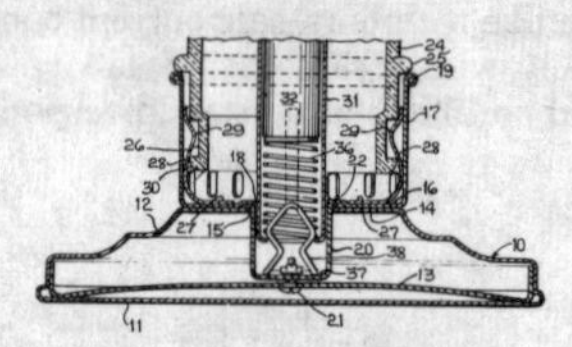

4. A candle lamp comprising a base having a top surface provided with a centrally located opening, a shade receiving cup mounted on and connected to said top surface and having an opening in registration with the opening of said top surface, a socket arranged in the base through said openings and carried by said top wall, a removable candle holding tube having an open end portion projected into said socket and including a candle projecting spring arranged in the tube and a spring member fixedly mounted in said socket and having arm portions extending upwardly and projecting into the open end of said tube into the adjacent end of said projecting spring and body for yieldable engagement with adjacent convolutions of said projecting spring, for detachably securing said projecting spring relative to said tube and said tube relative to said socket.

2,717,307
TUBULAR LUMINAIRE
Arthur M. Bjontegard, Marblehead, Mass., assignor to General Electric Company, a corporation of New York
Application August 18, 1950, Serial No. 180,220
2 Claims. (Cl. 240—51.11)

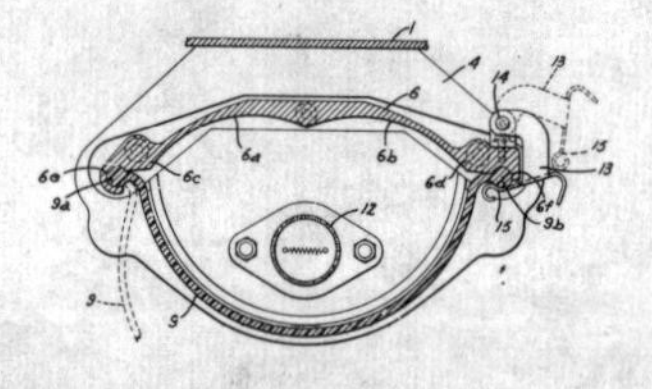

1. In a luminaire, an elongated reflector having a pair of parallel spaced-apart edges defining therebetween an open reflector mouth, each edge of said reflector being provided with a longitudinal slot and the slot in one of said edges being open and shallow and the slot in the other of said edges being circular in cross section and undercut to provide a side opening narrower than its diameter, and an elongated transparent cover having integrally formed on opposite edges thereof parallel beads spaced from one another by the spacing of said slots in said reflector edges, one of said beads having substantially the same surface configuration as said open slot in said one of said reflector edges and the other of said beads having substantially the same circular cross section as said undercut slot in said other of said reflector edges and being disposed in said undercut slot thereby hingedly to mount said cover for opening movement from its closed position in which its said one bead is seated in said shallow open slot of said reflector edge to close said reflector mouth.

2,717,308
LAMP GUARD CONSTRUCTION
Joseph D. Kevorkian, Philadelphia, Pa.
Application April 12, 1950, Serial No. 155,493
3 Claims. (Cl. 240—102)

1. In a lamp guard of the character described, a sheet metal reflector member having a central body portion of

substantially semi-cylindrical shape and opposite end portions disposed respectively at the top and base of the guard and of generally spherical contour, said central body portion having embossed therein a pair of axially spaced, circumferentially extending ribs projecting interiorly of the shell with the opposite ends of each rib respectively terminating short of the opposite vertical edges of said body portion, and a guard member formed of interconnected wire rod elements to form an open frame of a shape generally complemental to that of said reflector

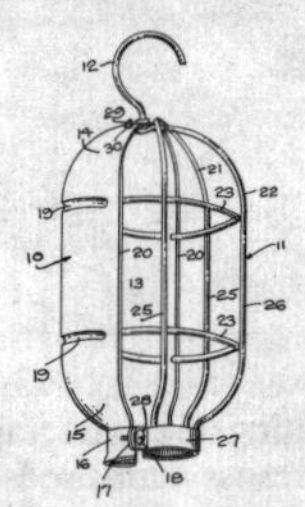

member, said wire rod elements including a pair of transversely curved wire rod elements axially spaced for coplanar disposition relatively to said internal ribs of the reflector shell with the opposite extremities of each of said transversely curved wire rod elements positioned inside of, and extending inwardly beyond, the vertical edges aforesaid of the reflector shell, whereby upon assembly of said reflector and guard members each of said transversely curved wire rod elements constitutes a circumferential continuation of an internal rib of the reflector shell.

2,717,309
RADIOSONDE WITH PROJECTILE MEANS TO CARRY IT ALOFT
Walter H. Campbell, Arlington, Va.
Application June 8, 1948, Serial No. 31,782
5 Claims. (Cl. 250—17)
(Granted under Title 35, U. S. Code (1952), sec. 266)

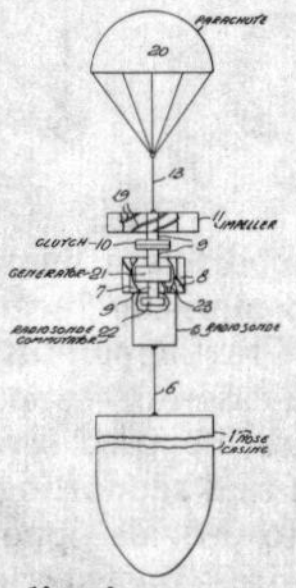

1. Apparatus for indicating meteorological characteristics of the upper atmosphere, comprising: a casing within which is placed a radiosonde, the radiosonde including a plurality of meteorological instruments and a radio transmitter for transmitting to a receiving station the indications of such instruments, switching means also in the radiosonde for sequentially connecting each of the instruments to the transmitter; projectile means for carrying the casing to the upper atmosphere; means attached to the casing for limiting its rate of descent during its return from the upper atmosphere, the last-named means including a parachute which is initially unopened; means associated with the casing for causing the parachute to open at a predetermined stage of the flight; and rotary-vane means included within the means for limiting the rate of descent of the casing, these rotary-vane means being operated by the passage of the latter through the air during the descent and serving also to actuate the switching means and to supply energy for operating the transmitter.

2,717,310
DIRECT CURRENT ELECTRONIC INTEGRATING SYSTEM
Thomas Ellis Woodruff, Los Angeles, Calif., assignor, by mesne assignments, to Hughes Aircraft Company, a corporation of Delaware
Application November 13, 1952, Serial No. 320,311
18 Claims. (Cl. 250—27)

1. An electronic integrating system for integrating a time-varying, direct-current potential, said system comprising: an electronic integrating network, including an integrating storage capacitor, whereby a time-varying, direct-current potential applied to said integrating network charges said integrating storage capacitor to develop an output voltage linearly proportional to the time integral of the time-varying, direct-current potential; a charge transfer capacitor; first means coupled to said capacitors and operable in response to said integrating storage capacitor being charged to a predetermined positive value, for substantially instantaneously charging said transfer capacitor to a predetermined negative value and for subsequently equalizing the charge of said capacitors, thereby to remove the positive charge across said integrating storage capacitor, said first means including a first output circuit for developing a positive pulse each time said integrating storage capacitor is charged to said positive value; and second means coupled to said capacitors and operable in response to said integrating storage capacitor being charged to a predetermined negative value, for substantially instantaneously charging said transfer capacitor to a predetermined positive value and for subsequently equalizing the charge of said capacitors, thereby to remove the negative charge across said integrating storage capacitor, said second means including a second output circuit for developing a positive pulse each time said integrating storage capacitor is charged to said negative value; the total number of said positive pulses from said first output circuit or from said second output circuit being linearly proportional to the integral with respect to time of the time-varying, direct-current potential.

2,717,311
SIMPLIFIED BINARY ADDER AND MULTIPLIER CIRCUIT
William A. Ogletree, Southampton, Pa., assignor to Philco Corporation, Philadelphia, Pa., a corporation of Pennsylvania
Application December 13, 1952, Serial No. 325,818
4 Claims. (Cl. 250—27)

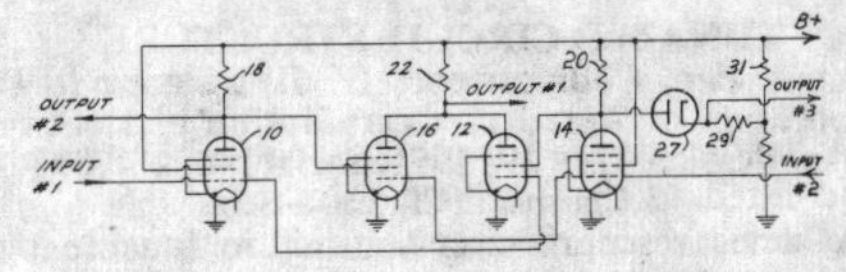

1. A circuit for providing indications of the correspondence of two bivalued signals, said circuit comprising: first and second electron tubes, each having at least an anode, a cathode, a control grid and a screen grid, a source of anode potential having at least a positive and a negative terminal, the cathodes of said first and second electron tubes being connected directly to said negative

terminal, an anode load resistor having one terminal thereof connected to said positive terminal and the other terminal thereof connected to both of said anodes, first and second signal inverter stages, each of said stages comprising an electron tube having at least an anode, a cathode and a control grid, means connecting said cathode to a point of fixed potential and an anode load resistor connecting said anode to a source of positive potential, the anode of said electron tube in said first inverter stage being connected directly to the screen grid of said first electron tube, the anode of said electron tube in said second inverter stage being connected directly to the screen grid of said second electron tube, means for supplying one of said bivalued signals to said control grid of said electron tube in said first inverter stage and to said control grid of said second electron tube, means for supplying the other bivalued signal to the control grid of said electron tube in said second inverter stage and to said control grid of said first electron tube, and means for deriving an output signal from the anode of at least one of said four electron tubes.

2,717,312
RADIO BEAM ANTENNA ARRANGEMENTS
Frank Howard Taylor, London, England, assignor to International Standard Electric Corporation, New York, N. Y., a corporation of Delaware
Application July 28, 1952, Serial No. 301,307
Claims priority, application Great Britain August 3, 1951
1 Claim. (Cl. 250—33)

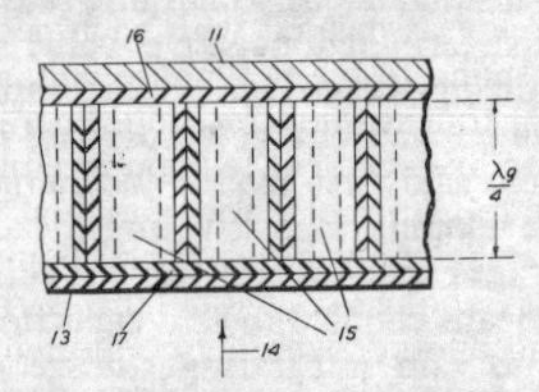

A directive radio antenna arrangement having means for producing a highly directive radiation beam, said means comprising an iris plate having an aperture centrally located and positioned forward with respect to the direction of radiation of said beam producing means, the surface of said plate facing said antenna being coated with an energy absorbing material, said plate being shaped and positioned so that the aperture passes radiated energy in a given direction, and to intercept and absorb energy radiated by said antenna in all other directions; said energy absorbing layer being in the form of a honeycomb structure and comprising a plurality of wave guide tubes closely spaced side by side with their longitudinal axes normal to the plate, each said tube having all its surfaces coated with high-loss dielectric material and having a guidelength substantially corresponding to an odd multiple of one quarter wavelength at the mean operating frequency, each of the two honeycomb surfaces defined by the ends of said tubes being closed by a respective thin layer of high-loss dielectric material.

2,717,313
TUNABLE CIRCUIT STRUCTURE
Wen Yuan Pan, Collingswood, N. J., assignor to Radio Corporation of America, a corporation of Delaware
Application May 29, 1951, Serial No. 228,891
2 Claims. (Cl. 250—36)

1. A series resonant circuit structure tunable over a frequency range of over two to one within the U. H. F. spectrum and comprising a first and a second rod disposed substantially parallel to each other to form a parallel wire transmission line, said second rod including an open loop intermediate its ends, a hollow tube of a material having a high dielectric constant, a first and a second metallic coating provided on the outer surface of said tube and spaced from each other substantially by the distance between said rods, said first coating being electrically connected to one end of said first rod and said second coating being electrically connected to one end of said second rod, a metallic core slideable in said tube, and means for moving said core into a first extreme position wherein said core extends within both of said coatings to provide effectively a pair of capacitors interconnected by

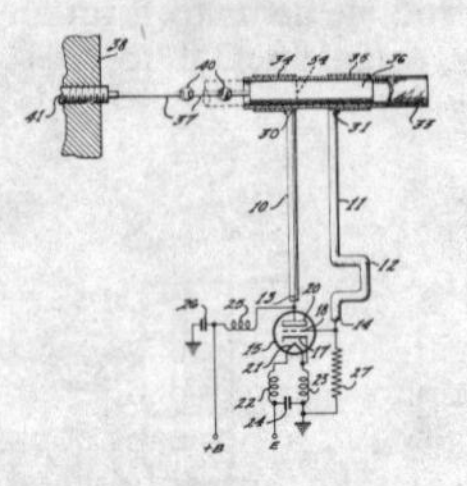

an inductor representing the inductance of said core and serially connected across said rods and extending the effective length of the parallel wire transmission line provided by said rods and into a second extreme position wherein said core extends within said first coating only and away from said second coating to provide a parallel wire transmission line open at its ends, said loop compensating for the electrical effect of said core on said first rod in its second extreme position.

2,717,314
DEVICE FOR SUPPORTING THE HEAD, AND HOLDING IT STATIONARY WHILE MAKING X-RAY PICTURES
James E. Delk, Sr., Atlanta, Ga.
Application January 23, 1952, Serial No. 267,857
11 Claims. (Cl. 250—50)

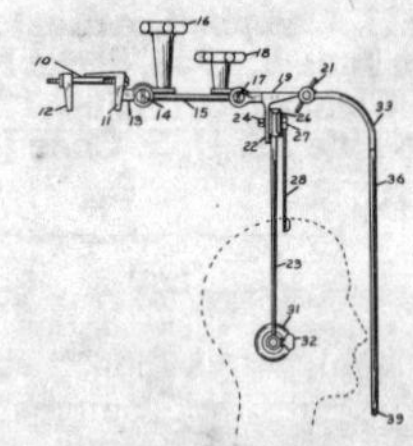

1. A device for supporting a human head while making X-ray pictures comprising means for engaging the film-cassette of an X-ray apparatus for mounting said device on said film-cassette, supporting means associated with said means, opposed spaced alignment arms pivotally mounted on said supporting means, means connected to said arms to provide opposed rotation of said arms, an alignment wire adjacent said supporting means in a plane midway between said arms for aligning the nose of a person whose head is engaged by said arms, and means for supporting said alignment wire.

2,717,315
X-RAY APPARATUS
Anthony Antal Nemet, Richmond-Surrey, and Matthew Berindei, Cheam-Surrey, England, assignors to Hartford National Bank and Trust Company, Hartford, Conn., as trustee
Application November 27, 1951, Serial No. 258,338
Claims priority, application Great Britain November 28, 1950
3 Claims. (Cl. 250—95)

1. X-ray apparatus comprising an X-ray tube having a given power rating, a source of high-tension potential connected to said tube, said potential source comprising an electric generator having a power rating of sufficient magnitude to supply said given power to said tube when

running substantially at its operating speed, a small electric motor coupled to said generator for driving the same, said motor having a power rating substantially below that of said generator whereby it can only drive said generator at its operating speed when the latter is

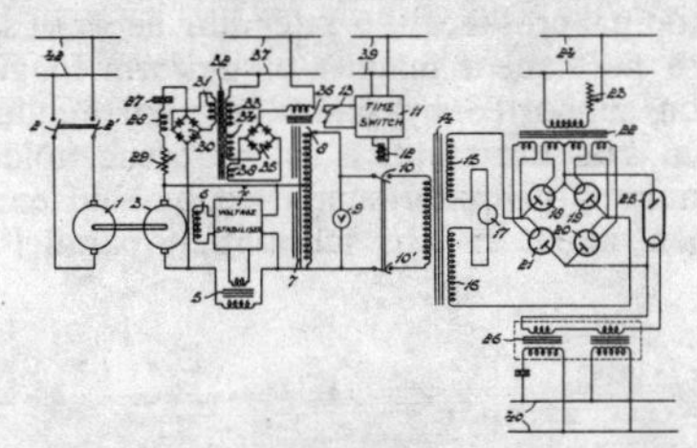

unloaded, means for separately energizing said motor when said generator is unloaded to bring said generator to its operating speed, and means coupling the output of said generator to said tube only when said generator is running at its operating speed.

2,717,316
PULSE LIMITER AND SHAPER
Richard Madey, Berkeley, Calif., assignor to the United States of America as represented by the United States Atomic Energy Commission
Application December 24, 1952, Serial No. 327,856
6 Claims. (Cl. 250—207)

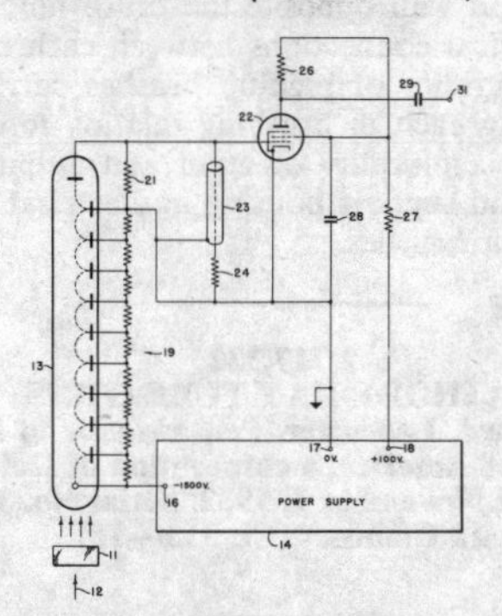

1. In a pulse limiter and shaper circuit for a photomultiplier tube circuit, the combination comprising a photomultiplier tube, means connected to the elements of said tube for impressing operating voltages, a length of coaxial cable with one end of the central conductor connected to the anode of said tube, the other end of the central conductor of said cable connected to a resistive impedance having a value such that voltage reflection occurs therefrom, and limiting means connected between the anode of said tube and an output terminal for limiting the amplitude of output pulses.

2,717,317
ENGINE TURNING MECHANISM
Sydney W. Scott, Peterborough, England, assignor to F. Perkins Limited, Peterborough, England
Application April 12, 1954, Serial No. 422,581
9 Claims. (Cl. 290—38)

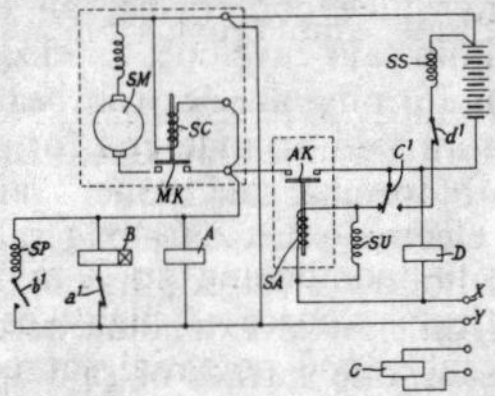

1. Engine turning mechanism comprising a starter motor, an auxiliary starting aid mechanism and control means comprising two electric relay members, one of which controls the circuit of the starting aid mechanism, but has its energising circuit controlled by the other relay, such other relay being operable when the voltage across the starter motor exceeds a predetermined value so as to render the starting aid mechanism inoperative, the relay controlling the starting aid mechanism being operable after a predetermined delay so as to actuate the starting aid mechanism when the voltage across the starter motor falls below the value required to operate said other relay.

2,717,318
OVERVOLTAGE DETECTION FOR ALTERNATING-CURRENT GENERATORS
Robert H. Keith and Clarence L. Mershon, Lima, Ohio, assignors to Westinghouse Electric Corporation, East Pittsburgh, Pa., a corporation of Pennsylvania
Application January 9, 1953, Serial No. 330,510
12 Claims. (Cl. 307—51)

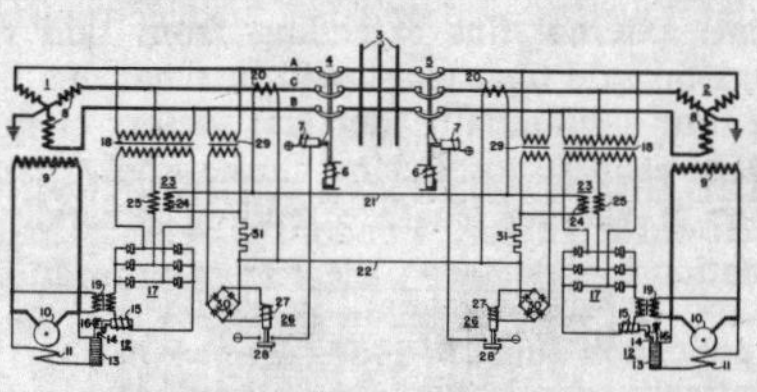

12. In a system comprising a plurality of alternating-current generators connected together for operation in parallel, a relay associated with each generator and connected to respond to the polyphase voltage of the generator, a current transformer for each generator energized by the generator load current, a loop circuit interconnecting said current transformers, and a mutual reactor for each generator having one winding connected in series with one phase of the voltage to which the corresponding relay responds and another winding connected to the loop circuit to be energized by a current proportional to the difference in the load currents of the generators.

2,717,319
METHOD AND APPARATUS FOR COOLING TRANSDUCERS
Francis P. Bundy, Alplaus, N. Y., assignor to General Electric Company, a corporation of New York
Application May 27, 1954, Serial No. 432,752
11 Claims. (Cl. 310—16)

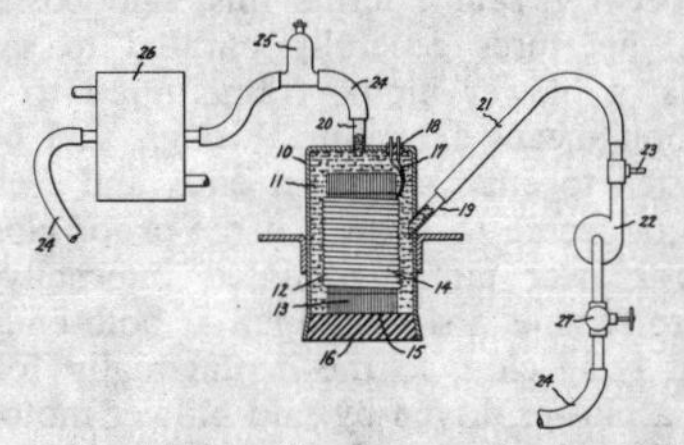

1. Apparatus for cooling a transducer having a transducer element, which comprises a housing to surround said element, and means for circulating an aerated acoustic cushioning fluid in said housing in direct thermal contact with said element.

2,717,320
HEAT EXCHANGER
Walter L. Shoulders and Robert R. Hayes, Cleveland, Ohio, assignors to Reliance Electric and Engineering Company
Application March 10, 1952, Serial No. 275,812
6 Claims. (Cl. 310—57)

4. A heat exchanger for a frame of a dynamoelectric machine with a base thereon and a shaft axis, said frame

having an opening near each end on the side thereof opposite said base with said openings having axes generally perpendicular to said shaft axis, said heat exchanger including, first and second duct ends, said duct ends having a substantially right angle bend between first and second portions thereof, said first portions being mountable on said frame openings and said second portions establishing a longitudinal axis substantially parallel to said shaft axis, said second portions extending toward one another and each having a female fitting on a circular opening therein, a plurality of substantially identical first duct sections each having a male and a female fitting on opposite ends of a circular opening therein, an adapter duct section having a male fitting on both ends of a circular opening therein, said plurality of first duct sections and said adapter duct section mating together and mating with the second portions of said duct ends to form a closed duct between said frame openings external of said frame, each of said first duct sections being formed as a ring to define the said circular opening and having internal and external fins extending from said ring, external fins on said duct ends, said internal fins being defined by four substantially identical sectors of ninety degrees each, each sector having substantially parallel internal fins of varying length extending generally toward

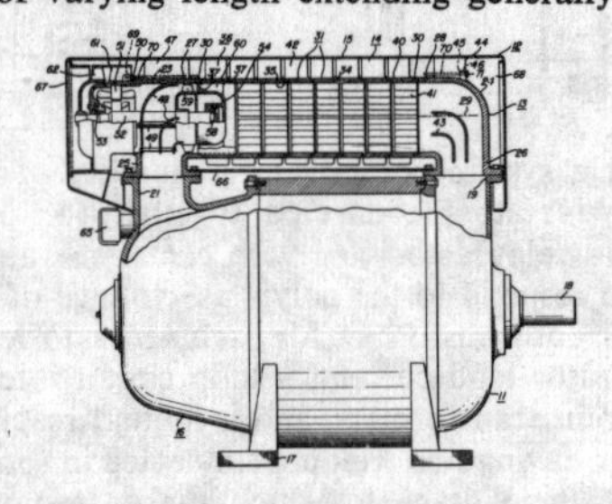

the center of said ring and terminating a short distance from the dividing line between sectors, said external fins all lying substantially parallel and terminating in the form of a heptagon, one side of said heptagon being the base formed by two oppositely extending external fins and disposed closest adjacent said frame, the two sides of said heptagon adjoining said base defining a forty-five degree angle relative to each other such that two complete heat exchangers may be paralleled on a single frame at an included angle of approximately forty-five degrees therebetween, bosses next adjacent the outer periphery of said duct ends between said external fins, said bosses having longitudinal apertures generally parallel to said longitudinal axis, a blower motor frame fitted to said first duct end, bolt means extending through said boss apertures to fasten together said duct ends and sections and blower motor frame to form a flame-proof closed duct, an electric blower motor mounted internally in said blower motor frame, a shroud open at both ends and enclosing and lying close to the heptagonally formed external fins, a blower driven by said blower motor and arranged to direct a stream of air within said shroud and between said external fins, and a second blower driven by said blower motor inside said closed duct to circulate air within said closed duct and said machine frame.

2,717,321
DYNAMOTOR
Harry C. Stearns, Glen Ellyn, Ill.
Application August 20, 1951, Serial No. 242,732
8 Claims. (Cl. 310—138)

7. In a dynamotor the combination comprising a support housing forming a stationary enclosing shell with a pair of oppositely disposed end walls; a shaft within said housing extending between and supported at the ends by said end walls; a stator armature mounted on said shaft with two sets of windings and a pair of commutators one for each set of windings that are disposed at opposite ends of the armature; a field member including a pair of end walls adjacent said commutators rotatably supported by and extending transversely of said shaft, a sleeve of uniform longitudinal cross section extending between said end walls, and a permanent magnet of uniform longitudinal cross section supported by said sleeve surrounding and spaced from said armature; a set of brush holders for each commutator pivotally supported by said end walls for movement about an axis substantialy parallel to said

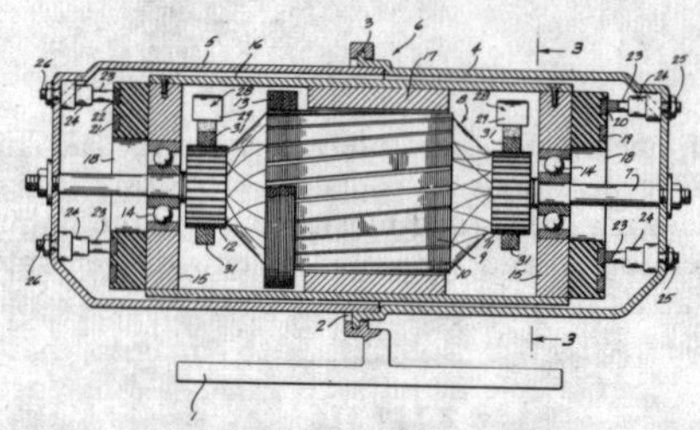

shaft with each extending circumferentially to each side of the pivotal support with a brush holding head to one side and a counterweight to the other side; brushes retained by said brush holding heads in commutating relation with said commutators; slip rings mounted on the side of each end wall opposite the brush holder supporting side; electrical connections between each brush and a slip ring, a plurality of pick-up brushes carried by said support housing each in brushing relation to one of said slip rings; and a plurality of input and output terminals mounted on said support housing in electrical connection with said pick-up brushes.

2,717,322
CATHODE RAY TUBE GUNS
David C. Ballard, Lancaster, Pa., assignor to Radio Corporation of America, a corporation of Delaware
Application November 1, 1952, Serial No. 318,191
10 Claims. (Cl. 313—76)

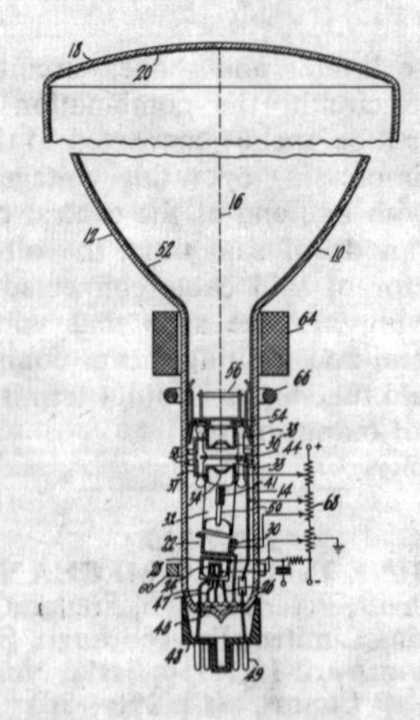

1. A cathode ray tube comprising, an enclosing envelope, a target within said envelope, an electron emitting cathode, means for forming an electron beam from electrons emanating from said cathode and for directing said beam along a path toward said target, said means including a hollow electrode spaced along said path from said cathode, said hollow electrode having an apertured transverse wall of thin metal extending across said path of said electron beam, one surface of said apertured wall being roughened to prevent evaporation of metal therefrom.

2,717,323
ELECTRON BEAM CENTERING APPARATUS
Burton R. Clay, Woodbury, N. J., assignor to Radio Corporation of America, a corporation of Delaware
Application March 23, 1954, Serial No. 418,021
15 Claims. (Cl. 313—77)

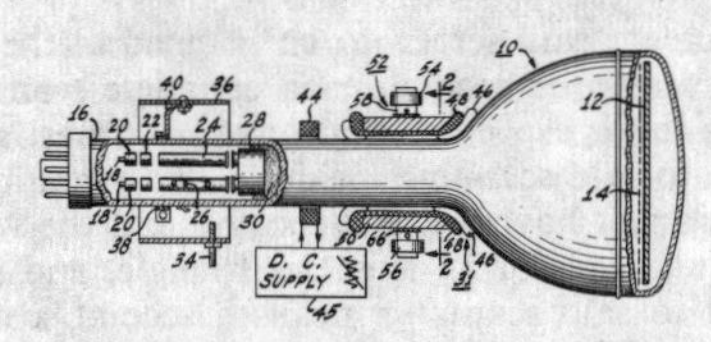

1. Beam centering apparatus for use in conjunction with a multi-beam kinescope of the type having a target, means for directing a plurality of electron beams toward such target and an electromagnetic deflection yoke for causing such beams to scan a raster, said centering apparatus comprising: a pair of permanent magnets; and means for supporting said magnets on diametrically opposed sides of such yoke in such manner that said magnets are rotatable about such yoke, said magnet supporting means having means permitting rotation of at least one of said magnets about an axis perpendicular to such beams.

2,717,324
CATHODE RAY TUBES
Ronald Charles Hall, Cowley, England, assignor to Electric & Musical Industries Limited, Hayes, England, a company of Great Britain
Application May 1, 1951, Serial No. 223,857
Claims priority, application Great Britain May 4, 1950
6 Claims. (Cl. 313—84)

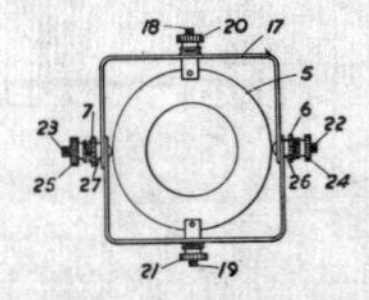

1. In combination with a cathode ray tube of the type having a neck portion along which a beam of electrons can be projected axially thereof, a magnetic focussing system for focussing said beam, means mounting said focussing system around said neck portion, substantially magnetically neutral ferromagnetic means disposed around said neck portion in a position physically spaced from said focussing system but magnetically coupled thereto, means mounting said ferromagnetic means independently of said focussing system mounting means, said means mounting said ferromagnetic means including means for effecting tilting movement of said ferromagnetic means so that said ferromagnetic means can be adjusted to positions other than normal to the axis of said neck portion to enable the direction of said beam to be varied subsequent to its passage through the magnetic field of said focussing system.

2,717,325
INDIRECTLY HEATED CATHODE, FOR CATHODE RAY TUBES IN PARTICULAR
Karl Gosslar, Oberesslingen, Germany, assignor to International Standard Electric Corporation, New York, N. Y., a corporation of Delaware
Application April 2, 1953, Serial No. 346,455
Claims priority, application Germany April 8, 1952
4 Claims. (Cl. 313—270)

1. A structure for securing a cathode electrode in fixed position regardless of thermal expansion effects, comprising a rod shaped cathode having a first rim at one end near the emitting surface, and a flanged second rim provided with an upturned groove at its other end, said second rim being integral with and turned up from said rod-shaped cathode, an insulating support member provided with an opening larger than the body of said rod but smaller than said first rim on one surface thereof, said cathode being seated in said opening with said first rim

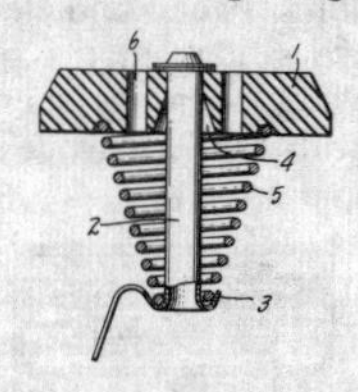

resting on said one surface, and a conically wound spring mounted in comparison with its smaller end in the groove on said second rim, and its other end pressing against the other surface of said insulating support.

2,717,326
ELECTRIC ARC FURNACE CONTROL SYSTEMS
Frank de la Roche Gunton, Rugby, England, assignor to General Electric Company, a corporation of New York
Application December 14, 1951, Serial No. 261,650
3 Claims. (Cl. 314—69)

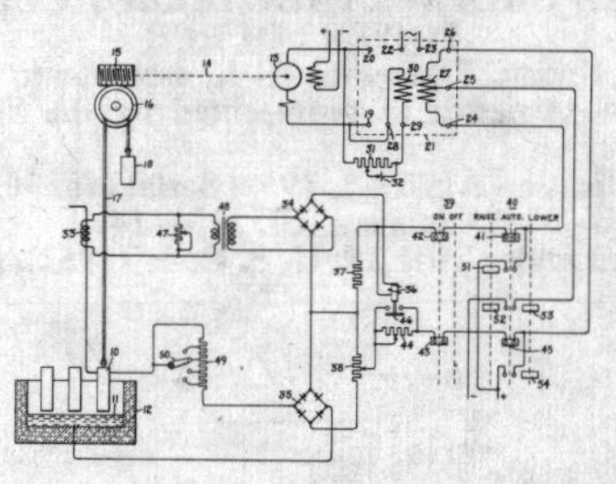

1. A control system for an electric arc furnace provided with a plurality of electrodes having connections to a source of alternating voltage, means for deriving a control voltage proportional to the current in one of said electrodes comprising a current transformer energized by said current and a rectifier having its input terminals connected to the secondary winding of said transformer and provided with output terminals, a first resistor connected across said output terminals, means for deriving a control voltage proportional to the voltage across the arc from said electrode to the charge in said furnace comprising a second rectifier provided with input terminals connected across said arc and provided with output terminals, a resistor connected across the output terminals of said second rectifier, an electric motor mechanically connected to drive said electrode and means responsive to the difference of said control voltages for selectively energizing said motor for raising and lowering said electrode comprising a magnetic amplifier having direct voltage output terminals connected to said motor, a main reactance winding connected to said output terminals, and a control circuit including said resistors connected in series relationship with their voltage drops opposing each other and a saturation control winding connected in series relationship with said resistors and mounted in inductive relationship with said reactance winding.

2,717,327
VELOCITY MODULATION DEVICES
Emile Touraton, Claude Dumousseau, and René Zwobada, Paris, France, assignors to International Standard Electric Corporation, New York, N. Y., a corporation of Delaware
Application January 15, 1948, Serial No. 2,355
Claims priority, application France January 17, 1947
7 Claims. (Cl. 315—6)

1. A high frequency electronic device comprising an elongated cathode electrode, an elongated reflector elec-

trode whose length runs in substantially the same direction as that of the cathode electrode, said electrodes being spaced apart a given distance throughout their length, wave guide lengths disposed adjacent each other in overlapping relation between said electrodes and spaced therefrom, said wave guide lengths being connected to each other and having openings therethrough substantially co-extensive with said electrodes and in alignment

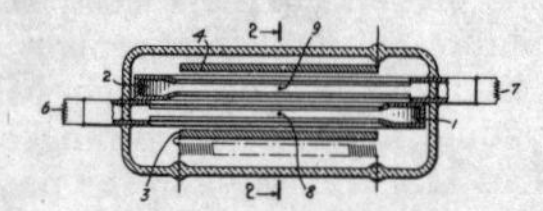

therewith, the transit time of an electron between corresponding points in said openings being a quarter of a period at the operating frequency, the distance between said corresponding points as measured through the connected wave guide lengths being effectively electrically a quarter wavelength at the same frequency, and means effectively terminating said wave guide lengths in their characteristic impedance, to thereby prevent standing waves.

2,717,328
PULSED HIGH VOLTAGE DIRECT CURRENT POWER SOURCE

Benjamin Kazan, Princeton, N. J., assignor to the United States of America as represented by the Secretary of the Army

Application August 4, 1952, Serial No. 302,635
5 Claims. (Cl. 315—14)
(Granted under Title 35, U. S. Code (1952), sec. 266)

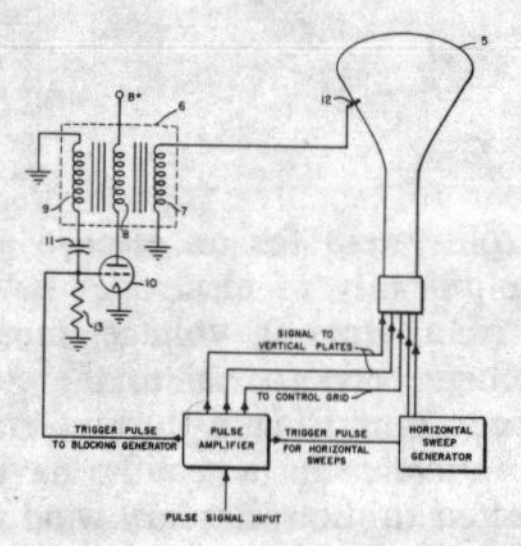

1. A high voltage direct current power supply for an electronically controlled load, comprising a blocking oscillator having a transformer and a grid controlled tube in its circuit, an additional high voltage winding on said transformer connected to intermittently apply the high voltage induced therein to said load, additional grid controlled power supply means across said load, trigger pulsing means and connections to apply the output thereof to trigger said blocking oscillator and the additional power supply means simultaneously.

2,717,329
TELEVISION SCAN SYSTEM

Charles H. Jones and Gordon S. Ley, Pittsburgh, Pa., assignors to Westinghouse Electric Corporation, East Pittsburgh, Pa., a corporation of Pennsylvania

Application September 19, 1950, Serial No. 185,678
2 Claims. (Cl. 315—24)

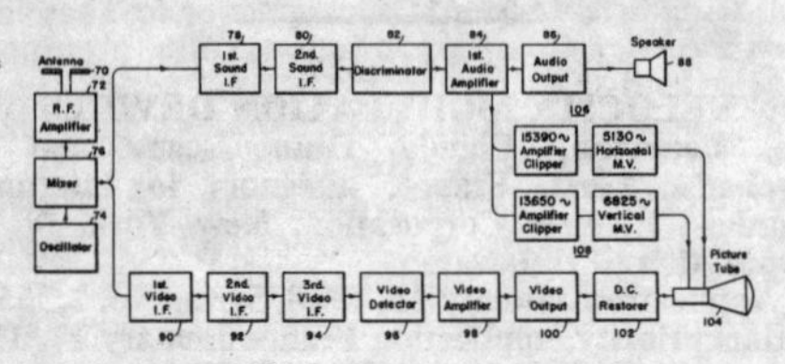

1. In combination a cathode ray tube having a sensitive screen and including scanning means adapted to have impressed thereon first and second right-angle components of a scanning parameter, corresponding respectively to the aspects a and b of the aspect ratio

$$\frac{a}{b}$$

of said screen; said screen to be scanned at the rate of r complete scans per second, each complete scan terminating along the b aspect of said screen n times; means for supplying to said scanning means, a first scanning parameter component having a frequency v and the wave form of the equal sides of an isosceles triangle, and means for supplying to said scanning means a second scanning parameter component at right angles to said first component having a frequency w and the wave form of the equal sides of an isosceles triangle, where v substantially equals nr and w substantially equals

$$n\frac{a}{b}r$$

and numerical n and

$$n\frac{a}{b}$$

are so related that they do not have a common factor.

2,717,330
CATHODE RAY TUBE SECTOR SELECTOR

Ralph E. Meagher, Champaign, Ill., and Chalmers W. Sherwin, Belmont, Mass., assignors, by mesne assignments, to the United States of America as represented by the Secretary of the Navy

Application January 21, 1946, Serial No. 642,469
9 Claims. (Cl. 315—24)

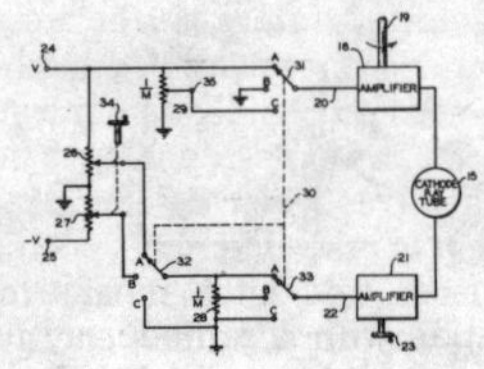

1. An electrical circuit comprising, a cathode ray tube; a radial sweep amplifier for said cathode ray tube having a first and second input wherein the sweep speed is controlled by the amplitude of an applied voltage, said first input being adapted to receive azimuth information; an off-center amplifier for said cathode ray tube having a first and second input, said first input being adapted to receive off-center direction information; a source of balanced voltage having first and second terminals; a first potentiometer connected between said first terminal and ground; a second potentiometer connected between said second terminal and ground; two tapped attenuators; a triple-pole triple-throw switch of positions A, B, and C where A is Off-center P. P. I., B is Spot and C is Centered P. P. I.; said elements being connected as follows; said second input of said radial sweep amplifier to the selector of the first section of said switch, contact A of said first section to the top of the first of said attenuators and to said first terminal of said source of balanced voltage, contact C of said first section to the tap of said first attenuator, said second input of said off-center amplifier to the selector of the second section of said switch, contact A of said second section to the top of the second of said attenuators and to the selector of the third section of said switch, contact B of said second section to the tap of said second attenuator, contact B of said first section, contacts C of said second and third sections, and the bottom of each of said attenuators being connected together and to ground, contact A of said third section to the sliding contact of said potentiometer connected to said first terminal of said source of balanced voltage, and contact B of said

third section to the sliding contact of said potentiometer connected to said second terminal of said source of balanced voltage, wherein said circuit enables switching to either of two cathode ray tube presentations, the second one of which will be magnified relative to the first by the ratio of said attenuators and displaced relative to the first to bring to the center of the second of said cathode ray tube presentations, the spot selected from said first presentation by adjustment of said first input to said off-center amplifier and said first and second potentiometers.

2,717,331
THERMOSTATICALLY OPERATED TIME DELAY SWITCHES
Jesse R. Hollins, Brooklyn, N. Y.
Application July 19, 1954, Serial No. 444,136
18 Claims. (Cl. 315—77)

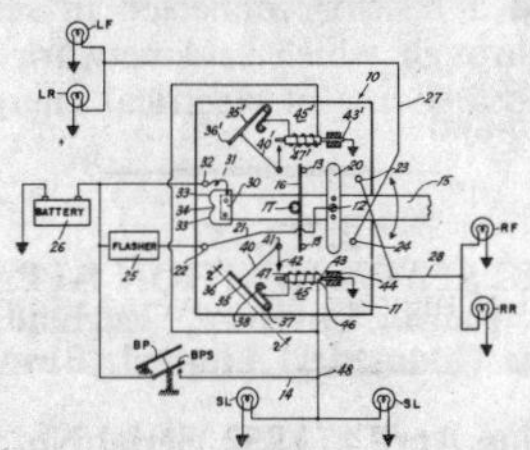

1. A time delay switch comprising, in combination, a switch operator movable between an operated position and a restored position; switch closure means operatively associated with said operator and operative, in one of said positions, to close a circuit controlled by said switch and, in the other of said positions, to open such circuit; means biasing said operator to the restored position; latch means operable to releasably latch said operator in the operated position, said latch means being biased to the operator-releasing position; electrically energized release means effective, when de-energized, to constrain said latch means to the latching position and, when energized, to release said latch means to the operator-releasing position; a normally open energizing circuit for said release means; a thermostatic device operable, when heated for a pre-set time, to close the energizing circuit of said release means; electric heating means for said device; and an energizing circuit for said heating means closed responsive to movement of said operator to the operated position; whereby, when said operator is moved to the operated position, said heating means is energized to heat said thermostatic device to close the energizing circuit of said release means after such pre-set time.

2,717,332
STARTING AND OPERATING CIRCUITS FOR FLUORESCENT LAMPS
William S. H. Hamilton, Larchmont, N. Y.
Application June 6, 1951, Serial No. 230,185
5 Claims. (Cl. 315—100)

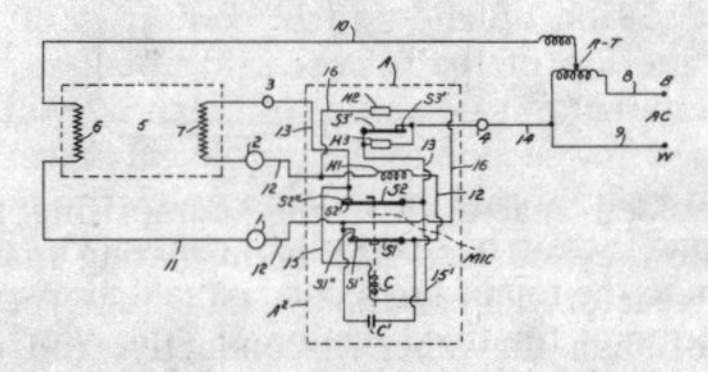

1. A system for supplying preheating and operating current to a discharge device comprising the combination of a supply line, a gaseous electric discharge device having two electrodes at least one of which is constructed to receive preheating current, a normally closed thermally operated switch and a normally closed electromagnetic switch, means including an inductive reactance device and the winding of said electromagnetic switch for connecting said electrodes with said line, a preheating circuit capable of producing firing temperature electrode heating, said circuit extending from one side of said line and including in series connection said inductive reactance device and the heater for said thermal switch and said electromagnetic switch and at least one of said electrodes and a current limiting device, said circuit continuing to the opposite side of said line, said electromagnetic switch having an armature of small inertia pivoted to be freely movable and said thermostatic switch being connected across said winding so that the opening of the thermal switch introduces said winding into the preheating circuit and causes a powerful and quick opening of the electromagnetic switch to cause the discharge device to fire by a high inductive voltage kick, and mechanical means operatively interconnecting said switches to hold the thermal switch open after the heater therefor cools, the operating current of said device continuing to energize said winding and hold both of said switches open during the operation of the discharge device.

2,717,333
STARTING AND OPERATING CIRCUIT FOR FLUORESCENT LAMP
William S. H. Hamilton, Larchmont, N. Y.
Application February 15, 1950, Serial No. 144,345
3 Claims. (Cl. 315—103)

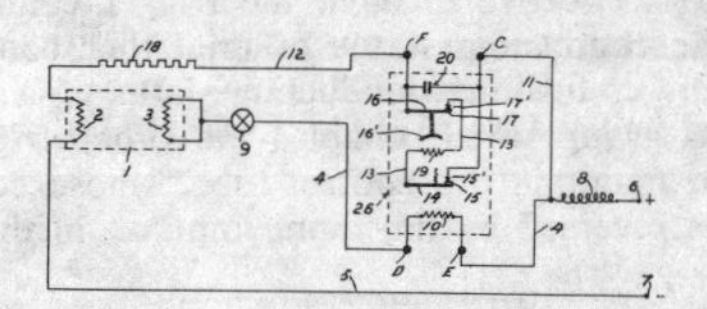

1. A system for supplying preheating and operating current to a discharge device comprising the combination of a supply circuit, a gaseous electric discharge device having two electrodes, at least one of which is constructed to receive preheating current, a relay having an operating winding and normally closed contacts, means including an inductive reactance device and said winding for connecting said electrodes with said supply circuit so that said winding is energized by the operating current of the discharge device after the device fires, a thermally operated switch having a heater and normally closed contacts, and a preheating circuit capable of producing firing temperature electrode heating, said circuit including in series connection said relay contacts and said thermal switch contacts and heater and the electrode to be preheated, said preheating circuit being connected across the discharge device and said winding in series connection so that the opening of said thermal switch causes said discharge device to fire by an inductive voltage kick and so that the opening of said relay contacts by the resulting establishment of the operating current through the discharge device prevents the re-establishment of the preheating circuit when said thermal switch cools and recloses.

2,717,334
ELECTRONIC COUNTERS
Joseph R. Desch, Dayton, Ohio, assignor to The National Cash Register Company, Dayton, Ohio, a corporation of Maryland
Application April 21, 1953, Serial No. 350,132
3 Claims. (Cl. 315—166)

1. In a device of the class described, the combination of a plurality of gaseous electron discharge devices; a first circuit connecting half of the devices into an operational group; a second circuit connecting the other half of the devices into another operational group; means connecting the devices in an operational series in which the conducting condition of the devices is advanced step-

by-step, each series connection extending from a discharge device in one group to a discharge device in the other group and enabling conduction in a device in one group to prepare a device in the other group for firing in response to an input impulse; a pair of control tubes, one related to each group of devices; means connecting each control tube to its related group of devices to control conduction in the devices of its related group; trigger connections between each group of devices and the control tube of the other group to enable conduction in each group of devices to control the conduction in the control tube related to the other group, whereby one control tube will be conducting and the other non-conducting according to which of the devices of the series is conducting; means to apply input impulses to the control tubes, each impulse reversing the conducting status of one of the control tubes, which in turn causes a change in the conducting status of a device in its related group, which change in the conducting status of the device in the group is effective through the trigger connection to reverse the conducting status of the other control tube and thereby change the conducting status of a device in its related group which enables the other trigger connection to maintain the control tube, whose conducting status was reversed by the input impulse, in that status.

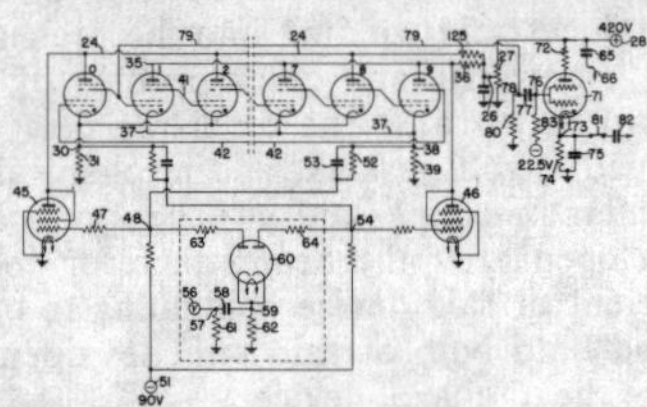

2,717,335
IGNITION SYSTEM

Marion W. Sims, Ezra C. Hill, and Aaron M. Krakower, Fort Wayne, Ind., assignors to General Electric Company, a corporation of New York

Application July 17, 1952, Serial No. 299,416
26 Claims. (Cl. 315—183)

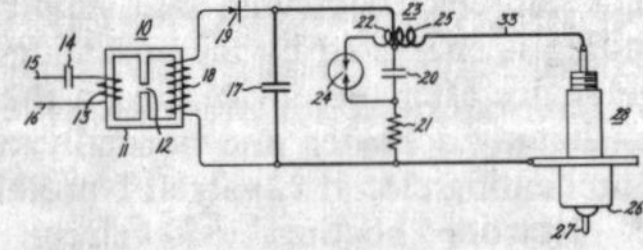

1. A capacitor discharge ignition system comprising, in combination, a source of unidirectional voltage capacitor-charging current, a main energy storage capacitor connected across said source, a triggering capacitor connected to be charged through a resistor by said source, a voltage step-up transformer having a low voltage primary winding and a high voltage secondary winding, a circuit completing triggering device connected to discharge said triggering capacitor through said low voltage primary winding when the voltage charge on said triggering capacitor attains a predetermined value, said transformer having at least said secondary winding connected directly in series with said main capacitor forming a discharge circuit, said discharge circuit being adapted to be connected directly to an ignition gap.

2,717,336
FLASHER CIRCUIT

Charles L. Craddock, North Hollywood, Calif., assignor to Michael Research Company, Inc., a corporation of California

Application May 8, 1953, Serial No. 353,736
6 Claims. (Cl. 315—183)

1. A flasher circuit of the class described which includes: a conductive device having the electrical characteristics of a glow discharge device; a control circuit comprising a resistor and a capacitor connected in parallel; a power utilization device connected in series with said conductive device; a reinforcing capacitor connected in parallel with the series circuit comprising said conductive device and said utilization device, forming a first series-parallel circuit; means connecting said control circuit in series with said first series-parallel circuit to form a second series-parallel circuit; a power capacitor connected in parallel with said series-parallel circuit to form a network; and a resistor connected in series with said network and through which said network is adapted to be connected to a source of electrical energy.

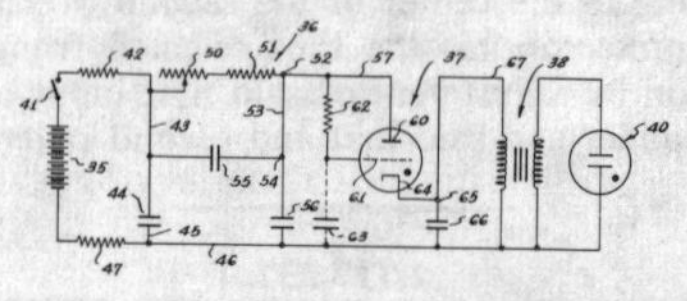

2,717,337
ELECTRIC SPARK IGNITION APPARATUS

John Andrew Laird, Coventry, England, assignor to Joseph Lucas (Industries) Limited, Birmingham, England

Application April 8, 1952, Serial No. 281,116
Claims priority, application Great Britain April 9, 1951
6 Claims. (Cl. 315—243)

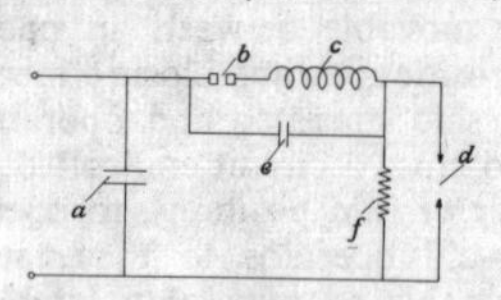

1. Electric spark gap ignition apparatus for the purpose specified comprising a discharge circuit including capacity, a control gap in said circuit adapted to control the discharge and an inductance in series with said capacity, and a condenser connected in parallel with the said control gap and at least a part of the inductance.

2,717,338
MOLDED REACTOR

Merritt W. Albright and John J. Foudy, Peabody, Mass., assignors to General Electric Company, a corporation of New York

Application April 29, 1954, Serial No. 426,470
3 Claims. (Cl. 317—13)

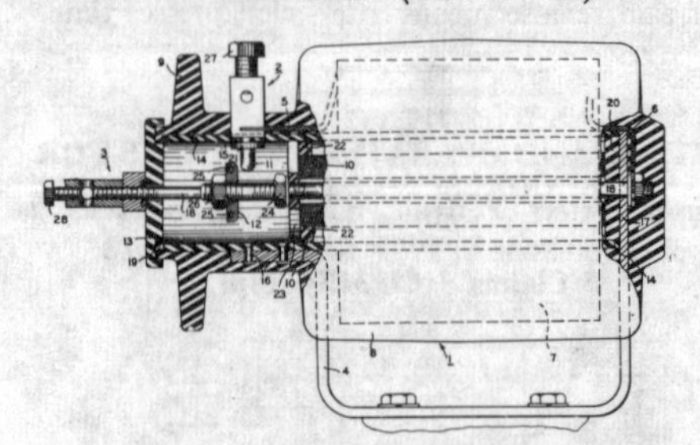

1. A molded reactor assembly comprising an open-ended hollow cylindrical tube of insulating material, an electrical reactor supported by said tube nearer one end thereof than the other, a conductive rod of length greater than the axial length of said tube disposed coaxially with said tube and with ends of said rod projecting axially outward beyond the ends of said tube, a first electrical terminal having a terminal block extending radially outward from said tube nearer said other end and a spark gap stud electrically and mechanically connected with said terminal block and extend-

ing radially inward in said tube nearer said other end for a distance less than the radial distance between said tube and said rod, the spacing between said rod and said stud being greater than spark gap lengths to be used with said reactor, electrical leads connecting said reactor between said terminal block and one projecting end of said rod at said one end of said tube, a spark gap nut axially movable along a threaded portion of said rod for adjustable axial positioning near said spark gap stud to form a spark gap therebetween of length shorter than said spacing, resilient molded material of high dielectric strength surrounding the outside of said tube and said leads and said reactor and closing said one end of said tube, said material bonding said tube and said one projecting end of said rod at said one end of said tube, an end cover slidable over the other projecting end of said rod for closing said other end of said tube, and a second electrical terminal connected with said other projecting end of said rod.

2,717,339
PANELBOARD CONSTRUCTION
Thomas F. Brown, Belleville, N. J., assignor to Westinghouse Electric Corporation, East Pittsburgh, Pa., a corporation of Pennsylvania
Application August 25, 1951, Serial No. 243,634
16 Claims. (Cl. 317—119)

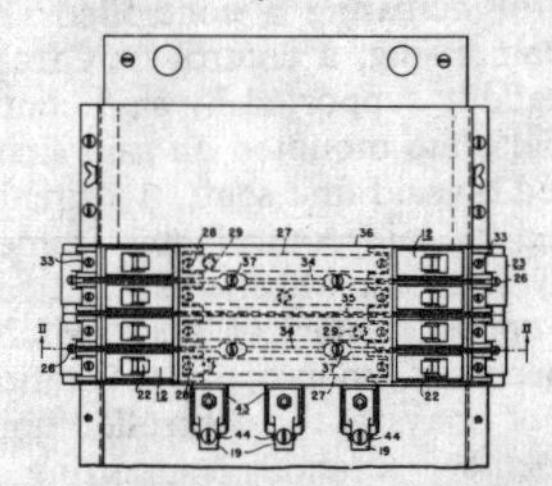

1. In a panelboard, in combination, a mounting pan, a unitary base removably secured to said pan for supporting a plurality of circuit interrupter units, said interrupter units having individual housings removably mounted on top of said base in oppositely disposed groups with an intervening space between opposite groups, each of said housings having a ventilating opening in the end thereof nearest said space, a barrier formed integrally with the base and disposed in said space longitudinally of the base, said openings being adjacent to the barrier to direct arc gases into said space, and a cover plate attached to the barrier, part of the barrier being reduced in height to provide space between the barrier and the cover plate for ventilating the housings through said openings.

2,717,340
DYNAMO REGULATORS
Albert Victor Waters, Perivale, Greenford, England, assignor to C. A. V. Limited, London, England
Application December 3, 1951, Serial No. 259,516
Claims priority, application Great Britain December 13, 1950
1 Claim. (Cl. 317—188)

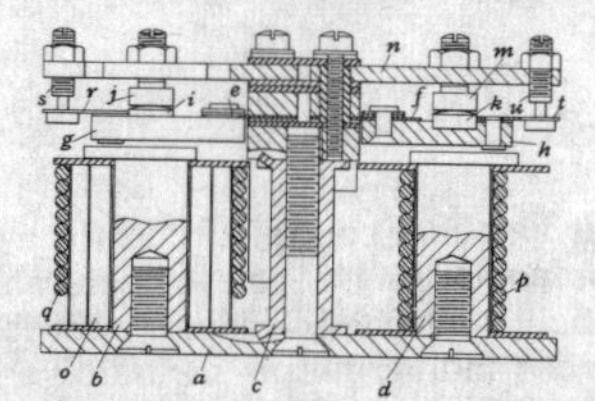

A dynamo regulator of the kind and for the purpose specified, having in combination, an iron base, three iron cores extending parallel with each other from the base with one of the cores situated between the others, a pair of independently movable contact-actuating armatures hingedly connected to the intermediate core and extending respectively over the outer ends of the end cores, a voltage winding on one of the end cores, a current winding on the other end core, and a supplementary current winding on the same core as the voltage winding, the arrangement being such that the magnetic fluxes created by all of the windings flow in the same direction through the intermediate core.

2,717,341
ASYMMETRICALLY CONDUCTIVE DEVICE
Harper Q. North, Los Angeles, Calif., assignor to General Electric Company, a corporation of New York
Application October 11, 1949, Serial No. 120,766
3 Claims. (Cl. 317—235)

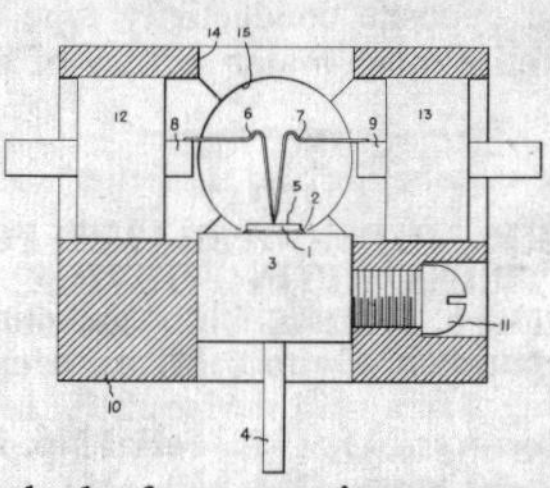

1. The method of constructing a current control device of the type employing a semi-conductor material with electrodes in contact with a face thereof comprising the steps of coating one electrode with a thin layer of a hardenable liquid insulating material, removing said coating from the tip of said electrode, positioning the tip of one of the electrodes on the surface of a semi-conductor member, positioning the tip of the other of the electrodes on the same surface of said semi-conductor member and moving one of said electrodes against the other of said electrodes so that said electrodes are maintained in insulatedly spaced relationship by said coating.

2,717,342
SEMICONDUCTOR TRANSLATING DEVICES
William G. Pfann, Basking Ridge, N. J., assignor to Bell Telephone Laboratories, Incorporated, New York, N. Y., a corporation of New York
Application October 28, 1952, Serial No. 317,191
3 Claims. (Cl. 317—235)

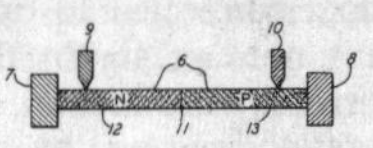

3. A semiconductor translating device comprising a filament of semiconductive material, two low-resistance contacts to said filament intermediate which contacts is a filamentary section of said body, which filamentary section is part P-type and part N-type, and at least two emitter contacts intermediate said low-resistance contacts at least one of which emitter contacts is substantially adjacent each low-resistance contact and in which there is a PN barrier intermediate said at least two emitter contacts.

2,717,343
P–N JUNCTION TRANSISTOR
Robert N. Hall, Schenectady, N. Y., assignor to General Electric Company, a corporation of New York
Original application November 18, 1952, Serial No. 321,262. Divided and this application June 25, 1954, Serial No. 439,319
5 Claims. (Cl. 317—235)

1. A power transistor comprising a semiconductor body having opposing major surfaces and having along a major dimension thereof a plurality of one conductivity-type

zones and a plurality of opposite conductivity-type zones, each said opposite conductivity-type zone being intermediate and integrally joined to two of said one conductivity-type zones to form respective P–N junctions therewith, a plurality of electrodes each making good

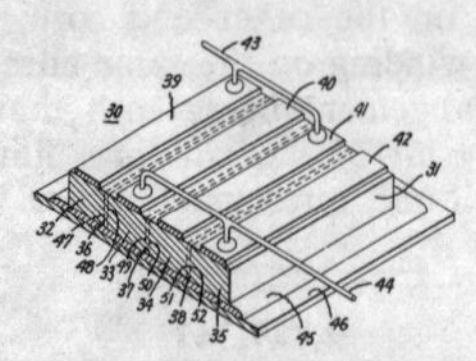

conductive contact with a respective one of said one conductivity-type zones along the same major surface of said body, and a further electrode consisting of an activator element for furnishing to said body conduction carriers of said opposite conductivity type and fused to and with the entire other major surface of said body.

2,717,344
ELECTRIC MOTOR OPERATED POWER TRANSMITTING DEVICE

George W. Jackson, Dayton, Ohio, assignor to General Motors Corporation, Detroit, Mich., a corporation of Delaware
Application April 11, 1952, Serial No. 281,844
4 Claims. (Cl. 318—31)

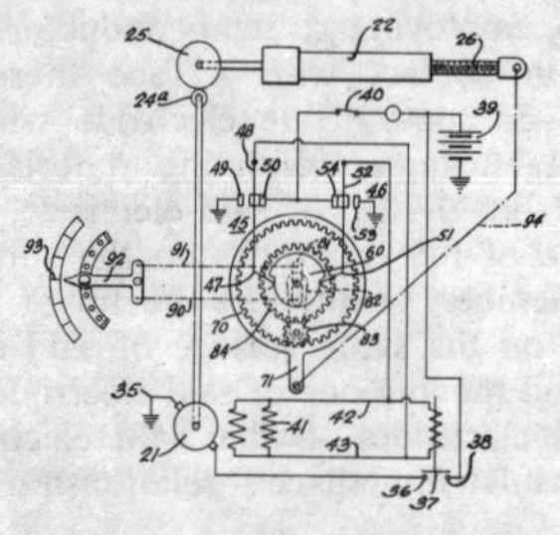

1. In combination with an actuator driven by a reversible electric motor for transmitting power in either direction, of a control mechanism manually operative to render the electric motor active in either direction and operative by the actuator to stop the electric motor, said control mechanism consisting of two electric switches in circuit with the motor, each switch having two stationary terminals and a movable contact, one stationary terminal of each switch being connected to one pole of a power source and normally not engaged by the movable contact of the respective switch, the second stationary terminal of each switch, normally engaged by the movable contact of said switch, being connected to the other pole of the power source, the movable contact of each switch being connected to the electric motor; a switch operating cam interposed between and engaged by the movable contacts of both switches, an arm rotatably supported coaxially of the cam, a manually operable lever connected to said arm for swinging it about its support in either direction of rotation; a gear train including a gear on the cam concentric therewith, a pinion rotatably mounted on the free end of said arm, and an internal ring gear mounted coaxially of the cam gear and meshing with the pinion; said cam being rotatable in one direction or the other for moving the movable contact of one or the other switch into engagement with its cooperating said one stationary terminal whereby the motor is energized to rotate in one or the other direction to the lever into any one of its plurality of selected positions swings the arm into predetermined corresponding positions, the swinging of the arm moving the pinion within the ring gear and thereby rolling it over the ring gear, the rotating pinion turning the cam gear and cam thereby causing said cam to actuate a switch into motor circuit closing position and causing the motor to operate in one direction; and a link attached to the actuator and operatively connected to the ring gear said link rotating the ring gear in response to and in accordance with the movement of the actuator, whereby the pinion is rotated and said pinion in turn rotates the cam gear and cam to normal position in which the both switches are in motor circuit opening positions and the motor rendered inactive.

2,717,345
BIDIRECTIONAL REMOTE ELECTRICAL CONTROL DEVICE

Dale L. Hileman, Cedar Rapids, Iowa, assignor to Collins Radio Company, Cedar Rapids, Iowa, a corporation of Iowa
Application April 8, 1953, Serial No. 347,519
9 Claims. (Cl. 318—129)

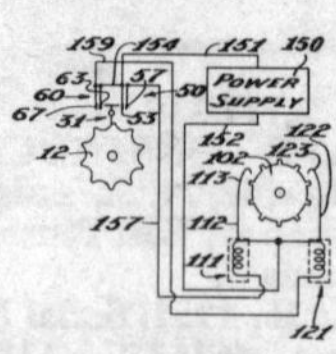

1. Means for actuating a controlled shaft at a controlled unit comprising, a control unit frame member, a first shaft rotatably supported by said control unit frame member, a dial plate mounted on said first shaft, a star-wheel mounted on said first shaft, a switching pawl rotatably supported by said control unit frame member and engageable by said star-wheel, first and second switches mounted on opposite sides of said switching pawl and engaged, respectively, upon opposite rotation of said star-wheel, a pair of relays at the controlled unit, a controlled unit frame member rotatably supporting the controlled shaft, a pair of ratchet pawls connected to said pair of relays, a pin-wheel mounted on said controlled shaft, said ratchet pawls engageable with said pin-wheel when their respective relays are energized, a power supply connected to the first and second switches and the first and second relays, and the first and second switches connected, respectively, to the first and second relays.

2,717,346
MAGNETIC SERVO-AMPLIFIER

Wilhelm A. Geyger, Takoma Park, Md., assignor to the United States of America as represented by the Secretary of the Navy
Application April 2, 1953, Serial No. 346,546
8 Claims. (Cl. 318—207)
(Granted under Title 35, U. S. Code (1952), sec. 266)

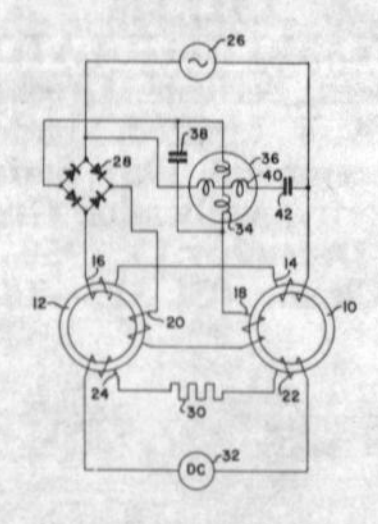

1. A magnetic amplifier control circuit for a two-phase induction motor comprising a pair of saturable reactor cores each having a controlled winding, a feedback winding and a control winding, means including a source of A.-C. potential and the controlled and feedback windings on said cores for causing said cores to saturate each on alternate half-cycles of said A.-C. potential, an energizing source for applying a control signal to said control wind-

ings, said control windings being arranged on said cores to effect push-pull operation in response to said control signal whereby the firing angles of said cores are differentially varied, means including a full-wave rectifier for applying the current flowing through the controlled windings on said cores to one field winding of said motor, and means for energizing the other field winding on said motor.

2,717,347
CIRCUITS FOR SUPPRESSING ARCING IN THE CONTROL OF CONDENSER-TYPE ELECTRIC MOTORS
Quentin G. Holtz, Pittsburgh, Pa., assignor to Gulf Research & Development Company, Pittsburgh, Pa., a corporation of Delaware
Application October 30, 1953, Serial No. 389,233
4 Claims. (Cl. 318—207)

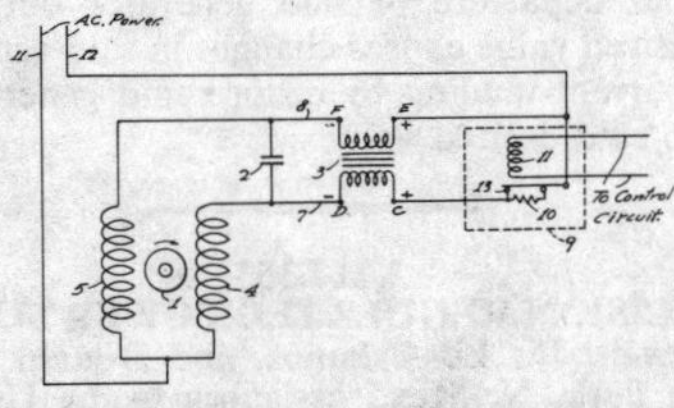

1. In a three-wire condenser-type motor circuit comprising two motor windings and a condenser all connected in a closed series circuit with a first lead wire connected to the junction of the two motor windings, a second lead wire connected to a junction of the condenser and a motor winding, and a third lead wire connected to the other junction of the condenser and a motor winding, and in which operating power is supplied to said first and second wires to effect motor rotation in one direction, or to said first and third wires to effect motor rotation in the other direction, or to said first wire and both said second and third wires to lock the motor, the improvement which comprises a transformer having windings which are respectively connected in series with the second and third wires leading to a junction of the condenser and a motor winding and with the transformer windings connected so that a voltage of one sign applied to the terminal of one transformer winding nearest a junction gives rise to a voltage of the same sign at the terminal of the other transformer winding nearest a junction.

2,717,348
DUAL FREQUENCY SINGLE PHASE ALTERNATING CURRENT MOTOR
Sol London, Fort Wayne, Ind., assignor to General Electric Company, a corporation of New York
Application July 10, 1953, Serial No. 367,213
6 Claims. (Cl. 318—221)

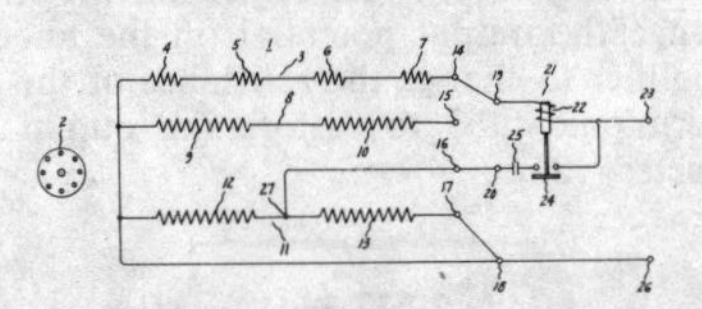

1. A dual frequency single phase alternating current motor comprising a first running winding forming four poles, a second running winding forming two poles, a starting winding having a plurality of sections forming two poles, and means for at times connecting said first running winding across a source of alternating current having a first frequency and said starting winding sections in parallel across said source, said starting winding poles being arranged to have the same polarity when so connected thereby to form consequent poles, and for at other times connecting said second running winding across a source of alternating current having a second frequency lower than said first frequency and said starting winding sections in series across said last named source, said starting winding poles being arranged to have opposite polarity when so connected.

2,717,349
SPEED CONTROL FOR INDUCTION MOTORS
William H. Lee, Norris, Tenn.
Application May 22, 1951, Serial No. 227,720
8 Claims. (Cl. 318—237)

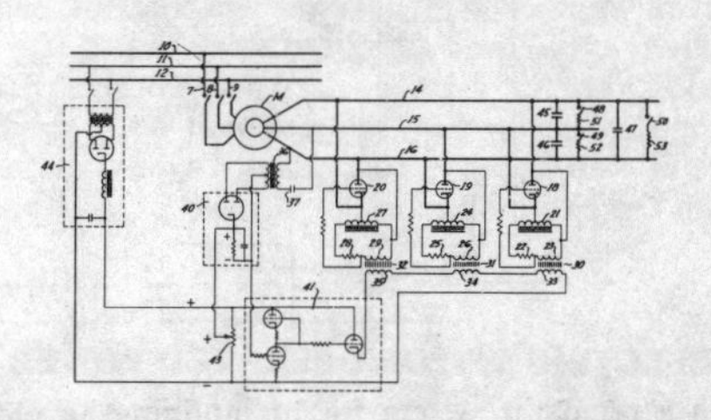

1. Apparatus for controlling the speed of a wound rotor induction motor comprising a plurality of thyratrons, means connecting the plate and cathode circuits of the thyratrons and the secondary windings of the motor whereby the thyratrons control the impedance of the motor secondary, phase shifting networks connected with the grids of each thyratron to control the firing thereof, each network including the A. C. winding of a saturable core reactor, the D. C. windings of the reactors being series connected, and circuits for supplying direct current to said D. C. winding to vary the magnetization of the cores of said reactors and thereby vary the reactance of said primaries and shift the phase of the grid voltage of said thyratrons, said direct current supply circuits including a transformer having its primary connected across at least one phase of the motor secondary, a condenser in series with said primary, a rectifier in circuit with the secondary of said transformer, and an amplifier for amplifying the direct current output of said rectifier.

2,717,350
GOVERNOR FOR ELECTRICAL MOTORS
Harrison D. Brailsford, Rye, N. Y.
Application February 25, 1953, Serial No. 338,829
11 Claims. (Cl. 318—254)

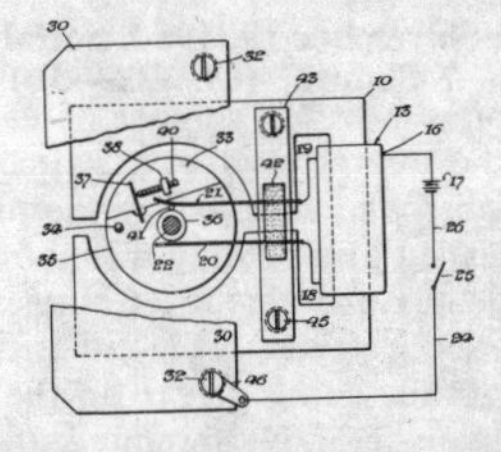

1. In a governor for an electrical motor of the class described and having energizing windings and a permanently magnetic rotor, in combination, a member pivotally mounted eccentrically of the motor shaft, resilient means normally holding said member in a predetermined position against the urge of centrifugal force and a contact carried by and rotating with said pivotally mounted member, said contact means cooperating with fixed contact members to together supply current to the motor energizing windings whereby the periods of closure of contacts are altered as the motor speed increases, and the speed is limited to a desired value.

2,717,351
REGULATORS
Carl A. Christian, Turtle Creek, and Slavo J. Murcek, White Oak, Pa., assignors to Westinghouse Electric Corporation, East Pittsburgh, Pa., a corporation of Pennsylvania
Application October 15, 1953, Serial No. 386,324
7 Claims. (Cl. 321—18)

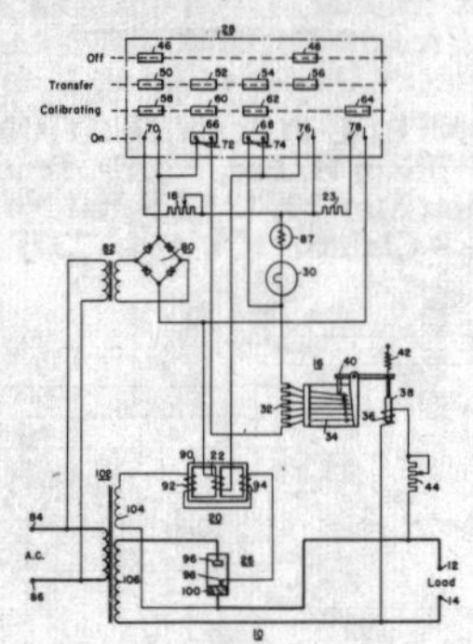

1. In a regulator system for maintaining an electrical quantity substantially constant, the combination comprising, means responsive to the magnitude of the electrical quantity for effecting a current flow in a given circuit proportional to the magnitude of the electrical quantity, means for manually adjusting the current flow in another circuit, control means for effecting a variation in the magnitude of the electrical quantity, switching means for selectively connecting, first, said given circuit in such circuit relationship with said control means so that said control means is responsive to the magnitude of the electrical quantity, second, for connecting said another circuit in parallel circuit relationship with said given circuit, and third, for disconnecting said given circuit from said control means so that said means for manually adjusting the current flow in said another circuit can be adjusted to control the magnitude of the current flow through said control means and thus the magnitude of said electrical quantity, and indicating means so connected to said given circuit and said another circuit as to indicate when the voltage across said given circuit is substantially equal to the voltage across said another circuit.

2,717,352
REGULATOR FOR D. C. GENERATOR
Morris Ribner, Silver Spring, Md., assignor to the United States of America as represented by the Secretary of the Navy
Application November 19, 1954, Serial No. 470,145
7 Claims. (Cl. 322—36)

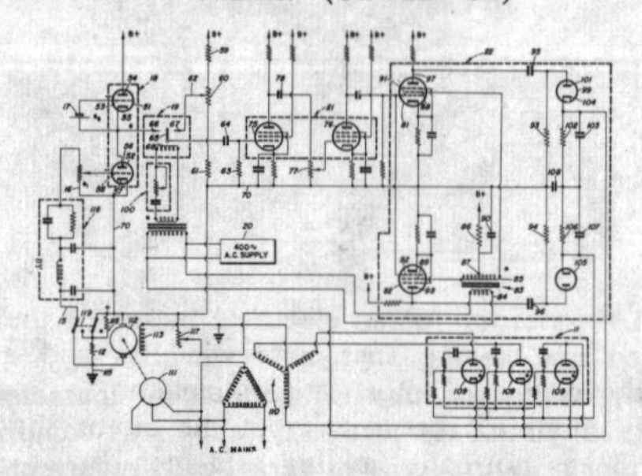

7. A regulated direct current generator comprising, a direct current generator having a field winding and providing a direct current output, means for supplying a reference voltage, a first source of alternating current, means for comparing said generator output with said reference voltage to provide a direct current error, means for converting said error from a direct to an alternating current in synchronism with current supplied by said first alternating current source, said converting means providing a phase shift with respect to said current from said alternating current source whenever said error signal indicates a change in said generator output from above a desired value to below said desired value, drift-free means for amplifying said converted error, a detector including means responsive to the amplitude and phase of said amplified converted error for reconverting said amplified converted error to a unidirectional control voltage, said control voltage having a magnitude proportional to said amplified converted error and a direction dependent upon the phase of said amplified converted error relative to the current from said first alternating current source, a second source of alternating current, and means controllable by said control voltage for converting current from said second source of alternating current into a direct current for exciting said generator field, so constructed and arranged that departure of said generator output voltage from a desired value causes changes in said generator field exciting current tending to restore said genertaor output voltage to said desired value.

2,717,353
PRECISION REGULATED POWER SUPPLY
Curtis Sewell, Jr., Los Alamos, and Donald M. Button, Mesilla Park, N. Mex., assignors to the United States of America as represented by the United States Atomic Energy Commission
Application October 27, 1954, Serial No. 465,160
1 Claim. (Cl. 323—22)

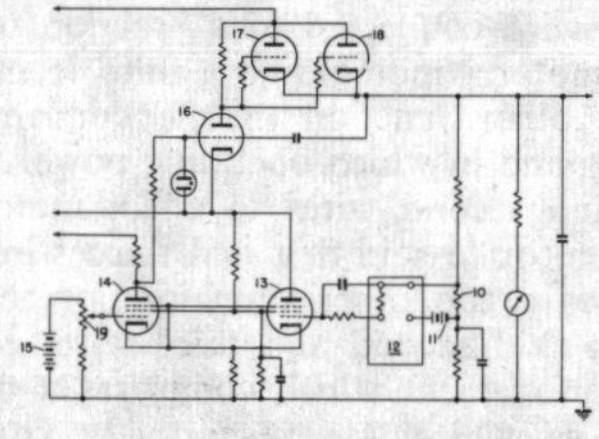

A precision voltage regulated power supply having at least one grid controlled thermionic tube in series therewith, a voltage divider connected across the output of the power supply, a standard cell employed as a reference voltage connected in series with a portion of the voltage divider and in polarity opposition to the voltage drop thereon, a direct current amplifier having its input connected in series with the standard cell and the resistance portion, a first amplifier tube connected to the output of the direct current amplifier, a second amplifier tube having its grid connected to a secondary reference voltage and being cathode-coupled to the first amplifier tube the second amplifier tube being anode-coupled to the grid of the grid controlled thermionic tube whereby a change in voltage across the resistance portion is amplified by the direct current amplifier, is impressed on the first amplifier to change the output potential on the anode of the second amplifier to change the resistance of the grid controlled thermionic tube to restore the output potential to the selected value.

2,717,354
VOLTAGE REGULATOR AND MEANS FOR OPERATING SAME
André Emile Jean Blévin, Paris, France
Application December 4, 1952, Serial No. 324,076
Claims priority, application France May 15, 1952
10 Claims. (Cl. 323—47)

1. A voltage-regulating device, comprising a discharge tube, circuit means feeding same with a voltage forming a predetermined fraction of the voltage to be regulated, an amplifying tube including a grid and an anode, means

wherethrough the discharge tube, upon discharge thereof, produces a negative bias on said grid of the amplifier tube, a circuit fed by the anode of the amplifier tube, a beating relay fed by said anode circuit, said beating relay including two blades moving between two predetermined extreme positions and occupying said positions respectively when the voltage feeding the discharge tube rises above and sinks below a predetermined normal value slightly beneath the firing potential of the discharge tube, an

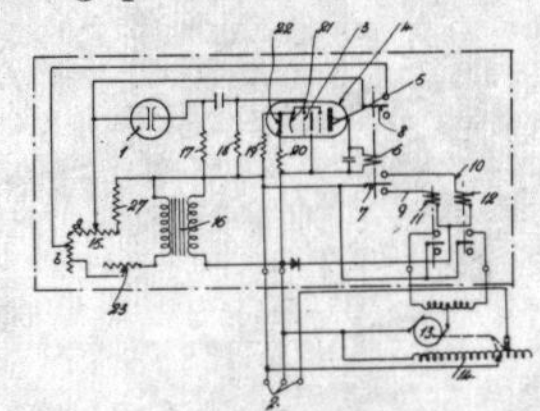

auxiliary relay adapted to be energized by one of said blades in one of its extreme positions, means controlled by said auxiliary relay and adapted to modify the voltage to be regulated, a potentiometric resistance inserted in the circuit means feeding voltage to the discharge tube, and normally open short-circuiting means connected across at least part of said potentiometric resistance and adapted to be closed by the second blade in one of its extreme positions.

2,717,355
AMPLIFIER LIMIT CIRCUIT
Victor J. Louden, Schenectady, N. Y., assignor to General Electric Company, a corporation of New York
Application August 18, 1952, Serial No. 304,892
22 Claims. (Cl. 323—64)

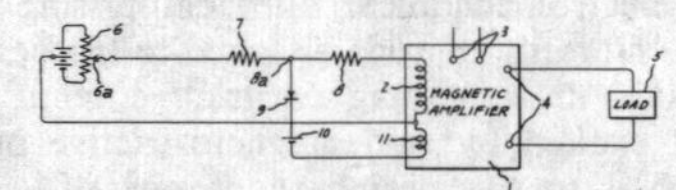

1. In a magnetic amplifier having a direct current signal input winding, an amplifier output limiting circuit including a unilateral conducting device and a source of bias voltage connected in series circuit relation across said input winding, and a compensating winding connected in series circuit relation in said limiting circuit and magnetically coupled with said input winding to control said amplifier.

2,717,356
TEMPERATURE AND VOLTAGE CONTROL CAPACITORS
James H. Foster, Erie, Pa., assignor to Erie Resistor Corporation, Erie, Pa., a corporation of Pennsylvania
Application March 28, 1951, Serial No. 217,959
3 Claims. (Cl. 323—74)

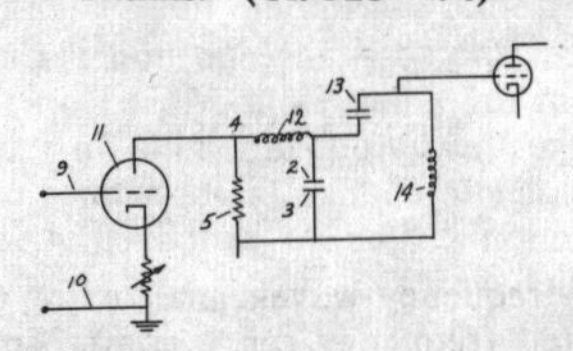

1. A capacitor unit for frequency control and the like comprising a dielectric body having both a voltage and temperature coefficient of capacity and having a Curie point in the region of the normal ambient temperature, a resistor material on the dielectric, a pair of capacitor electrodes on the dielectric, a source of control voltage variable in accordance with a quantity to be controlled, and connections for impressing the control voltage across the resistor and across the capacitor electrodes, said resistor being related to the thermal capacity of the dielectric such that the dielectric quickly assumes a temperature corresponding to the control voltage and the temperature rise of the dielectric supplements the change in capacity due to the control voltage.

2,717,357
VARIABLE ELECTRIC IMPEDANCE
Edward Lewis Mather, Liverpool, and Reginald James Cheetham, Prescot, England, assignors to British Insulated Callender's Cables Limited, London, England, a British company
Application March 2, 1953, Serial No. 339,714
Claims priority, application Great Britain March 10, 1952
7 Claims. (Cl. 323—77)

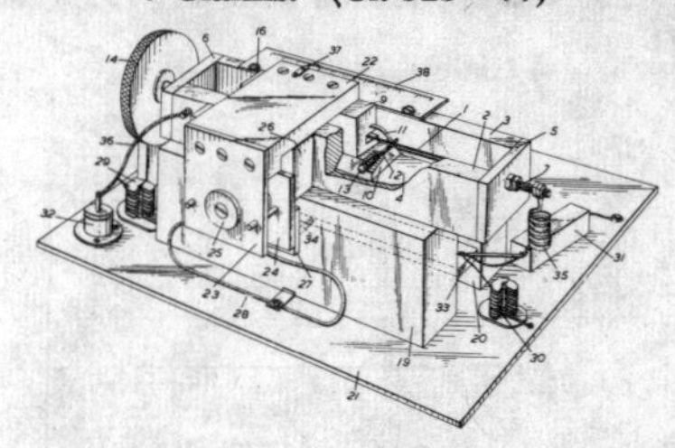

1. A two terminal variable impedance network which behaves substantially as a pure resistance throughout its range of adjustment comprising two terminals, a fixed non-inductive resistor and a variable inductive resistor electrically connected in series between the two terminals, a variable capacitor connected across the two terminals, a variable capacitor connected across the fixed resistor, adjusting means for each of the capacitors and a mechanical coupling between both said adjusting means and the variable resistor, the values of the elements being so chosen and the mechanical coupling being so constructed that for all adjustments

$$\sqrt{\frac{L}{2C}}$$

is substantially equal to the resistance (R) of the fixed non-inductive resistor, L being the value of the inductance of the variable resistor and C being the value of the capacitance of each of the two capacitors, R being much greater than the maximum values of the reactive and resistive components of the variable resistor.

2,717,358
ELECTRICAL SYSTEM
Allen C. Munster, Hatboro, Pa., assignor to Philco Corporation, Philadelphia, Pa., a corporation of Pennsylvania
Application November 4, 1949, Serial No. 125,623
10 Claims. (Cl. 324—68)

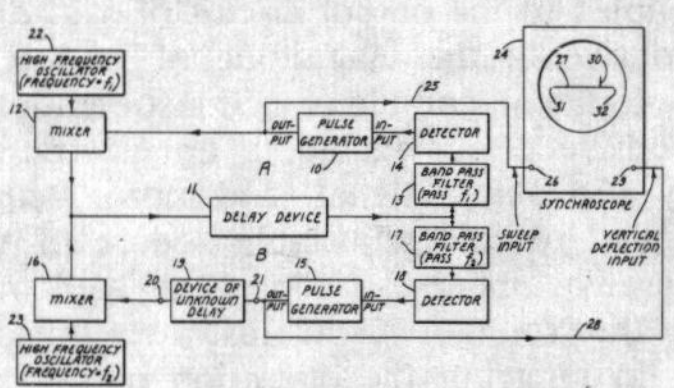

10. Apparatus for measuring the delay characteristics of electrical circuit elements, said apparatus comprising: a generator of a first series of periodically recurrent time-spaced pulse signals, a generator of a second series of periodically recurrent time-spaced pulse signals, means for utilizing a circuit element whose delay characteristic is to be measured to cause the period of recurrence of pulses in one of said series to differ from the period of recurrence of pulses in the other of said series by an amount equal substantially to the delay inherent in said

circuit element, and means for observing the change which occurs in the time-spacing of pulses in one of said series relative to pulses in the other of said series during intervals of duration substantially greater than the period of recurrence of pulses in either of said series.

2,717,359
MEASURING APPARATUS
Rudolf F. Wild, Wilmington, Del., assignor to Minneapolis-Honeywell Regulator Company, Minneapolis, Minn., a corporation of Delaware
Original application January 10, 1947, Serial No. 721,419. Divided and this application June 17, 1950, Serial No. 168,667
4 Claims. (Cl. 324—98)

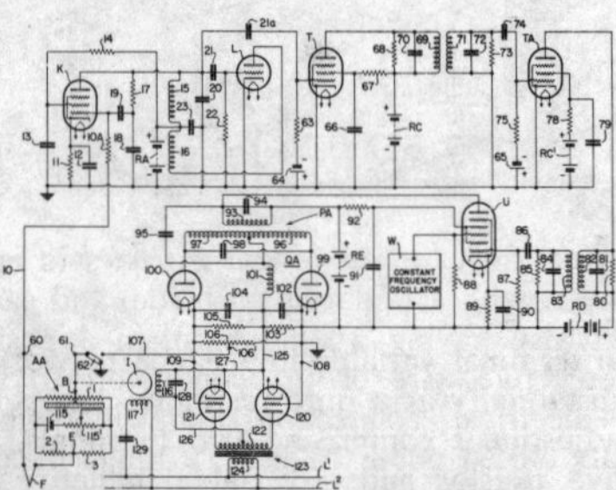

4. Apparatus responsive to the magnitude of a small unidirectional voltage, comprising signal producting means including a frequency determining portion and an output portion and operative to produce in said output portion a high frequency signal of a frequency determined by said frequency determining portion, reactance tube means including a control circuit and an output circuit and having an amplification factor which is dependent upon the magnitude of a unidirectional voltage applied to said control circuit, circuit means adapted to connect a source of unidirectional voltage to said control circuit, whereby the amplification factor of said reactance tube means is controlled in accordance with the magnitude of the last mentioned voltage, said circuit means including a potentiometric circuit including an adjustable source of voltage adapted to be connected in series voltage opposition with the first mentioned source of voltage to said control circuit, means connecting said output circuit into said frequency determining portion, whereby the frequency of said high frequency signal has a value dependent upon that of said last mentioned voltage, a balanced frequency discriminator having an input circuit coupled to said output portion and having an output circuit and operative to produce in the last mentioned output circuit an output signal having a magnitude dependent upon the frequency of said high frequency signal, and a responsive indicating device coupled to the output circuit of said discriminator and responsive to the magnitude of said output signal and hence to the magnitude of said last mentioned voltage, said responsive device including an electric motor controlled by said output signal, said motor being mechanically coupled to said adjustable source of voltage and being operative under the control of said output signal to adjust the voltage of the last mentioned source as necessary to interrupt the operation of said motor.

2,717,360
DOUBLE L MIXER
George W. Price, Cedar Rapids, Iowa, assignor to Collins Radio Co., Cedar Rapids, Iowa, a corporation of Iowa
Application June 29, 1951, Serial No. 234,419
3 Claims. (Cl. 333—6)

2. A wave guide mixer comprising, first, second, and third wave guide sections with the first and second wave guide sections joined so as to have their top walls in a common plane and to form an L, the third wave guide section mounted to the first and second wave guide sections at right angles to the common plane, and with the edge of the third wave guide section off-set a constant

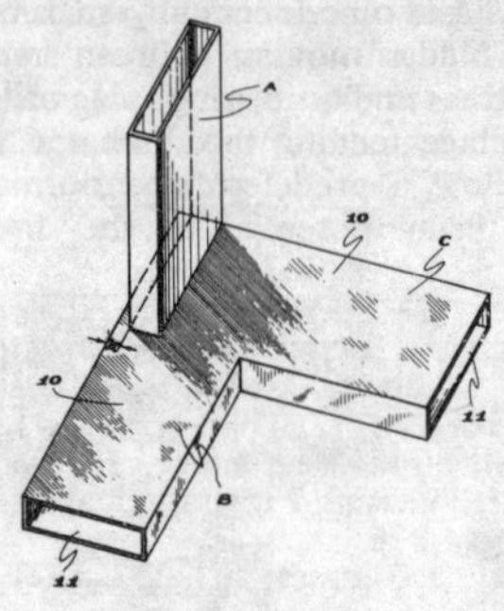

distance "*x*" from the edge of the first wave guide section where *x* is less than one-half the width of the first wave guide section.

2,717,361
MECHANICAL FILTERS
Melvin L. Doelz, Los Angeles, Calif., assignor to Collins Radio Company, Cedar Rapids, Iowa, a corporation of Iowa
Application September 24, 1951, Serial No. 248,011
4 Claims. (Cl. 333—71)

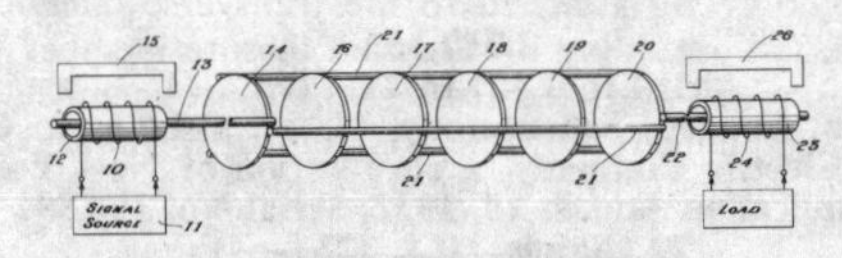

1. An electromechanical filter comprising a plurality of discs, a plurality of rods attached to the peripheries of each of said discs, a magnetostrictive input means, an input rod excited by said magnetostrictive input means and attached to the periphery of one of said discs, a magnetostrictive output means, and a magnetostrictive output rod connected to the periphery of another one of said discs and couple to the magnetostrictive output means.

2,717,362
HIGH-FREQUENCY WAVE-SIGNAL TUNING DEVICE
Meyer Press, Flushing, N. Y., assignor to Hazeltine Research, Inc., Chicago, Ill., a corporation of Illinois
Application May 2, 1950, Serial No. 159,423
5 Claims. (Cl. 333—82)

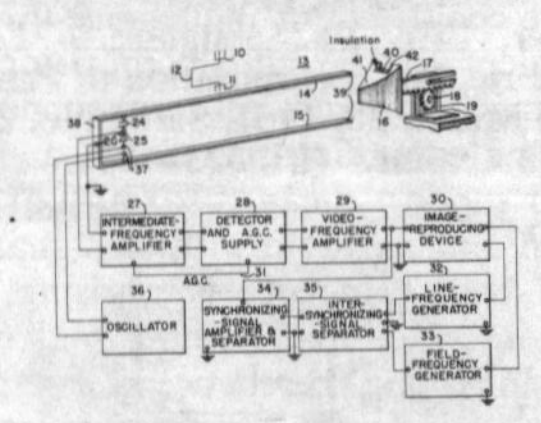

1. A high-frequency wave-signal tuning device tunable over a selected frequency range comprising: a high-frequency wave-signal transmission line including a pair of elongated conductors having substantially uniform spacing therebetween and having an open end portion presenting an impedance discontinuity; and reactive tuning means having a tapered end portion disposed longitudinally of and longitudinally displaceable along said end portion of said conductors in spaced relation with said conductors and at least semi-disengageable therefrom for tuning said transmission line over said selected range, said tapered portion being effective to reduce the spacing between engaged portions of said conductors and said tuning means

with displacements of said tuning means toward said conductors, and said tuning means and said conductors comprising an effective condenser having a value which increases with said displacements toward said conductors and providing for said transmission line a frequency-displacement tuning characteristic which decreases in frequency with said displacements toward said conductors.

2,717,363
RESONATOR TUNER
Merle R. Hubbard, Cedar Rapids, Iowa, assignor to Collins Radio Company, Cedar Rapids, Iowa, a corporation of Iowa
Original application September 18, 1952, Serial No. 310,289. Divided and this application March 4, 1954, Serial No. 414,070
3 Claims. (Cl. 333—82)

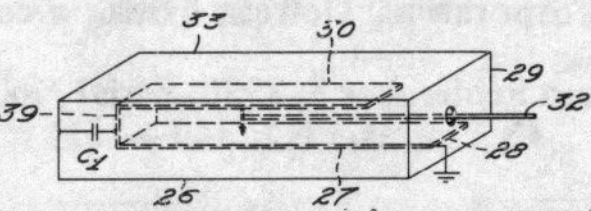

2. A cavity resonator comprising, a container member of conducting material, a conducting closure member connected across said container member, a first conductor with one end attached to said closure member, a transverse conducting portion connected to the other end of said first conducting member, a second conducting member with one end attached to the transverse conducting member and extending toward said closure member with its opposite end insulated from said closure member, and a shorting plunger movable longitudinally of said first and second conductors and engageable therewith.

2,717,364
TEMPERATURE COMPENSATION OF PERMEABILITY TUNED CIRCUITS
David M. Hodgin and Leo V. Mifflin, Cedar Rapids, Iowa, assignors to Collins Radio Company, Cedar Rapids, Iowa, a corporation of Iowa
Application February 5, 1951, Serial No. 209,498
5 Claims. (Cl. 336—30)

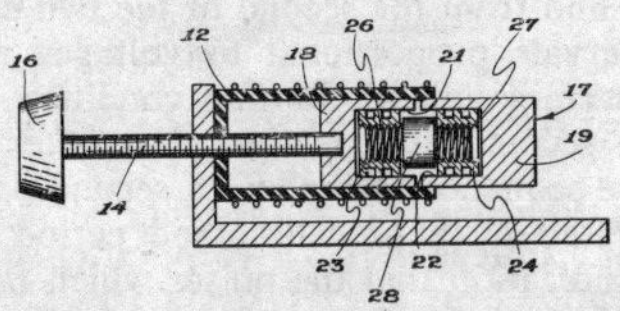

2. A movable core inductor with a substantially constant temperature coefficient of inductance throughout the range of movement of the core comprising, a support means, an inductance coil and coil form supported on said support means and extending therefrom, a shaft extending through said support means coaxially of said coil, a core mounted on and movable with said shaft axially of and within said coil, said core comprising two spaced cylindrical portions, a coupler joining said two portions but maintaining an air gap therebetween, said portions and said coupler having different temperature coefficients of expansion whereby said air gap varies with temperature.

2,717,365
ELECTRICAL OUTLET RECEPTACLE HAVING INSULATION PIERCING MEANS FOR AN ELECTRICAL CORD
Arthur Greenbaum, Tuckahoe, N. Y., assignor to Academy Electrical Products Corp., New York, N. Y., a corporation of New York
Application November 13, 1951, Serial No. 255,987
11 Claims. (Cl. 339—99)

1. An electrical outlet receptacle for use with quick-connect and disconnect contact blades of a conventional cord plug comprising, in combination, two molded members; one of which is a base, having a conductor-receiving groove, dimensioned in cross-section for a close-fit with a flat parallel two-wire conductor adapted to supply current to said receptacle, said groove being open at one end to receive the conductor, and closed at its other end to form a stop, against which seats the extremity of the conductor; the other molded member of which is a cover for closing the base, and said cover having apertures through which the contact blades of the aforesaid cord plug are adapted to be thrust; a pair of conductive parts fixed in the cover, each having a

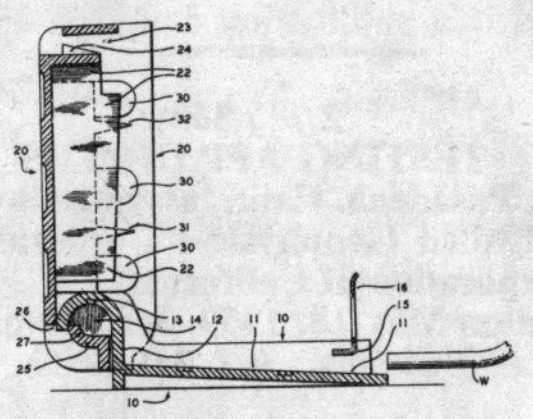

resilient contact finger in alignment with a said aperture, for engagement with a said contact blade, and each conductive part also having a rigid barb, in a position to pierce a respective wire of the conductor, to make a permanent electrical connection therewith, when the cover is fully closed upon the base; a hinge means, pivotally connecting the cover on the base, adjacent the closed end of said groove, and including trunnion means concealed when the cover is closed on the base; and detent latching means, enclosed by the cover, and engageable with detent means carried by the base, to lock the cover and base together in fully closed position.

2,717,366
HIGH VOLTAGE RECTIFIER TUBE MOUNTING
William H. Summerer, Chicago, Ill., assignor to Admiral Corporation, Chicago, Ill., a corporation of Delaware
Application September 13, 1952, Serial No. 309,530
6 Claims. (Cl. 339—126)

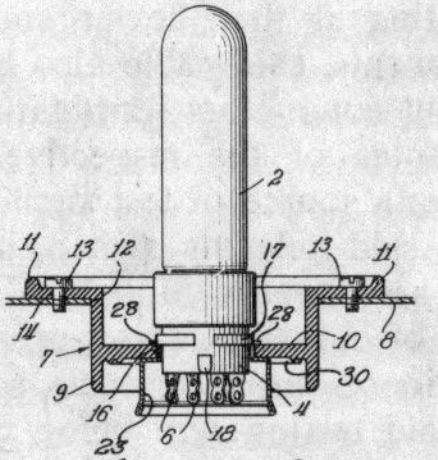

1. A high voltage tube mounting comprising an insulating member for connection to the chassis and having a tube socket receiving opening therein, a tube socket disposed in said opening and a high voltage shield connected to said member extending around and enclosing the tube socket and means on said shield engaging the tube socket to lock it in place.

2,717,367
MOULDED COVER JACK ASSEMBLY
George O. Puerner, Indianapolis, Ind., assignor to P. R. Mallory & Co., Inc., Indianapolis, Ind., a corporation of Delaware
Application January 27, 1954, Serial No. 406,439
15 Claims. (Cl. 339—185)

1. A jack comprising a plastic moulded shell including an annular side wall having a plurality of converging slots extending along the length thereof, a substantially closed back wall, an open end opposite said back wall, a pair of contact springs, each having a surrounding integral supporting frame, said frames having solder lugs penetrating said back wall and being held in said slots along said

annular wall the length thereof, and a cap over said shell for closing said open end, said cap having a central aper-

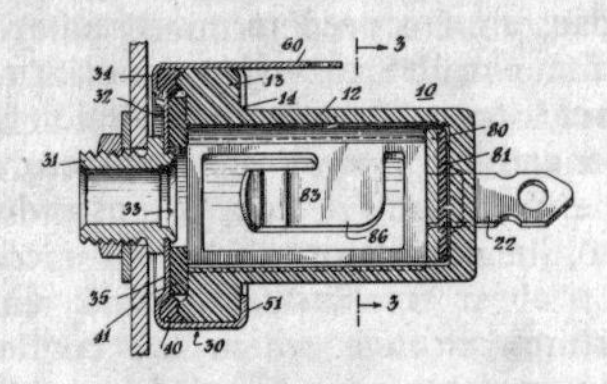

ture adapted to allow a plug to pass therethrough and to be tightly gripped by said frames.

2,717,368
TESTING APPARATUS
Merrill Swan, Pasadena, Calif., assignor, by mesne assignments, to United Geophysical Corporation, Pasadena, Calif., a corporation of California
Application May 18, 1953, Serial No. 355,508
9 Claims. (Cl. 340—15)

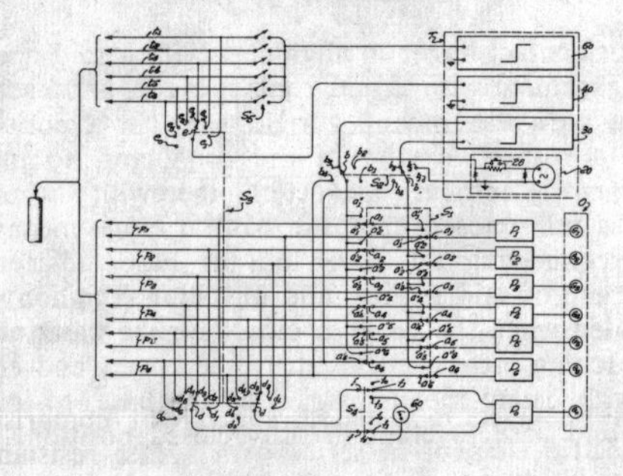

1. In combination: a series of hydrophones each comprising a sealed case, each hydrophone having a piezoelectric detector element and a preamplifier mounted therein, the detector element of each hydrophone being connected to the input of the preamplifier of said hydrophone; a multiple-conductor cable interconnecting said hydrophones, said cable having a plurality of test conductors terminating at their lower ends at the respective detector elements, said cable also having a plurality of pairs of output conductors terminating at their lower ends at the outputs of the respective preamplifiers; a measuring device; a source of test signals; and a channel selector switch, said selector switch including a movable testing arm associated with a plurality of testing contacts and a pair of movable output arms associated with a plurality of pairs of output contacts, said movable arms being ganged, said testing arm being connected to said source of test signals, said output arms being connected to said measuring device, said plurality of test conductors terminating at their upper ends at corresponding testing contacts, said plurality of pairs of output conductors terminating at their upper ends at corresponding pairs of output contacts, whereby said measuring device and said signal source may be selectively connected respectively to the outputs of said amplifiers and to the detector elements at the respective inputs thereof.

2,717,369
PRESSURE-SENSITIVE DEEP WELL SEISMOGRAPH DETECTOR
Thomas Bardeen, Fox Chapel, and Roger W. Williams, Penn Township, Allegheny County, Pa., assignors to Gulf Research & Development Company, Pittsburgh, Pa., a corporation of Delaware
Application July 31, 1952, Serial No. 301,908
3 Claims. (Cl. 340—17)

1. A hydrophone for use in wells, comprising a housing defining a first chamber having an opening formed therein, first flexible diaphragm means covering said opening, flexible dome means covering one face of said flexible diaphragm means for protecting the same and defining therebetween a protective hollow second chamber, liquid of low compressibility in said protective hollow

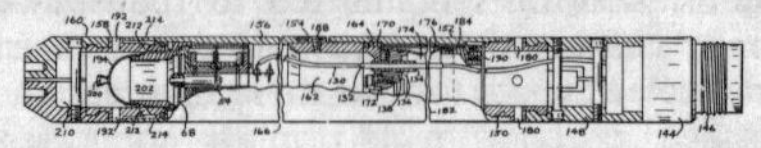

chamber, aperture means formed in said flexible dome means, plug means for closing said aperture means, and gas outlet means forming an outlet for any gases trapped between said dome means and said housing, for allowing said gases to escape into said well.

2,717,370
TRANSMITTING SYSTEM AND METHOD
Charles A. Piper, Detroit, Mich., assignor to Bendix Aviation Corporation, Detroit, Mich., a corporation of Delaware
Application September 8, 1950, Serial No. 183,717
13 Claims. (Cl. 340—151)

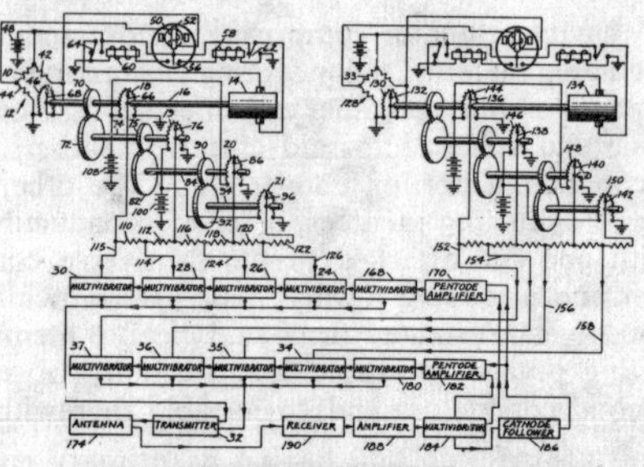

1. In combination, means for determining the values of a plurality of conditions, means for expressing each determination by a plurality of digital values and for providing voltages proportional to the value of each digit in a determination, means operative by a first interrogation pulse to produce a pair of coded pulses separated by a predetermined period of time to provide an indication that the pulses which follow represent a first condition, means for producing a sequence of pulses spaced from one another and from the second of the two coded pulses by time intervals proportional to voltages representing a digital determination of the first condition, means operative by a second interrogation pulse after the end of the first pulse sequence to produce a second pair of coded pulses separated by a predetermined period of time to provide an indication that the pulses which follow represent a second condition, means for producing a sequence of pulses spaced from one another and spaced from the second pulse in the second coded pair by time intervals proportional to the voltages representing a determination of the second condition, means for transmitting the different pulse sequences, and means operative at a removed station to decode the pulse sequences and provide a record of the variable conditions.

2,717,371
TONE CONTROL SYSTEM FOR CONTROLLING A REMOTE STATION
Warren B. Bruene, Cedar Rapids, Iowa, assignor to Collins Radio Company, Cedar Rapids, Iowa, a corporation of Iowa
Application August 9, 1951, Serial No. 241,100
3 Claims. (Cl. 340—170)

1. A tone control system for controlling a controlled station from a remote position comprising, a power station supplying power to the control and the controlled stations, a first transformer located at the control station with its primary connected to the power station, the secondary of said first transformer having its midpoint connected to ground, first and second switching

means connected to opposite sides of the secondary of the first transformer, a second transformer located at the control station, a primary of the second transformer connected to the first and second switching means, a pair of communication lines connected to the secondary of said second transformer and extending from the control to the controlled stations, a third transformer located at the controlled station with its primary connected to the pair of communication lines, the midpoint of the secondary of the third transformer connected to ground, two pairs of electron tubes at the controlled station with the control grids of the first pair of electron tubes connected to one side of the secondary of said third transformer, the control grids of the second pair of electron tubes connected to the opposite side of the secondary of the third transformer, a fourth transformer located at the controlled station and having its primary connected to the power station, a first secondary of the fourth transformer with one end connected to the plate of one of the first pair of electron tubes, and the opposite end connected to the plate of one of the second pair of electron tubes, a second secondary of said fourth transformer with one end connected to the plate of the other of said first pair of electron tubes and the opposite end connected to the plate of the other of said second pair of electron tubes, a first relay connected between ground and the mid-point of the first secondary of the said fourth transformer, a second relay connected between ground and the mid-point of the second secondary of said fourth transformer, and the said first and second relays furnishing control signals.

2,717,372
FERROELECTRIC STORAGE DEVICE AND CIRCUIT

John R. Anderson, Berkeley Heights, N. J., assignor to Bell Telephone Laboratories, Incorporated, New York, N. Y., a corporation of New York

Application November 1, 1951, Serial No. 254,245

13 Claims. (Cl. 340—173)

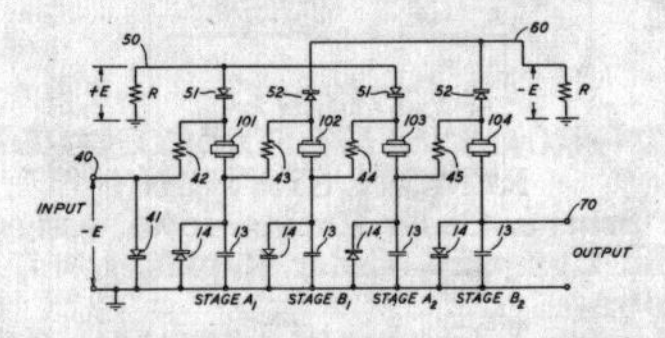

1. A memory circuit comprising a condenser having a dielectric of a ferroelectric material, said material having a hysteresis loop with a high ratio between the slopes of the side portions of the loop and the slopes of the top and bottom portions of the loop, means applying a pulse to said condenser to polarize said material to one point along said hysteresis loop, means applying storage pulses to said condenser, said storage pulses of one polarity reversing the polarization of said material to another point on said hysteresis loop, means applying a read out pulse to said condenser of a predetermined polarity and of a magnitude sufficient to cause said material to traverse a portion of said loop, and means receiving an output pulse from said condenser on application thereto of said read out pulse, the magnitude of said output pulse depending on the slope of the portion of said hysteresis loop traversed by said material on application of said read out pulse whereby the number is stored in said condenser by polarizing said material to one of said points along said loop.

2,717,373
FERROELECTRIC STORAGE DEVICE AND CIRCUIT

John R. Anderson, Berkeley Heights, N. J., assignor to Bell Telephone Laboratories, Incorporated, New York, N. Y., a corporation of New York

Application December 14, 1951, Serial No. 261,665

13 Claims. (Cl. 340—173)

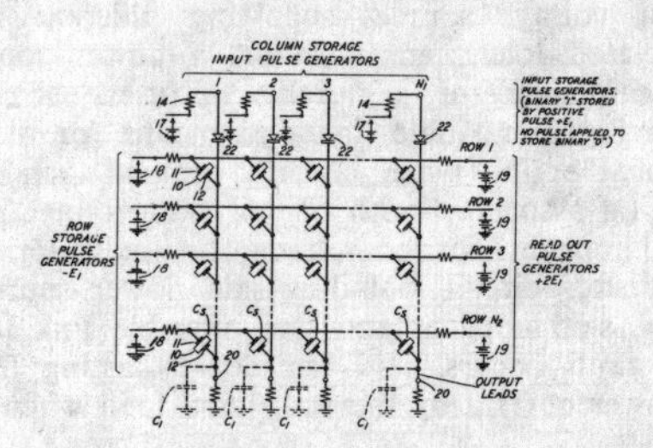

1. A ferroelectric data storage circuit comprising a ferroelectric element in series with a first resistance, a condenser and a second resistance in series and shunting the first resistance, means for applying across the first resistance a first voltage pulse of one polarity and of selected magnitude and simultaneously applying across the element and the first resistance a second voltage pulse of the selected magnitude and of the opposite polarity, and means for thereafter applying across the element and the first resistance a third voltage pulse of the one polarity and of twice the selected magnitude.

2,717,374
DEFLECTION VOLTAGE GENERATOR

Gifford E. White, Woodland Hills, Calif., assignor to Sperry Rand Corporation, a corporation of Delaware

Original application April 30, 1942, Serial No. 441,188. Divided and this application March 25, 1948, Serial No. 17,042

12 Claims. (Cl. 340—212)

5. In combination with a spiral scanning device, apparatus for producing deflecting potentials adapted to spirally scan the beam of a cathode ray tube synchronously with said device, comprising means for varying the amplitude of an alternating electromotive force synchronously with each scanning cycle of said device, a first modulator for modulating said electromotive force by a signal of frequency corresponding to the rate of rotation of said device, a second modulator for modulating said electromotive force by a signal having a phase quadrature relation to said first modulating signal, and means for demodulating the outputs of said modulators to produce spiral deflecting potentials.

2,717,375
FAULT INDICATOR
Samuel Lubkin, Brooklyn, N. Y., assignor to Underwood Corporation, New York, N. Y., a corporation of Delaware
Application December 31, 1953, Serial No. 401,559
9 Claims. (Cl. 340—250)

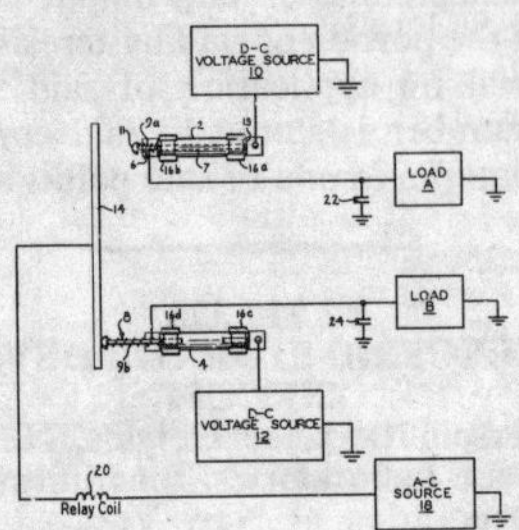

1. In electrical apparatus for indicating the occurrence of a fault which causes an overload on one of a plurality of voltage sources supplying different voltages to associated loads, an indicator, a power source for feeding said indicator, a plurality of fuses, each of said fuses having a movable link and means for displacing said movable link when a fault occurs, each of said fuses being associated with one of said voltage sources, and an impeder for completing an indicator circuit, said indicator circuit including said power source, said indicator, said impeder and said movable link in series when a fault occurs, said impeder bypassing the indicator power current by the associated load when a fault occurs.

2,717,376
TRAFFIC SIGNAL LIGHT
Ralph H. Carpenter, Mitchell, Ind., and Steve V. Palladino, Bradenton, Fla.
Application December 8, 1952, Serial No. 324,704
1 Claim. (Cl. 340—382)

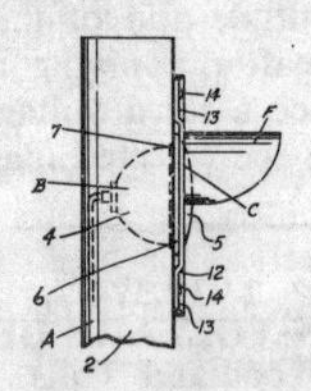

A traffic signal light comprising a housing having a front wall which throughout its entire area is flat and is provided with a circular opening therein, a light associated with the housing opening and comprising a casing having a front end of circular shape in cross sectional configuration, the light being a hermetically sealed unitary structure comprising at the front end of a casing a lens and interiorly of the casing a filament and a reflector, the light casing adjacent the front end thereof provided with a circumferential outwardly extending flat flange giving to that portion of the casing a diameter greater than the diameter of the circular opening in the housing front wall, the light casing extending through the housing front wall opening and having its flange in abutment with the front wall of the housing, a flat comparatively thin plate associated with the light casing and having therein a circular opening of lesser diameter than the diameter of the light casing flange and through which the light lens extends outwardly, the plate abutting the outer face of the light casing flange and having at each of its corners an offset ear, said ears being offset in respect to the rest of the plate a distance equal to the thickness of the light casing flange, said ears being flat and abutting the front flat wall of the housing, a sun visor immovably secured to the plate and extending around approximately one-third of the perimeter of the plate opening and extending outwardly in respect thereto, the plate being square, means passing through the plate ears for detachably securing the plate to the housing and clamping the flange of the light casing against the housing front wall to immovably secure the light casing in place, and the plate being selectively positionable of the housing to selectively position the sun visor at different quadrants of the circular opening in the housing front wall.

2,717,377
AUTOMATIC AMPLITUDE CANCELLATION IN MOVING TARGET INDICATOR
Homer G. Tasker, Van Nuys, and Verland A. Olson, Canoga Park, Calif., assignors to Gilfillan Bros., Inc., Los Angeles, Calif., a corporation of California
Application February 21, 1951, Serial No. 212,146
13 Claims. (Cl. 343—7.7)

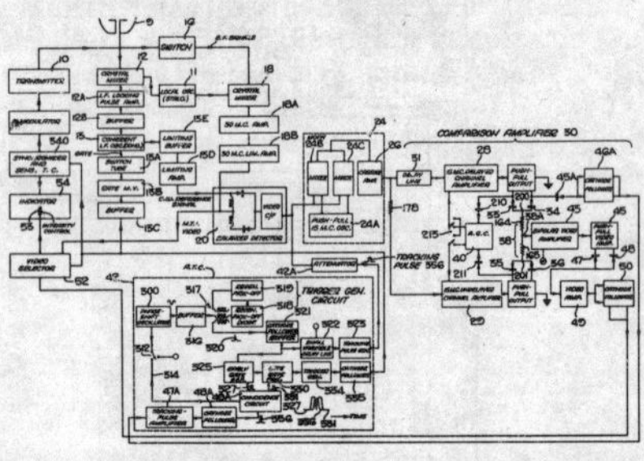

1. In a moving target indicating system of the character described wherein it is desired to eliminate indications produced by fixed targets, the improvement which resides in providing means for passing a train of video signals through a delayed channel amplifier, means for passing the same train of video signals through an undelayed channel amplifier, a cancellation network coupled to said delayed and undelayed channel amplifiers for developing a resultant voltage in said network which is equal to the difference in voltage appearing in said channel amplifiers, indicating means responsive to the output from said cancellation network, means deriving a control voltage in accordance with the intensity of a signal in one of said channels, and means automatically controlling the gain in said one channel in accordance with said control voltage.

2,717,378
DISTANCE MEASURING DEVICES
Floyd T. Wimberly, Watertown, Mass., assignor to Raytheon Manufacturing Company, Newton, Mass., a corporation of Delaware
Application August 2, 1952, Serial No. 302,300
14 Claims. (Cl. 343—14)

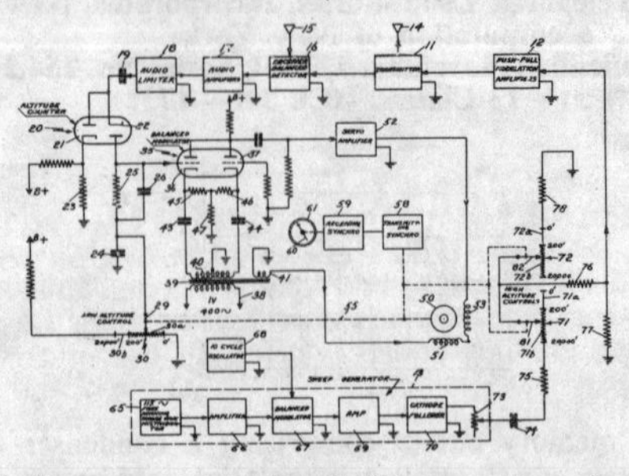

1. A frequency modulated distance measuring device for measuring the distance between a body carrying said device and a reflecting surface comprising a transmitter adapted to transmit a wave toward said reflecting surface, first means including a sweep generator for frequency modulating said transmitted wave, second means responsive to said transmitted wave and the wave reflected from

said surface for deriving a difference frequency wave, third means energized by said difference frequency wave for producing a voltage proportional to the frequency of said difference frequency wave, fourth means responsive to said voltage for energizing a servomotor, a distance indicator having a movable pointer, means including said servomotor for driving said pointer linearly with distance for distances less than a predetermined amount, and means for driving said pointer substantially inversely as the logarithm of said distance for distances greater than said predetermined amount.

2,717,379
RADIO NAVIGATION
Charles William Earp, London, England, assignor to International Standard Electric Corporation, New York, N. Y., a corporation of Delaware
Application September 25, 1952, Serial No. 311,435
Claims priority, application Great Britain October 29, 1951
11 Claims. (Cl. 343—106)

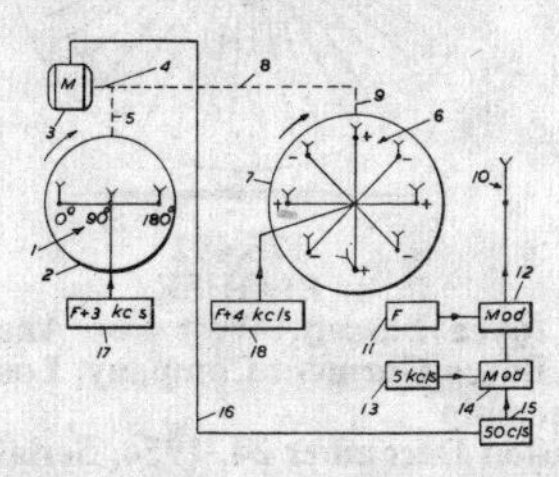

1. A radio navigation system comprising means at a receiver for deriving respective trains of pulses from two synchronously rotating radiation patterns having a single lobe and a plurality of lobes respectively, means for relatively adjusting the phases of the two trains so that single pulses of one train overlap corresponding single pulses of the other train, and means for applying only the resultant coincidental pulses for phase comparison with a reference wave held in synchronism with the rotation of said patterns.

2,717,380
ANTENNA SYSTEM FOR MEASURING LOW ELEVATION ANGLES
Frederick E. Brooks, Jr., Austin, Tex., assignor, by mesne assignments, to the United States of America as represented by the Secretary of the Navy
Application September 16, 1952, Serial No. 309,833
15 Claims. (Cl. 343—113)

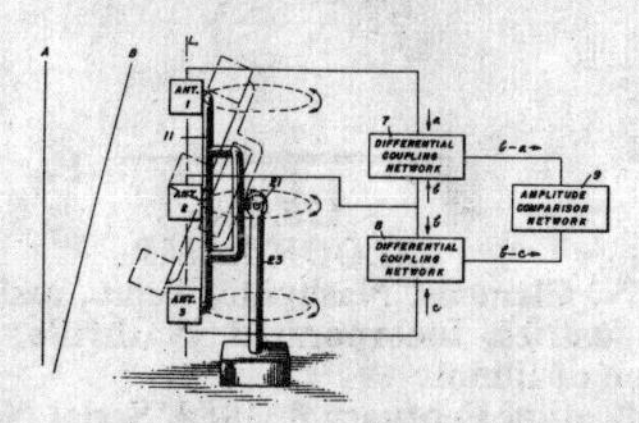

1. A direction-finding system for separately measuring the angle-of-arrival of one of two simultaneously received waves of the same frequency comprising the combination of a plurality of energy-detecting elements movable in unison in the plane in which the angle-of-arrival is to be measured and spaced along a line parallel to the wave front of the expected waves, a first pair of said energy-detecting elements differentially connected to cancel out signal components originating from a wave front to which the detecting elements are aligned, a second similar pair of energy-detecting elements differentially connected to cancel out the signal components originating from a wave front to which they are aligned, and means coupled to said two pairs of energy-detecting elements for indicating when the amplitude of the resultant output of said differentially connected pairs of energy-detecting elements are equal.

2,717,381
THERMOELECTRICALLY POWERED CONTROL DEVICE FOR WATER HEATERS AND THE LIKE
Russell B. Matthews, Wauwatosa, Wis., assignor to Milwaukee Gas Specialty Company, Milwaukee, Wis., a corporation of Wisconsin
Application September 13, 1951, Serial No. 246,464
15 Claims. (Cl. 236—21)

1. Condition responsive apparatus for controlling flow of fuel comprising, in combination, a control body for attachment to a heater and having a fuel inlet and a fuel outlet, a valve for controlling the flow of fuel, an electromagnetic operator for actuating said valve, a thermoelectric generator affording a source of electric energy for energizing said operator and actuating said valve, said electromagnetic operator comprising an electromagnet in circuit with said thermoelectric generator and an armature adapted to be attracted to said electromagnet when the latter is energized, connections between said armature and said valve comprising a spring arm which stores up energy imparted by attraction of said armature until the stored energy is sufficient to overcome the sealing force of the valve, and condition responsive means comprising an enclosure expansible and contractible with changes in a condition and contacts encapsulated within said enclosure for relative movement with expansion and contraction of said enclosure, said contacts being in circuit with said thermoelectric generator for controlling the supply of electric energy to said operator thereby controlling actuation of said valve in response to changes in the condition, said condition responsive means being carried by said control body to be held in heat conductive relation to an outer surface of an imperforate wall portion of the heater by attachment of said control body to said heater.

DESIGNS

SEPTEMBER 6, 1955

175,488
CLAMP
Melvin C. Anderson and Sidney F. Hanks, Strasburg, Colo.
Application June 22, 1955, Serial No. 36,623
Term of patent 14 years
(Cl. D54—13)

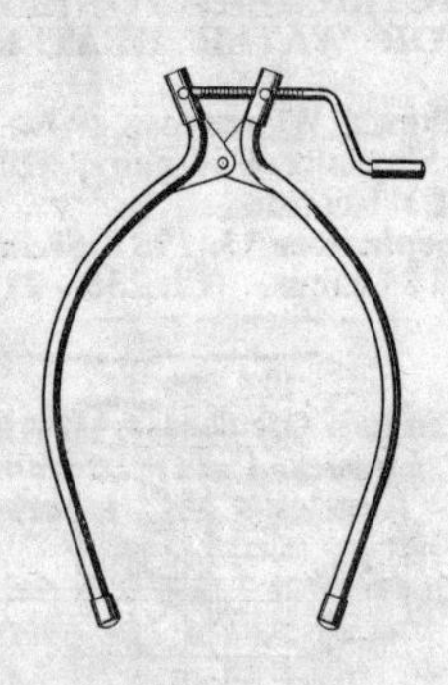

175,489
ANTENNA
George B. Arnold, Coral Gables, Fla.
Application February 28, 1955, Serial No. 34,737
Term of patent 7 years
(Cl. D26—14)

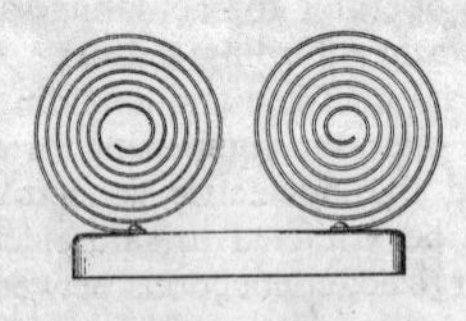

175,490
SOUND TRANSLATING MACHINE
Leo J. Aucoin, Bantam, Conn., assignor to The Gray Manufacturing Company, a corporation of Connecticut
Application December 15, 1954, Serial No. 33,568
Term of patent 14 years
(Cl. D26—14)

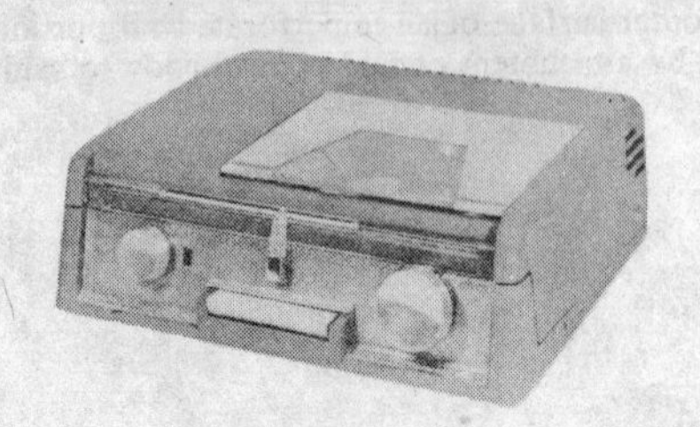

175,491
PLATE OR SIMILAR ARTICLE
A. Baker Barnhart, New York, N. Y.
Application February 28, 1955, Serial No. 34,768
Term of patent 7 years
(Cl. D44—15)

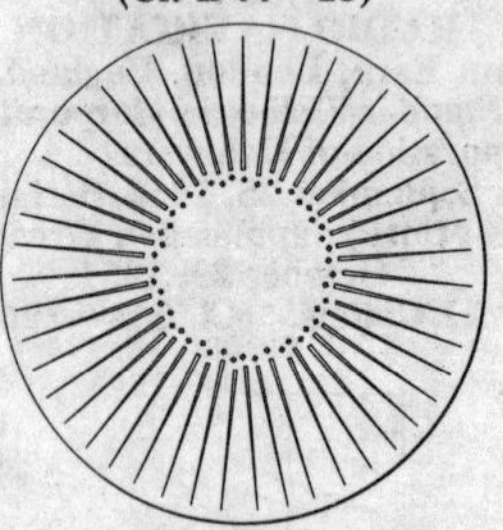

175,492
BLOUSE
Jacqueline Joyce Bassett, West Los Angeles, Calif., assignor to Barco Garment Company, Los Angeles, Calif., a copartnership
Application December 24, 1954, Serial No. 33,667
Term of patent 7 years
(Cl. D3—25)

175,493
TUBE UNION WRENCH
Allen J. Brame, Los Angeles, Calif., assignor to Tubing Appliance Co., Inc., Los Angeles, Calif., a corporation of California
Application June 7, 1955, Serial No. 36,395
Term of patent 14 years
(Cl. D54—16)

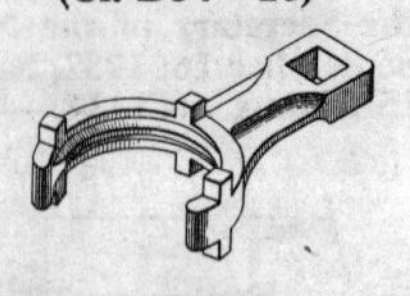

175,494
FLOOR LAMP
Charles W. Clemens, Nashville, Tenn., assignor to Aladdin Industries, Incorporated, Nashville, Tenn., a corporation of Illinois
Application February 8, 1954, Serial No. 28,902
Term of patent 7 years
(Cl. D48—20)

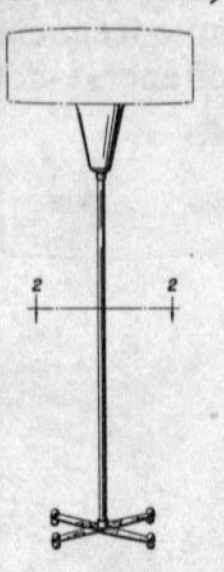

175,495
SUCTION CLEANER
Auguste Conord, Paris, France, assignor to André Conord, Paris, France
Application April 29, 1955, Serial No. 35,745
Claims priority, application France December 23, 1954
Term of patent 14 years
(Cl. D9—2)

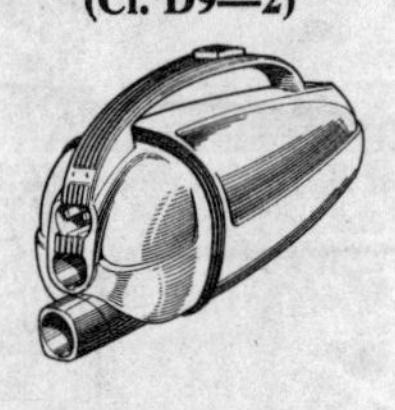

175,496
AUTOGRAPHIC REGISTER
John T. Davidson and Paul E. Simmons, Dayton, Ohio
Application November 3, 1953, Serial No. 27,442
Term of patent 14 years
(Cl. D64—11)

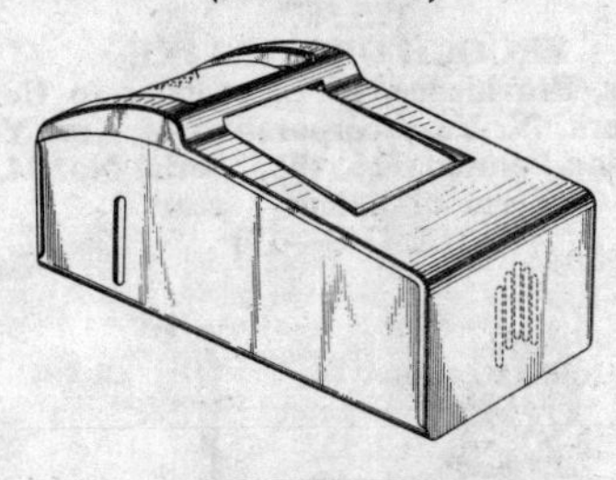

175,497
PORTABLE CABLE VULCANIZER
William G. Driemeyer, Webster Groves, and Gustav C. Thym, Lemay, Mo., assignors to Joy Manufacturing Company, Pittsburgh, Pa., a corporation of Pennsylvania
Application January 3, 1955, Serial No. 33,784
Term of patent 14 years
(Cl. D81—1)

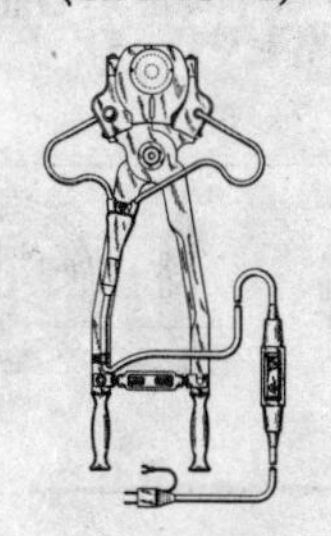

175,498
COMBINED SALAD BOWL AND STAND
Anthony Easton, New York, N. Y.
Application February 10, 1955, Serial No. 34,446
Term of patent 14 years
(Cl. D44—10)

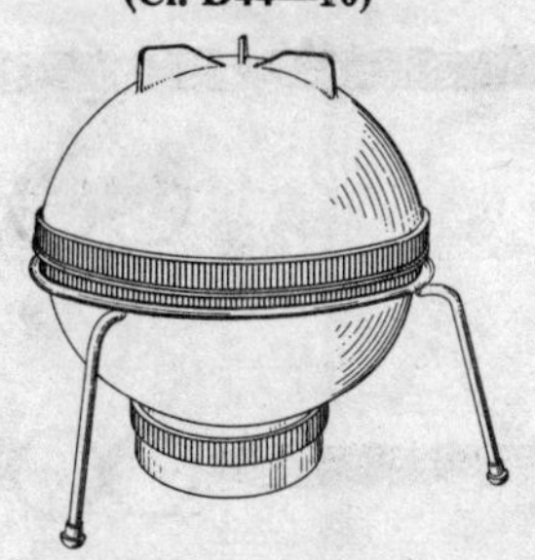

175,499
CEREAL BOWL
Walter J. Else, Everett, Wash.
Application May 17, 1955, Serial No. 36,060
Term of patent 14 years
(Cl. D44—15)

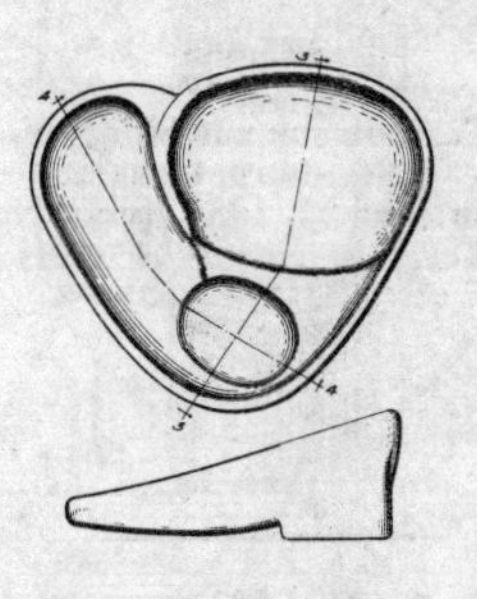

175,500
LADY'S SHOE OR SIMILAR ARTICLE
Jules Fern, Los Angeles, Calif., assignor to Fern Shoe Company, Los Angeles, Calif., a corporation of California
Application September 3, 1954, Serial No. 32,152
Term of patent 3½ years
(Cl. D7—7)

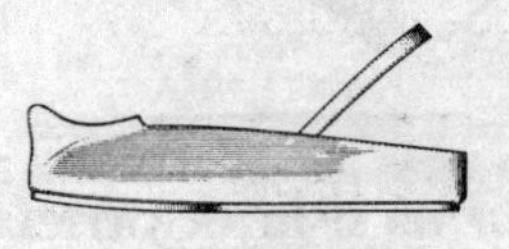

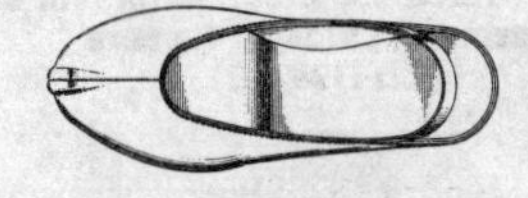

175,501
CLOCK
Robert O. Fletcher, Ashland, Mass., assignor to General Electric Company, a corporation of New York
Application September 30, 1954, Serial No. 32,495
Term of patent 7 years
(Cl. D42—7)

175,502
GAME BOARD
Joseph Follettie, Sacramento, Calif.
Application January 27, 1955, Serial No. 34,266
Term of patent 7 years
(Cl. D34—8)

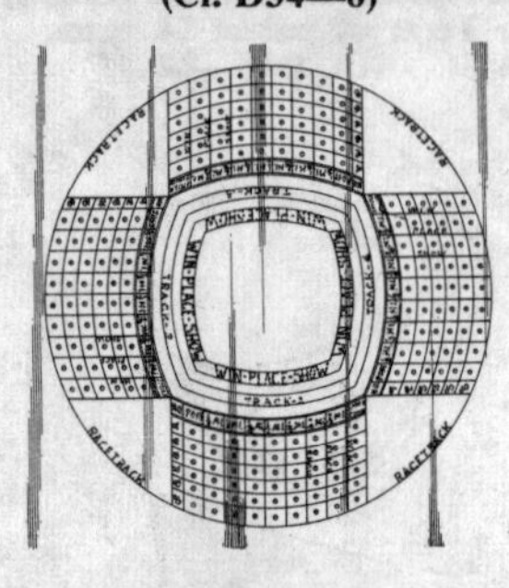

175,503
HOUSING FOR AN ANAMORPHOSER
Robert E. Gottschalk and John R. Moore, Los Angeles, Calif., assignors to Panavision, Inc., a corporation of California
Application January 17, 1955, Serial No. 34,008
Term of patent 14 years
(Cl. D57—1)

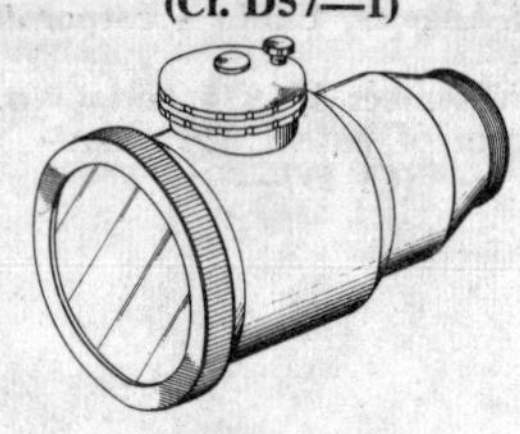

175,504
BADGE OR SIMILAR ARTICLE
Millard Joseph Hines, St. Petersburg, Fla.
Application March 25, 1954, Serial No. 29,695
Term of patent 14 years
(Cl. D29—2)

175,505
PRESSURE-SENSITIVE ADHESIVE TAPE DISPENSER AND APPLIER
Walter C. Larsen, Minneapolis, Minn., assignor to Minnesota Mining & Manufacturing Company, St. Paul, Minn., a corporation of Delaware
Application April 27, 1954, Serial No. 30,217
Term of patent 14 years
(Cl. D74—1)

175,506
SEWING MACHINE HEAD
Clarence Karstadt, Santa Barbara, Calif., assignor to Birtman Electric Company, a corporation of Illinois
Application November 19, 1952, Serial No. 22,369
Term of patent 14 years
(Cl. D70—1)

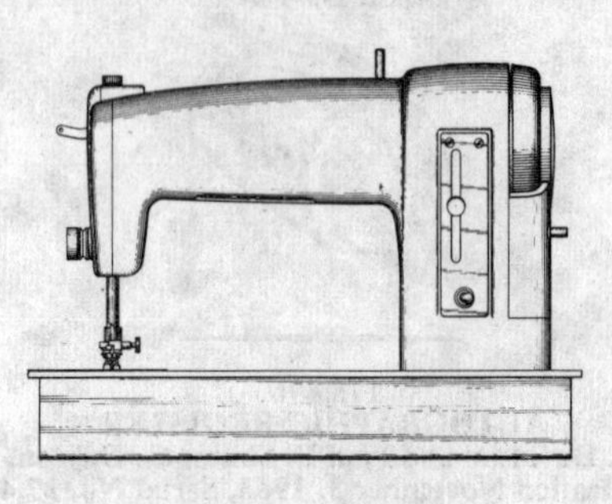

175,507
BROOCH OR THE LIKE
Adolph Katz, Providence, R. I., assignor to Coro, Inc., New York, N. Y., a corporation of New York
Application February 25, 1955, Serial No. 34,715
Term of patent 7 years
(Cl. D45—19)

175,508
CHARM BRACELET OR THE LIKE
Adolph Katz, Providence, R. I., assignor to Eastern Jewelry Mfg. Co. Inc., New York, N. Y., a corporation of New York
Application March 8, 1955, Serial No. 34,922
Term of patent 7 years
(Cl. D45—4)

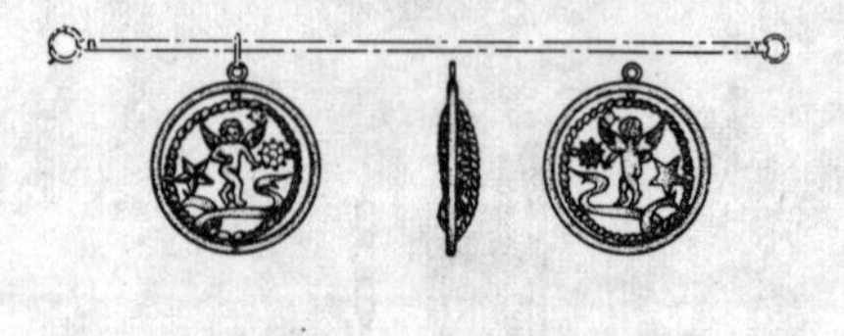

175,509
BRACELET OR THE LIKE
Adolph Katz, Providence, R. I., assignor to Coro, Inc., New York, N. Y., a corporation of New York
Application March 8, 1955, Serial No. 34,932
Term of patent 7 years
(Cl. D45—4)

175,510
NECKLACE OR THE LIKE
Adolph Katz, Providence, R. I., assignor to Coro, Inc., New York, N. Y., a corporation of New York
Application March 8, 1955, Serial No. 34,933
Term of patent 7 years
(Cl. D45—16)

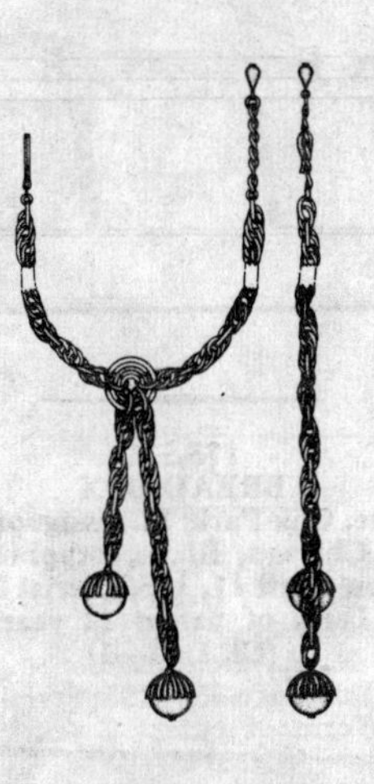

175,511
FLASHLIGHT
Edward C. Klotz, Jr., Milwaukee, Wis., assignor to Ray-O-Vac Company, Madison, Wis., a corporation of Wisconsin
Application November 24, 1954, Serial No. 33,227
Term of patent 14 years
(Cl. D48—24)

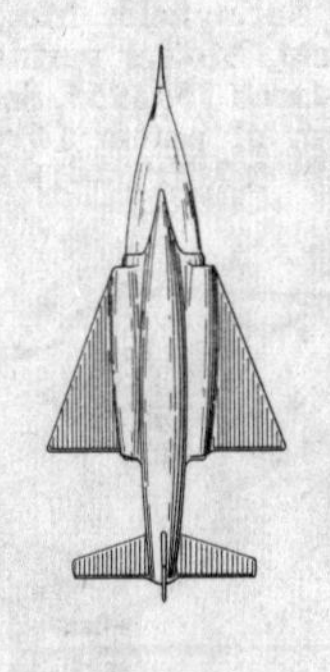

175,512
FLASHLIGHT
Edward C. Klotz, Jr., Milwaukee, Wis., assignor to Ray-O-Vac Company, a corporation of Wisconsin
Application November 24, 1954, Serial No. 33,228
Term of patent 14 years
(Cl. D48—24)

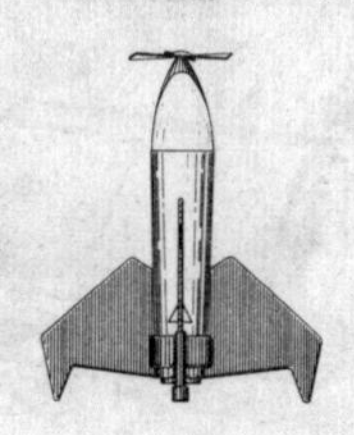

175,513
FLASHLIGHT
Edward C. Klotz, Jr., Milwaukee, Wis., assignor to Ray-O-Vac Company, Madison, Wis., a corporation of Wisconsin
Application November 24, 1954, Serial No. 33,229
Term of patent 14 years
(Cl. D48—24)

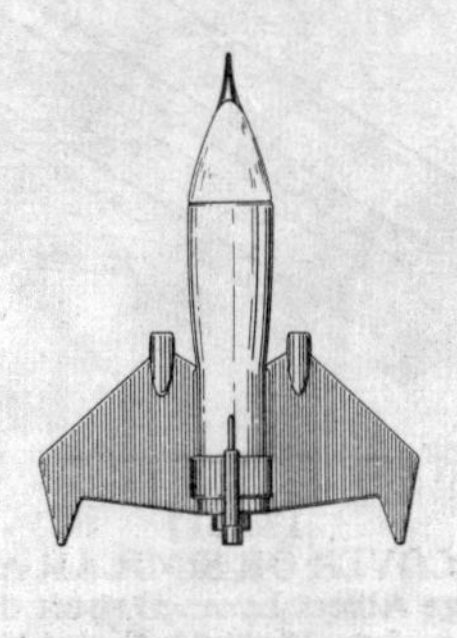

175,514
FLASHLIGHT
Edward C. Klotz, Jr., Milwaukee, Wis., assignor to Ray-O-Vac Company, Madison, Wis., a corporation of Wisconsin
Application November 24, 1954, Serial No. 33,230
Term of patent 14 years
(Cl. D48—24)

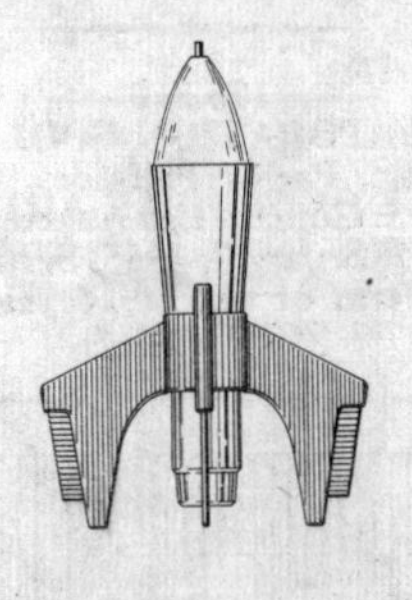

175,515
COMBINATION KITCHEN AND SEWING MACHINE CABINET
Carl J. Lerch, Cleveland, Ohio, assignor to White Sewing Machine Corporation, Lakewood, Ohio, a corporation of Delaware
Application August 6, 1952, Serial No. 20,922
Term of patent 14 years
(Cl. D33—12)

175,516
SPICE RACK
Thompson A. Lively, San Lorenzo, Calif.
Application March 31, 1955, Serial No. 35,302
Term of patent 14 years
(Cl. D33—3)

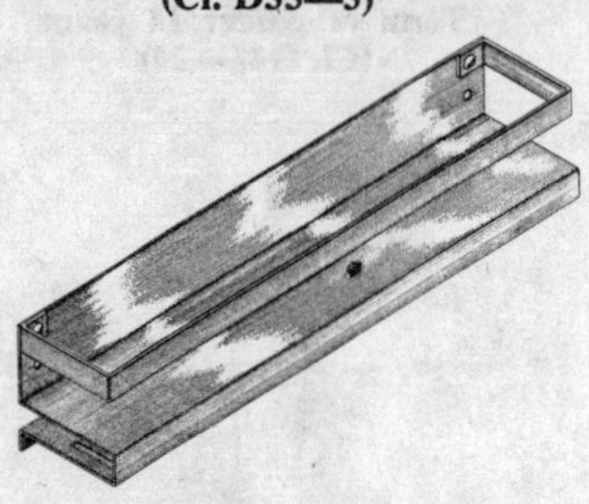

175,517
WHEEL COVER OR SIMILAR ARTICLE
George Albert Lyon, Detroit, Mich.
Application July 17, 1953, Serial No. 26,013
Term of patent 14 years
(Cl. D14—30)

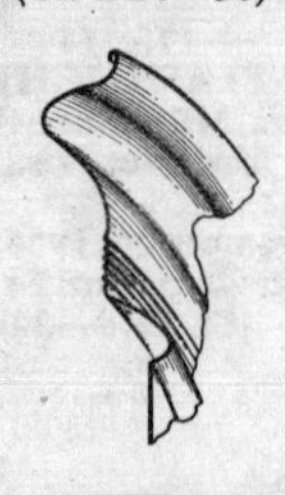

175,518
EVAPORATION COOLER
Thomas B. Martin, Pacific Palisades, Calif., assignor to Harry S. Guthait, Los Angeles, Calif.
Application February 7, 1955, Serial No. 34,392
Term of patent 14 years
(Cl. D62—4)

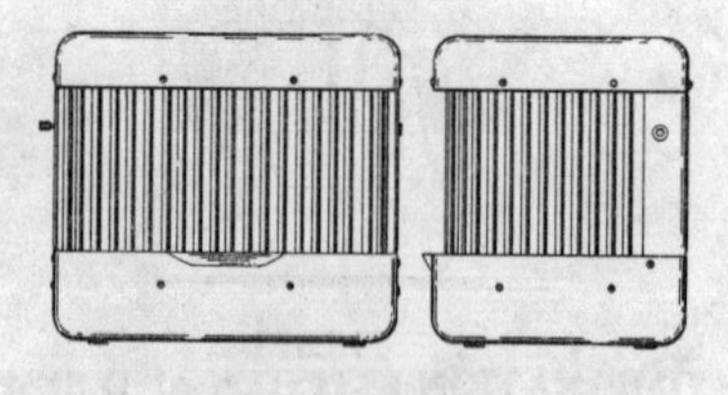

175,519
ORNAMENTAL PLATE
Colin Edmonds Pascal, Kings Norton, Birmingham, England, assignor to Lloyd, Pascal and Company Limited, Birmingham, England, a British company
Application November 9, 1954, Serial No. 33,038
Claims priority, application Great Britain June 3, 1954
Term of patent 14 years
(Cl. D29—20)

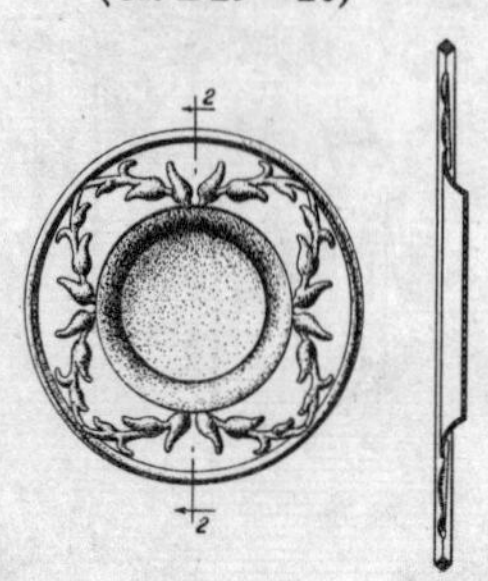

175,520
HUMIDIFIER UNIT
Milton A. Powers and Robert C. Champlin, Detroit, Mich., assignors to Skuttle Manufacturing Company, Milford, Mich., a corporation of Michigan
Application June 17, 1954, Serial No. 31,032
Term of patent 14 years
(Cl. D62—4)

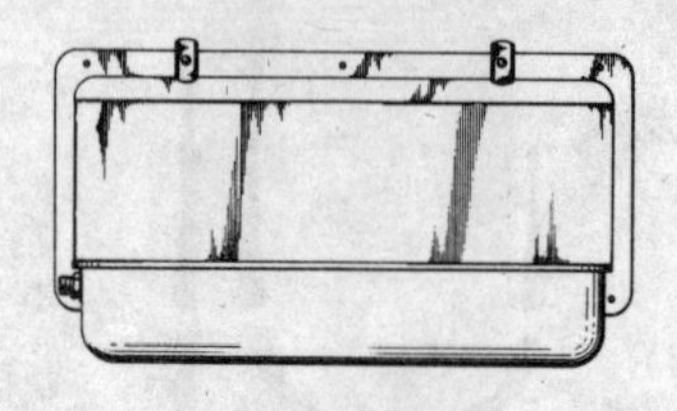

175,521
BREAD BOX
Jean O. Reinecke, Oak Park, Ill., assignor to Federal Tool Corporation, Chicago, Ill., a corporation of Illinois
Application April 11, 1955, Serial No. 35,441
Term of patent 14 years
(Cl. D44—1)

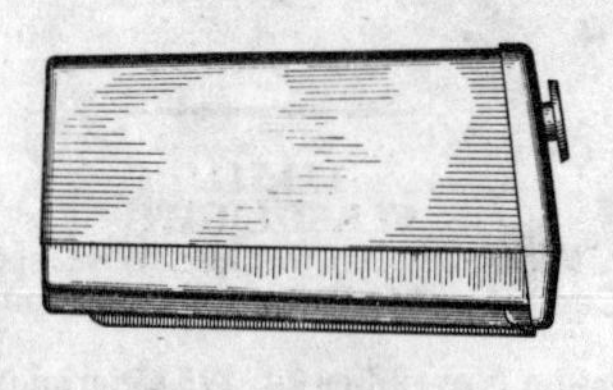

175,522
BOAT
Horace H. Roby, Springfield, Mo., assignor to Roby-Hutton, Springfield, Mo., a partnership of Missouri
Application March 15, 1955, Serial No. 35,051
Term of patent 14 years
(Cl. D71—1)

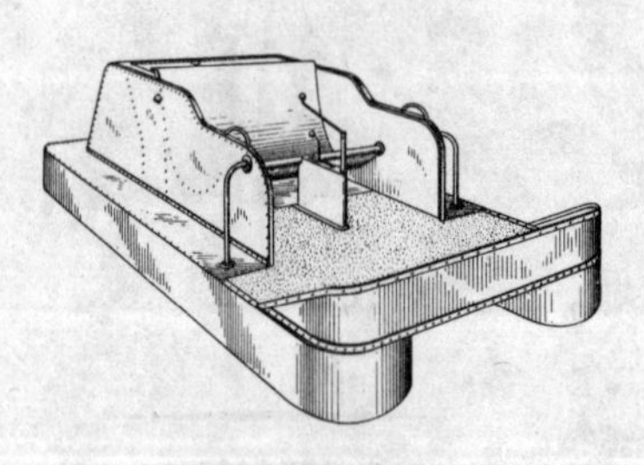

175,523
DICTATION MACHINE
Emil C. Steinbach, Chicago, Ill., assignor to Charles P. Peirce, Wilmette, Ill.
Application March 8, 1954, Serial No. 29,407
Term of patent 7 years
(Cl. D26—14)

175,524
LOOPED PILE CARPET OR SIMILAR ARTICLE
George V. Uihlein, Worcester, Mass., assignor to M. J. Whittall Associates, Inc., Worcester, Mass.
Application July 9, 1954, Serial No. 31,365
Term of patent 7 years
(Cl. D92—4)

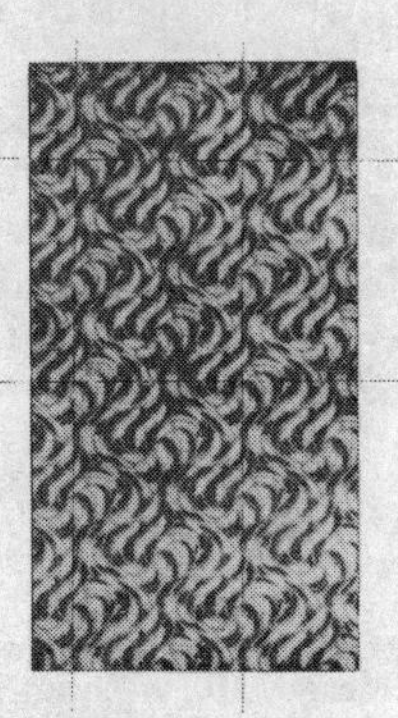

175,525
XYLOPHONE
Harry Zimmerman, New York, N. Y.
Application April 6, 1955, Serial No. 35,392
Term of patent 14 years
(Cl. D56—1)

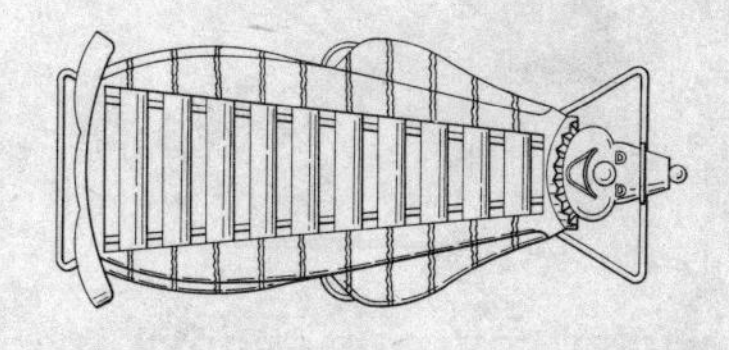

LIST OF REISSUE PATENTEES

TO WHOM

PATENTS WERE ISSUED ON THE 6TH DAY OF SEPTEMBER, 1955

NOTE.—Arranged in accordance with the first significant character or word of the name (in accordance with city and telephone directory practice).

Eckstein, Ernest E. Re. 24,058, Cl. 294—110.
Killough, Jack D.: *See*—
Pinion, Thomas J., and Killough. Re. 24,059.
Owens-Corning Fiberglass Corp.: *See*—
Russell, Robert C. Re. 24,060.
Pinion, Thomas J., and J. D. Killough; said Pinion assor to said Killough. Re. 24,059, Cl. 16—153.
Russell, Robert G., to Owens-Corning Fiberglass Corp. Re. 24,060, Cl. 49—17.

LIST OF PLANT PATENTEES

Howard & Smith: *See*—
Howard, Arthur P. 1,417.
Howard, Arthur P., to Howard & Smith. 1,417, Cl. 47—61.
Pawla, Edith W. 1,418, Cl. 47—60.

LIST OF DESIGN PATENTEES

Aladdin Industries, Inc.: *See*—
Clemens, Charles W. 175,494.
Anderson, Melvin C., and S. F. Hanks. 175,488, Cl. D54—13.
Arnold, George B. 175,489, Cl. D26—14.
Aucoin, Leo J., to The Gray Mfg. Co. 175,490, Cl. D26—14.
Barco Garment Co.: *See*—
Bassett, Jacqueline J. 175,492.
Barnhart, A. Baker. 175,491, Cl. D44—15.
Bassett, Jacqueline J., to Barco Garment Co. 175,492, Cl. D3—25.
Birtman Electric Co.: *See*—
Karstadt, Clarence. 175,506.
Brame, Allen J., to Tubing Appliance Co., Inc. 175,493, Cl. D54—16.
Champlin, Robert C.: *See*—
Powers, Milton A., and Champlin. 175,520.
Clemens, Charles W., to Aladdin Industries, Inc. 175,494, Cl. D48—20.
Conord, André: *See*—
Conord, Auguste. 175,495.
Conord, Auguste, to André Conord. 175,495, Cl. D9—2.
Coro, Inc.: *See*—
Katz, Adolph. 175,507.
Katz, Adolph. 175,509.
Katz, Adolph. 175,510.
Davidson, John T., and P. E. Simmons. 175,496, Cl. D64—11.
Driemeyer, William G., and G. C. Thym, to Joy Mfg. Co. 175,497, Cl. D81—1.
Eastern Jewelry Mfg. Co. Inc.: *See*—
Katz, Adolph. 175,508.
Easton, Anthony. 175,498, Cl. D44—10.
Else, Walter J. 175,499, Cl. D44—15.
Federal Tool Corp.: *See*—
Reinecke, Jean O. 175,521.
Fern, Jules, to Fern Shoe Co. 175,500, Cl. D7—7.
Fern Shoe Co.: *See*—
Fern, Jules. 175,500.
Fletcher, Robert O., to General Electric Co. 175,501, Cl. D42—7.
Follettie, Joseph. 175,502, Cl. D34—8.
General Electric Co.: *See*—
Fletcher, Robert O. 175,501.
Gottschalk, Robert E., and J. R. Moore, to Panavision, Inc. 175,503, Cl. D57—1.
Gray Mfg. Co., The: *See*—
Aucoin, Leo J. 175,490.
Guthait, Harry S.: *See*—
Martin, Thomas B. 175,518.
Hanks, Sidney F.: *See*—
Anderson, Melvin C., and Hanks. 175,488.
Hines, Millard J. 175,504, Cl. D29—2.
Joy Mfg. Co.: *See*—
Driemeyer, William G., and Thym. 175,497.
Karstadt, Clarence, to Birtman Electric Co. 175,506, Cl. D70—1.
Katz, Adolph, to Coro, Inc. 175,507, Cl. D45—19.
Katz, Adolph, to Eastern Jewelry Mfg. Co. Inc. 175,508, Cl. D45—4.
Katz, Adolph, to Coro, Inc. 175,509, Cl. D45—4.
Katz, Adolph, to Coro, Inc. 175,510, Cl. D45—16.
Klotz, Edward C., Jr., to Ray-O-Vac Co. 175,511, Cl. D48—24.
Klotz, Edward C., Jr., to Ray-O-Vac Co. 175,512, Cl. D48—24.
Klotz, Edward C., Jr., to Ray-O-Vac Co. 175,513, Cl. D48—24.
Klotz, Edward C., Jr., to Ray-O-Vac Co. 175,514, Cl. D48—24.
Larsen, Walter C., to Minnesota Mining & Mfg. Co. 175,505, Cl. D74—1.
Lerch, Carl J., to White Sewing Machine Corp. 175,515, Cl. D33—12.
Lively, Thompson A. 175,516, Cl. D33—3.
Lloyd, Pascal and Co. Ltd.: *See*—
Pascal, Colin E. 175,519.
Lyon, George A. 175,517, Cl. D14—30.
Martin, Thomas B., to Harry S. Guthait. 175,518, Cl. D62—4.
Minnesota Mining & Mfg. Co.: *See*—
Larsen, Walter C. 175,505.
Moore, John R.: *See*—
Gottschalk, Robert E., and Moore. 175,503.
Panavision, Inc.: *See*—
Gottschalk, Robert E., and Moore. 175,503.
Pascal, Colin E., to Lloyd, Pascal and Co. Ltd. 175,519, Cl. D29—20.
Peirce, Charles P.: *See*—
Steinbach, Emil C. 175,523.
Powers, Milton A., and R. C. Champlin, to Skuttle Mfg. Co. 175,520, Cl. D62—4.
Ray-O-Vac Co.: *See*—
Klotz, Edward C., Jr. 175,511.
Klotz, Edward C., Jr. 175,512.
Klotz, Edward C., Jr. 175,513.
Klotz, Edward C., Jr. 175,514.
Reinecke, Jean O., to Federal Tool Corp. 175,521, Cl. D44—1.
Roby, Horace H., to Roby-Hutton. 175,522, Cl. D71—1.
Roby-Hutton: *See*—
Roby, Horace H. 175,522.
Simmons, Paul E.: *See*—
Davidson, John T., and Simmons. 175,496.
Skuttle Mfg. Co.: *See*—
Powers, Milton A., and Champlin. 175,520.
Steinbach, Emil C., to C. P. Peirce. 175,523, Cl. D26—14.
Thym, Gustav C.: *See*—
Driemeyer, William G., and Thym. 175,497.
Tubing Appliance Co., Inc.: *See*—
Brame, Allen J. 175,493.
Uihlein, George V., to M. J. Whittall Associates, Inc. 175,524, Cl. D92—4.
White Sewing Machine Corp.: *See*—
Lerch, Carl J. 175,515.
Whittall, M. J., Associates, Inc.: *See*—
Uihlein, George V. 175,524.
Zimmerman, Harry. 175,525, Cl. D56—1

LIST OF PATENTEES

TO WHOM

PATENTS WERE ISSUED ON THE 6TH DAY OF SEPTEMBER, 1955

NOTE.—Arranged in accordance with the first significant character or word of the name (in accordance with city and telephone directory practice).

Abbott Laboratories: *See*—
Ryan, George R. 2,716,982.
Windischman, Edward F., Hartop, and Ryan. 2,716,983.
Abell, Harry D. and H. E., and N. R. Heald, to Fischer & Christen. 2,717,153, Cl. 270—81.
Abell, Homer E.: *See*—
Abell, Harry D. and H. E., and Heald. 2,717,153.
Academy Electrical Products Corp.: *See*—
Greenbaum, Arthur. 2,717,365.
Adams, David M. 2,717,010, Cl. 143—43.
Adler, Joseph I., Jr. 2,716,828, Cl. 41—13.
Admiral Corp.: *See*—
Summerer, William H. 2,717,366.
Agar, Kenneth C., to Clinton Machine Co. 2,717,150, Cl. 261—64.
Agriculture, United States of America as represented by the Secretary of: *See*—
McKinney, Leonard L., Setzkorn, and Uhing. 2,717,263.
Rose, William G., and Lundgren. 2,717,194.
Air Preheater Corp., The: *See*—
Rosenberg, George E. 2,716,884.
Akai, Saburo. 2,716,898, Cl. 74—194.
Akshun Mfg. Co.: *See*—
Lees, Gerald M. 2,716,869.
Albright, Merritt W., and J. J. Foudy, to General Electric Co. 2,717,338, Cl. 317—13.
Alford Cartons: *See*—
Vines, Oscar L. 2,717,115.
Vines, Oscar L. 2,717,116.
Alloy Steel and Metals Co.: *See*—
Francis, Paul R. 2,716,824.
Amendola, James V. 2,716,847, Cl. 51—187.
American Can Co.: *See*—
Hebert, Harold C. 2,717,089.
American Cyanamid Co.: *See*—
Garbarini, John J. 2,717,267.
Hook, Edwin O., and Beegle. 2,717,241.
Hultquist, Martin E., and Roth. 2,717,250
Kienle, Roy H., Eastes, and Cooke. 2,717,246.
Westlake, Harry E., Jr., and Hardy. 2,717,255.
American Machine and Foundry Co.: *See*—
Waddington, William H. 2,717,085.
American Oil Co., The: *See*—
Lutz, Irvin H., Collier, and Bellman. 2,717,231.
American Seating Co.: *See*—
Hoven, Alfred C., and Nordmark. 2,717,026.
Kilmer, Simon T. 2,716,774.
Morgan, Erving B., Jurewicz, and Plachecki. 2,717,187.
Amerman, Jack A., to Emsco Mfg. Co. 2,717,184, Cl. 308—187.
Ames, James G. 2,717,050, Cl. 183—4.1.
Ampro Corp.: *See*—
Bauman, Harold W. 2,717,281.
Anderson, Axel, to Sundstrand Magnetic Products Co. 2,717,080, Cl. 210—1.5.
Anderson, David E., to Thompson Products, Inc. 2,717,149, Cl. 261—18.
Anderson, John R., to Bell Telephone Laboratories, Inc. 2,717,372, Cl. 340—173.
Anderson, John R., to Bell Telephone Laboratories, Inc. 2,717,373, Cl. 340—173.
Anderson, Katherine B. 2,717,175, Cl. 299—83.
Anderson, Walter E., to Globe-Union Inc. 2,717,273, Cl. 136—168.
Andres, Stanley G., to Research Corp. 2,717,051, Cl. 183—7.
Andrew, James M. 2,717,145, Cl. 254—98.
Andrews, Joseph W., and O. R. Schuler, to Calumet & Hecla, Inc. 2,717,072, Cl. 205—3.
Antonidis, John E., to General Motors Corp. 2,716,895, Cl. 74—7.
Apra Precipitator Corp.: *See*—
Valvo, Joseph F. 2,717,052.
Arcturus Mfg. Corp.: *See*—
Sheehan, James W. 2,717,138.
Armstrong, Bruce, to Jackson and Church Co. 2,716,926, Cl. 92—26.
Armstrong, Bruce, to Jackson and Church Co. 2,717,195, Cl. 8—156.
Armstrong, Raymond E. 2,716,840, Cl. 46—142.
Army, United States of America as represented by the Secretary of the: *See*—
Bankes, Willard G., Jr. 2,717,057.
Kazan, Benjamin. 2,717,328.
Arneson, Edwin L., to Morris Paper Mills. 2,717,097, Cl. 220—113.
Arneson, Edwin L., to Morris Paper Mills. 2,717,098, Cl. 220—113.
Aro Equipment Corp.: *See*—
Noonan, Richard E. 2,716,812.
Arone, Nicholas F., to General Electric Co. 2,717,216, Cl. 117—75.
Arvan, Peter G.: *See*—
Jones, Otha C., and Arvan. 2,717,198.
Asaff, Annis G., to Callaghan Hession Corp. 2,717,245, Cl. 260—31.2.
Auger, Harold. 2,717,087, Cl. 214—16.1.
Bader, William, to Wesson Multicut Co. 2,716,799, Cl. 29—96.
Bader, William, to Wesson Multicut Co. 2,716,800, Cl. 29—96.
Badger, Everett H., Jr., C. P. Dahl, and J. W. Overbeke, to The Parker Appliance Co. 2,716,999, Cl. 137—235.
Bailey, Alton E., to National Cylinder Gas Co. 2,717,202, Cl. 23—283.
Baker, David B.: *See*—
Bechman, William O., Land, Baker, and Henning. 2,716,907.
Baker, Donald S.: *See*—
Blackhall, James M., Baker, and Dahlman. 2,717,278.
Baldwin, Hornsby S. 2,717,019, Cl. 152—226.
Balentine, Conrad J., to General Electric Co. 2,717,294, Cl. 200—150.
Bales, Max G., to General Motors Corp. 2,717,286, Cl. 200—31.
Ballard, David C., to Radio Corp. of America. 2,717,322, Cl. 313—76.
Bankes, Willard G., Jr., to the United States of America as represented by the Secretary of the Army. 2,717,057, Cl. 188—5.
Barber Machinery Ltd.: *See*—
Malo, John H. 2,717,090.
Bardeen, Thomas, and R. W. Williams, to Gulf Research & Development Co. 2,717,369, Cl. 340—17.
Barkham, Joseph, to Chisholm, Boyd & White Co. 2,716,900, Cl. 74—520.
Barrett, Roger M. 2,717,131, Cl. 244—6.
Bartelt Engineering Co.: *See*—
Harker, Charles B. 2,716,795.
Baskerville, Ralph: *See*—
Woodward, Lewis J., and Baskerville. 2,717,298.
Battles, William M. 2,717,007, Cl. 139—192.
Bauerlein, Carl C., to The Dole Valve Co. 2,716,996, Cl. 137—98.
Baugh, Everett L., and D. D. Wallace, to General Motors Corp. 2,716,995, Cl. 137—87.
Bauman, Harold W., to Ampro Corp. 2,717,281, Cl. 179—100.2.
Baumheckel, Ralph E., to Link-Belt Co. 2,717,185, Cl. 308—187.2.
Beard, Ernest G. 2,716,919, Cl. 88—16.6.
Beare, Charles H., to General Motors Corp. 2,716,778, Cl. 18—53.
Bechman, William O., H. A. Land, D. B. Baker, and W. W. Henning, to International Harvester Co. 2,716,907, Cl. 74—720.5.
Bednarek, Richard E.: *See*—
Titus, Charles H., and Bednarek. 2,717,293.
Beecham Research Laboratories Ltd.: *See*—
Holland, David O. 2,717,252.
Holland, David O. 2,717,253.
Beegle, Lindley C.: *See*—
Hook, Edwin O., and Beegle. 2,717,241.
Beldt, Lauren F., and D. M. Lewis, to Collins Radio Co. 2,716,896, Cl. 74—10.2.
Bell, Edward R.: *See*—
Rust, Frederick F., and Bell. 2,717,264.
Bell Telephone Laboratories, Inc.: *See*—
Anderson, John R. 2,717,372.
Anderson, John R. 2,717,373.
Lovell, Clarence A. 2,717,280.
Matlack, Richard C., and Metzger. 2,717,279.
Pfann, William G. 2,717,342.
Bellamy, John C., to Cook Electric Co. 2,717,120, Cl. 235—61.
Bellman, David J.: *See*—
Lutz, Irvin H., Collier, and Bellman. 2,717,231.
Beloit Iron Works: *See*—
Goodwillie, John E. 2,717,037.
Bemis Bro. Bag Co.: *See*—
Kindseth, Harold V. 2,716,852.
Bendix Aviation Corp.: *See*—
Block, Arnold H. 2,716,862.
Piper, Charles A. 2,717,370.
Presnell, Frank G. 2,716,945.
Stuart, Alfred A. 2,716,891.
Bennett, John W.: *See*—
Egee, Walter W., and Bennett. 2,717,302.
Berindei, Matthew: *See*—
Nemet, Anthony A., and Berindei. 2,717,315.
Berry, Morris J. 2,717,015, Cl. 150—34.
Betts, George E., Jr., E. L. Blekfeld, and A. T. McCanner, Jr., to the United States of America as represented by the Secretary of the Navy. 2,716,959, Cl. 114—241.

Be Vant Mining & Refining Corp.: *See*—
Edwards, Enoch F. 2,717,205.
Biber, Albert, to Gulf Research & Development Co. 2,716,915 Cl. 82—34.
Biblis, William B. 2,716,750, Cl. 1—49.8.
Biehn, Gerald L., to Westinghouse Electric Corp. 2,716,868, Cl. 62—6.
Biehn, Gerald L., to Westinghouse Electric Corp. 2,716,870, Cl. 62—115.
Bierman, William. 2,716,980, Cl. 128—44.
Biersack, Joseph: *See*—
Sihler, Immanuel, and Biersack. 2,717,287.
Bindler, Jakob, to J. R. Geigy A. G. 2,717,270, Cl. 260—584.
Bingham-Herbrand Corp., The: *See*—
Skareen, Willard C. 2,716,902.
Binstock, Martin H., and H. A. Steinherz, to Westinghouse Electric Corp. 2,717,223, Cl. 148—122.
Birch, Stanley F., F. A. Fidler, and R. A. Dean, to The British Petroleum Co. Ltd. 2,717,269, Cl. 260—582.
Birdsall, Edwin H., to General Dynamics Corp. 2,716,893, Cl. 74—5.
Bitzer, M. A.: *See*—
Winkelsträter, Fritz. 2,716,836.
Bjontegard, Arthur M., to General Electric Co. 2,717,307, Cl. 240—51.11.
Blackhall, James M., D. S. Baker, and P. O. Dahlman, to The North Electric Mfg. Co. 2,717,278, Cl. 179—16.
Blekfeld, Elmer L.: *See*—
Betts, George E., Jr., Blekfeld, and McCanner. 2,716,959.
Bletz, Edward, to Stevens Mfg. Co., Inc. 2,717,289, Cl. 200—138.
Blévin, André E. J. 2,717,354, Cl. 323—47.
Bloch, Herman S., and G. L. Hervert, to Universal Oil Products Co. 2,717,243, Cl. 252—138.
Block, Arnold H., to Bendix Aviation Corp. 2,716,862, Cl. 60—39.28.
Bloomfield Industries, Inc.: *See*—
Meara, William F. 2,717,306.
Bluestein, Ben A., to General Electric Co. 2,717,257, Cl. 260—448.2.
Blumstein, Albert, to Cavitron Corp. 2,716,849, Cl. 51—187.
Bourner, Howard L., to Temco, Inc. 2,716,820, Cl. 34—82.
Bourquin, Jean-Pierre: *See*—
Stoll, Arthur, and Bourquin. 2,717,251.
Boyd, Charles J. 2,717,172, Cl. 294—86.
Brailsford, Harrison D. 2,717,350, Cl. 318—254.
Brand, R. A., & Co. Ltd.: *See*—
Feasey, John. 2,717,017.
Brandon, Clarence W., 14⅙% to N. A. Hardin, 14⅙% to H. H. Wright, 14⅙% to C. H. Newton, and 15% to H. B. Jackson. 2,716,958, Cl. 114—74.
Braum, Blanchard K. 2,716,816, Cl. 32—40.
Brenfleck, Gene. 2,716,991, Cl. 134—182.
Brennan, Joseph B. 2,716,790, Cl. 22—73.
Breslow, Donald M., and O. S. Schesvold, to D. M. Breslow, d. b. a. The Curtition Co. 2,717,033, Cl. 160—84.
Bridgeport Fabrics, Inc.: *See*—
Naramore, Harold B. 2,716,788.
Bristol Aeroplane Co. Ltd., The: *See*—
Whitehead, Frederick W. 2,716,924.
British Insulated Callender's Cables Ltd.: *See*—
Dixon, Henry J., Chapman, and Harwood. 2,717,301.
Mather, Edward L., and Cheetham. 2,717,357.
British Petroleum Co. Ltd., The: *See*—
Birch, Stanley F., Fidler, and Dean. 2,717,269.
Brooks, Frederick E., Jr., to the United States of America as represented by the Secretary of the Navy. 2,717,380, Cl. 343—113.
Brown, Alexander M. 2,716,798, Cl. 29—34.
Brown, Alfred E., to The Gillette Co. 2,717,228, Cl. 167—87.1.
Brown, Cicero C. 2,717,041, Cl. 166—186.
Brown, Eugene L. B. 2,716,871, Cl. 62—146.
Brown, Gerald T., to Peter Spence and Sons Ltd. 2,717,197, Cl. 23—88.
Brown, Thomas F., to Westinghouse Electric Corp. 2,717,339, Cl. 317—119.
Bruene, Warren B., to Collins Radio Co. 2,717,371, Cl. 340—170.
Brundrett, George A., to General Motors Corp. 2,717,058, Cl. 188—88.
Brunswick-Balke-Collender Co., The: *See*—
Huck, William F. 2,717,155.
Buchmann, Gerhard, to International Standard Electric Corp. 2,717,047, Cl. 181—32.
Bundy, Francis P., to General Electric Co. 2,717,319, Cl. 310—16.
Burden, Martin L. 2,716,830, Cl. 43—42.06.
Burns, Beeler D. 2,716,856, Cl. 56—30.
Burns, Jabez, & Sons, Inc.: *See*—
Kopf, Joseph L. 2,716,936.
Bush, George L., to The Teleregister Corp. 2,717,086, Cl. 214—11.
Button, Donald M.: *See*—
Sewell, Curtis, Jr., and Button. 2,717,353.
Byers, Robert C., to the United States of America as represented by the Secretary of the Navy. 2,716,873, Cl. 64—11.
C. A. V. Ltd.: *See*—
Howe, Frank R. 2,716,901.
Waters, Albert V. 2,717,340.
Cairelli, Eremeldo: *See*—
Rundblad, George J., and Cairelli. 2,717,099.
Caldwell, Clarence H., and W. K. Caldwell. 2,716,770, Cl. 15—126.
Caldwell, Walter K.: *See*—
Caldwell, Clarence H., and W. K. 2,716,770.
California Research Corp.: *See*—
Foehr, Edward G. 2,717,242.
Callaghan Hession Corp.: *See*—
Asaff, Annis G. 2,717,245.
Calumet & Hecla, Inc.: *See*—
Andrews, Joseph W., and Schuler. 2,717,072.
Campbell, David H., to Ethyl Corp. 2,717,199, Cl. 23—154.
Campbell, George E., to The Wheland Co. 2,717,186, Cl. 309—3.
Campbell, Walter H. 2,717,309, Cl. 250—17.
Campfield, Arthur E., and W. Davis, to Arthur E. Campfield, Inc. 2,716,992, Cl. 135—3.
Campfield, Arthur E., Inc.: *See*—
Campfield, Arthur E., and Davis. 2,716,992.
Carp, Arthur. 2,716,807, Cl. 30—15.5.
Carpenter, Ralph H., and S. V. Palladino. 2,717,376, Cl. 340—382.
Carruth, Herman A.: *See*—
Williamson, Marshall I., and Carruth. 2,717,074.
Carton, Jean. 2,717,151, Cl. 265—7.
Casanovas, Michel. 2,717,174, Cl. 299—24.
Case Co.: *See*—
Case, Larue R. 2,716,811.
Case, Ethel: *See*—
Case, Rogers. 2,717,274.
Case, J. I., Co.: *See*—
Greeson, Clarett B. 2,717,136.
Case, Larue R., to Case Co. 2,716,811, Cl. 30—47.
Case, Rogers, deceased; E. Case, executrix, to Transandean Associates, Inc. 2,717,274, Cl. 174—33.
Castelli, Joseph L.: *See*—
Manheim, Theodore B., and Castelli. 2,716,961.
Caterpillar Tractor Co.: *See*—
King, Ralph J., and Eyman. 2,716,970.
Ramsel, Charles A. 2,717,067.
Cavitron Corp.: *See*—
Blumstein, Albert. 2,716,849.
Chapman, Joseph E. G.: *See*—
Dixon, Henry J., Chapman, and Harwood. 2,717,301.
Chappell, Ralph R. 2,717,192, Cl. 346—139.
Cheetham, Reginald J.: *See*—
Mather, Edward L., and Cheetham. 2,717,357.
Chemical Elaborations, Inc.: *See*—
Cole, Robert M. 2,717,262.
Chicopee Mfg. Corp.: *See*—
Simon, Simon A., and Ruhf. 2,717,193.
Chiksan Co.: *See*—
Hedden, Robert R. 2,717,166.
Chisholm, Boyd & White Co.: *See*—
Barkham, Joseph. 2,716,900.
Chloride Electrical Storage Co. Ltd., The: *See*—
Jay, James B., White, and Hunt. 2,717,003.
Christian, Carl A., and S. J. Murcek, to Westinghouse Electric Corp. 2,717,351, Cl. 321—18.
Christner, Robert M. 2,717,221, Cl. 117—127.
Cincinnati Milling Machine Co., The: *See*—
Roehm, Erwin G. 2,716,925.
Clark, Robert H. 2,717,113, Cl. 226—125.
Clark, Virgil R.: *See*—
Streich, Philip A., Clark, and Tait. 2,716,776.
Clay, Burton R., to Radio Corp. of America. 2,717,323, Cl. 313—77.
Clement, Joseph J.: *See*—
Wilson, Ashley F., and Clement. 2,717,000.
Clift, Gilbert B.: *See*—
McGay, John B., and Clift. 2,716,860.
Clinton Machine Co.: *See*—
Agar, Kenneth C. 2,717,150.
Cockrell, Jesse S. 2,716,772, Cl. 15—306.
Codrick, Thomas H. 2,716,993, Cl. 135—4.
Cole, Robert M., to Chemical Elaborations, Inc. 2,717,262, Cl. 260—468.
Collier, Oran W.: *See*—
Lutz, Irvin H., Collier, and Bellman. 2,717,231.
Collins Radio Co.: *See*—
Price, George W. 2,717,360.
Beldt, Lauren F., and Lewis. 2,716,896.
Bruene, Warren B. 2,717,371.
Doelz, Melvin L. 2,717,361.
Hileman, Dale L. 2,717,345.
Hodgin, David M., and Mifflin. 2,717,364.
Hubbard, Merle R. 2,717,363.
Smith, Roy A. 2,716,887.
Wulfsberg Arthur H. 2,716,897.
Collins, Tappan, to National Steel Corp. 2,717,060, Cl. 189—1.
Columbia-Southern Chemical Corp.: *See*—
Neubauer, Joseph A., Strain, and Kung. 2,717,238.
Commonwealth Color & Chemical Co.: *See*—
Fronmuller, George D. 2,717,240.
Compton, William H., J. M. Duff and J. K. Smith, to The Reliance Electric & Engineering Co. 2,716,859, Cl. 57—100.
Comstock, Millard F. 2,716,988, Cl. 131—242.
Condon, Fred E.: *See*—
Vaughan, William E., and Condon. 2,717,248.
Contois, Leo L., Jr., to Monsanto Chemical Co. 2,717,247, Cl. 260—78.5.
Converse, John O., to Nichols Engineering Co. 2,716,949, Cl. 104—32.
Cook, Curtiss L., to Deere & Co. 2,717,071, Cl. 198—189.
Cook Electric Co.: *See*—
Bellamy, John C. 2,717,120.
Cooke, Theodore F.: *See*—
Kienle, Roy H., Eastes, and Cooke. 2,717,246.
Cooper, James M., to General Electric Co. 2,717,132, Cl. 244—77.
Corn, John A. 2,717,014, Cl. 150—4.
Cornelius, Nelson F. 2,717,178, Cl. 299—97.
Cory Corp.: *See*—
Karlen, Harvey R. 2,717,303.
Costello, James R., Jr.: *See*—
Toy, Arthur D. F., and Costello. 2,717,249.
Crabtree, David, & Sons Ltd.: *See*—
Felton, William. 2,717,117.

Craddock, Charles L., to Michael Research Co., Inc. 2,717,336, Cl. 315—183.
Cranston, Hoy A., to C. S. Miner. 2,717,211, Cl. 99—54.
Crehan, William J.: See—
 Talmey, Paul, and Crehan. 2,717,218.
Crookston, Robert R., to Esso Research and Engineering Co. 2,716,997, Cl. 137—102.
Crutis, Arvel C., to Pipelife, Inc. 2,717,038, Cl. 166—1.
Culbertson, John L., to International Telephone and Telegraph Corp. 2,717,284, Cl. 179—175.2.
Cultiguard Shovel Co.: See—
 Demorest, Leroy A. 2,716,934.
 Demorest, Leroy A. 2,716,935.
Cuny, Ernest A. 2,716,948, Cl. 103—161.
Curtition Co., The: See—
 Breslow, Donald M., and Schesvold. 2,717,033.
Dahl, Carl P.: See—
 Badger, Everett H., Jr., Dahl, and Overbeke. 2,716,999.
Dahlman, Per O.: See—
 Blackhall, James M., Baker, and Dahlman. 2,717,278.
Daimler-Benz Aktiengesellschaft: See—
 Nallinger, Fredrich K. H. 2,717,045.
Damm, Felix H. H., of 40% to G. Seemann. 2,717,031, Cl. 158—27.4.
Danault, Clinton L.: See—
 Ford, James G., and Denault. 2,717,219.
Dardani, Edward V., to The Heim Co. 2,717,006, Cl. 139—151.
Daudelin, Roland G.: See—
 Kent, Raymond C., Jr., and Daudelin. 2,716,957.
Davis, Harold G. 2,716,984, Cl. 128—225.
Davis, Hubert G., and T. P. Wilson, to Union Carbide and Carbon Corp. 2,717,259, Cl. 260—449.6.
Davis, Hubert G., T. P. Wilson, and A. N. Kurtz, to Union Carbide and Carbon Corp. 2,717,260, Cl. 260—449.6.
Davis, Michael Z. 2,716,767, Cl. 15—21.
Davis, Ralph P., to Walworth Co. 2,716,789, Cl. 22—10.
Davis, Walter: See—
 Campfield, Arthur E., and Davis. 2,716,992.
Dawson, Helen L., and M. D. Wheatley. 2,717,227, Cl. 167—74.
Dean, Ronald A.: See—
 Birch, Stanley F., Fidler, and Dean. 2,717,269.
De Bonville, Arthur L., and J. S. Swercewski. 2,716,834, Cl. 43—55.
Deere & Co.: See—
 Cook, Curtiss L. 2,717,071.
 Sorensen, Knud B., and Peirson. 2,716,855.
Delk, James E., Sr. 2,717,314, Cl. 250—50.
Demetry, Thomas A.: See—
 Shoup, Carl A. 2,717,190.
Demorest, Leroy A., to Cultiguard Shovel Co. 2,716,934, Cl. 97—204.
Demorest, Leroy A., to Cultiguard Shovel Co. 2,716,935, Cl. 97—204.
Denman Enterprises, Ltd.: See—
 Stuart, John C. K. 2,716,843.
Deremer, Floyd E., to Oldberg Mfg. Co. 2,717,048, Cl. 181—61.
Desch, Joseph R., to The National Cash Register Co. 2,717,334, Cl. 315—166.
Desi, Paul F. 2,716,973, Cl. 124—1.
De Witte, Marinus C. 2,717,210, Cl. 99—11.
Diamond, Richard: See—
 Thomas, Clifford A. 2,717,159.
Dickinson, Charles H.: See—
 Timson, Ernest A., and Dickinson. 2,716,942.
Dieterich, Joseph O. E. 2,717,158, Cl. 273—141.
Dixon, Henry J., J. E. G. Chapman, and F. A. Harwood, to British Insulated Callender's Cables Ltd. 2,717,301, Cl. 219—12.
Dobias, George L. 2,717,020, Cl. 153—32.
Doelz, Melvin L., to Collins Radio Co. 2,717,361, Cl. 333—71.
Dole Valve Co., The: See—
 Bauerlein, Carl C. 2,716,996.
Dolliver, Morris A., and S. Semenoff, to Olin Mathieson Chemical Corp. 2,717,236, Cl. 204—75.
Doman, Glidden S., to Doman Helicopters, Inc. 2,716,889, Cl. 73—147.
Doman Helicopters, Inc.: See—
 Doman, Glidden S. 2,716,889.
Douglas, Peyton W., to Easy Washing Machine Corp. 2,717,135, Cl. 248—20.
Dr. Ing. H. C. F. Porsche K.-G.: See—
 Rabe, Karl. 2,717,173.
Driscoll, William F. 2,717,069, Cl. 194—9.
Duerksen, Arnold, to Super Mold Corp. of California. 2,717,022, Cl. 154—9.
Duff, John M.: See—
 Compton, William H., Duff, and Smith. 2,716,859.
Dumousseau, Claude: See—
 Touraton, Emile, Dumousseau, and Zwobada. 2,717,327.
Dupin, Eugene A. 2,717,032, Cl. 158—132.
Du Pont, E. I., de Nemours and Co.: See—
 Fay, Robert E., Jr. 2,717,220.
 Krespan, Carl G. 2,717,261.
 Noddin, George A., and Spaeth. 2,717,204.
 Sullivan, David J. 2,717,217.
Durex, S. A.: See—
 Lang, Heinrich. 2,716,969.
Dusing & Hunt, Inc.: See—
 Dusing, Leon F., and Saino. 2,717,062.
Dusing, Leon F., and F. H. Saino; said Dusing assor. to Dusing & Hunt, Inc., and said Saino assor. to F. H. Saino Mfg. Co. 2,717,062, Cl. 189—46.
Dylewski, Stanley A. 2,717,157, Cl. 273—135.
Earp, Charles W., to International Standard Electric Corp. 2,717,379, Cl. 343—106.
Eastes, John W.: See—
 Kienle, Roy H., Eastes, and Cooke. 2,717,246.
Eastman, Fred C., to United Shoe Machinery Corp. 2,716,764, Cl. 12—1.
Easy Washing Machine Corp.: See—
 Douglas, Peyton W. 2,717,135.
Eaton Mfg. Co.: See—
 Pipes, George R., and Findley. 2,717,179.
Edwards, Enoch F., to Be Vant Mining & Refining Corp. 2,717,205, Cl. 75—5.
Egee, Walter W., to Fletcher Works Inc. 2,717,126, Cl. 242—39.
Egee, Walter W., and J. W. Bennett, to Fletcher Works Inc. 2,717,302, Cl. 219—19.
Electric & Musical Industries Ltd.: See—
 Hall, Ronald C. 2,717,324.
 Holman, Herbert E., and Newton. 2,716,921.
Elliott, James M., and D. R. Green. 2,716,781, Cl. 19—165.
Ely, Charles M., to National Distillers Products Corp. 2,717,208, Cl. 99—4.
Ely, Charles M., and S. Schott, to National Distillers Products Corp. 2,717,209, Cl. 99—4.
Empire Brushes, Inc.: See—
 Schwartz, Harold H., and Tucker. 2,716,768.
Emsco Mfg. Co.: See—
 Amerman, Jack A. 2,717,184.
Engelder, Arthur E. 2,717,100, Cl. 222—5.
Erie Resistor Corp.: See—
 Foster, James H. 2,717,356.
Eriksson, Maurits S. 2,716,757, Cl. 4—166.
Esso Research and Engineering Co.: See—
 Crookston, Robert R. 2,716,997.
Ethyl Corp.: See—
 Campbell, David H. 2,717,199.
 Shapiro, Hymin. 2,717,206.
 Trotter, Percy W. 2,717,233.
 Trotter, Percy W. 2,717,272.
Eyman, Earl D.: See—
 King, Ralph J., and Eyman. 2,716,970.
Ezzell, Louie M.: See—
 Owens, Edward O., and Ezzell. 2,716,885.
Farbenfabriken Bayer Aktiengesellschaft: See—
 Rempel, Nikolaus. 2,717,237.
Farny, Paul, and E. Weidmann. 2,716,972, Cl. 123—90.
Faulkner, Clarence F., to Puritan Cordage Mills, Inc. 2,717,215, Cl. 117—7.
Faulkner, Harry. 2,716,880, Cl. 67—7.1.
Faximile, Inc.: See—
 Hester, Frank A. 2,717,191.
Fay, Robert E., Jr., to E. I. du Pont de Nemours and Co. 2,717,220, Cl. 117—126.
Feasey, John, to R. A. Brand & Co. Ltd. 2,717,017, Cl. 150—52.
Fegan, Thomas G. 2,716,783, Cl. 20—55.
Fejmert, Bernhard V.: See—
 Fejmert, Erik V. and B. V. 2,717,147.
Fejmert, Erik V. and B. V.; said B. V. Fejmert assor. to said E. V. Fejmert. 2,717,147, Cl. 259—178.
Felton, William, ½ to David Crabtree & Sons Ltd. 2,717,117, Cl. 28—49.
Ferber, Robert, to Usines Gallus Societe Anonyme. 2,717,134, Cl. 248—11.
Ferris, Walter, to The Oilgear Co. 2,716,944, Cl. 103—2.
Fetterle, Marcel: See—
 Huguenin, Pierre, and Fetterle. 2,716,829.
Fidler, Frederick A.: See—
 Birch, Stanley F., Fidler, and Dean. 2,717,269.
Fielden, Fred, and F. Murphy, to Thomas Robinson & Son Ltd. 2,717,124, Cl. 241—60.
Findlay, Robert A., to Phillips Petroleum Co. 2,717,229, Cl. 196—14.15.
Findley, Howard J.: See—
 Pipes, George R., and Findley. 2,717,179.
Firma A. J. Tröster: See—
 Weitz, Otto. 2,717,105.
Fischer, Georg, Aktiengesellschaft: See—
 Wernli, Max. 2,716,905.
Fischer, Paul W., and J. W. Scheffel, to Union Oil Co. of California. 2,717,239, Cl. 252—8.5.
Fisher & Christen: See—
 Abell, Harry D. and H. E., and Heald. 2,717,153.
Fitler, Lester D., to Rome Cable Corp. 2,716,818. Cl. 33—131.
Fleischer, Kurt W.: See—
 Hess, Frederic O., and Fleischer. 2,716,968.
Fletcher, George, & Co. Ltd.: See—
 Murray, Cecil W. 2,717,142.
Fletcher Works Inc.: See—
 Egee, Walter W. 2,717,126.
 Egee, Walter W., and Bennett. 2,717,302.
 Schaum, Fletcher. 2,716,858.
Fluor Corp., Ltd., The: See—
 Langford, John A. 2,717,049.
Focke, Gustavus K. 2,716,911, Cl. 81—23.
Foehr, Edward G., to California Research Corp. 2,717,242, Cl. 252—49.6.
Foglio, James J. 2,716,755, Cl. 2—279.
Foley, Robert T., and O. P. McCarty, to General Electric Co. 2,717,296, Cl. 200—166.
Ford Alexander Corp., The: See—
 Gieske, William R. 2,717,039.
Ford, James G., and C. L. Denault, to Westinghouse Electric Corp. 2,717,219, Cl. 117—126.
Ford, Wayne B. 2,716,815, Cl. 32—32.
Foster, James H., to Erie Resistor Corp. 2,717,356, Cl. 323—74.
Foudy, John J.: See—
 Albright, Merritt W., and Foudy. 2,717,338.
Fouré, Claude: See—
 Reingold, Lucien, and Fouré. 2,716,863.

Francis, Paul R., to Alloy Steel and Metals Co. 2,716,824, Cl. 37—147.
Franklin, Edna M. 2,716,817, Cl. 33—15.
Friedman, Joseph D.: *See*—
Groth, Fred A. 2,717,035.
Friess, Hans, to Klangfilm Gesellschaft mit beschraenkter Haftung. 2,717,283, Cl. 179—100.3.
Frink, Russell E., and P. Olsson, to Westinghouse Electric Corp. 2,717,292, Cl. 200—144.
Fronmuller, George D., to Commonwealth Color & Chemical Co. 2,717,240, Cl. 252—8.6.
Fry, Millard E., to General Motors Corp. 2,717,291, Cl. 200—140.
Funk, Roger S. 2,717,110, Cl. 224—42.45.
Gable, Myron W., to Shell Development Co. 2,717,095, Cl. 220—26.
Gaidos, Alonzo F. 2,716,923, Cl. 89—140.
Gainsburgh, Charles B. 2,716,756, Cl. 2—321.
Garbarini, John J., to American Cyanamid Co. 2,717,267, Cl. 260—534.
Garmgard Products Co.: *See*—
Schmidt, John B., and Hanson. 2,717,160.
Gartner, Stanley J., and H. J. Zwald, to Sylvania Electric Products Inc. 2,717,092, Cl. 214—658.
Geigy, J. R., A. G.: *See*—
Bindler, Jakob. 2,717,270.
Geller, Julius, and H. Ratte, to Rutgerswerke-Aktiengesellschaft. 2,717,232, Cl. 202—40.
General American Transportation Corp.: *See*—
Talmey, Paul, and Crehan. 2,717,218.
General Dynamics Corp.: *See*—
Birdsall, Edwin H. 2,716,893.
General Electric Co.: *See*—
Albright, Merritt W., and Foudy. 2,717,338.
Arone, Nicholas F. 2,717,216.
Balentine, Conrad J. 2,717,294.
Bjontegard, Arthur M. 2,717,307.
Bluestein, Ben A. 2,717,257.
Bundy, Francis P. 2,717,319.
Cooper, James M. 2,717,132.
Foley, Robert T., and McCarty. 2,717,296.
Gunton, Frank D. 2,717,326.
Hall, Robert N. 2,717,343.
Heintz, Milton L. 2,717,288.
Kantor, Simon W. 2,717,258.
London, Sol. 2,717,348.
Louden, Victor J. 2,717,355.
North, Harper Q. 2,717,341.
Prober, Maurice. 2,717,235.
Silva, William C. 2,716,866.
Sims, Marion W., Hill, and Krakower. 2,717,335.
Titus, Charles H., and Bednarek. 2,717,293.
Woodward, Lewis J., and Baskerville. 2,717,298.
General Motors Corp.: *See*—
Antonidis, John E. 2,716,895.
Bales, Max G. 2,717,286.
Baugh, Everett L., and Wallace. 2,716,995.
Beare, Charles H. 2,716,778.
Brundrett, George A. 2,717,058.
Fry, Millard E. 2,717,291.
Harris, Edward P. 2,716,787.
Jackson, George W. 2,717,344.
Jacobs, James W. 2,716,867.
Marvin, John T. 2,717,295.
Shaw, Willard C. 2,716,803.
Stickel, Carl A. 2,716,865.
Gerth, Roy J. 2,716,846, Cl. 51—128.
Geyger, Wilhelm A., to the United States of America as represented by the Secretary of the Navy. 2,717,346, Cl. 318—207.
Gieske, William R., to The Ford Alexander Corp. 2,717,039, Cl. 166—65.
Gilardi, William D. 2,717,111, Cl. 224—45.
Gilfillan Bros., Inc.: *See*—
Tasker, Homer G., and Olson. 2,717,377.
Gill, William M., and E. D. Vold. 2,716,882, Cl. 70—159.
Gillette Co., The: *See*—
Brown, Alfred E. 2,717,228.
Glass, Arthur A. 2,716,831, Cl. 43—42.31.
Glass, Marvin I.: *See*—
Young, George G. 2,716,839.
Globe-Union Inc.: *See*—
Anderson, Walter E. 2,717,273.
Goda, Lawrence B., Sr. 2,717,177, Cl. 299—86.
Godbey, John W.: *See*—
McMichael, Charles E., Godbey, and Zehnder. 2,717,256.
Goddard, Esther C.: *See*—
Goddard, Robert H. 2,717,182.
Goddard, Robert H., deceased; E. C. Goddard, executrix, ½% to The Daniel and Florence Guggenheim Foundation. 2,717,182, Cl. 308—9.
Goodwillie, John E., to Beloit Iron Works. 2,717,037, Cl. 164—65.
Goodyear, James W. 2,716,861, Cl. 60—13.
Gordon, Herman, deceased; S. S. Gordon, executrix. 2,716,753, Cl. 2—195.
Gordon, Selma S.: *See*—
Gordon, Herman. 2,716,753.
Gosslar, Karl, to International Standard Electric Corp. 2,717,325, Cl. 313—270.
Gottstein, Adolph J. 2,717,171, Cl. 294—16.
Graflex, Inc.: *See*—
Smith, Clarence E. 2,716,929.
Graham Mfg. Co., Inc., The: *See*—
Viall, Richmond, Jr., Peterson, and Johnson. 2,716,845.
Grant, Harry C., Jr., and F. B. Parsons, to Specialties Development Corp. 2,717,042, Cl. 169—11.
Grant, Harry C., Jr. 2,716,821, Cl. 37—43.
Green, Dick R.: *See*—
Elliott, James M., and Green. 2,716,781.
Greenbaum, Arthur, to Academy Electrical Products Corp. 2,717,365, Cl. 339—99.
Greer, Carl S., Jr., to Tranter Mfg. Inc. 2,716,802, Cl. 29—157.3.
Greeson, Clarett B., to J. I. Case Co. 2,717,136, Cl. 248—75.
Gregory, James. 2,717,133, Cl. 244—148.
Grimes, Claude E. 2,717,009, Cl. 140—2.
Groenendal, Ubbo W. 2,717,084, Cl. 211—20.
Grogan, David H. 2,717,137, Cl. 248—99.
Grote, Henry W.: *See*—
Murray, Maurice J., Haensel, and Grote. 2,717,230.
Groth, Fred A., to F. A. Groth and J. D. Friedman. 2,717,035, Cl. 160—178.
Grounds, Henry. 2,716,932, Cl. 97—10.
Guerinet, Roger J., to Societe Stapfer & Cie, s. a. r. l. 2,716,910, Cl. 81—3.8.
Guerrini, Vincenzo: *See*—
Pasquini, Vittorio, and Guerrini. 2,716,952.
Guggenheim, Daniel and Florence, Foundation, The: *See*—
Goddard, Robert H. 2,717,182.
Gulbrandsen, Helge, to United Shoe Machinery Corp. 2,716,766, Cl. 12—36.
Gulf Research & Development Co.: *See*—
Bardeen, Thomas, and Williams. 2,717,369.
Biber, Albert. 2,716,915.
Holtz, Quentin G. 2,717,347.
Pigott, Reginald J. S. 2,716,914.
Gunton, Frank D., to General Electric Co. 2,717,326, Cl. 314—69.
Guthrie, James M. 2,717,305, Cl. 219—38.
Gwaltney, Eugene C., Jr. 2,717,122, Cl. 236—44.
Hacker, George H. 2,716,864, Cl. 61—16.
Hadley, Wilfred N., to Parks & Woolson Machine Co. 2,716,797, Cl. 26—29.
Haensel, Vladimir: *See*—
Murray, Maurice J., Haensel, and Grote. 2,717,230.
Hagen, Norbert. 2,716,777, Cl. 18—19.
Haines, William C. 2,716,954, Cl. 112—176.
Hajecate, Thomas H., to The Light House, Inc. 2,716,758, Cl. 9—8.
Hall, Michael F. 2,717,148, Cl. 261—16.
Hall, Robert N., to General Electric Co. 2,717,343, Cl. 317—235.
Hall, Ronald C., to Electric & Musical Industries Ltd. 2,717,324, Cl. 313—84.
Hamilton, William S. H. 2,717,332, Cl. 315—100.
Hamilton, William S. H. 2,717,333, Cl. 315—103.
Hammer, Oscar. 2,717,106, Cl. 222—318.
Haneberg, Raymond A.: *See*—
Hobing, John W. 2,716,879.
Haneberg, Richard H.: *See*—
Hobing, John W. 2,716,879.
Hanford, William E., to The M. W. Kellogg Co. 2,717,200, Cl. 23—190.
Hansen, Hans P. 2,716,903, Cl. 74—557.
Hanson, Harold J.: *See*—
Schmidt, John B., and Hanson. 2,717,160.
Hardin, N. A.: *See*—
Brandon, Clarence W. 2,716,958.
Hardy, James A., to Schwitzer-Cummins Co. 2,716,946, Cl. 103—37.
Hardy, William B.: *See*—
Westlake, Harry E., Jr., and Hardy. 2,717,255.
Harker, Charles B., to Bartelt Engineering Co. 2,716,795, Cl. 24—255.
Harris, Edward P., to General Motors Corp. 2,716,787, Cl. 20—69.
Harris, Estelle F. 2,717,036, Cl. 160—354.
Harris-Seybold Co.: *See*—
Wood, James R. 2,717,154.
Hart, Charles C. 2,716,808, Cl. 30—16.
Hartford National Bank and Trust Co.: *See*—
Nemet, Anthony A., and Berindei. 2,717,315.
Van Laar, Jan A. W. 2,717,203.
Zalm, Pieter, and van den Boomgaard. 2,717,244.
Hartop, William L., Jr.: *See*—
Windischman, Edward F., Hartop, and Ryan. 2,716,983.
Hartzell Industries, Inc.: *See*—
Johnston, Danal W. 2,716,975.
Harwood, Frederick A.: *See*—
Dixon, Henry J., Chapman, and Harwood. 2,717,301.
Hattman, Frederick A., to United States Steel Corp. 2,716,941, Cl. 101—40.
Hayden, Vernon H., and M. G. Leonard, to Westinghouse Electric Corp. 2,717,275, Cl. 174—37.
Hayes, Robert R.: *See*—
Shoulders, Walter L., and Hayes. 2,717,320.
Hazeltine Research, Inc.: *See*—
Press, Meyer. 2,717,362.
Heald, Norman R.: *See*—
Abell, Harry D., and H. E., and Heald. 2,717,153.
Hebert, Harold C., to American Can Co. 2,717,089, Cl. 214—83.26.
Hedden, Robert R., to Chiksan Co. 2,717,166, Cl. 285—10.
Heim Co., The: *See*—
Dardani, Edward V. 2,717,006.
Heiniger, Wilfred, to Paillard S. A. 2,717,055, Cl. 185—39.
Heintz, Milton L., to General Electric Co. 2,717,288, Cl. 200—97.
Heizer, Edward J., to Specialties Development Corp. 2,717,128, Cl. 242—72.
Henderson, Minnis W. 2,717,096, Cl. 220—40.
Henning, William W.: *See*—
Bechman, William O., Land, Baker, and Henning. 2,716,907.
Hensgen, Bernard T., and J. C. V. Bosch, A. M. Lederer, and P. M. Wood, to Swift & Co. 2,717,212, Cl. 99—116.
Hervert, George L.: *See*—
Bloch, Herman S., and Hervert. 2,717,243.

Hess, Frederic O., and K. W. Fleischer, to Selas Corp. of America. 2,716,968, Cl. 122—510.
Hester, Frank A., to Faximile, Inc. 2,717,191, Cl. 346—101.
Hetherington, Alexander C., to The M. W. Kellogg Co. 2,717,023, Cl. 154—33.1.
Hileman, Dale L., to Collins Radio Co. 2,717,345, Cl. 318—129.
Hilgert, Adolph J., and R. B. Matthews, to Milwaukee Gas Specialty Co. 2,717,123, Cl. 236—75.
Hill, Charles, and C. K. John. 2,716,875, Cl. 64—29.
Hill, Ezra C.: See—
Sims, Marion W., Hill, and Krakower. 2,717,335.
Hill, Glenn M. 2,717,103, Cl. 222—100.
Hirsch, Harold S. 2,716,754, Cl. 2—227.
Hoagland, Arns. 2,717,304, Cl. 219—32.
Hobing, John W., to R. A. and R. H. Haneberg. 2, 716,879, Cl. 67—6.1.
Hock, Lester R. 2,717,064, Cl. 189—46.
Hodgin, David M., and L. V. Mifflin, to Collins Radio Co. 2,717,364, Cl. 336—30.
Hoeselbarth, Frank W. E., to C. H. Masland & Sons. 2,717,005, Cl. 139—39.
Holcroft & Co.: See—
Joy, Clarence L. 2,716,989.
Holland, David O., to Beecham Research Laboratories Ltd. 2,717,252, Cl. 260—309.5.
Holland, David O., to Beecham Research Laboratories Ltd. 2,717,253, Cl. 260—309.5.
Hollins, Jesse R. 2,717,331, Cl. 315—77.
Holman, Herbert E., and G. C. Newton, to Electric & Musical Industries Ltd. 2,716,921, Cl. 88—18.4.
Holtz, Quentin G., to Gulf Research & Development Co. 2,717,347, Cl. 318—207.
Hook, Edwin O., and L. C. Beegle, to American Cyanamid Co. 2,717,241, Cl. 252—32.7.
Hopkins, Roscoe B., ½ to A. L. Ketchum. 2,717,152, Cl. 267—20.
Hoppes, Lloyd G., to L. Wilkinson. 2,717,104, Cl. 222—177.
Horvath, Artemy A. 2,717,207, Cl. 99—2.
Hosemaster Machine Co. Ltd.: See—
Wickardt, Kurt W. 2,716,877.
Hosmer, Chester C.: See—
Launder, Ernie L., and Hosmer. 2,716,822.
Hoven, Alfred C., and W. E. Nordmark, to American Seating Co. 2,717,026, Cl. 155—85.
Howe, Frank R., to C. A. V. Ltd. 2,716,901, Cl. 74—526.
Howe, Robert F. 2,716,963, Cl. 120—24.
Hubbard, Merle R., to Collins Radio Co. 2,717,363, Cl. 333—82.
Hubert, Clarence A., and J. F. Ziskal, to International Harvester Co. 2,716,966, Cl. 121—40.
Huck, William F., to The Brunswick-Balke-Collender Co. 2,717,155, Cl. 273—49.
Huebner Co., The: See—
Huebner, William C. 2,716,826.
Huebner, William C., to The Huebner Co. 2,716,826, Cl. 41—1.
Hughes Aircraft Co.: See—
Woodruff, Thomas E. 2,717,310.
Huguenin & Cie: See—
Huguenin, Pierre, and Fetterle. 2,716,829.
Huguenin, Pierre, Bienne, and M. Fetterle, to Huguenin & Cie. 2,716,829, Cl. 41—35.
Hulse, Alexander B., Jr. 2,717,140, Cl. 248—223.
Hultquist, Martin E., and B. Roth, to American Cyanamid Co. 2,717,250, Cl. 260—251.5.
Hunt, Joseph A.: See—
Jay, James B., White, and Hunt. 2,717,003.
Illinois Tool Works: See—
Nielsen, Moritz H. 2,716,761.
Poupitch, Ougljesa J. 2,716,760.
Stern, William. 2,716,762.
International Business Machines Corp.: See—
Luhn, Hans P. 2,717,121.
International Electronics Co., The: See—
Reed, Frank J. 2,717,282.
International Harvester Co.: See—
Bechman, William O., Land, Baker, and Henning. 2,716,907.
Hubert, Clarence A., and Ziskal. 2,716,966.
International Standard Electric Corp.: See—
Buchmann, Gerhard. 2,717,047.
Earp, Charles W. 2,717,379.
Gosslar, Karl. 2,717,325.
Taylor, Frank H. 2,717,312.
Touraton, Emile, Dumousseau, and Zwobada. 2,717,327.
International Telephone and Telegraph Corp.: See—
Culbertson, John L. 2,717,284.
Isacco, Vittorio. 2,717,043, Cl. 170—135.4.
Jackson and Church Co.: See—
Armstrong, Bruce. 2,716,926.
Armstrong, Bruce. 2,717,195.
Jackson, George W., to General Motors Corp. 2,717,344, Cl. 318—31.
Jackson, Harvey B.: See—
Brandon, Clarence W. 2,716,958.
Jacobi, Werner, to Siemens & Halske. 2,717,299, Cl. 201—63.
Jacobs, James W., to General Motors Corp. 2,716,867, Cl. 62—4.
Janik, Anton J., to The Ridge Tool Co. 2,716,947, Cl. 103—138.
Jay, James B., A. J. White, and J. A. Hunt, to The Chloride Electrical Storage Co. Ltd. 2,717,003, Cl. 137—625.34.
Jelinek, Ulrich, to The M. W. Kellogg Co. 2,717,024, Cl. 154—33.1.
Jelinek, Ulrich, to The M. W. Kellogg Co. 2,717,025, Cl. 154—33.1.
Jewell, Bernard W., to The O. A. Sutton Corp., Inc. 2,717,139, Cl. 248—208.
Johns, Charles K.: See—
Hill, Charles, and Johns. 2,716,875.
Johnson, Charles B., to Woodall Industries, Inc. 2,716,804, Cl. 29—453.
Johnson, Francis M. 2,717,130, Cl. 244—2.
Johnson, Herbert L., to Sun Oil Co. 2,717,265, Cl. 260—504.
Johnson, Roger H.: See—
Viall, Richmond, Jr., Peterson, and Johnson. 2,716,845.
Johnston, Danal W., to Hartzell Industries, Inc. 2,716,975, Cl. 126—110.
Johnston, Mary A. 2,716,950, Cl. 107—47.
Jonas, Samuel S. 2,717,029, Cl. 155—179.
Jones, Charles H., and G. S. Ley, to Westinghouse Electric Corp. 2,717,329, Cl. 315—24.
Jones, Leo D., to The Sharples Corp. 2,717,119, Cl. 233—21.
Jones, Otha C., and P. G. Arvan, to Monsanto Chemical Co. 2,717,198, Cl. 23—106.
Joy, Clarence L., to Holcroft & Co. 2,716,989, Cl. 134—66.
Jurewicz, Bruno S.: See—
Morgan, Erving B., Jurewicz, and Plachecki. 2,717,187.
Kantor, Simon W., to General Electric Co. 2,717,258, Cl. 260—448.2.
Karlen, Harvey R., to Cory Corp. 2,717,303, Cl. 219—20.
Katz, Robert. 2,717,061, Cl. 189—46.
Katz, Samuel W., and N. T. Salmon. 2,716,990, Cl. 134—155.
Kazan, Benjamin, to the United States of America as represented by the Secretary of the Army. 2,717,328, Cl. 315—14.
Keefe, John S. 2,716,987, Cl. 131—237.
Keene Edward W. W. 2,717,083, Cl. 210—199.
Keith, Robert H., and C. L. Mershon, to Westinghouse Electric Corp. 2,717,318, Cl. 307—51.
Kelem, Howard. 2,716,751, Cl. 1—215.
Keller, Harry. 2,716,928, Cl. 93—58.5.
Kellermann, Willy. 2,716,842, Cl. 46—208.
Kellogg, M. W., Co., The: See—
Hanford, William E. 2,717,200.
Hetherington, Alexander C. 2,717,023.
Jelinek, Ulrich. 2,717,024.
Jelinek, Ulrich. 2,717,025.
Passino, Herbert J. 2,717,201.
Kemper, Walter O. 2,716,940, Cl. 101—6.
Kenimer, Loy F. 2,716,775, Cl. 16—138.
Kent, Raymond C., Jr., and R. G. Daudelin, to the United States of America as represented by the Secretary of the Navy. 2,716,957, Cl. 114—23.
Ketchum, Art L.: See—
Hopkins, Roscoe B. 2,717,152.
Kevorkian, Joseph D. 2,717,308, Cl. 240—102.
Kienle, Roy H., J. W. Eastes, and T. F. Cooke, to American Cyanamid Co. 2,717,246, Cl. 260—37.
Kilmer, Simon T., to American Seating Co. 2,716,774, Cl. 16—42.
Kinderman, Walter J., to Yarnall-Waring Co. 2,716,892, Cl. 73—293.
Kindseth, Harold V., to Bemis Bro. Bag Co. 2,716,852, Cl. 53—148.
King, Carl B. 2,716,837, Cl. 46—44.
King, Ralph J., and E. D. Eyman, to Caterpillar Tractor Co. 2,716,970, Cl. 123—41.73.
Klamp, Paul, to Mechanical Handling Systems Inc. 2,716,965, Cl. 121—38.
Klangfilm Gesellschaft mit Beschraenkter Haftung: See—
Friess, Hans. 2,717,283.
Knasko, Joseph J. 2,716,998, Cl. 137—231.
Kopf, Joseph L., to Jabez Burns & Sons, Inc. 2,716,936, Cl. 99—236.
Koval, Walter C. 2,716,810, Cl. 30—47.
Krakower, Aaron M.: See—
Sims, Marion W., Hill, and Krakower. 2,717,335.
Kraus, Charles E. 2,716,801, Cl. 29—105.
Kresban, Carl G., to E. I. du Pont de Nemours and Co. 2,717,261, Cl. 260—465.8.
Krøyer, Karl K. K. 2,716,792, Cl. 22—209.
Kubota, Hidetsugu. 2,717,044, Cl. 170—165.
Kulicke, Frederick W., Jr., to Proctor Electric Co. 2,716,825, Cl. 38—79.
Kung, Frederick E.: See—
Neubauer, Joseph A., Strain, and Kung. 2,717,238.
Kurtz, Abraham N.: See—
Davis, Hubert G., Wilson, and Kurtz. 2,717,260.
Kuss, Ferdinand. 2,717,188, Cl. 312—323.
Kuyper, Peter H., to Rolscreen Co. 2,716,784, Cl. 20—56.4.
Labuza, Walter. 2,717,144, Cl. 254—16.
Lado, Anthony L., to Pettibone New York Corp. 2,716,899, Cl. 74—385.
Laird, John A., to Joseph Lucas (Industries) Ltd. 2,717,337, Cl. 315—243.
Land, Harry A.: See—
Bechman, William O., Land, Baker, and Henning. 2,716,907.
Lang, Heinrich, to Durex, S. A. 2,716,969, Cl. 123—41.16.
Langford, John A., to The Fluor Corp., Ltd. 2,717,049, Cl. 183—2.
Langhein, Walter H. D., to I. F. F. E. Pauls. 2,716,953, Cl. 112—158.
Launder, Ernie L., and C. C. Hosmer. 2,716,822, Cl. 37—142.
Lederer, Albrecht M.: See—
Hensgen, Bernard T., Vanden Bosch, Lederer, and Wood. 2,717,212.
Lee, William H. 2,717,349, Cl. 318—237.
Leerkamp, Herbert H., to J. A. Rockwood and H. H. Leerkamp. 2,716,913, Cl. 82—12.
Lees, Gerald M., to Akshun Mfg. Co. 2,716,869, Cl. 62—107.
Leighton, Charles R. and N. D. 2,717,076, Cl. 209—138.
Leighton, Charles R. and N. D. 2,717,077, Cl. 209—139.
Leighton, Noel D.: See—
Leighton, Charles R. and N. D. 2,717,076.
Leighton, Charles R. and N. D. 2,717,077.

Lemieux, Raymond U., to National Research Council. 2,717,266, Cl. 260—530.
Leonard, Merrill G.: *See*—
Hayden, Vernon H., and Leonard. 2,717,275.
Lerner, George. 2,716,814, Cl. 31—22.
Lesjak, Babette V.: *See*—
Lesjak, Michael. 2,716,931.
Lesjak, Michael, deceased; B. V. Lesjak, administratrix. 2,716,931, Cl. 95—90.5.
Letterman, Charles B., to Underwood Corp. 2,717,070, Cl. 197—114.
Levi, Cass B., to The McNally-Pittsburg Mfg. Corp. 2,717,078, Cl. 209—172.5.
Levi, Cass B., to The McNally Pittsburg Mfg. Corp. 2,717,079, Cl. 209—172.5.
Lewis, Dean M.: *See*—
Beldt, Lauren F., and Lewis. 2,716,896.
Lewyt Corp.: *See*—
Meyerhoefer, Carl E. 2,716,773.
Ley, Gordon S.: *See*—
Jones, Charles H., and Ley. 2,717,329.
Liebert, Wilhelmus A. J. 2,717,046, Cl. 181—31.
Light House, Inc., The: *See*—
Hajecate, Thomas H. 2,716,758.
Lindbloom, Franz. 2,717,169, Cl. 292—144.
Link-Belt Co.: *See*—
Baumheckel, Ralph E. 2,717,185.
Lisk-Savory Corp.: *See*—
Wilson, Ashley F., and Clement. 2,717,000.
Livingston, Harry F. 2,717,141, Cl. 248—278.
Loewe, Siegmund. 2,716,962, Cl. 116—124.4.
Loftin, John A. 2,716,841, Cl. 46—191.
London, Sol, to General Electric Co. 2,717,348, Cl. 318—221.
Louden, Victor J., to General Electric Co. 2,717,355, Cl. 323—64.
Lovell, Clarence A., to Bell Telephone Laboratories, Inc. 2,717,280, Cl. 179—90.
Loyles, Rudolph O., and J. W. Walker. 2,716,977, Cl. 126—360.
Lubkin, Samuel, to Underwood Corp. 2,717,375, Cl. 340—250.
Lucas, Joseph, (Industries) Ltd.: *See*—
Laird, John A. 2,717,337.
Lucien, Rene, to Societe d'Inventions Aeronautiques et Mecaniques S. I. A. M. 2,717,002, Cl. 137—620.
Luhn, Hans P., to International Business Machines Corp. 2,717,121, Cl. 235—145.
Lundberg, Roland O. 2,716,908, Cl. 76—40.
Lundgren, Harold P.: *See*—
Rose, William G., and Lundgren. 2,717,194.
Lutz, Irvin H., O. W. Collier, and D. J. Bellman, to The American Oil Co. 2,717,231, Cl. 196—50.
Lyon, George A. 2,717,059, Cl. 188—264.
Madey, Richard, to the United States of America as represented by the United States Atomic Energy Commission. 2,717,316, Cl. 250—207.
Maitland, Charles L. 2,716,912, Cl. 81—86.
Major Leather Goods Mfg. Co.: *See*—
Wolf, Harold H. 2,716,985.
Malick, Franklin S., to the United States of America as represented by the Secretary of the Navy. 2,717,066, Cl. 192—84.
Mallory, P. R., & Co., Inc.: *See*—
Puerner, George O. 2,717,367.
Malo, John H., to Barber Machinery Ltd. 2,717,090, Cl. 214—131.
Malone, Carl E., 50% to H. M. Sutton. 2,716,809, Cl. 30—30.
Maloney-Crawford Tank and Mfg. Co.: *See*—
Wilson, Samuel A. 2,717,081.
Mamere, Joseph P.: *See*—
Merlin, Alfred G., and Mamere. 2,716,759.
Manheim, Theodore B., and J. L. Castelli. 2,716,961, Cl. 116—124.
Marden, John W.: *See*—
Nagy, Rudolph, and Marden. 2,717,234.
Marotta, Ralph, and C. R. Martinson, to Monsanto Chemical Co. 2,717,214, Cl. 106—228.
Marson, Samuel. 2,716,930, Cl. 95—64.
Martin, Paul H. 2,717,163, Cl. 280—477.
Martin, Philip W. 2,716,890, Cl. 73—151.
Martinson, Carl R.: *See*—
Marotta, Ralph, and Martinson. 2,717,214.
Marvin, John T., to General Motors Corp. 2,717,295, Cl. 200—159.
Masland, C. H., & Sons: *See*—
Hoeselbarth, Frank W. E. 2,717,005.
Mather, Edward L., and R. J. Cheetham, to British Insulated Callender's Cables Ltd. 2,717,357, Cl. 323—77.
Matlack, Richard C., and F. W. Metzger, to Bell Telephone Laboratories, Inc. 2,717,279, Cl. 179—17.
Matthews, Russell B.: *See*—
Hilgert, Adolph J., and Matthews. 2,717,123.
Matthews, Russell B., to Milwaukee Gas Specialty Co. 2,717,381, Cl. 236—21.
Mautner, Steven E., to Skydyne, Inc. 2,717,093, Cl. 217—56.
McCann, Kelly F., to Warner Lewis Co. 2,717,082, Cl. 210—169.
McCanner, Arthur T., Jr.: *See*—
Betts, George E., Jr., Blekfeld, and McCanner. 2,716,959.
McCargar, George L. 2,717,143, Cl. 249—19.
McCarty, Orin P.: *See*—
Foley, Robert T., and McCarty. 2,717,296.
McConnell, Albert L., and R. W. Medeiros, to Scott Paper Co. 2,717,224, Cl. 154—117.
McCumber, Forest H. 2,716,960, Cl. 115—41.
McDonald, Seymour F. 2,717,129, Cl. 242—99.
McGay, John B., and G. B. Clift, to Rockwell Mfg. Co. 2,716,860, Cl. 60—7.
McKinney, Leonard L., E. A. Setzkorn, and E. H. Uhing, to the United States of America as represented by the Secretary of Agriculture. 2,717,263, Cl. 260—471.
McMichael, Charles E., J. W. Godbey, and V. L. Zehnder, to National Cylinder Gas Co. 2,717,256, Cl. 260—428.
McNally-Pittsburg Mfg. Corp., The: *See*—
Levi, Cass B. 2,717,078.
Levi, Cass B. 2,717,079.
Meagher, Ralph E., and C. W. Sherwin, to the United States of America as represented by the Secretary of the Navy. 2,717,330, Cl. 315—24.
Meara, William F., to Bloomfield Industries, Inc. 2,717,306, Cl. 240—13.
Mechanical Handling Systems Inc.: *See*—
Klamp, Paul. 2,716,965.
Medeiros, Robert W.: *See*—
McConnell, Albert L., and Medeiros. 2,717,224.
Melvin, Francis F. 2,716,857, Cl. 56—400.17.
Merlin, Alfred G., and J. P. Mamere, to The National Screw & Mfg. Co. 2,716,759, Cl. 10—26.
Mershon, Clarence L.: *See*—
Keith, Robert H., and Mershon. 2,717,318.
Metzger, Frederick W.: *See*—
Matlack, Richard C., and Metzger. 2,717,279.
Meyer, Everet E. 2,717,164, Cl. 280—491.
Meyerhoefer, Carl E., to Lewyt Corp. 2,716,773, Cl. 15—369.
Meyers, George F. 2,716,844, Cl. 51—34.
Michael Research Co., Inc.: *See*—
Craddock, Charles L. 2,717,336.
Mifflin, Leo V.: *See*—
Hodgin, David M., and Mifflin. 2,717,364.
Milano, Cesare A. 2,716,937, Cl. 99—306.
Miles, John R., to M. S. Wolk. 2,716,918, Cl. 88—1.
Miller, Robert E., to Monsanto Chemical Co. 2,717,254, Cl. 260—332.3.
Milwaukee Gas Specialty Co.: *See*—
Hilgert, Adolph J., and Matthews. 2,717,123.
Matthews, Russell B. 2,717,381.
Thornbery, John H. 2,717,290.
Miner, Carl S.: *See*—
Cranston, Hoy A. 2,717,211.
Minneapolis-Honeywell Regulator Co.: *See*—
Wild, Rudolf F. 2,717,359.
Minnesota and Ontario Paper Co.: *See*—
Stevens, Gilbert. 2,717,213.
Minnie, Raymond J., III. 2,716,832, Cl. 43—43.12
Mixter, Leona M. 2,716,827, Cl. 41—12.
Moletz, George J., Jr., and B. Pollack. 2,717,107, Cl. 222—334.
Molyneux, John C. 2,717,016, Cl. 150—40.
Monfils, Napoleon A.: *See*—
Quinn, Edward, and Monfils. 2,716,765.
Monsanto Chemical Co.: *See*—
Contois, Leo L., Jr. 2,717,247.
Jones, Otha C., and Arvan. 2,717,198.
Marotta, Ralph, and Martinson. 2,717,214.
Miller, Robert E. 2,717,254.
Moore, John A. 2,716,786, Cl. 20—62.
Moore, George A. 2,717,094, Cl. 220—4.
Moore, Robert J.: *See*—
Wachter, Aaron, and Moore. 2,717,196.
More, Agnes M. 2,716,981, Cl. 128—163.
Morgan, Erving B., B. S. Jurewicz, and A. J. Plachecki, to American Seating Co. 2,717,187, Cl. 311—106.
Morley, Herbert F. 2,717,088, Cl. 214—16.1.
Morris, Joseph L. 2,716,878, Cl. 66—120.
Morris Paper Mills: *See*—
Arneson, Edwin L. 2,717,097.
Arneson, Edwin L. 2,717,098.
Moss, Charles W. 2,717,008, Cl. 139—336.
Mullin, Henry A. 2,716,823, Cl. 37—145.
Munster, Allen C., to Philco Corp. 2,717,358, Cl. 324—68.
Murcek, Slavo J.: *See*—
Christian, Carl A., and Murcek. 2,717,351.
Murphy, Frank: *See*—
Fielden, Fred, and Murphy. 2,717,124.
Murray, Cecil W., to George Fletcher & Co. Ltd. 2,717,142, Cl. 249—3.
Murray, Maurice J., V. Haensel, and H. W. Grote, to Universal Oil Products Co. 2,717,230, Cl. 196—50.
Nagy, Rudolph, and J. W. Marden, to the United States of America as represented by the United States Atomic Energy Commission. 2,717,234, Cl. 204—10.
Nallinger, Friedrich K. H. to Daimler-Benz Aktiengesellschaft. 2,717,045, Cl. 180—1.
Naramore, Harold B., to Bridgeport Fabrics, Inc. 2,716,788, Cl. 20—69.
National Cash Register Co., The: *See*—
Desch, Joseph R. 2,717,334.
National Cylinder Gas Co.: *See*—
Bailey, Alton E. 2,717,202.
McMichael, Charles E., Godbey, and Zehnder. 2,717,256.
National Distillers Products Corp.: *See*—
Ely, Charles M. 2,717,208.
Ely, Charles M., and Schott. 2,717,209.
National Folding Box Co., Inc.: *See*—
Williamson, Marshall I., and Carruth. 2,717,074.
National Marking Machine Co., The: *See*—
Sutton, Gerald J. 2,716,748.
National Research Council: *See*—
Lemieux, Raymond U. 2,717,266.
National Research Development Corp.: *See*—
Rowe, David S. 2,716,886.
National Screw & Mfg. Co., The: *See*—
Merlin, Alfred G., and Mamere. 2,716,759.
National Steel Corp.: *See*—
Collins, Tappan. 2,717,060.

Navy, United States of America as represented by the Secretary of the: *See*—
Betts, George E., Jr., Blekfeld, and McCanner. 2,716,959.
Brooks, Frederick E., Jr. 2,717,380.
Byers, Robert C. 2,716,873.
Geyger, Wilhelm A. 2,717,346.
Kent, Raymond C., Jr., and Daudelin. 2,716,957.
Malick, Franklin S. 2,717,066.
Meagher, Ralph E., and Sherwin. 2,717,330.
Nichols, Nathaniel B., and Rider. 2,716,894.
Ribner, Morris. 2,717,352.
Nelson, Erdick H. 2,717,065, Cl. 190—57.
Nelson, George E. 2,717,156, Cl. 273—134.
Nemet, Anthony A., and M. Berindei, to Hartford National Bank and Trust Co., trustee. 2,717,315, Cl. 250—95.
Neubauer, Joseph A., F. Strain, and F. E. Kung, to Columbia-Southern Chemical Corp. 2,717,238, Cl. 204—163.
Newton, Catherine H.: *See*—
Brandon, Clarence W. 2,716,958.
Newton, George C.: *See*—
Holman, Herbert E., and Newton. 2,716,921.
Nichols Engineering Co.: *See*—
Converse, John O. 2,716,949.
Nichols, Nathaniel B., and C. H. Rider, to the United States of America as represented by the Secretary of the Navy. 2,716,894, Cl. 74—5.41.
Nielson, Moritz H., to Illinois Tool Works. 2,716,761, Cl. 10—155.
Noddin, George A., and C. P. Spaeth, to E. I. du Pont de Nemours & Co. 2,717,204, Cl. 52—2.
Noonan, Richard E., to Aro Equipment Corp. 2,716,812, Cl. 30—94.
Nordmark, Walter E.: *See*—
Hoven, Alfred C., and Nordmark. 2,717,026.
North Electric Mfg. Co., The: *See*—
Blackhall, James M., Baker, and Dahlman. 2,717,278.
North, Harper Q., to General Electric Co. 2,717,341, Cl. 317—235.
Office National d'Etudes et de Recherches Aeronautiques (O. N. E. R. A.): *See*—
Reingold, Lucien, and Foure. 2,716,863.
Ogletree, William A., to Philco Corp. 2,717,311, Cl. 250—27.
Oilgear Co., The: *See*—
Ferris, Walter. 2,716,944.
Oldberg Mfg. Co.: *See*—
Deremer, Floyd E. 2,717,048.
Olin Mathieson Chemical Corp.: *See*—
Dolliver, Morris A., and Semenoff. 2,717,236.
Olson, Verland A.: *See*—
Tasker, Homer G., and Olson. 2,717,377.
Olsson, Paul: *See*—
Frink, Russell E., and Olsson. 2,717,292.
Orlowsky, Emilie L.: *See*—
Sihler, Immanuel, and Biersack. 2,717,287.
Orr, James V. 2,717,161, Cl. 280—36.
Osrow, Harold: *See*—
Osrow, Leonard, and H. Osrow. 2,717,176.
Osrow, Leonard, and H. Osrow. 2,717,176, Cl. 299—84.
Overbeke, John W.: *See*—
Badger, Everett H., Jr., Dahl, and Overbeke. 2,716,999.
Owens, Edward O., and L. M. Ezzell. 2,716,885, Cl. 73—48.
Page, John S., to Page Oil Tools Inc. 2,717,004, Cl. 137—637.
Page Oil Tools, Inc.: *See*—
Page, John S. 2,717,004.
Paillard S. A.: *See*—
Heiniger, Wilfred. 2,717,055.
Paish, Harold P. S., to Henry Sykes Ltd. 2,717,040, Cl. 166—157.
Palladino, Steve V.: *See*—
Carpenter, Ralph H., and Palladino. 2,717,376.
Pan, Wen Y., to Radio Corporation of America. 2,717,313, Cl. 250—36.
Parham, Charles L., Jr. 2,717,114, Cl. 229—14.
Parke, Davis & Co.: *See*—
Rebstock, Mildred C., and Pfeiffer. 2,717,268.
Parker Appliance Co., The: *See*—
Badger, Everett H., Jr., and Dahl, and Overbeke. 2,716,999.
Parks & Woolson Machine Co.: *See*—
Hadley, Wilfred N. 2,716,797.
Parry, Frank, to The Singer Mfg. Co. 2,716,956, Cl. 112—230.
Parsons, Floyd B.: *See*—
Grant, Harry C., Jr., and Parsons. 2,717,042.
Paschal, Arthur E. 2,717,068, Cl. 193—2.
Pasquini, Vittorio, and V. Guerrini, to Vittorio Necchi S. p. A. 2,716,952, Cl. 112—2.
Passino, Herbert J., to The M. W. Kellogg Co. 2,717,201, Cl. 23—190.
Patterson, Cecil R. 2,717,168, Cl. 292—70.
Pauls, Ingolf F. F. E.: *See*—
Langhein, Walter H. D. 2,716,953.
Paulsen, Harold A. 2,716,782, Cl. 20—1.
Peirson, Walter R.: *See*—
Sorensen, Knud B., and Peirson. 2,716,855.
Percival, Isabel C., and J. E. Tierney. 2,717,170, Cl. 292—325.
Perkins, F., Ltd.: *See*—
Scott, Sydney W. 2,717,317.
Perlmutter, Samuel. 2,716,793, Cl. 24—84.
Perrault, Ainslie. 2,717,001, Cl. 137—514.
Person, Lars B., and C. H. Von Sivers, R. K. Wadö, and K. R. Wickmam, to Telefonaktiebolaget L. M. Ericsson. 2,717,277, Cl. 179—15.6.
Petersen, Alfred A., to Prat Daniel Corp. 2,717,054, Cl. 183—81.
Peterson, Aldon M. 2,716,833, Cl. 43—44.88.
Peterson, Russell I., Jr.: *See*—
Viall, Richmond, Jr., Peterson, and Johnson. 2,716,845.
Pettibone New York Corp.: *See*—
Lado, Anthony L. 2,716,899.
Pfann, William G., to Bell Telephone Laboratories, Inc. 2,717,342, Cl. 317—235.
Pfeiffer, Elizabeth L.: *See*—
Rebstock, Mildred C., and Pfeiffer. 2,717,268.
Philco Corp.: *See*—
Munster, Allen C. 2,717,358.
Ogletree, William A. 2,717,311.
Phillips Petroleum Co.: *See*—
Findlay, Robert A. 2,717,229.
Phinn, George A. 2,717,011, Cl. 144—93.
Piana, Angelo. 2,716,916, Cl. 84—360.
Pigott, Reginald J. S., to Gulf Research & Development Co. 2,716,914, Cl. 82—34.
Pinatelli, Thomas. 2,716,976, Cl. 126—116.
Pipelife, Inc.: *See*—
Crutis, Arvel C. 2,717,038.
Piper, Charles A., to Bendix Aviation Corp. 2,717,370, Cl. 340—151.
Pipes, George R., and H. J. Findley, to Eaton Mfg. Co. 2,717,179, Cl. 301—37.
Plachecki, Alexander J.: *See*—
Morgan, Erving B., Jurewicz, and Plachecki. 2,171,187.
Pollack, Bennie: *See*—
Moletz, George J., Jr., and Pollack. 2,717,107.
Poupitch, Ougljesa J., to Illinois Tool Works. 2,716,760, Cl. 10—155.
Pouret, Pierre. 2,716,979, Cl. 128—2.1.
Prat Daniel Corp.: *See*—
Petersen, Alfred A. 2,717,054.
Presnell, Frank G., to Bendix Aviation Corp. 2,716,945, Cl. 103—4.
Press, Meyer, to Hazeltine Research, Inc. 2,717,362, Cl. 333—82.
Price, George W., to Collins Radio Co. 2,717,360, Cl. 333—6.
Prober, Maurice, to General Electric Co. 2,717,235, Cl. 204—59.
Proctor Electric Co.: *See*—
Kulicke, Frederick W., Jr. 2,716,825.
Puerner, George O., to P. R. Mallory & Co., Inc. 2,717,367, Cl. 339—185.
Pullman-Standard Car Mfg. Co.: *See*—
Test, Ellis W. 2,716,872.
Puritan Cordage Mills, Inc.: *See*—
Faulkner, Clarence F. 2,717,215.
Quercetti, Alessandro. 2,716,838, Cl. 46—105.
Quinn, Edward, and N. A. Monfils, to United Shoe Machinery Corp. 2,716,765, Cl. 12—1.
Rabe, Karl, to Dr. Ing. H. C. F. Porsche K.-G. 2,717,173, Cl. 296—28.
Radio Corporation of America: *See*—
Ballard, David C. 2,717,322.
Clay, Burton R. 2,717,323.
Pan, Wen Y. 2,717,313.
Schroeder, Alfred C. 2,717,276.
Ralston, John D. 2,717,112, Cl. 226—116.
Ramsel, Charles A., to Caterpillar Tractor Co. 2,717,067, Cl. 192—93.
Ratte, Heinrich: *See*—
Geller, Julius, and Ratte. 2,717,232.
Rayburn, Vincent A., to Western Electric Co., Inc. 2,717,125, Cl. 242—25.
Rayburn, Vincent A., to Western Electric Co., Inc. 2,717,127, Cl. 242—45.
Rayburn, Vincent A., to Western Electric Co., Inc. 2,717,285, Cl. 191—1.
Raytheon Mfg. Co.: *See*—
Wimberly, Floyd T. 2,717,378.
Rebstock, Mildred C., and E. L. Pfeiffer, to Parke, Davis & Co. 2,717,268, Cl. 260—562.
Reed, Frank J., to The International Electronics Co. 2,717,282, Cl. 179—100.2.
Reed, Macdonald S. 2,716,805, Cl. 29—548.
Reed, Macdonald S. 2,716,806, Cl. 29—548.
Reingold, Lucien, and C. Fouré, to Office National d'Etudes et de Recherches Aeronautiques (O. N. E. R. A.). 2,716,863, Cl. 60—39.28.
Reitzes, Charles. 2,716,850, Cl. 51—211.
Reliance Electric and Engineering Co.: *See*—
Shoulders, Walter L., and Hayes. 2,717,320.
Reliance Electric & Engineering Co., The: *See*—
Compton, William H., Duff, and Smith. 2,716,859.
Rempel, Erika G.: *See*—
Rempel, Nikolaus. 2,717,237.
Rempel, Nikolaus, deceased; E. G. Rempel, admin., to Farbenfabriken Bayer Aktiengesellschaft. 2,717,237, Cl. 204—101.
Research Corp.: *See*—
Andres, Stanley G. 2,717,051.
Wintermute, Harry A. 2,717,053.
Reynolds, William G. 2,716,779, Cl. 19—131.
Ribner, Morris, to the United States of America as represented by the Secretary of the Navy. 2,717,352, Cl. 322—36.
Rider, Clinton H.: *See*—
Nichols, Nathaniel B., and Rider. 2,716,894.
Ridge Tool Co., The: *See*—
Janik, Anton J. 2,716,947.
Rives, Halcolm D., to Swingspout Measure Co. 2,717,102, Cl. 222—89.
Robertson, Elvin W. 2,716,752, Cl. 1—260.
Robinson, Thomas, & Son Ltd.: *See*—
Fielden, Fred, and Murphy. 2,717,124.
Rockwell Mfg. Co.: *See*—
McGay, John B., and Clift. 2,716,860.
Rockwood, John A.: *See*—
Leerkamp, Herbert H. 2,716,913.
Roehm, Erwin G., to The Cincinnati Milling Machine Co. 2,716,925, Cl. 90—13.5.
Rolatape, Inc.: *See*—
Staples, James D. 2,716,819.
Rolscreen Co.: *See*—
Kuyper, Peter H. 2,716,784.

Svenska Skofabrikantforeningen : *See*—
Rylander, Uno B. 2,717,021.
Svenson, Ernest J. 2,716,888, Cl. 73—116.
Swan, Merrill, to United Geophysical Corp. 2,717,368, Cl. 340—15.
Swanson, Kenneth P., to Textile Engineering Corp. 2,716,780, Cl. 19—142.
Swercewski, Joseph S.: *See*—
De Bonville, Arthur L., and Swercewski. 2,716,834.
Swift & Co.: *See*—
Hensgen, Bernard T., Vanden Bosch, Lederer, and Wood. 2,717,212.
Swingspout Measure Co.: *See*—
Rives, Halcolm D. 2,717,102.
Sykes, Allen H. 2,716,971, Cl. 123—46.
Sykes, Henry, Ltd.: *See*—
Paish, Harold P. S. 2,717,040.
Sylvan, Joseph. 2,717,063, Cl. 189—46.
Sylvania Electric Products Inc.: *See*—
Gartner, Stanley J., and Zwald. 2,717,092.
Tackaberry, Middleton J. 2,716,964, Cl. 120—89.
Tait-Clark-Streich Machiney Corp.: *See*—
Streich, Philip A., Clark, and Tait. 2,716,776.
Tait, Emmitte P.: *See*—
Streich, Philip A., Clark, and Tait. 2,716,776.
Talmey, Paul, and W. J. Crehan, to General American Transportation Corp. 2,717,218, Cl. 117—97.
Tasker, Homer G., and V. A. Olson, to Gilfillan Bros., Inc. 2,717,377, Cl. 343—7.7.
Taylor, Frank H., to International Standard Electric Corporation. 2,717,312, Cl. 250—33.
Tea, Clark A. 2,716,883, Cl. 73—11.
Teague, Walter D., and S. D. Wassyng, to Servel, Inc. 2,717,189, Cl. 312—351.
Telefonaktiebolaget L. M. Ericsson: *See*—
Person, Lars B., Von Sivers, Wadö, and Wickman. 2,717,277.
Teleregister Corp., The: *See*—
Bush, George L. 2,717,086.
Temco, Inc.: *See*—
Bourner, Howard L. 2,716,820.
Terrill, John R. 2,716,881, Cl. 70—14.
Test, Ellis W., to Pullman-Standard Car Mfg. Co. 2,716,872, Cl. 62—171.
Textile Engineering Corp.: *See*—
Swanson, Kenneth P. 2,716,780.
Thatcher, Ralph H. 2,717,027, Cl. 155—115.
Thomas, Clifford A., ½ to R. Diamond. 2,717,159, Cl. 280—12.
Thompson Products, Inc.: *See*—
Anderson, David E. 2,717,149.
Thornbery, John H., to Milwaukee Gas Specialty Co. 2,717,290, Cl. 200—140.
Tierney, Joseph E.: *See*—
Percival, Isabel C., and Tierney. 2,717,170.
Timmerbeil, Ewald R. 2,716,749, Cl. 1—49.
Timson, Ernest A., and C. Hillingdon. 2,716,942, Cl. 101—144.
Titus, Charles H., and R. E. Bednarek, to General Electric Co. 2,717,293, Cl. 200—150.
Torricelli Creations, Inc.: *See*—
Torricelli, Ugo. 2,716,978.
Torricelli, Ugo. 2,716,994.
Torricelli, Ugo, to Torricelli Creations, Inc. 2,716,978, Cl. 128—2.
Torricelli, Ugo, to Torricelli Creations, Inc. 2,716,994, Cl. 135—33.
Touraton, Emile, C. Dumousseau, and R. Zwobada, to International Standard Electric Corp. 2,717,327, Cl. 315—6.
Toy, Arthur D. F., and J. R. Costello, Jr., to Victor Chemical Works. 2,717,249, Cl. 260—247.5.
Transandean Associates, Inc.: *See*—
Case, Rogers. 2,717,274.
Tranter Mfg. Inc.: *See*—
Greer, Carl S., Jr. 2,716,802.
Troppe, Frederick J. 2,716,917, Cl. 84—400.
Trotter, Percy W., to Ethyl Corp. 2,717,233, Cl. 202—57.
Trotter, Percy W., to Ethyl Corp. 2,717,272, Cl. 260—648.
Tucker, Jean S.: *See*—
Schwartz, Harold H., and Tucker. 2,716,768.
Turner, Wesley. 2,716,771, Cl. 15—136.
Tyne, George H. 2,717,300, Cl. 219—1.
Uhing, Eugene H.: *See*—
McKinney, Leonard L., Setzkorn, and Uhing. 2,717,263.
Ulfves, Carl A. 2,716,851, Cl. 51—232.
Underwood Corp.: *See*—
Letterman, Charles B. 2,717,070.
Lubkin, Samuel. 2,717,375.
Union Carbide and Carbon Corp.: *See*—
Davis, Hubert G., and Wilson. 2,717,259.
Davis, Hubert G., Wilson, and Kurtz. 2,717,260.
Union Oil Co. of California: *See*—
Fischer, Paul W., and Scheffel. 2,717,239.
United Geophysical Corp.: *See*—
Swan, Merrill. 2,717,368.
United Shoe Machinery Corp.: *See*—
Eastman, Fred C. 2,716,764.
Gulbrandsen, Helge. 2,716,766.
Quinn, Edward, and Monfils. 2,716,765.
Senfleben, Paul W. 2,716,763.
United States Atomic Energy Commission, United States of America as represented by the: *See*—
Madey, Richard. 2,717,316.
Nagy, Rudolph, and Marden. 2,717,234.
Sewell, Curtis, Jr., and Button. 2,717,353.
Vandenberg, Leonard B. 2,716,943.
United States Steel Corp.: *See*—
Hattman, Frederick A. 2,716,941.
Shiflet, Harry A. 2,716,967.
Universal Oil Products Co.: *See*—
Bloch, Herman S., and Hervert. 2,717,243.
Murray, Maurice J., Haensel, and Grote. 2,717,230.
Usines Gallus Societe Anonyme: *See*—
Ferber, Robert. 2,717,134.
Valvo, Joseph F., to Apra Precipitator Corp. 2,717,052, Cl. 183—7.
Vandenberg, Leonard B., to the United States of America as represented by the United States Atomic Energy Commission. 2,716,943, Cl. 103—1.
Van den Boomgard, Jan: *See*—
Zalm, Pieter, and van den Boomgaard. 2,717,244.
Vanden Bosch, Johan C.: *See*—
Hensgen, Bernard T., Vanden Bosch, Lederer, and Wood. 2,717,212.
Van Handel, Ambrose B. 2,717,101, Cl. 222—80.
Van Laar, Jan A. W., to Hartford National Bank and Trust Co., trustee. 2,717,203, Cl. 41—42.
Van Zwalenburg, Benjamin R. 2,717,013, Cl. 144—253.
Vaughan, William E., and F. E. Condo, to Shell Development Co. 2,717,248, Cl. 260—92.8.
Ver Haigh, Frank W. 2,716,951, Cl. 110—18.
Viall, Richmond, Jr., R. I. Peterson, Jr., and R. H. Johnson, to The Graham Mfg. Co., Inc. 2,716,845, Cl. 51—48.
Victor Chemical Works: *See*—
Toy, Arthur D. F., and Costello. 2,717,249.
Villemure, Joseph. 2,717,028, Cl. 155—124.
Vines, Oscar L., to Alford Cartons. 2,717,115, Cl. 229—28.
Vines, Oscar L., to Alford Cartons. 2,717,116, Cl. 229—28.
Vittorio Necchi S. p. A.: *See*—
Pasquini, Vittorio, and Guerrini. 2,716,952.
Vold, Earl D.: *See*—
Gill, William M., and Vold. 2,716,882.
Von Sivers, Carl H.: *See*—
Person, Lars B., Von Sivers, Wadö, and Wickman. 2,717,277.
Wachter, Aaron, and R. J. Moore, to Shell Development Co. 2,717,196, Cl. 21—2.5.
Waddington, William H., to American Machine and Foundry Co. 2,717,085, Cl. 211—74.
Wadö, Reid K.: *See*—
Person, Lars B., Von Sivers, Wadö, and Wickman. 2,717,277.
Wagner, Donald D. 2,717,018, Cl. 150—52.
Walker, John W.: *See*—
Loyles, Rudolph O., and Walker. 2,716,977.
Walker, Neville E. 2,717,297, Cl. 200—166.
Wallace, De Loss D.: *See*—
Baugh, Everett L., and Wallace. 2,716,995.
Walsh, Bernard A. 2,717,109, Cl. 224—5.
Walter, Hellmuth, to Worthington Corp. 2,717,118, Cl. 230—116.
Walters, Albert F. 2,717,162, Cl. 280—150.
Walworth Co.: *See*—
Davis, Ralph P. 2,716,789.
Warner Lewis Co.: *See*—
McCann, Kelly F. 2,717,082.
Wassyng, Seymour D.: *See*—
Teague, Walter D., and Wassyng. 2,717,189.
Waters, Albert V., to C. A. V. Ltd. 2,717,340, Cl. 317—188.
Weidmann, Ernst: *See*—
Farny, Paul, and Weidmann. 2,716,972.
Weitz, Otto, to Firma A. J. Tröster. 2,717,105, Cl. 222—283.
Wernli, Max, to Georg Fischer Aktiengesellschaft. 2,716,905, Cl. 74—665.
Wesson Multicut Co.: *See*—
Bader, William. 2,716,799.
Bader, William. 2,716,800.
Western Electric Co., Inc.: *See*—
Rayburn, Vincent A. 2,717,125.
Rayburn, Vincent A. 2,717,127.
Rayburn, Vincent A. 2,717,285.
Westinghouse Electric Corp.: *See*—
Biehn, Gerald L. 2,716,868.
Biehn, Gerald L. 2,716,870.
Binstock, Martin H., and Steinherz. 2,717,223.
Brown, Thomas F. 2,717,339.
Christian, Carl A., and Murcek. 2,717,351.
Ford, James G., and Denault. 2,717,219.
Frink, Russell E., and Olsson. 2,717,292.
Hayden, Vernon H., and Leonard. 2,717,275.
Jones, Charles H., and Ley. 2,717,329.
Keith, Robert H., and Mershon. 2,717,318.
Santini, Danilo, and Suozzo. 2,717,056.
Westlake, Harry E., Jr., and W. B. Hardy, to American Cyanamid Co. 2,717,255, Cl. 260—381.
Wheatley, Max D.: *See*—
Dawson, Helen L., and Wheatley. 2,717,227.
Wheland Co., The: *See*—
Campbell, George E. 2,717,186.
White, Albert J.: *See*—
Jay, James B., White, and Hunt. 2,717,003.
White, Gifford E., to Sperry Rand Corp. 2,717,374, Cl. 340—212.
Whitehead, Frederick W., to The Bristol Aeroplane Co. Ltd. 2,716,924, Cl. 90—13.
Wickardt, Kurt W., to Hosemaster Machine Co. Ltd. 2,716,877, Cl. 66—96.
Wickman, Klas R.: *See*—
Person, Lars B., Von Sivers, Wadö, and Wickman. 2,717,277.
Wild, Rudolf F., to Minneapolis-Honeywell Regulator Co. 2,717,359, Cl. 324—98.
Wiley, Evans C. 2,716,874, Cl. 64—23.
Wilkinson, Lester: *See*—
Hoppes, Lloyd G. 2,717,104.
Williams, Albert E. 2,717,225, Cl. 154—128.

Williams, Roger W. : *See*—
Bardeen, Thomas, and Williams. 2,717,369.
Williamson, Marshall I., and H. A. Carruth, to National Folding Box Co., Inc. 2,717,074, Cl. 206—58.
Wilson, Ashley F., and J. J. Clement, to Lisk-Savory Corp. 2,717,000, Cl. 137—343.
Wilson, Fred E. : *See*—
Wray, George O., and Wilson. 2,716,974.
Wilson-Jones Co. : *See*—
Rundblad, George J., and Cairelli. 2,717,099.
Wilson, Samuel A., to Maloney-Crawford Tank and Mfg. Co. 2,717,081, Cl. 210—52.5.
Wilson, Thomas P. : *See*—
Davis, Hubert G., and Wilson. 2,717,259.
Davis, Hubert G., Wilson, and Kurtz. 2,717,260.
Wimberly, Floyd T., to Raytheon Mfg. Co. 2,717,378, Cl. 343—14.
Windischman, Edward F., W. L. Hartop, Jr., and G. R. Ryan, to Abbott Laboratories. 2,716,983, Cl. 128—221.
Winkelsträter, Fritz, to M. A. Bitzer. 2,716,836, Cl. 43—146.
Wintermute, Harry A., to Research Corp. 2,717,053, Cl. 183—7.
Witt, Frank. 2,716,986, Cl. 131—88.
Wolf, Harold H., to Major Leather Goods Mfg. Co. 2,716,985, Cl. 129—1.
Wolf, Victor, Ltd. : *See*—
Rowe, Richard. 2,717,271.
Wolk, Michael S. : *See*—
Miles, John R. 2,716,918.
Wood, James R., to Harris-Seybold Co. 2,717,154, Cl. 271—12.
Wood, Peirce M. : *See*—
Hensgen, Bernard T., Vanden Bosch, Lederer, and Wood. 2,717,212.
Woodall Industries, Inc. : *See*—
Johnson, Charles B. 2,716,804.
Woodruff, Thomas E., to Hughes Aircraft Co. 2,717,310, Cl. 250—27.
Woodward, Lewis J., and R. Baskerville, to General Electric Co. 2,717,298, Cl. 200—168.
Worth, William E. 2,717,167, Cl. 292—30.
Worthington Corp. : *See*—
Walter, Hellmuth. 2,717,118.
Wray, George O., and F. E. Wilson. 2,716,974, Cl. 126—109.
Wright, Hazel H. : *See*—
Brandon, Clarence W. 2,716,958.
Wulfsberg, Arthur H., to Collins Radio Co. 2,716,897, Cl. 74—10.54.
Yarnall-Waring Co. : *See*—
Kinderman, Walter J. 2,716,892.
Young, Douglas, Inc. : *See*—
Young, Lewis D. 2,717,073.
Young, George G., to M. I. Glass. 2,716,839, Cl. 46—135.
Young, Lewis D., to Douglas Young, Inc. 2,717,073, Cl. 206—45.34.
Zalm, Pieter, and J. van den Boomgaard, to Hartford National Bank and Trust Co., trustee. 2,717,244, Cl. 252—301.6.
Zehnder, Victor L. : *See*—
McMichael, Charles E., Godbey, and Zehnder. 2,717,256.
Zelenay, Ludwig. 2,716,794, Cl. 24—113.
Ziskal, Joseph F. : *See*—
Hubert, Clarence A., and Ziskal. 2,716,966.
Zublin, John A. 2,717,146, Cl. 255—28.
Zwald, Henry J. : *See*—
Gartner, Stanley J., and Zwald. 2,717,092.
Zwobada, René : *See*—
Touraton, Emile, Dumousseau, and Zwobada. 2,717,327.

CLASSIFICATION OF PATENTS

ISSUED SEPTEMBER 6, 1955

NOTE.—First number=class, second number=subclass, third number=patent number

1— 2: 2,716,748
49: 2,716,749
49.8: 2,716,750
215: 2,716,751
260: 2,716,752
2— 195: 2,716,753
227: 2,716,754
279: 2,716,755
321: 2,716,756
4— 166: 2,716,757
8— 111: 2,717,193
112: 2,717,194
156: 2,717,195
9— 8: 2,716,758
10— 26: 2,716,759
155: 2,716,760
2,716,761
2,716,762
12— 1: 2,716,763
2,716,764
2,716,765
36: 2,716,766
15— 21: 2,716,767
119: 2,716,768
2,716,769
126: 2,716,770
136: 2,716,771
306: 2,716,772
369: 2,716,773
16— 42: 2,716,774
138: 2,716,775
153: Re.24,059
17— 2: 2,716,776
18— 19: 2,716,777
53: 2,716,778
19— 131: 2,716,779
142: 2,716,780
165: 2,716,781
20— 1: 2,716,782
55: 2,716,783
56.4: 2,716,784
57.5: 2,716,785
62: 2,716,786
69: 2,716,787
2,716,788
21— 2.5: 2,717,196
22— 10: 2,716,789
73: 2,716,790
200: 2,716,791
209: 2,716,792
23— 88: 2,717,197
106: 2,717,198
154: 2,717,199
190: 2,717,200
2,717,201
283: 2,717,202
24— 84: 2,716,793
113: 2,716,794
255: 2,716,795
25— 83: 2,716,796
26— 29: 2,716,797
28— 49: 2,717,117
29— 34: 2,716,798
96: 2,716,799
2,716,800
105: 2,716,801
157.3: 2,716,802
205: 2,716,803
453: 2,716,804
548: 2,716,805
2,716,806
30— 15.5: 2,716,807
16: 2,716,808
30: 2,716,809
47: 2,716,810
2,716,811
94: 2,716,812
241: 2,716,813
31— 22: 2,716,814
32— 32: 2,716,815
40: 2,716,816
33— 15: 2,716,817
131: 2,716,818
141: 2,716,819
34— 82: 2,716,820
37— 43: 2,716,821
142: 2,716,822
145: 2,716,823
147: 2,716,824
38— 79: 2,716,825
41— 1: 2,716,826
12: 2,716,827
13: 2,716,828
35: 2,716,829
42: 2,717,203
43—42.06: 2,716,830
42.31: 2,716,831
43.12: 2,716,832
44.88: 2,716,833
55: 2,716,834
43— 95: 2,716,835
146: 2,716,836
46— 44: 2,716,837
105: 2,716,838
135: 2,716,839
142: 2,716,840
191: 2,716,841
208: 2,716,842
47— 60: P.P.1,418
61: P.P.1,417
49— 1: 2,716,843
17: Re.24,060
51— 34: 2,716,844
48: 2,716,845
128: 2,716,846
187: 2,716,847
2,716,848
2,716,849
211: 2,716,850
232: 2,716,851
52— 2: 2,717,204
53— 148: 2,716,852
55— 84: 2,716,853
89: 2,716,854
126: 2,716,855
56— 30: 2,716,856
400.17: 2,716,857
57— 80: 2,716,858
100: 2,716,859
60— 7: 2,716,860
13: 2,716,861
39.28: 2,716,862
2,716,863
61— 16: 2,716,864
62— 4: 2,716,865
2,716,866
2,716,867
6: 2,716,868
107: 2,716,869
115: 2,716,870
146: 2,716,871
171: 2,716,872
64— 11: 2,716,873
23: 2,716,874
29: 2,716,875
66— 9: 2,716,876
96: 2,716,877
120: 2,716,878
67— 6.1: 2,716,879
7.1: 2,716,880
70— 14: 2,716,881
159: 2,716,882
73— 11: 2,716,883
12: 2,716,884
48: 2,716,885
53: 2,716,886
69: 2,716,887
116: 2,716,888
147: 2,716,889
151: 2,716,890
194: 2,716,891
293: 2,716,892
74— 5: 2,716,893
5.41: 2,716,894
7: 2,716,895
10.2: 2,716,896
10.54: 2,716,897
194: 2,716,898
385: 2,716,899
520: 2,716,900
526: 2,716,901
541: 2,716,902
557: 2,716,903
574: 2,716,904
665: 2,716,905
677: 2,716,906
720.5: 2,716,907
75— 5: 2,717,205
167: 2,717,206
76— 40: 2,716,908
107: 2,716,909
81— 3.8: 2,716,910
23: 2,716,911
86: 2,716,912
82— 12: 2,716,913
34: 2,716,914
2,716,915
84— 360: 2,716,916
400: 2,716,917
88— 1: 2,716,918
16.6: 2,716,919
2,716,920
18.4: 2,716,921
28: 2,716,922
89— 140: 2,716,923
90— 13: 2,716,924
13.5: 2,716,925
92— 26: 2,716,926
44: 2,716,927
93— 58.5: 2,716,928
95— 34: 2,716,929
64: 2,716,930
90.5: 2,716,931
97— 10: 2,716,932
26: 2,716,933
204: 2,716,934
2,716,935
99— 2: 2,717,207
4: 2,717,208
2,717,209
11: 2,717,210
54: 2,717,211
116: 2,717,212
236: 2,716,936
306: 2,716,937
408: 2,716,938
2,716,939
101— 6: 2,716,940
40: 2,716,941
144: 2,716,942
103— 1: 2,716,943
2: 2,716,944
4: 2,716,945
37: 2,716,946
138: 2,716,947
161: 2,716,948
104— 32: 2,716,949
106— 214: 2,717,213
228: 2,717,214
107— 47: 2,716,950
110— 18: 2,716,951
112— 2: 2,716,952
158: 2,716,953
176: 2,716,954
210: 2,716,955
230: 2,716,956
114— 23: 2,716,957
74: 2,716,958
241: 2,716,959
115— 41: 2,716,960
116— 124: 2,716,961
124.4: 2,716,962
117— 7: 2,717,215
75: 2,717,216
76: 2,717,217
97: 2,717,218
126: 2,717,219
2,717,220
127: 2,717,221
120— 24: 2,716,963
89: 2,716,964
121— 38: 2,716,965
40: 2,716,966
122— 499: 2,716,967
510: 2,716,968
123—41.16: 2,716,969
41.73: 2,716,970
46: 2,716,971
90: 2,716,972
124— 1: 2,716,973
126— 109: 2,716,974
110: 2,716,975
116: 2,716,976
360: 2,716,977
128— 2: 2,716,978
2.1: 2,716,979
44: 2,716,980
163: 2,716,981
214: 2,716,982
221: 2,716,983
225: 2,716,984
129— 1: 2,716,985
131— 88: 2,716,986
237: 2,716,987
242: 2,716,988
134— 66: 2,716,989
155: 2,716,990
182: 2,716,991
135— 3: 2,716,992
4: 2,716,993
33: 2,716,994
136— 168: 2,717,273
137— 87: 2,716,995
98: 2,716,996
102: 2,716,997
231: 2,716,998
235: 2,716,999
343: 2,717,000
514: 2,717,001
620: 2,717,002
625.34: 2,717,003
637: 2,717,004
139— 39: 2,717,005
151: 2,717,006
192: 2,717,007
336: 2,717,008
140— 2: 2,717,009
143— 43: 2,717,010
144— 93: 2,717,011
162: 2,717,012
144— 253: 2,717,013
148— 12: 2,717,222
122: 2,717,223
150— 4: 2,717,014
34: 2,717,015
40: 2,717,016
52: 2,717,017
2,717,018
152— 226: 2,717,019
153— 32: 2,717,020
48: 2,717,021
154— 9: 2,717,022
33.1: 2,717,023
2,717,024
2,717,025
117: 2,717,224
128: 2,717,225
155— 85: 2,717,026
115: 2,717,027
124: 2,717,028
179: 2,717,029
188: 2,717,030
158— 27.4: 2,717,031
132: 2,717,032
160— 84: 2,717,033
176: 2,717,034
178: 2,717,035
354: 2,717,036
164— 65: 2,717,037
166— 1: 2,717,038
65: 2,717,039
157: 2,717,040
186: 2,717,041
167— 46: 2,717,226
74: 2,717,227
87.1: 2,717,228
169— 11: 2,717,042
170—135.4: 2,717,043
165: 2,717,044
174— 33: 2,717,274
37: 2,717,275
178— 5.4: 2,717,276
179— 15.6: 2,717,277
16: 2,717,278
17: 2,717,279
90: 2,717,280
100.2: 2,717,281
2,717,282
100.3: 2,717,283
175.2: 2,717,284
180— 1: 2,717,045
181— 31: 2,717,046
32: 2,717,047
61: 2,717,048
183— 2: 2,717,049
4.1: 2,717,050
7: 2,717,051
2,717,052
2,717,053
81: 2,717,054
185— 39: 2,717,055
187— 29: 2,717,056
188— 5: 2,717,057
88: 2,717,058
264: 2,717,059
189— 1: 2,717,060
46: 2,717,061
2,717,062
2,717,063
2,717,064
190— 57: 2,717,065
191— 1: 2,717,285
192— 84: 2,717,066
93: 2,717,067
193— 2: 2,717,068
194— 9: 2,717,069
196—14.15: 2,717,229
50: 2,717,230
2,717,231
197— 114: 2,717,070
198— 184: 2,717,071
200— 31: 2,717,286
48: 2,717,287
97: 2,717,288
138: 2,717,289
140: 2,717,290
2,717,291
144: 2,717,292
150: 2,717,293
2,717,294
159: 2,717,295
166: 2,717,296
2,717,297
168: 2,717,298
201— 63: 2,717,299
202— 40: 2,717,232
57: 2,717,233
204— 10: 2,717,234
59: 2,717,235
75: 2,717,236
204— 101: 2,717,237
163: 2,717,238
205— 3: 2,717,072
206—45.34: 2,717,073
58: 2,717,074
82: 2,717,075
209— 138: 2,717,076
139: 2,717,077
172.5: 2,717,078
2,717,079
210— 1.5: 2,717,080
52.5: 2,717,081
169: 2,717,082
199: 2,717,083
211— 20: 2,717,084
74: 2,717,085
214— 11: 2,717,086
16.1: 2,717,087
2,717,088
83.26: 2,717,089
131: 2,717,090
132: 2,717,091
658: 2,717,092
217— 56: 2,717,093
219— 1: 2,717,300
12: 2,717,301
19: 2,717,302
20: 2,717,303
32: 2,717,304
38: 2,717,305
220— 4: 2,717,094
26: 2,717,095
40: 2,717,096
113: 2,717,097
2,717,098
221— 293: 2,717,099
222— 5: 2,717,100
80: 2,717,101
89: 2,717,102
100: 2,717,103
177: 2,717,104
283: 2,717,105
318: 2,717,106
334: 2,717,107
223— 91: 2,717,108
224— 5: 2,717,109
42.45: 2,717,110
45: 2,717,111
226— 116: 2,717,112
125: 2,717,113
229— 14: 2,717,114
28: 2,717,115
2,717,116
230— 116: 2,717,118
233— 21: 2,717,119
235— 61: 2,717,120
145: 2,717,121
236— 21: 2,717,381
44: 2,717,122
75: 2,717,123
240— 13: 2,717,306
51.11: 2,717,307
102: 2,717,308
241— 60: 2,717,124
242— 25: 2,717,125
39: 2,717,126
45: 2,717,127
72: 2,717,128
99: 2,717,129
244— 2: 2,717,130
6: 2,717,131
77: 2,717,132
148: 2,717,133
248— 11: 2,717,134
20: 2,717,135
75: 2,717,136
99: 2,717,137
183: 2,717,138
208: 2,717,139
223: 2,717,140
278: 2,717,141
249— 3: 2,717,142
19: 2,717,143
250— 17: 2,717,309
27: 2,717,310
2,717,311
33: 2,717,312
36: 2,717,313
50: 2,717,314
95: 2,717,315
207: 2,717,316
252— 8.5: 2,717,239
8.6: 2,717,240
32.7: 2,717,241
49.6: 2,717,242
138: 2,717,243
301.6: 2,717,244
254— 16: 2,717,144
98: 2,717,145
255— 28: 2,717,146
259— 178: 2,717,147
260— 31.2: 2,717,245
37: 2,717,246
78.5: 2,717,247
92.8: 2,717,248
247.5: 2,717,249
251.5: 2,717,250
293.4: 2,717,251
309.5: 2,717,252
2,717,253
332.3: 2,717,254
381: 2,717,255
428: 2,717,256
448.2: 2,717,257
2,717,258
449.6: 2,717,259
2,717,260
465.8: 2,717,261
468: 2,717,262
471: 2,717,263
488: 2,717,264
504: 2,717,265
530: 2,717,266
534: 2,717,267
562: 2,717,268
582: 2,717,269
584: 2,717,270
616: 2,717,271
648: 2,717,272
261— 16: 2,717,148
18: 2,717,149
64: 2,717,150
265— 7: 2,717,151
267— 20: 2,717,152
270— 81: 2,717,153
271— 12: 2,717,154
273— 49: 2,717,155
134: 2,717,156
135: 2,717,157
141: 2,717,158
280— 12: 2,717,159
20: 2,717,160
36: 2,717,161
150: 2,717,162
477: 2,717,163
491: 2,717,164
281— 16: 2,717,165
285— 10: 2,717,166
290— 38: 2,717,317
292— 30: 2,717,167
70: 2,717,168
144: 2,717,169
325: 2,717,170
294— 16: 2,717,171
86: 2,717,172
110: Re.24,058
296— 28: 2,717,173
299— 24: 2,717,174
83: 2,717,175
84: 2,717,176
86: 2,717,177
97: 2,717,178
301— 37: 2,717,179
302— 48: 2,717,180
59: 2,717,181
307— 51: 2,717,318
308— 9: 2,717,182
77: 2,717,183
187: 2,717,184
187.2: 2,717,185
309— 3: 2,717,186
310— 16: 2,717,319
57: 2,717,320
138: 2,717,321
311— 106: 2,717,187
312— 323: 2,717,188
351: 2,717,189
313— 76: 2,717,322
77: 2,717,323
84: 2,717,324
270: 2,717,325
314— 69: 2,717,326
315— 6: 2,717,327
14: 2,717,328
24: 2,717,329
2,717,330
77: 2,717,331
100: 2,717,332
103: 2,717,333
166: 2,717,334
183: 2,717,335
2,717,336
243: 2,717,337
316— 2: 2,717,190
317— 13: 2,717,338
119: 2,717,339
188: 2,717,340
235: 2,717,341
2,717,342
2,717,343

318— 31: 2,717,344
129: 2,717,345
207: 2,717,346
2,717,347
221: 2,717,348
237: 2,717,349
254: 2,717,350

321— 18: 2,717,351
322— 36: 2,717,352
323— 22: 2,717,353
47: 2,717,354
64: 2,717,355
74: 2,717,356
77: 2,717,357

324— 68: 2,717,358
98: 2,717,359
333— 6: 2,717,360
71: 2,717,361
82: 2,717,362
2,717,363
336— 30: 2,717,364

339— 99: 2,717,365
126: 2,717,366
185: 2,717,367
340— 15: 2,717,368
17: 2,717,369
151: 2,717,370
170: 2,717,371

340— 173: 2,717,372
2,717,373
212: 2,717,374
250: 2,717,375
382: 2,717,376
343— 7.7: 2,717,377
14: 2,717,378

343— 106: 2,717,379
113: 2,717,380
346— 101: 2,717,191
139: 2,717,192

Classification of Designs

D 3—25: Des. 175,491
D 7— 7: Des. 175,500
D 9— 2: Des. 175,495
D14—30: Des. 175,517
D26—14: Des. 175,489
Des. 175,490
Des. 175,523

D29— 2: Des. 175,504
20: Des. 175,519
D33— 3: Des. 175,516
12: Des. 175,515
D34— 5: Des. 175,502
D42— 7: Des. 175,501
D44— 1: Des. 175,521

D44—10: Des. 175,498
15: Des. 175,492
Des. 175,499
D45— 4: Des. 175,508
Des. 175,509
16: Des. 175,510
19: Des. 175,507

D48—20: Des. 175,494
24: Des. 175,511
Des. 175,512
Des. 175,513
Des. 175,514
D54—13: Des. 175,488
16: Des. 175,493

D56— 1: Des. 175,525
D57— 1: Des. 175,503
D62— 4: Des. 175,518
Des. 175,520
D64—11: Des. 175,496
D70— 1: Des. 175,506
D71— 1: Des. 175,522

D74— 1: Des. 175,505
D81— 1: Des. 175,497
D92— 4: Des. 175,524

OFFICIAL GAZETTE ✦ UNITED STATES PATENT OFFICE
September 6, 1955 Volume 698 Number 1

TRADEMARKS

NOTICES

Trademark Suits

Notices under 15 U. S. C. 1116 ; Trademark Act of July 5, 1946

TM 305,331 ("News-Week" and design), TM 385,859 (Newsweek The Magazine of News Significance), TM 603,656 (Newsweek), Weekly Publications, Inc., Weekly magazine, filed July 5, 1955, D. C., S. D. N. Y., Doc. 102/6, *Weekly Publications, Inc.* v. *The Hagedorn Publishing Co., Inc.*

TM 354,199 (The Lone Ranger Magazine), Trojan Publishing Corp., Periodical publication; TM 365,670, TM 507,040 (The Lone Ranger), The Lone Ranger, Inc., Cartoon or comic strip series; TM 527,916, same, Radio broadcasts, filed Apr. 28, 1953, D. C., S. D. N. Y., Doc. 84/329, *The Lone Ranger, Inc.* v. *M. A. Henry Co., Inc., et al.* Stipulation and order of discontinuance July 6, 1955.

TM 365,670. (See TM 354,199.)

TM 385,859. (See TM 305,331.)

TM 507,040. (See TM 354,199.)

TM 527,916. (See TM 354,199.)

TM 603,656. (See TM 305,331.)

Trademark Rules

New Rules of Practice in Trademark Cases became effective on August 15, 1955. Copies are available without charge upon request addressed to the U. S. Patent Office. Distribution is generally limited to one copy per person, and in no event will more than five copies be furnished without charge on a single order. Quantities in excess of five copies may be purchased from the U. S. Patent Office at $.15 each. Copies of this edition will not be sold by the Superintendent of Documents.

CONDITION OF TRADEMARK APPLICATIONS AS OF AUGUST 12, 1955

Total number of applications awaiting action (excluding renewals and republications)	11, 037
Date of oldest new application	Feb. 1, 1955
Date of oldest amended application	Mar. 14, 1955

MERCHANT, JOHN, Executive Examiner TRADEMARK EXAMINING DIVISIONS, EXAMINERS AND TRADEMARK CLASSES UNDER EXAMINATION	Oldest Application	
	New	Amended
I. STERBA, J. R., Classes 4, 5, 12, 13, 14, 16, 19, 21, 23, 24, 25, 26, 27, 28, 30, 33, 35, 44, 52	2-2-55	4-27-55
II. SHRYOCK, R. F. (Acting), Classes 2, 6, 18, 22, 46, 51 and Service Mark Classes 100, 101, 102, 103, 104, 105, 106, 107	2-1-55	3-14-55
III. WENDT, C. M. (Acting), Classes 1, 3, 7, 8, 9, 10, 11, 15, 17, 20, 29, 31, 32, 34, 36, 37, 38, 39, 40, 41, 42, 43, 45, 47, 48, 49, 50.	2-10-55	4-7-55
Renewals (All Classes)	7-1-55	7-11-55
Republications (All Classes)	6-11-55	7-4-55

Applications Filed During Week Ended August 12, 1955—424

Registrations Issued	348—No. 611,601 to No. 611,948
Renewals Issued	48

MARKS PUBLISHED FOR OPPOSITION

The following marks are published in compliance with section 12(a) of the Trademark Act of 1946. Notice of opposition under section 13 may be filed within thirty days of this publication. See Rules 20.1 to 20.5.

As provided by section 31 of said act, a fee of twenty-five dollars must accompany each notice of opposition.

CLASS 1

SN 666,597. The Ransom & Randolph Company, Toledo, Ohio. Filed May 18, 1954.

R&R

Applicant claims ownership of Reg. No. 300,906.

For Refractory Investment Materials for Casting Metals, Alloys and the Like; Plaster Materials for Molds for Casting Dies, Models and the Like; and Bonding, Coating, and Pattern Materials for Casting Molds.

Use since Mar. 1, 1910.

CLASS 2

SN 669,512. Waldorf Paper Products Company, St. Paul, Minn. Filed July 6, 1954.

For Cartons of Paper Board and Corrugated Paper Board.

Use since Sept. 8, 1953.

SN 681,480. Atlanta Paper Company, Atlanta, Ga. Filed Feb. 11, 1955.

CLUSTER-PAK

For Folding Paperboard Cartons, Corrugated Paperboard Shipping Containers, and Carry-Home Cartons of Paper and Paperboard.

Use since Jan. 12, 1955.

SN 683,165. American Can Company, New York, N. Y. Filed Mar. 10, 1955.

For Metal Cans.

Use since Feb. 17, 1955.

SN 686,426. Cohoes Carrybag Company, Inc., Cohoes, N. Y. Filed Apr. 28, 1955.

COINVELOPE

For Paper Merchandise Envelopes.

Use since Apr. 6, 1955.

CLASS 4

SN 670,281. Milton J. Gordon, d. b. a. M. J. Gordon Company, Glassport, Pa. Filed July 20, 1954.

For Cleaning and Polishing Liquid Preparation for Automobiles.

Use since June 21, 1954.

SN 676,956. Gulf Oil Corporation, Pittsburgh, Pa. Filed Nov. 19, 1954.

The drawing is lined for orange and blue. Applicant claims ownership of Reg. No. 224,356.

For Cleaning and Polishing Liquid for Furniture, Woodwork and Household Dusting.

Use since on or about Sept. 23, 1924.

SN 680,177. Cadie Chemical Products, Inc., New York, N. Y. Filed Jan. 20, 1955.

♊ GEMINI ♊

For Mitts Impregnated With a Silver Cleaning and Polishing Preparation.

Use since Dec. 9, 1954.

CLASS 5

SN 670,635. Vernon Chemical & Manufacturing Corp., Mount Vernon, N. Y. Filed July 26, 1954.

For Pressure Sensitive Tapes—Namely, Drafting Tape, Economy Cloth Tape, Fiberglass Reinforced Packaging Tape, Freezer Tape, High Temperature Tape, Home Workshop Tape, Paper Masking Tape, Shoe Covering Tape, Vinyl Plastic Tape and Waterproof Cloth Tape.
Use since Aug. 17, 1951.

CLASS 6

SN 653,581. Mathieson Chemical Corporation, New York, N. Y., now by merger and change of name Olin Mathieson Chemical Corporation. Filed Sept. 22, 1953.

SWEETA

For Saccharin.
Use since Aug. 19, 1953.

SN 653,600. United Aniline Co., Boston, Mass. Filed Sept. 22, 1953.

UNITOL

For Liquid Penetrants as a Wetting Agent Comprising an Alkyl-Aryl Sulphonate for the Specific Use in the Dyeing, Fulling, and Carbonizing Operations in the Textile Industry.
Use since January 1950.

SN 657,210. L. M. Gould, d. b. a. L. M. Gould Co., San Francisco, Calif. Filed Dec. 1, 1953.

TOILETABS

For Chemical Pellets for Water Conditioning for Use in Flush Tanks, Toilet Bowls, Septic Tanks, and Sewer Pipes.
Use since Apr. 30, 1945.

SN 657,211. L. M. Gould, d. b. a. L. M. Gould Co., San Francisco, Calif. Filed Dec. 1, 1953.

Sta-Cleen

For Chemical Pellets for Water Conditioning for Use in Flush Tanks, Toilet Bowls, Septic Tanks, and Sewer Pipes.
Use since Sept. 28, 1946.

SN 671,250. Geigy Chemical Corporation, New York, N. Y. Filed Aug. 6, 1954.

GY-FUME

For Insecticides and Nematocides.
Use since Oct. 25, 1949.

SN 671,465. Rhode Island Laboratories, Inc., West Warwick, R. I. Filed Aug. 10, 1954. Sec. 2(f).

VIOLITE

Applicant claims ownership of Reg. Nos. 399,944, 444,337, and 509,634.
For Luminous Fluorescent, Phosphorescent Radioactive Pigments.
Use since Dec. 26, 1941.

SN 676,892. Carmen L. Hueston, d. b. a. C. L. Hueston Company, Detroit, Mich. Filed Nov. 18, 1954.

THERMOTUNG

For Treated Tung Oil for General Industrial Use.
Use since Sept. 3, 1954.

SN 679,563. Crawford, Keen & Cia. Sociedad de Responsabilidad Limitada, Buenos Aires, Argentina. Filed Jan. 7, 1955.

For Casein and Other Derivatives of Milk.
Use since in the year 1926.

SN 680,496. Virginia Smelting Company, West Norfolk, Va. Filed Jan. 25, 1955. Sec. 2(f) as to "Virginia."

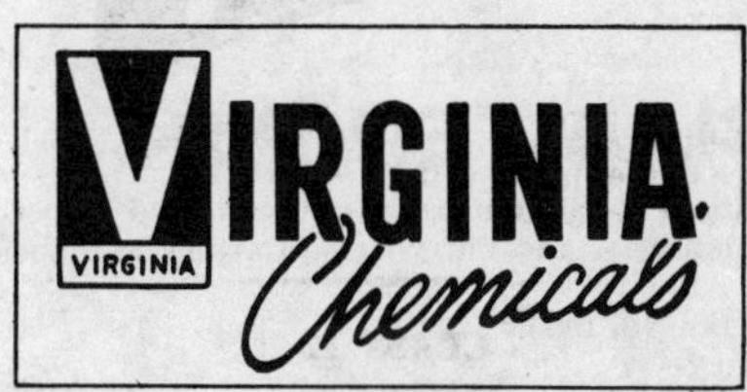

Exclusive right to the use of the word "Chemicals" is disclaimed apart from the mark as shown. Applicant claims ownership of Reg. Nos. 526,525 and 554,280.
For Aerosol Insecticides, Aerosol Fungicides, Aerosol Deodorants, Fungicides, Insecticides, Insecticide and Fungicide Mixtures, Methyl Chloride, Halogenated Methane Refrigerants, Sodium Hydrosulfite, Sulfur Dioxide, Zinc Hydrosulfite, Zinc Sulfate, and Zinc Carbonate.
Use since Jan. 10, 1951, on sodium hydrosulfite; and Sept. 7, 1941, as to "Virginia."

CLASS 8

SN 670,441. Smokador Manufacturing Co., Inc., Bloomfield, N. J. Filed July 22, 1954.

SPIN-ADOR

For Desk Type Ash Receivers.
Use since May 21, 1954.

CLASS 10

SN 643,383. The Stadler Fertilizer Company, Cleveland, Ohio. Filed Mar. 9, 1953.

The words "Stadler's" and "Soil-Bil-Der" are disclaimed apart from the mark as shown. Applicant claims ownership of Reg. Nos. 61,315, 556,897, and others.
For Lawn and Tree Food.
Use since Nov. 1, 1932.

SN 675,019. Smith-Douglass Company, Incorporated, Norfolk, Va. Filed Oct. 18, 1954.

For Fertilizers.
Use since in the spring of 1947.
Subj. to Intf. with SN 675,053.

CLASS 12

SN 676,240. Simpson Logging Company, Seattle, Wash. Filed Nov. 8, 1954.

For Standard Mill Products To Be Used as Construction Materials.
Use since Sept. 16, 1954.

SN 676,430. The Verticel Company, Englewood, Colo. Filed Nov. 10, 1954.

VERTICEL

For Structural and Insulating Panel Core Materials of the Cellular Honeycomb Type.
Use since on or about July 1, 1945.

SN 678,084. Edmund C. Barbera, d. b. a. Iritox Chemical Co., New York, N. Y. Filed Dec. 10, 1954.

BRICKALL

For Waterproof Brick Facing in Panel Form for Application to Any Surface.
Use since June 14, 1954.

SN 680,344. Air-O-Therm Application Co., Inc., Des Plaines, Ill. Filed Jan. 24, 1955.

Jet-Sulation

For Shredded Insulation Material Such as Dry Mineral and/or Asbestos Fibers.
Use since June 1951.

SN 680,345. Air-O-Therm Application Co., Inc., Des Plaines, Ill. Filed Jan. 24, 1955.

AIR·O·THERM

For Shredded Insulation Material Such as Dry Mineral and/or Asbestos Fibers.
Use since June 1951.

SN 680,632. Storm Flooring Co., Inc., New York, N. Y. Filed Jan. 27, 1955. Sec. 2(f) as to "Continuous Strip" and "Ironbound."

No claim is made to exclusive right to use of the words "Hardwood Floor" or to the representation of wooden floor elements per se apart from the mark as shown. The drawing is lined to indicate the color yellow. Applicant claims ownership of Reg. Nos. 309,312, 512,568, and 565,243.
For Wooden Floors.
Use since at least as early as March 1953.

SN 684,817. M. H. Detrick Company, Chicago, Ill. Filed Apr. 4, 1955.

DECRETE

For Castable Insulation Material for Heat Enclosures.
Use since July 6, 1951.

SN 684,918. American Polyglas Corporation, Carlstadt, N. J. Filed Apr. 5, 1955.

TRANSLUX

For Corrugated Panels of Plastic Material.
Use since October 1953.

SN 685,000. The Philip Carey Manufacturing Company, Cincinnati, Ohio. Filed Apr. 6, 1955.

ALLTEMP

For Industrial Insulation.
Use since Mar. 14, 1955.

SN 685,059. United States Plywood Corporation, New York, N. Y. Filed Apr. 6, 1955.

WEAVETEX

For Plywood.
Use since Jan. 1, 1955.

SN 685,179. The Flintkote Company, New York, N. Y. Filed Apr. 8, 1955.

For Plastic Composition Flooring Tile.
Use since on or about Jan. 6, 1955.

SN 685,180. The Flintkote Company, New York, N. Y. Filed Apr. 8, 1955.

modnar

For Plastic Composition Flooring Tile.
Use since on or about Nov. 30, 1954.

SN 685,213. Elwin G. Smith & Company, Pittsburgh, Pa. Filed Apr. 8, 1955.

EGSCO

For Metal Roofing and Siding and Flooring; Insulated Panel Walls; Metal, Composition and Plastic Sheeting.
Use since January 1954.

SN 685,258. Consolidated General Products Inc., Houston, Tex. Filed Apr. 11, 1955.

REXALUM

For Siding.
Use since on or before June 30, 1954.

SN 685,363. The Flintkote Company, New York, N. Y. Filed Apr. 12, 1955.

KOLORDRIFT

Applicant claims ownership of Reg. No. 248,278.
For Asbestos-Cement Shingles.
Use since Feb. 21, 1954.

SN 685,416. Win-Chek, Garfield, N. J. Filed Apr. 12, 1955.
For Storm Doors and Windows.

WIN-CHEK

Use since Nov. 2, 1953.

SN 685,448. Kentile, Inc., Brooklyn, N. Y. Filed Apr. 13, 1955.

ROYAL KENFLOR

Applicant claims ownership of Reg. Nos. 336,151, 596,234, and others.
For Vinyl Plastic Material in Roll and/or Sheet Form and/or Tiles for Construction Purposes for Covering Floors and Walls, and for Covering Table Tops, Sink Tops, Counter Tops, and Stair Treads.
Use since Mar. 22, 1955.

SN 685,558. Childers Manufacturing Co., Houston, Tex. Filed Apr. 15, 1955.

For Plastic Coated Aluminum Building Panels.
Use since Aug. 2, 1954.

CLASS 13

SN 648,092. Fabricated Products Company, West Newton, Pa. Filed June 2, 1953.

The representation of the goods apart from the mark as shown in the drawing is hereby disclaimed.
For Pre-Assembled Unit, Consisting of a Fastener, Threaded or Non-Threaded, Having Assembled to It at the Factory an Appropriate Washer for Fastening Articles Including Sheet Metal Panels Together.
Use since July 1, 1951.

SN 667,181. Tinnerman Products Inc., Cleveland, Ohio. Filed May 26, 1954. Sec. 2(f).

For Locking Clips.
Use since on or about Dec. 15, 1932.

SN 667,445. The Reichert Float & Manufacturing Company, Toledo, Ohio. Filed June 1, 1954.

SPIN SEAT

For Tank Balls.
Use since May 14, 1954.

SN 671,210. Standard Pressed Steel Co., Jenkintown, Pa. Filed Aug. 5, 1954. Sec. 2(f).

HALLOWELL

Applicant claims ownership of Reg. No. 511,641.
For Solid Steel Collars.
Use since in the year 1912.

SN 674,608. Moist O'Matic, Incorporated, Riverside, Calif. Filed Oct. 11, 1954.

Moist O' Matic

For Irrigation Devices, More Specifically: Irrigation Systems Including Soil Moisture Sensing Devices for Initiating and Terminating Irrigation Cycles, Program Timers for Controlling a Series of Irrigation Outlets, Vacuum Responsive Valves and Remote Control Valves, and Other Components of Irrigation Systems.
Use since Sept. 22, 1952.

SN 675,469. Ruez M. Dodge, d. b. a. The Ruez Company, Charleston, W. Va. Filed Oct. 26, 1954.

SNUGGY

For End Play Take-Up Washer for Automobile Steering Assemblies and Gearshift Control Boxes.
Use since Sept. 15, 1951.

SN 675,571. McKinney Manufacturing Company, Pittsburgh, Pa. Filed Oct. 27, 1954. Sec. 2(f) as to "McKinney."

The words "Home & Hobby Hardware" are disclaimed apart from the mark as shown.
For Shelf Hardware—Namely, Hinges, Hinge Hasps, Safety Hasps, Door and Drawer Pulls, Corner Irons, Mending Plates, Barrel Bolts, Sash Lifts, Screen Hangers and Turn Buttons.
Use since Sept. 23, 1954; and since Jan. 1, 1873, as to "McKinney."

SN 678,382. The Eastern Foundry Company, Boyertown, Pa. Filed Dec. 15, 1954.

Applicant claims ownership of Reg. Nos. 176,579 and 179,575.
For Metal Soil Pipe and Fittings.
Use since Aug. 23, 1922.

SN 678,742. The Palnut Company, Irvington, N. J. Filed Dec. 21, 1954.

PUSHNUT

Applicant claims ownership of Reg. No. 324,612.
For Sheet Metal Fasteners.
Use since about June 9, 1952.

SN 679,897. Damar Products, Inc., Newark, N. J. Filed Jan. 14, 1955.

Mrs. Damar's

Applicant claims ownership of Reg. No. 563,095.
For Casters, Perforated Containers for Boiling Rice and the Like, Protective Cooking Supports for Cooking Utensils, Steam Baskets, Trivets, Hot Plates, and Metal Plates for Cleaning Silver.
Use since 1952.

SN 680,213. Paul H. Miller & Son, Indianapolis, Ind. Filed Jan. 20, 1955. Sec. 2(f).

MILLER - MAID

For Cabinet Drawer Slides.
Use since 1934.

SN 680,390. Keystone Steel & Wire Company, Peoria, Ill. Filed Jan. 24, 1955.

KEYLINE

For Metal Mesh Fencing.
Use since Jan. 3, 1955.

SN 680,982. The Union Chain and Manufacturing Company, Sandusky, Ohio. Filed Feb. 2, 1955.

UNION

Applicant claims ownership of Reg. Nos. 139,416, 517,339, and others.
For Chain and Chains for General Use, and Parts Thereof and Attachments Therefor.
Use since Mar. 14, 1914.

SN 680,994. Anti-Corrosive Metal Products Co., Inc., Schodack, N. Y. Filed Feb. 3, 1955.

nylo-fast

For Machine Screws Made of Non-Corrosive Plastic Material.
Use since Nov. 1, 1954.

SN 681,282. W. H. Maze Company, Peru, Ill. Filed Feb. 8, 1955.

STORM GUARD

For Nails.
Use since Jan. 7, 1955.

CLASS 14

SN 664,567. Titan Metal Manufacturing Company, Bellefonte, Pa. Filed Apr. 14, 1954.

TITAN

Applicant claims ownership of Reg. No. 120,800.
For Bronzes, Round, Square, and Hexagonal Bronze Bars, Rolled and Cold Drawn Bronze Bars; Brass Rods, Naval Brass, Muntz's Metal, and Other Brasses; Corrosion Resistant Nonferrous Alloys Consisting Primarily of Copper, Zinc, and Nickel and Sold in the Form of Extruded and Cold Drawn Rods and Special Shapes; Metal Welding Rods; Brass Forgings and Brass Castings; Brass and Bronze Rods, Shapes, and Forgings, Brass Wire and Automatic Screw Machine Parts Unfinished and in the Rough.
Use since June 1, 1915.

SN 681,394. Arcos Corporation, Philadelphia, Pa. Filed Feb. 10, 1955.

ELECTROPAK

For Consumable Arc Welding Electrodes and Gas Welding Rods.
Use since Dec. 8, 1954.

SN 682,797. Alfried Krupp von Bohlen und Halbach, d. b. a. Fried. Krupp, Essen, Germany. Filed Mar. 4, 1955.

WIDIA

Applicant claims ownership of German Reg. Nos. 351,828 and 470,849, dated May 6, 1926 and Nov. 20, 1934, respectively.
For Metallic Carbides and Their Alloys, Metal Shapes for General Use and for Hard Surfacing Operations, Hard Metal Alloys in the Form of Blocks, Bars, Rods, Powder and Granules, Wires, Heat and Corrosion Resistant Alloys, Wear-Resistant Alloys.
Subj. to Intf. with SN 673,763 and SN 673,764.

CLASS 15

SN 661,647. The Lubrizol Corporation, Wickliffe, Ohio. Filed Feb. 25, 1954.

The drawing is lined for orange and blue. Applicant claims ownership of Reg. Nos. 275,737 and 427,464.
For Lubricating Compositions and Lubricating Composition Compounds.
Use since on or about Dec. 31, 1934.

SN 670,750. Muench-Kreuzer Candle Co., Inc., Syracuse, N. Y. Filed July 28, 1954.

DECO-KIT

For Candles.
Use since May 16, 1954.

SN 670,753. Muench-Kreuzer Candle Co., Inc., Syracuse, N. Y. Filed July 28, 1954. Sec. 2(f).

FUNERALITE

Applicant claims ownership of Reg. No. 353,710.
For Candles for Use With Mortuary Lamps.
Use since Apr. 15, 1927.

CLASS 16

SN 659,580. Gunther Wagner, Hannover, Germany. Filed Jan. 15, 1954.

Eilido

Applicant claims ownership of German Reg. No. 138,350, dated Dec. 27, 1910.
For Artists' Water Colors and Oil Colors.

SN 670,757. Protective Coatings, Inc., Tampa, Fla. Filed July 28, 1954.

For Vinyl Plastic Movie Screen Coating.
Use since Mar. 11, 1954.

SN 684,620. American Zinc Sales Company, St. Louis, Mo. Filed Mar. 31, 1955.

The words "Zinc Oxide" are disclaimed apart from the mark as a whole. Applicant claims ownership of Reg. No. 323,490.

For Co-Fumed 35% Leaded Zinc Oxide, Characterized by Its Low Oil Absorption.

Use since May 11, 1935.

SN 684,621. American Zinc Sales Company, St. Louis, Mo. Filed Mar. 31, 1955.

The words "Zinc Oxide" are disclaimed apart from the mark as a whole. Applicant claims ownership of Reg. No. 323,490.

For 35% Leaded Zinc Oxide Partially Blended, Also Characterized by Its Medium Oil Absorption.

Use since May 11, 1935.

SN 684,622. American Zinc Sales Company, St. Louis, Mo. Filed Mar. 31, 1955.

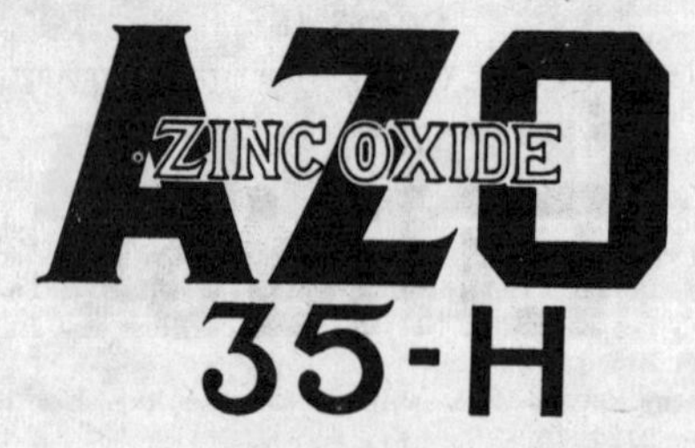

The words "Zinc Oxide" are disclaimed apart from the mark as a whole. Applicant claims ownership of Reg. No. 323,490.

For 35% Leaded Zinc Oxide Partially Blended, Also Characterized by Its High Oil Absorption.

Use since May 11, 1935.

SN 685,289. Napko Paint & Varnish Works, Houston, Tex. Filed Apr. 11, 1955.

For Paints.

Use since Mar. 1, 1955.

SN 685,594. M & H Chemical Products, Inc., Seattle, Wash. Filed Apr. 15, 1955.

ANKORBOND

For Sealers and Coatings, and Particularly That Type Used on Wood Construction and Wood Veneer.

Use since on or about Feb. 14, 1955.

SN 686,347. Alkydol Laboratories, Inc., Cicero, Ill. Filed Apr. 27, 1955.

JEL-O-MER

For Liquid Alkyd Resin and Solutions Thereof Both for Use as Coating Compositions and as Vehicles in Making Paints and Enamels.

Use since Feb. 15, 1955.

SN 686,368. Illinois Farm Supply Co., Chicago, Ill. Filed Apr. 27, 1955. Sec. 2(f).

SOYOIL

Applicant claims ownership of Reg. Nos. 331,916 and 349,119.

For Ready Mixed Paints and Paint Enamels.

Use since October 1931.

SN 686,521. American-Marietta Company, Chicago, Ill. Filed Apr. 29, 1955.

Jelled MAGIC

Applicant claims the exclusive right to the use of the word "Jelled" as a part of its trademark, but not otherwise.

For Thixotropic Gel Paint.

Use since on or about Mar. 7, 1955.

CLASS 17

SN 665,766. General Cigar Co., Inc., New York, N. Y. Filed May 5, 1954.

"WHITEY" WHITE OWL

Applicant claims ownership of Reg. Nos. 150,823 and 599,881.

For Cigars.

Use since on or about Mar. 24, 1954.

CLASS 18

SN 647,903. Blax, Inc., New York, N. Y. Filed May 29, 1953.

For Analgesic, Anti-Rheumatic, Anti-Neuritic, and Anti-Arthritic Composition.

Use since Oct. 16, 1951.

SN 651,957. Soluble Products Co., Brooklyn, N. Y. Filed Aug. 17, 1953.

SOLPROGEL

For Soluble Gel Base for Use With Chemotherapeutic and Medicinal Agents.
Use since September 1949.

SN 660,128. Galt Chemical Products Limited, Galt, Ontario, Canada. Filed Jan. 26, 1954.

AROMATEUM

For Concentrated Ground Mixture of Botanical Roots, Seeds and Herbs, Used as an Additive for Animal Feed.
Use since Feb. 26, 1921.

SN 664,202. Allcock Manufacturing Company, Ossining, N. Y. Filed Apr. 9, 1954. Sec. 2(f) as to "Brandreth's."

Applicant claims ownership of Reg. Nos. 80,454, 438,234, and others.
For Pharmaceutical and Medicinal Preparation Indicated in and for the Treatment of Liver Complaint, Constipation, Dyspepsia, Stomach Disorders, Headaches, and for All Such Other Analogous, Similar or related Human Ailments.
Use since 1835.

SN 677,940. Cosmetics Inc., Bound Brook, N. J. Filed Dec. 8, 1954.

The drawing is lined for the color brown but color is not claimed as a feature of the mark.
For Medicated Powder Base, To Aid in Clearing Skin Blemishes.
Use since Sept. 1, 1953.

SN 681,235. Weymouth Laboratories, North Weymouth, Mass. Filed Feb. 7, 1955.

For Liniment Used in the Treatment of Muscular Aches and Pains.
Use since October 1947.

SN 681,250. James B. Buman, Jr., d. b. a. Buman Pharmaceuticals, Oakland, Calif. Filed Feb. 8, 1955.

BASPERSEE

For Pharmaceutical Preparations for Relief of Pain Associated With Arthritis.
Use since Nov. 19, 1954.

SN 681,305. Schering Corporation, Bloomfield, N. J. Filed Feb. 8, 1955.

NERAVAL

For Anesthetic Preparation.
Use since Nov. 9, 1954.

SN 681,313. Vick Chemical Company, New York, N. Y. Filed Feb. 8, 1955.

THERMORUB

For Analgesic Preparation for External Use.
Use since Jan. 20, 1955.

SN 681,314. Vick Chemical Company, New York, N. Y. Filed Feb. 8, 1955.

THERMOTRATE

For Analgesic Preparation for External Use.
Use since Jan. 20, 1955.

SN 681,327. Abbott Laboratories, North Chicago, Ill. Filed Feb. 9, 1955.

CALCIDRINE

Applicant claims ownership of Reg. No. 346,822.
For Sedative and Expectorant Preparation for the Symptomatic Treatment of Coughs.
Use since Jan. 10, 1955.

SN 681,333. B. F. Ascher & Company, Inc., Kansas City, Mo. Filed Feb. 9, 1955.

SEROLFIA

For Medicinal Preparation Useful as a Sedative, To Reduce Hypertension, Anxiety, Nervousness, To Slow the Pulse and To Lower Blood Pressure.
Use since Feb. 2, 1955.

SN 681,586. Frederick Trout Company, Incorporated, Atlanta, Ga. Filed Feb. 14, 1955.

DUO - GESIC

For Analgesic-Antidepressant Preparation.
Use since Dec. 13, 1954.

SN 681,587. Frederick Trout Company, Incorporated, Atlanta, Ga. Filed Feb. 14, 1955.

METH-AVERIN

For Antitussive-Expectorant Preparation.
Use since Dec. 13, 1954.

CLASS 19

SN 688,093. Studebaker-Packard Corporation, Detroit, Mich. Filed May 23, 1955.

EXECUTIVE

For Automotive Vehicles—Namely, Passenger Cars and Parts Thereof.
Use since Apr. 19, 1955.

CLASS 20

SN 671,011. Armstrong Cork Company, Lancaster, Pa. Filed Aug. 3, 1954.

ROYALTONE

For Felt Base Floor Coverings.
Use since July 20, 1954.

SN 680,050. Vinyl Plastics, Inc., Sheboygan, Wis. Filed Jan. 17, 1955.

TERRALAST

For Flexible Plastic Tiles and Sheets for Covering Floors, Stairs, Walls, Counter Tops, and Desk and Table Tops.
Use since Apr. 22, 1954.

SN 680,297. The Pantasote Company, Passaic, N. J. Filed Jan. 21, 1955.

LIFE-FACE

For Wall Covering, Consisting of Pressure Sensitive Backed Vinyl, Sold in Bolts or Sheets.
Use since Oct. 22, 1954.

SN 680,298. The Pantasote Company, Passaic, N. J. Filed Jan. 21, 1955.

LIFE-TAC

For Wall Covering, Comprising Pressure Sensitive Backed Vinyl Sold in Bolts or Sheets.
Use since Oct. 22, 1954.

CLASS 21

SN 640,756. Radio Ceramics Corporation, Angola, Ind. Filed Jan. 14, 1953.

"THERMFLEX"

For Printed Electrical Circuit Heaters.
Use since July 26, 1951.

SN 666,827. Compagnie Generale de Telegraphie Sans Fil., Paris, France. Filed May 21, 1954.

Priority under Sec. 44(d). French application filed Mar. 24, 1954, Reg. No. 440,432, dated Mar. 24, 1954.
For Electrical, Radio-Electric, Electronic, Optical and Acoustical Apparatus for Telecommunications for Radio and Television Transmission and Reception, Radar, Telephotography and Facsimile, Control and Regulation of Machines, Meters, Computors and Calculators, Direction, Distance and Altitude Indicators and Finders, Sounding Apparatus; Replacement Parts for the Said Apparatus—Namely, Tubes, Crystals, Condensers, Resistors, Transistors, Thermistors; Conducting, Semi-Conducting, Magnetic and Insulating Materials for Use in the Operation and Manufacture of Such Apparatus; Machines and Equipment for Processing and Manufacturing Such Materials.

SN 669,630. Ram Electronics, Inc., Irvington-on-Hudson, N. Y. Filed July 8, 1954.

For Radio and Television Components.
Use since Nov. 26, 1948.

SN 677,031. Bliss Electronic Corp., Sussex, N. J. Filed Nov. 22, 1954.

Applicant disclaims the representation of the goods apart from the mark as shown.
For Automatic Dial-Telephone Exchange System Apparatus, Components Thereof, and Parts Therefor.
Use since Aug. 19, 1954.

SN 680,547. Pittsburgh Plate Glass Company, Pittsburgh, Pa. Filed Jan. 26, 1955.

PITTCOMATIC

For Electrically Operated Door Hinges, Door Operators and Parts Therefor.
Use since Oct. 27, 1950.

SN 680,874. Devry Technical Institute, Inc., Chicago, Ill. Filed Feb. 1, 1955.

The words "Electronics" and "Television" are disclaimed apart from the mark as shown.
For Electronic Circuit Components—Namely, Resistors, Capacitors, Sockets, Tubes, Transformers, Switches, Transducers, Meters, Cabinets, Panels, Terminals and Clips, Chassis Knobs, Test Prods, and Wire.
Use since Jan. 2, 1954.

SN 680,957. International Resistance Company, Philadelphia, Pa. Filed Feb. 2, 1955. Sec. 2(f).

Applicant claims ownership of Reg. No. 580,110.
For Package Containing Parts for Electronic Controls Used in Radio, Television, and Electronic Circuits.
Use since Nov. 1, 1949.

SN 682,161. Metallizing Engineering Co. Inc., Westbury, N. Y. Filed Feb. 23, 1955.

METCO

Applicant claims ownership of Reg. Nos. 358,092, 552,362, and others.
For Electric Apparatus of the Power Source Type—Namely, Transformers and Motor Generators for Electric Grinders and Electric Control Equipment—Namely, Ignition Control Equipment, Transformers, Spark Gaps, Motors, and Switches.
Use since Dec. 30, 1942.

SN 682,475. Redmond Company, Inc., Owosso, Mich. Filed Feb. 28, 1955.

TRI FLUX

For Electrical Motors.
Use since December 1952.

SN 682,476. Redmond Company, Inc., Owosso, Mich. Filed Feb. 28, 1955.

Applicant claims ownership of Reg. No. 423,433.
For Electrical Motors.
Use since January 1947.

SN 682,746. Metals & Controls Corporation, Attleboro, Mass. Filed Mar. 3, 1955.

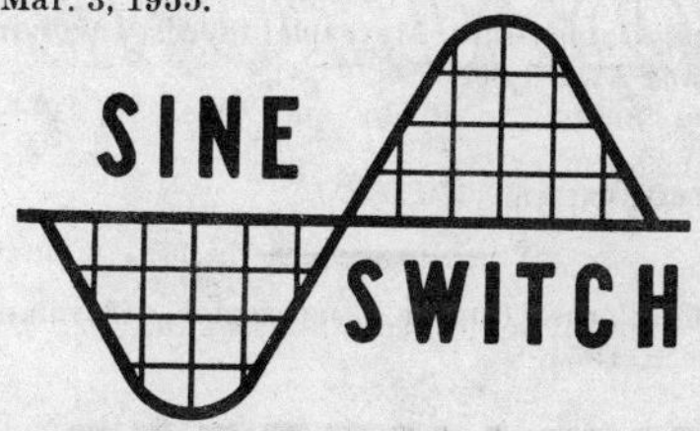

No registration rights are claimed for the word "Switch" apart from the mark as shown.
For Electrical Switches.
Use since on or before Sept. 21, 1954.

SN 683,347. Boom Electric Corporation, Chicago, Ill. Filed Mar. 14, 1955.

No claim is made to the exclusive use of the word "Sound" apart from the mark shown.
For Electrical Sound Producing and Reproducing Equipment—Namely, Public Address Systems, Interoffice Telephone Communication Systems, Paging Systems, Music Distribution Systems, Magnetic Sound Recording Equipment, and Parts for Each Thereof.
Use since Jan. 17, 1955.

CLASS 22

SN 670,589. Luc-Ton Corporation, Kansas City, Mo. Filed July 26, 1954.

For Board Game Constituting a Modern Adaptation of Chess.
Use since Dec. 1, 1953.
Subj. to Intf. with SN 675,293.

SN 675,293. J. Ben Lieberman, d. b. a. Orinda Enterprises, Orinda, Calif. Filed Oct. 22, 1954.

For Board Game With Moveable Pieces Involving Simulated Opposing Armed Services.

Use since Sept. 23, 1954; and Dec. 15, 1944, as to "Strategy."

Subj. to Intf. with SN 670,589.

SN 676,265. True Temper Corporation, Cleveland, Ohio. Filed Nov. 8, 1954.

MUSKETEER

For Fishing Rods.

Use since Aug. 1, 1953.

CLASS 23

SN 645,345. Snap-On Tools Corporation, Kenosha, Wis. Filed Apr. 15, 1953.

WEDGE-GRIP

For Open End Wrenches and Socket Wrenches.

Use since Feb. 10, 1953.

SN 655,778. Heinrich Flottmann G. m. b. H., Bochum, Germany. Filed Nov. 3, 1953.

The term "Flottmann" is disclaimed apart from the mark as shown. Applicant claims ownership of German Reg. No. 642,941, dated Aug. 21, 1953.

For Machines for Mining Purposes—Namely, Reciprocating Conveyor Motors, Reverse Current Motors, Driving Motors, Coal Cutting Machines, Coal Mining Machines, Charging Machines, Drill Sharpening Machines, Upsetting Machines, Spiral Drilling Machines, Grinding Machines, and Air Coduit Ventilators; Compressed-Air Hammers—Namely, Drilling Hammers, Cutting Hammers, Splitting Hammers, Spade Hammers, Stone Splitting Hammers, Riveting Hammers, Caulking Hammers, and Rammers; Auxiliary Implements for Drills and Accessories—Namely, Advance Props, Stretching Columns, Tripods, Boring Cars, Bore Crowns, Boring Rods, Rock Borers, Coal Borers, Pointed Chisels, Spades, Gads, and Compressed-Air Fittings; Conveying Implements—Namely, Conveyer Chutes and Conveyer Bands; Compressed-Air Producing Machines—Namely, Stationary and Portable Compressors; Combustion Engines; and Parts of the Various Machines and Implements.

SN 656,272. The Standard Register Company, Dayton, Ohio. Filed Nov. 12, 1953.

CARBOMATIC

For Autographic Registers and Carbon Feeds Therefor.

Use since July 1953.

SN 664,663. Farmgard Products Company, Minneapolis, Minn. Filed Apr. 16, 1954.

FARMGARD

For Tractor Canopy Consisting of a Metal Framework Attached to a Tractor and Utilizing a Canvas Top.

Use since on or about Dec. 1, 1953.

SN 669,786. Metalastik Limited, Leicester, England. Filed July 12, 1954.

SPHERILASTIK

Applicant claims ownership of British Reg. No. 721,373, dated Aug. 31, 1953.

For Vibration or Shock Absorption Load-Bearing Mountings.

SN 672,585. Clark Cutler McDermott Co., Franklin, Mass. Filed Sept. 1, 1954.

For Vibration Absorbing Pads for Supporting Machines Which Develop Vibrations When in Use.

Use since Mar. 5, 1954.

SN 675,579. New Hermes Laminating Machine Company, New York, N. Y. Filed Oct. 27, 1954.

For Machines for Laminating Identification Cards, Photographs or Like Products.

Use since Sept. 9, 1954.

SN 685,425. J. I. Case Company, Racine, Wis. Filed Apr. 13, 1955.

The drawing is lined for black.

For Tractors.

Use since Dec. 15, 1954.

CLASS 26

SN 646,292. Lear, Incorporated, Grand Rapids, Mich. Filed May 1, 1953.

DYNA-FLUX

For Flux Valve for Detecting the Intensity and Sense of the Earth's Magnetic Field and Translating the Same Into a Voltage of a Magnitude and Phase Corresponding to Said Intensity and Sense and Which Devices Include an Arrangement of Coils Energized by a Source of Alternating Current and, When Used Aboard Craft, for Example, an Airplane, Capable of Interaction With the Relatively Straight Flux Lines Representing the Earth's Magnetic Field To Provide Variation in the Magnitude and Phase in the Currents in the Coils Which Is Representative of the Position of the Craft With Reference to the Earth's Field.

Use since Apr. 1, 1953.

SN 651,508. Empire Machinery Company, Odessa, Tex. Filed Aug. 7, 1953.

The representation of a dial and associated hand are disclaimed apart from the mark as shown.

For Tension Measuring Means for Tong Arms.

Use since May 14, 1953.

SN 660,713. Deltronic Corporation, Los Angeles, Calif. Filed Feb. 8, 1954.

No claim is made to the words "Plug Gauges" nor to the representation of the goods apart from the mark as shown.

For Plug Gauges.

Use since May 20, 1953.

SN 666,822. Aurora Industries, Inc., Chicago, Ill. Filed May 21, 1954.

COLORAMA

For Projection Screens for the Display of Light-Projected Motion and Still Pictures.

Use since on or about Mar. 15, 1954.

SN 669,752. Decade Instrument Co., Caldwell, N. J. Filed July 12, 1954.

DECA-SWEEP

For Decade-Switched Sweeping Oscillator for Producing Frequencies in the Mega-Cycle Range for Use in Radio and Television Equipment.

Use since during May 1952.

SN 672,545. Taber Instrument Corporation, North Tonawanda, N. Y. Filed Aug. 31, 1954. Sec. 2(f).

Taber

For Measuring Instruments, Comprising Instruments for Measuring Stiffness or Qualities of Flexure; Instruments for Measuring Resistance to Abrasion; Instruments for Measuring Hardness and Scratch Resistance; and Electrical Apparatus for Process Control and Research Purposes Comprising Electrical Transmitters.

Use since on or about Mar. 28, 1932, on instruments for measuring stiffness or flexural qualities of materials.

SN 672,828. Magnaflux Corporation, Chicago, Ill. Filed Sept. 7, 1954.

SEDAC

For Electronic Instruments for Locating Fissures, Seams and Voids, and for Determining Their Depth, in Metallic Surfaces.

Use since Aug. 20, 1954.

SN 674,116. Wilhelm Witt, d. b. a. Iloca Camera, Hamburg, Germany. Filed Sept. 30, 1954.

Quick B

For Photographic Cameras.

Use since Apr. 26, 1952.

Subj. to Intf. with SN 664,053.

SN 675,322. Spar Engineering & Development, Inc., Wyncote, Pa. Filed Oct. 22, 1954.

For Oscillograph Recording Accessories, Aircraft Control Systems and Components Thereof, Impulse Timers, and Automatic Sprinkling Apparatus Including Moisture Sensing Elements and Electronic Controls.

Use since Feb. 18, 1952.

Subj. to Intf. with SN 666,196.

SN 681,985. Ernst U. Wilhelm Bertram, Fabrik Phototechnischer Messgeraete. Munich-Pasing, Germany. Filed Feb. 21, 1955.

Bewi

Applicant claims ownership of German Reg. No. 436,844, dated Apr. 18, 1931.

For Automatic Exposure Meters for Use in Connection With Cameras and Parts Thereof.

Use since December 1930.

SN 682,067. Robertshaw-Fulton Controls Company, Greensburg, Pa. Filed Feb. 21, 1955.

GRIDTROL

For Thermostats, Thermostatic Valves and Thermostatic Controls for Gas Range Surface Burners.
Use since Jan. 25, 1955.

SN 684,499. Marie Thoresen, d. b. a. Thoresen Direct Sales, New York, N. Y. Filed Mar. 29, 1955.

POWERHOUSE

For Binoculars and Field Glasses.
Use since Mar. 14, 1955.

SN 684,558. The Haloid Company, Rochester, N. Y. Filed Mar. 30, 1955.

DAINITE

For Photographic Paper.
Use since on or about Mar. 1, 1955.

SN 684,721. Eastman Kodak Company, Rochester, N. Y. Filed Apr. 1, 1955.

Showtime

For Motion Picture Projectors.
Use since Mar. 14, 1955.

SN 685,504. Ledoux & Company, Teaneck, N. J. Filed Apr. 14, 1955.

OXYATOR

For Machines for the Determination of Oxygen in Metals by the Bromination-Carbon Reduction Technique.
Use since Mar. 17, 1955.

CLASS 27

SN 661,914. Harry Shragie, d. b. a. Ritex Watch Company, New York, N. Y. Filed Mar. 2, 1954.

SHRAGIE

For Watches in Cases, Watch Movements, Combined Watches and Bracelets, Combined Watches and Watch Bands.
Use since January 1950.

SN 671,956. The Parker Pen Company, Janesville, Wis. Filed Aug. 19, 1954. Sec. 2(f).

PARKER

Applicant claims ownership of Reg. Nos. 408,997 and 509,034.
For Clocks.
Use since on or about Jan. 1, 1927.

SN 682,913. Granat Bros., San Francisco, Calif. Filed Mar. 7, 1955.

Navette

For Watches.
Use since on or about Feb. 18, 1955.

CLASS 28

SN 670,834. House of Rousseau, Limited, New York, N. Y. Filed July 30, 1954.

For Costume Jewelry for Personal Wear or Adornment—Namely, Earrings, Bracelets, and Necklaces.
Use since Oct. 19, 1951.

SN 671,561. Feature Ring Co., Inc., New York, N. Y. Filed Aug. 12, 1954.

JEWEL LOCK

The word "Lock" appearing on the drawing, is disclaimed apart from the mark as shown.
For Finger Rings and Finger Ring Mountings Made of Precious Metal.
Use since on or about July 28, 1954.

SN 672,390. Feature Ring Co., Inc., New York, N. Y. Filed Aug. 30, 1954.

EVER-GLOW

For Finger Rings and Finger Ring Mountings Made of Precious Metal.
Use since on or about Aug. 1, 1954.

SN 679,091. Fithian, Sun Valley, Calif. Filed Dec. 29, 1954.

For Jewelry—Namely, Bracelets, Earrings, Charms, Cuff Links, Tie Bars, Tie Pins, Sets Consisting of Matching Belt Buckle and Cuff Links and Key Chains.
Use since June 25, 1954.

SN 682,812. M & M Mfg. Co., Providence, R. I. Filed Mar. 4, 1955.

For Jewelry—Namely, Pendants, Necklaces, Bracelets, Pins, Earrings, Brooches, Chokers, Buckles, Ladies' Rings, and Ladies' Cuff Links.
Use since on or about Dec. 5, 1954.

SN 682,912. Granat Bros., San Francisco, Calif. Filed Mar. 7, 1955.

Navette

For Finger Rings.
Use since on or about Feb. 18, 1955.

CLASS 29

SN 678,946. Nelson E. Abrahamsen, Cleveland, Ohio. Filed Dec. 27, 1954.

For Scrub Brush for Spectacles.
Use since Nov. 24, 1954.

CLASS 30

SN 668,523. Booths and Colcloughs Limited, Stoke-on-Trent, England. Filed June 21, 1954. Sec. 2(f).

colclough

For Chinaware—Namely, Tea, Dinner, and Tableware.
Use since as early as the year 1937.

SN 668,524. Booths and Colcloughs Limited, Stoke-on-Trent, England. Filed June 21, 1954. Sec. 2(f).

For Chinaware—Namely, Tea, Dinner, and Tableware.
Use since as early as the year 1900.

SN 673,453. Iroquois China Company, Solvay, N. Y. Filed Sept. 20, 1954.

carrara
MODERN BY
IROQUOIS

Applicant disclaims any right to the descriptive word "Modern." Applicant claims ownership of Reg. No. 537,079.
For China Tableware.
Use since Aug. 25, 1954.

CLASS 31

SN 666,545. Sparkler Mfg. Co., Mundelein, Ill. Filed May 17, 1954.

SPARKLER

Applicant claims ownership of Reg. No. 507,573.
For Domestic, Industrial, and Automotive, Filters for Liquids, and Parts Therefor.
Use since Jan. 11, 1929.

SN 673,670. M & A, Richmond Hill, N. Y. Filed Sept. 23, 1954.

"VABA"

For Filters for Tropical Fish Aquariums.
Use since May 25, 1952.

SN 677,861. British Filters Limited, Maidenhead, England. Filed Dec. 7, 1954.

ALBION

For Beer Polishing Filter Presses.
Use since Feb. 28, 1954.

CLASS 32

SN 641,821. The Wright Line, Inc., Worcester, Mass., by change of name from Wright and Company, Incorporated. Filed Feb. 5, 1953. Sec. 2(f) as to "Wright Line."

For Punched Card Business Machine Accessory Equipment Consisting of Cabinets, Tubs and Files for Storing File Cards, Cabinets and Sections for Storing Business Machine Equipment, and Card File Trays, Racks, Tubs and Cabinets for Storing Card File Trays and Boxes.
Use since 1937.

SN 670,495. Myers-Spalti Manufacturing Company, Houston, Tex. Filed July 23, 1954.

Colony Arts Group

For Sofas, Settees, Chairs, Rockers, Tables, Beds, Dressers, Chests, Chests on Chests, Desks, Vanities, Sideboards, Hutches, Night Stands, Mirrors, Benches, and Dressing Tables.
Use since Apr. 6, 1954.

SN 670,624. Streater Industries, Inc., Spring Park, Minn. Filed July 26, 1954.

FLEX = ORAMA

For Store Fixtures—Namely, Display Cases, Display Racks, Display Counters, and Display Stands.
Use since July 15, 1954.

SN 671,209. Standard Pressed Steel Co., Jenkintown, Pa. Filed Aug. 5, 1954. Sec. 2(f) as to "Hallowell."

HALLOWELL

Applicant claims ownership of Reg. No. 533,959.
For Adjustable Steel Shelving for Cabinets or Other Temporary Use.
Use since July 1, 1954.

SN 672,240. L. C. Phenix Co., Inc., Los Angeles, Calif. Filed Aug. 25, 1954.

For Chairs.
Use since Jan. 5, 1950.

SN 678,080. American Fixture and Manufacturing Co., d. b. a. Chromcraft, Division of American Fixture and Manufacturing Co., St. Louis, Mo., now by change of name to Chromcraft Corporation. Filed Dec. 10, 1954.

For End Tables, Corner Tables, Cocktail Tables, Game Tables, and Desks.
Use since July 17, 1954.

SN 678,774. Douglas Furniture Corporation, Chicago, Ill. Filed Dec. 22, 1954.

Applicant claims ownership of Reg. No. 587,945.
For Kitchen and Dinette Tables and Chairs.
Use since Sept. 20, 1954.

SN 681,399. Balcrank, Inc., Cincinnati, Ohio. Filed Feb. 10, 1955.

MIRRO GLEAM

For Metal Furniture—Namely, Chairs and Chaises.
Use since Feb. 7, 1955.

SN 681,675. CCH Products Company, Chicago, Ill. Filed Feb. 15, 1955.

Desk-Sider

For Mobile Bookracks.
Use since Nov. 26, 1954.

SN 681,713. Sleep Promotion Laboratory, Inc., New York, N. Y. Filed Feb. 15, 1955.

REJUVENATOR

For Mattresses.
Use since Jan. 27, 1955.

CLASS 33

SN 683,488. Corning Glass Works, Corning, N. Y. Filed Mar. 15, 1955. Sec. 2(f).

CORNING

Applicant claims ownership of the marks disclosed in expired Reg. Nos. 6,799, 545,056, and is also the owner of Reg. No. 545,056.
For Tableware, Kitchenware, Tumblers, Pitchers, Jars, Bottles, Bowls, Globes, Vessels, Tubing, Cane, Panels, Optical and Ophthalmic Shapes and Lens Blanks, and Glass Shapes in Flat, Solid, Hollow, Cylindrical, Tubular, or Sagged Form for Use in the Industrial Arts.
Use since October 1878.

CLASS 34

SN 666,021. The Hobart Manufacturing Company, Troy, Ohio. Filed May 10, 1954.

Kitchenaire

Applicant claims ownership of Reg. No. 442,096.
For Plate Warmers and Dish Dryers Based Upon Application of Heat.
Use since May 28, 1952.

SN 666,497. Thomas O. Marini, d. b. a. Romac Industries, Camden, N. J. Filed May 17, 1954.

SUNTEMP

For Baseboard Radiators, Air Distributing Ducts, Convectors and Their Parts and Radiator Coverings for Use With Home and Industrial Hot Water and Hot Air Heating Systems.
Use since Apr. 28, 1954.

SN 670,138. Walker Machine & Foundry Corporation, Roanoke, Va. Filed July 16, 1954.

C
H
A
R
C
O
A
L
K
I
N
G

For Charcoal Grills.
Use since in or about August 1953.

SN 674,539. Thermal Research & Engineering Corporation, Conshohocken, Pa. Filed Oct. 8, 1954.

For Burners, Air Heaters, Heat Exchangers, and Furnaces.
Use since Mar. 16, 1954.

SN 680,374. George Van Dyck, d. b. a. Therm-Mite Boiler Company, Oswego, Oreg. Filed Jan. 24, 1955.

THERM-MITE

For Commercial and Residential Oil or Gas Fired Boilers for Steam or Hot Water Heating Systems.
Use since Nov. 9, 1954.

CLASS 35

SN 662,764. Sure-Seal Equipment Company, Portland, Oreg. Filed Mar. 16, 1954.

For Grease Retainer Rings.
Use since Dec. 30, 1953.

SN 668,416. The Uni-Shim Company, Atlanta, Ga. Filed June 17, 1954.

UNI–SHIM

For Shims for Use in the Repair of Automobiles.
Use since on or about May 26, 1954.

SN 681,708. Phoenix Manufacturing Company, Joliet, Ill. Filed Feb. 15, 1955.

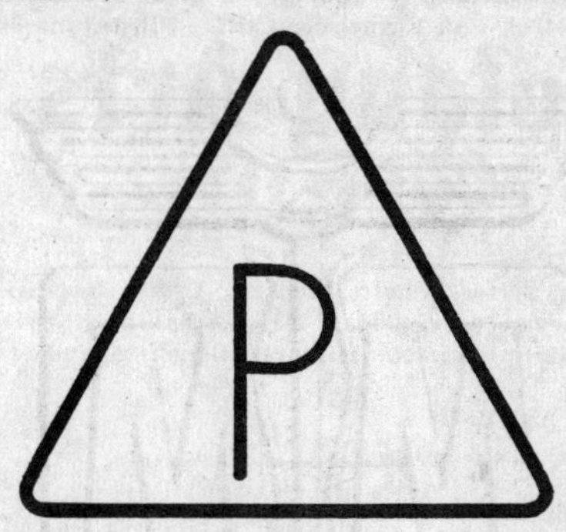

For Rubber Tires and Rubber Hose.
Use since 1949 on rubber tires.

SN 687,157. Wichita Clutch Company, Inc., Wichita Falls, Tex. Filed May 9, 1955. Sec. 2(f).

WICHITA

Applicant claims ownership of Reg. No. 597,037.
For Rotary Fluid Seals.
Use since December 1948.

SN 687,185. Crane Packing Company, Chicago, Ill. Filed May 10, 1955.

FreeFlow

For Gaskets.
Use since on or about Mar. 11, 1955.

SN 687,408. The Polymer Corporation, Reading, Pa. Filed May 12, 1955.

For Synthetic Resin Tubing.
Use since Mar. 24, 1955.

CLASS 36

SN 671,780. Romar Recording Studios, Glendale, N. Y. Filed Aug. 16, 1954.

ROMAR

For Mechanically Grooved Phonograph Records.
Use since January 1954.

CLASS 37

SN 649,563. Richard V. Bibbero, d. b. a. Medical Management Control, San Francisco, Calif. Filed June 30, 1953.

For Bookkeeping, Accounting, and General Office Forms and Ledgers.
Use since June 10, 1953.

SN 661,688. Richard O. Buck, d. b. a. A. J. Buck & Son, Baltimore, Md. Filed Feb. 26, 1954.

BUCK STERI - BAG

No claim is made to the words "Buck" and "Bag" apart from the mark shown in the drawing.
For Paper Bags.
Use since June 15, 1953.

SN 662,970. The Champion Paper and Fibre Company, Hamilton, Ohio. Filed Mar. 30, 1954.

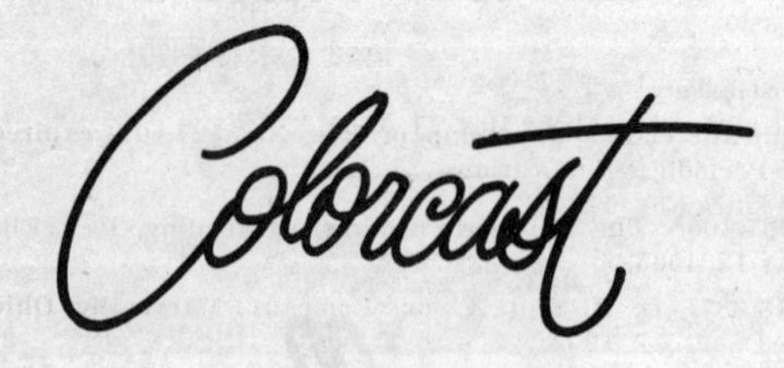

For Coated Printing Paper.
Use since on or about Apr. 19, 1950.

SN 665,456. Valley Paper Company, Holyoke, Mass. Filed Apr. 29, 1954.

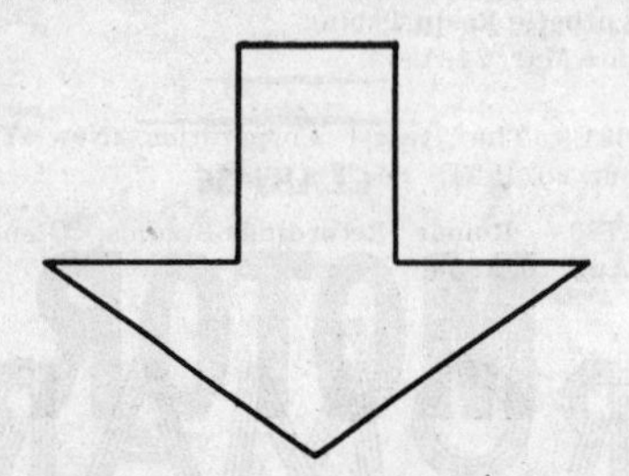

Applicant claims ownership of Reg. No. 384,485.
For Writing, Typewriting, and Printing Paper.
Use since at least as early as June 1940.

SN 665,787. Scripto, Inc., Atlanta, Ga. Filed May 5, 1954.

ADGIF

Applicant claims ownership of Reg. No. 311,755.
For Mechanical Pencils, Leads and Erasers for Mechanical Pencils, Mechanical Typewriter Erasers (in the Form of Mechanical Pencils) and Eraser Refills Therefor, Ink Erasers, Fountain Pens, Ball Point Pens, and Refill Units for Ball Point Pens, Consisting of Ink Supply With Ball Point Tip in Writing Attachment.
Use since Nov. 25, 1931.

SN 665,836. The Mead Corporation, Dayton, Ohio. Filed May 6, 1954.

MOISTRITE

For Paper—Namely, Uncoated Offset and Bond, Including Mimeo, Duplicator, and Ledger Paper.
Use since March 1939.

SN 667,128. Keuffel & Esser Company, Hoboken, N. J. Filed May 26, 1954.

CRYSTALENE

For Tracing Paper.
Use since April 1954.

SN 668,958. Wilson Jones Company, Chicago, Ill. Filed June 25, 1954. Sec. 2(f).

PERFECTION

Applicant claims ownership of Reg. No. 400,249.
For Manually Operated Moisteners for Gummed Tape, Gummed Envelope Flaps, and the Like.
Use since Apr. 23, 1934.

SN 670,804. Royal Lace Paper Works, Inc., Brooklyn, N. Y. Filed July 29, 1954. Sec. 2(f).

DOUBL-EDGE

Applicant claims ownership of Reg. No. 318,501, expired.
For Paper Products—Namely, Shelf Paper, Lace Paper, Doilies, Napkins, Towels, Tissue Paper, and Fancy Printed Lace Paper in Rolls and Sheets.
Use since Aug. 26, 1933.

SN 672,139. Brown Paper Industries, Incorporated, West Monroe, La. Filed Aug. 24, 1954.

BAYOU KRAFT

Without waiver of common law rights, the word "Kraft" is disclaimed apart from the mark as shown.
For Wrapping Paper.
Use since June 1, 1954.

SN 673,513. Frank Novelty Co., New York, N. Y. Filed Sept. 21, 1954.

For Leather Book Covers for the Ecclesiastical Trade.
Use since July 1, 1951.

SN 679,956. White Washburne Corporation, Hinsdale, N. H. Filed Jan. 14, 1955.

PURA SNOW

Applicant claims ownership of Reg. No. 155,119.
For Toilet Tissue and Paper Towels.
Use since Dec. 21, 1917, on toilet tissue.

SN 685,068. Norman Harrower, d. b. a. Linton Brothers & Company, Fitchburg, Mass. Filed Apr. 22, 1955.

Toreador Cover

The word "Cover" is disclaimed apart from the mark as shown. Applicant claims ownership of the mark shown in Reg. No. 317,847.
For Cover Stock.
Use since Jan. 1, 1930.

CLASS 38

SN 645,591. The Reardon Company, St. Louis, Mo. Filed Apr. 20, 1953.

For Color Guides in Booklet Form for Use by Painters in Selecting and Matching Paint Colors.
Use since Nov. 25, 1952.

SN 662,434. The Flow Publishing Company, Cleveland, Ohio. Filed Mar. 11, 1954. Sec. 2(f).

Applicant claims ownership of Reg. Nos. 415,565 and 506,210.
For Publication Published Monthly or From Time to Time, the Subject Matter of Which Integrates Materials Directed to the Handling of Equipment Into the Flow of Production.
Use since on or about Feb. 3, 1945.

SN 672,377. Bishop Publishing Company, Chicago, Ill. Filed Aug. 30, 1954.

Color-Bax

For Modern Color Background Displays Printed on Paper and Cards.
Use since July 28, 1954.

SN 672,453. Standard Oil Company of California, Wilmington, Del. Filed Aug. 30, 1954.

LUBRIGRAM

Applicant claims ownership of Reg. No. 308,465.
For Periodic Publications Distributed From Time to Time.
Use since November 1946.

SN 672,621. Thomas Ashwell & Company, Inc., New York, N. Y. Filed Sept. 2, 1954. Sec. 2(f) as to "Exporters' Encyclopaedia."

For Periodical Publications in the Nature of an Annual Export Shipper's Guide.
Use since January 1904.

SN 672,622. Thomas Ashwell & Company, Inc., New York, N. Y. Filed Sept. 2, 1954. Sec. 2(f).

EXPORT TRADE
and SHIPPER

Applicant claims ownership of Reg. No. 127,807, expired.
For Periodical Publications.
Use since Nov. 11, 1935.

SN 673,387. O. M. Scott & Sons Company, Marysville, Ohio. Filed Sept. 17, 1954.

"YARD BIRD"

For Newspaper Column, and Mailing Pieces.
Use since Aug. 24, 1954.

SN 675,631. The Hearst Corporation, New York, N. Y. Filed Oct. 28, 1954. Sec. 2(f).

MOTOR

Applicant claims ownership of Reg. Nos. 124,102 and 192,880.
For Magazine Published Monthly and/or at Other Intervals.
Use since October 1903.

SN 680,092. Paper Cup and Container Institute, Inc., New York, N. Y. Filed Jan. 18, 1955. Sec. 2(f).

SINGLE SERVICE NEWS

For Public Health News Magazine.
Use since Oct. 1, 1943.

SN 680,093. Paper Cup and Container Institute, Inc., New York, N. Y. Filed Jan. 18, 1955. Sec. 2(f).

HEALTH OFFICERS
News
Digest

For Public Health News Magazine.
Use since Feb. 1, 1936.

SN 680,393. Lane-Wells Company, Los Angeles, Calif. Filed Jan. 24, 1955.

Tomorrow's tools—today!

Applicant claims ownership of Reg. No. 319,715.
For Periodical Publication.
Use since Mar. 20, 1938.

CLASS 39

SN 658,901. I. Gendelman & Son, Philadelphia, Pa. Filed Jan. 4, 1954.

WINDSOR
TOGS

The word "Togs" is disclaimed apart from the mark as shown on the drawing.
For Infants' and Children's Sportswear—Namely, Overalls, Overall Sets, Jackets, Jodhpur Sets, Slacks, Eton Suits, Boxer Shorts, Girls' Shorts, Playsuits, Snowsuits, Pedal Pushers and Sets of Shirts and Shorts.
Use since Dec. 2, 1953.

SN 664,036. Abe Rivetz, d. b. a. A. Rivetz Co., Boston, Mass. Filed Apr. 6, 1954.

THE NECKTIE BELT

No claim is made for the exclusive use of the word "Belt" except as a part of the entire mark as shown.
For Waist Belts Having Conventional Means for Adjustably Securing the Ends Together.
Use since Jan. 29, 1954.

SN 667,473. Seymour Troy, New York, N. Y. Filed June 1, 1954.

Seymour
Originals

Applicant claims ownership of Reg. No. 399,089.
For Ladies' Shoes.
Use since May 5, 1954.

SN 668,047. New England Overall Co., Inc., also d. b. a. Bilt-Well, Boston, Mass. Filed June 10, 1954.

For Dungarees.
Use since Mar. 8, 1954.

SN 668,115. The Olen Company, Mobile, Ala. Filed June 11, 1954.

The drawing is lined for the color red.
For Men's, Women's, and Children's Work Clothing—Namely, Shirts, Pants, Gloves, Dungarees, Overalls, Jumpers, Shoes, Caps, Socks, and Belts.
Use since Oct. 20, 1953.

SN 668,585. Ben Rosenberg Shoe Co., St. Louis, Mo. Filed June 21, 1954.

Thunderbird

For Men's, Women's, and Children's Moccasins and Shoes.
Use since June 4, 1954.

SN 668,758. Margaret B. Karcher, Playa Del Rey, Calif. Filed June 23, 1954.

Hi-Chair-Safe

For Combined Bib and Restraining Jacket for Small Children.
Use since on or about June 9, 1954.

SN 669,145. Max Rosen, d. b. a. Western Slipper Mfg. Co., Chicago, Ill. Filed June 29, 1954.

KIDLINE

For Adults' and Infants' Shoes and Slippers.
Use since Aug. 15, 1950.

SN 669,191. The Havre de Grace Hosiery Mill, Inc., Havre de Grace, Md. Filed June 30, 1954.

For Women's Hosiery.
Use since Apr. 26, 1954.

SN 670,259. Tru Balance Corsets Inc., New York, N. Y. Filed July 19, 1954.

DIABOLO

For Ladies' Corsets, Girdles, Foundation Garments, Corselettes, Brassieres, Bathing Suits, Ladies' Underwear and Shoes Made of Leather and a Combination of Leather and Fabrics.
Use since May 17, 1954.

SN 670,456. M. Beckerman & Sons, Inc., New York, N. Y. Filed July 23, 1954.

For Women's, Infants', Children's, and Boys' Shoes.
Use since January 1952.

SN 670,460. Campus Sweater & Sportswear Company, d. b. a. Campus Sweater and Sportswear Co., Cleveland, Ohio. Filed July 23, 1954.

Doeskin Interlock

For Men's and Boys' Outer Shirts Made in Whole or in Substantial Part of a Knitted Fabric Made Entirely of Cotton Fibers.
Use since on or about July 9, 1954.

SN 670,911. Dietz & Company, Inc., East Norwalk, Conn. Filed Aug. 2, 1954.

For Boys' Sport Jackets, Slacks, and Pants.
Use since June 2, 1954.

SN 671,036. Jay Manufacturing Company, Incorporated, Roxbury, Mass. Filed Aug. 3, 1954.

Jeansters

For Jeans, Outer Shorts, and Outer Shirts.
Use since May 14, 1954.

SN 671,102. The Jones Trouser Company, Inc., Philadelphia, Pa. Filed Aug. 4, 1954.

For Men's and Boys' Slacks.
Use since May 29, 1954.

SN 671,194. Nippon Gomu Kabushiki Kaisha, Chiyoda-ku, Tokyo, Japan. Filed Aug. 5, 1954.

AIRESIN

Priority under Sec. 44(d). Japanese application filed June 24, 1954, Reg. No. 461,851, dated Mar. 9, 1955.
For Sandals and Slippers.

SN 672,560. M. Beckerman & Sons, Inc., New York, N. Y. Filed Sept. 1, 1954.

GLIDE RITE

For Women's, Infants', Children's, and Boys' Shoes.
Use since Aug. 11, 1954.

SN 673,897. The Cashmere Corporation of America, Cleveland, Ohio. Filed Sept. 28, 1954.

Applicant claims ownership of Reg. No. 502,901.
For Men's, Women's, and Children's Knitted Woolen Sweaters, Cardigans, and Pullovers.
Use since June 21, 1954.

SN 673,963. Bambi Footwear Inc., Brooklyn, N. Y. Filed Sept. 29, 1954.

For Women's and Children's Shoes.
Use since July 15, 1954.

SN 674,035. L. Isaacson & Sons, New York, N. Y. Filed Sept. 9, 1954.

For Wearing Apparel, Consisting of Boys' Shirts.
Use since 1933.

SN 674,532. Stan's Shoe Outlet, Oakland, Calif. Filed Oct. 8, 1954.

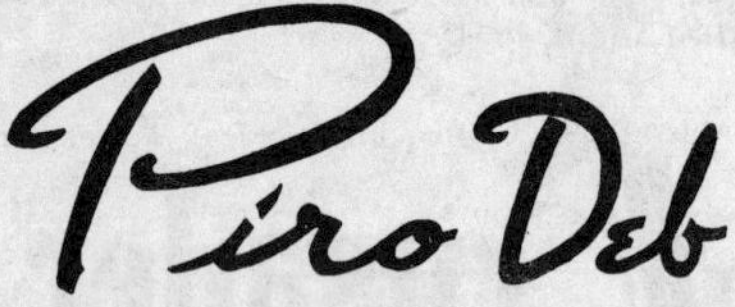

For Shoes.
Use since Dec. 30, 1953.

SN 674,533. Stan's Shoe Outlet, Oakland, Calif. Filed Oct. 8, 1954.

For Shoes.
Use since Apr. 27, 1953.

SN 675,282. Robert G. Hildebrand, d. b. a. R. G. Hildebrand Co., St. Louis, Mo. Filed Oct. 22, 1954.

Applicant claims ownership of Reg. No. 512,305.
For Unit Assemblies of Cut-Out Leather Parts for Making a Pair of Infants' or Young Children's Shoes by Hand.
Use since December 1945.

SN 675,698. Gross Galesburg Co., Galesburg, Ill. Filed Oct. 29, 1954.

For Clothing for Men, Women, and Children—Namely, Cowboy Jeans, Sport Shirts, Sport Coats, and Overalls.
Use since Aug. 12, 1954

SN 676,370. Fowlkes Hat Co., Denver, Colo. Filed Nov. 10, 1954.

OPEN RANGE

For Hats and Caps for Adults and Children.
Use since Oct. 29, 1954.

SN 676,839. David E. Schwab & Company, Inc., New York, N. Y. Filed Nov. 17, 1954.

Applicant claims ownership of Reg. Nos. 185,804, 365,348, and others.
For Handkerchiefs.
Use since July 2, 1928.

SN 678,257. Poplar Textile, Inc., Philadelphia, Pa. Filed Dec. 13, 1954.

For Hosiery, Sweaters, for Use by Men, Women, and Children.
Use since Aug. 1, 1949.

SN 678,609. Don Juan Corporation of California, Los Angeles, Calif. Filed Dec. 20, 1954.

La Scala

For Men's and Boys' Shirts.
Use since Oct. 6, 1954.

SN 678,779. Andrew Geller, Inc., Brooklyn, N. Y. Filed Dec. 22, 1954.

Andrew Geller
EXQUISITE FOOTWEAR

The words "Exquisite Footwear" are disclaimed apart from the mark as shown. Applicant claims ownership of Reg. No. 306,943, expired.
For Ladies' and Misses' Shoes.
Use since June 1, 1924; and since Nov. 1, 1907, as to "Andrew Geller."

SN 678,991. The Kendall Company, Walpole, Mass. Filed Dec. 27, 1954.

Applicant claims ownership of Reg. No. 358,738.
For Diapers.
Use since Dec. 12, 1947.

SN 679,137. The Cashmere Corporation of America, Cleveland, Ohio. Filed Dec. 30, 1954.

Hadley Caprice

Applicant claims ownership of Reg. No. 502,901.
For Men's, Women's, and Children's Knitted Woolen Sweaters, Cardigans, and Pullovers.
Use since Dec. 15, 1954.

SN 679,361. The Ward Stilson Co., Anderson, Ind. Filed Jan. 3, 1955.

For Ladies', Misses', and Junior Misses' Dresses.
Use since Dec. 1, 1954.

SN 679,385. Gloria Gloves, Inc., New York, N. Y. Filed Jan. 4, 1955.

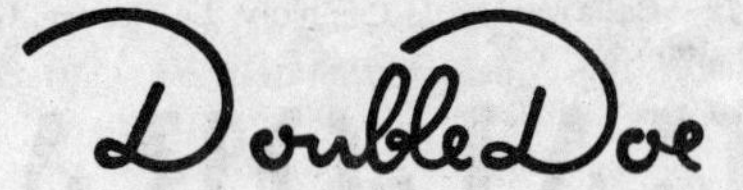

For Ladies' Gloves.
Use since Nov. 12, 1954.

SN 680,370. The Walker T. Dickerson Company, Columbus, Ohio. Filed Jan. 24, 1955. Sec. 2(f) as to "Dickerson."

No claim is made to the words "Fashioned to Fit" apart from the mark as shown. Applicant claims ownership of Reg. Nos. 274,781 and 555,075.
For Women's and Misses' Shoes.
Use since Dec. 17, 1954; and since February 1930 as to "Dickerson."

SN 680,540. Lumbard-Watson Company, Auburn, Maine. Filed Jan. 26, 1955.

scampers

For Sock Linings and Shoes.
Use since Dec. 31, 1954.
Subj. to Intf. with SN 687,287.

CLASS 42

SN 666,517. Charles Pindyck, Inc., New York, N. Y. Filed May 17, 1954.

Tubmates

For Towels, Wash Cloths, and Sets of Towels and Wash Cloths.
Use since Aug. 14, 1950.

SN 666,795. Stanhill Fashions Inc., New York, N. Y. Filed May 20, 1954.

Silva-Duv

For Fabric Made Into Ladies' Suits and Coats.
Use since Apr. 29, 1954.

SN 668,230. Eugene T. Barwick, d. b. a. E. T. Barwick Mills, Chamblee, Ga. Filed June 15, 1954.

THE
Brookwood

For Textile Rugs and Carpeting.
Use since on or about Apr. 19, 1954.

SN 668,469. Roanoke Mills Company, Roanoke Rapids, N. C. Filed June 18, 1954.

SKIP CORD

For Piece Goods for Wearing Apparel and Piece Goods for Home Furnishings.
Use since Apr. 14, 1954.

SN 669,107. D. B. Fuller & Co., Inc., New York, N. Y. Filed June 29, 1954.

LORALETTE

For Textile Fabrics in the Piece of Cotton, Rayon, Synthetic Fibres, and Mixtures Thereof.
Use since Jan. 19, 1953.

SN 670,488. Lincoln Fabrics Co. Inc., New York, N. Y. Filed July 23, 1954.

LincoRama

For Textile Fabrics Made of Cotton, Rayon, and Synthetic Fibres.
Use since May 12, 1954.

SN 676,656. Pacific Mills, New York, N. Y. Filed Nov. 15, 1954.

PACIDRIL

For Shoe Lining Fabrics.
Use since June 24, 1954.

SN 678,750. Staple Coat Co. Inc., New York, N. Y. Filed Dec. 21, 1954.

For Fabrics Made Into Ladies' and Misses' Coats and Toppers.
Use since Oct. 6, 1954.

SN 679,197. Liberty & Co. Limited, London, England. Filed Nov. 30, 1954.

Applicant claims ownership of British Reg. No. 725,765, dated Jan. 15, 1954.
For Bed Sheets, Blankets, Bedspreads, Quilts, Table Cloths, and Place Mats and Piece Goods of Cotton, Wool, Silk, and Synthetic Fibers.

SN 680,168. Tootal Broadhurst Lee Company Limited, Manchester, England. Filed Jan. 19, 1955.

TOOTAFA

For Piece Goods Made Wholly or Mainly of Cotton.
Use since July 13, 1954.

SN 680,353. The Barbizon Corporation, New York, N. Y. Filed Jan. 24, 1955.

BLENDAIRE

For Woven Fabric Made of Nylon, Cotton, and Other Synthetic Fibers, and Combinations Thereof.
Use since Oct. 15, 1954.

SN 681,577. Celanese Corporation of America, New York, N. Y. Filed Feb. 14, 1955.

CELACOT

For Textile Fabrics Made of Synthetic Fibers, Alone and in Admixture With Cotton, Wool or Silk.
Use since Jan. 29, 1955.

SN 681,671. Callaway Mills Company, La Grange, Ga. Filed Feb. 15, 1955.

VALPOINT

Applicant claims ownership of Reg. No. 566,659.
For Textile Carpeting.
Use since Jan. 24, 1955.

SN 681,672. Callaway Mills Company, La Grange, Ga. Filed Feb. 15, 1955.

CALORA

For Textile Carpeting.
Use since Jan. 24, 1955.

SN 681,673. Callaway Mills Company, La Grange, Ga. Filed Feb. 15, 1955.

CALMONT

For Textile Carpeting.
Use since Jan. 24, 1955.

CLASS 43

SN 677,181. Tissages de Soieries Reunis (T. S. R.), Paris, France. Filed Nov. 23, 1954.

MOUSSENYL

Applicant claims ownership of French Reg. No. 28,503, dated Nov. 26, 1953.
For Thread and Yarn.

CLASS 44

SN 668,915. O. E. M. Corporation, East Norwalk, Conn. Filed June 25, 1954. Sec. 2(f).

CLEERLITE

For Oxygen Tent Canopies.
Use since 1947.

SN 670,134. Dr. E. Ulhorn & Co., Wiesbaden-Biebrich, Germany. Filed July 16, 1954.

BETASTRAL

Applicant claims ownership of German Reg. No. 646,968, dated Oct. 30, 1953.

For Applicators or Vials for Radiation Therapy.

SN 675,604. Stanley Webster Laboratories, Inc., Elmhurst, Ill. Filed Oct. 27, 1954.

"COR-VOX"

For Electrical Hearing Aids.

Use since Sept. 29, 1954.

SN 676,747. The J. Bird Moyer Company, Inc., Philadelphia, Pa. Filed Nov. 16, 1954.

MOYCO

Applicant claims ownership of Reg. No. 393,579.

For Dental Waxes, Impression Materials Made With Powders or With Compounds and Plastic Denture Materials.

Use since Feb. 12, 1914.

SN 678,106. William L. Gould, d. b. a. Research Supplies, Albany, N. Y. Filed Dec. 10, 1954.

WANDEX

For Vaginal Applicator.

Use since on or about Dec. 17, 1945.

SN 678,156. Sears, Roebuck and Co., Chicago, Ill. Filed Dec. 10, 1954.

Ann Barton

For Sanitary Goods for Women (Namely, Napkins, Briefs, Panties, Tampons, Belts, Shields).

Use since on or about Jan. 18, 1952.

SN 679,005. Samuel Mushlin, Quincy, Mass. Filed Dec. 27, 1954.

Sonartherm

For Medical Equipment for Generating High Frequency Sonic Waves for Treatment of Muscular and/or Joint Conditions.

Use since Dec. 16, 1954.

CLASS 45

SN 660,936. Fred Fear & Co., Brooklyn, N. Y. Filed Feb. 11, 1954.

SCHOONER

For Imitation Fruit Syrup for Making Soft Drinks.

Use since approximately in the year 1942.

SN 665,111. Hoffman Beverage Company, Newark, N. J. Filed Apr. 23, 1954.

Streamline

For Soft Drinks.

Use since Jan. 15, 1954.

Subj. to Intf. with SN 669,178.

SN 666,057. Red Owl Stores, Inc., Hopkins, Minn. Filed May 10, 1954.

RED OWL

Applicant claims ownership of Reg. Nos. 602,779 and 606,801.

For Carbonated Beverages Sold as Soft Drinks and for Use as Mixers.

Use since April 1922.

SN 666,756. Hoffman Beverage Company, Newark, N. J. Filed May 20, 1954. Sec. 2(f) as to "Hoffman."

HOFFMAN
Streamline

For Soft Drinks.

Use since Jan. 15, 1954; and since 1911 as to "Hoffman."

Subj. to Intf. with SN 669,178.

SN 666,860. John Molaison, New Orleans, La. Filed May 21, 1954. Sec. 2(f).

All wording except "Molay" is disclaimed apart from the mark as shown. The drawing is lined for orange.

For Non-Alcoholic Cocktail Mix Used in Mixing Alcoholic Drinks.

Use since Jan. 1, 1948.

SN 668,500. Flotill Products, Inc., Stockton, Calif. Filed June 18, 1954.

The picture and name shown are that of "Tillie Lewis" (Mrs. Meyer L. Lewis), whose consent is of record. Exclusive registration rights in the words "Tasti-Diet" are not claimed apart from the rest of the mark, all common law rights being retained.

For Low Calorie Soft Drinks.

Use since Mar. 10, 1954.

SN 669,178. Dolly Fruit Products Corp., Brooklyn, N. Y. Filed June 30, 1954.

Stream-Line
Rickey

Applicant disclaims the word "Rickey" apart from the mark as shown.

For Lime Flavored Fountain Syrups Used in Making Soft Drinks.

Use since Jan. 15, 1954.

Subj. to Intf. with SN 665,111 and 666,756.

SN 676,991. Orange Smile Sirup Company, St. Louis, Mo. Filed Nov. 19, 1954.

The drawing is lined for red. The spoken feature of applicant's trademark are the words "Cheer Up." Applicant claims ownership of Reg. No. 340,923.

For Soft Drink and Syrups and Extracts Used in Making Same.

Use since Aug. 20, 1954.

SN 678,431. W. R. Beyer, d. b. a. Beyer and Company, Odessa, Tex. Filed Dec. 16, 1954.

WILD CAT

For Soft Drinks.

Use since Aug. 1, 1952.

SN 679,531. Mission Bottling Company of Buffalo, Inc., Buffalo, N. Y. Filed Jan. 6, 1955. Sec. 2(f).

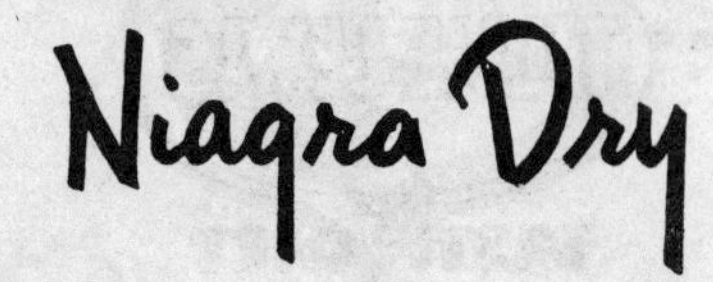

For Ginger Ale and Club Soda.

Use since on or about June 26, 1939.

CLASS 46

SN 604,061. Acme-Evans Company, Inc., Indianapolis, Ind. Filed Sept. 27, 1950. Sec. 2(f) as to "O-So-Fine."

The words "Kansas," "Quality Guaranteed," "Hard Wheat," and "Flour" are disclaimed apart from the mark shown.

For Wheat Flour.

Use since Jan. 1, 1905.

SN 650,889. John Engelhorn & Sons, Newark, N. J. Filed July 27, 1953. Sec. 2(f) as to "Engelhorn's."

For Smoked Pork Shoulder Butt.

Use since Mar. 15, 1953; since 1875 as to "Engelhorn's"; and since 1946 as to design of piglets.

SN 652,197. The E. Kahn's Sons Co., Cincinnati, Ohio. Filed Aug. 24, 1953.

BOUQUET

For Fresh Meat for Human Consumption—Namely, Beef Carcasses, Veal Carcasses, and Lamb Carcasses.

Use since on or about May 21, 1948.

SN 654,870. International Packers Limited, Chicago, Ill. Filed Oct. 15, 1953.

INTERPACK

Applicant claims ownership of Reg. Nos. 505,233 and 504,790.

For Canned Corned Beef, Beef and Sheep Sausage Casings.

Use since Oct. 23, 1951, on sausage casings.

SN 661,644. Liberty Fish Co., Philadelphia, Pa. Filed Feb. 25, 1954.

For Frozen Rock Lobster Tails.
Use since on or about Dec. 11, 1953.

SN 661,997. Farmers Cooperative Creamery Association, Slater, Iowa. Filed Mar. 4, 1954.

For Butter.
Use since 1924.

SN 662,508. L. W. Bonsib, d. b. a. Bonsib Drive-Ins, Denver, Colo. Filed Mar. 12, 1954.

Bonburger

Applicant claims ownership of Reg. No. 380,927.
For Prepared Frozen Beef and Sandwiches Made Therefrom.
Use since on or about Aug. 15, 1953.

SN 663,255. Chip Steak Company, Oakland, Calif. Filed Mar. 25, 1954.

For Frozen Fresh Beef Wafer Steaks.
Use since on or about Jan. 15, 1953.

SN 663,310. Rockwood & Co., Brooklyn, N. Y. Filed Mar. 25, 1954.

CHOCOLATE CRUSTIES

For Chocolate Candy Pieces.
Use since Apr. 24, 1953.

SN 665,599. Interstate Bakeries Corporation, Kansas City, Mo. Filed May 3, 1954.

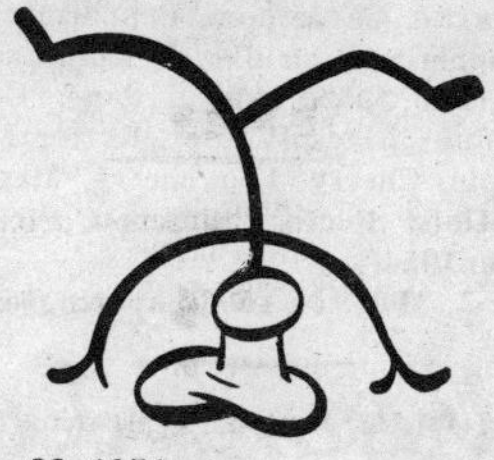

For Bread.
Use since Mar. 22, 1954.

SN 666,090. Andersen's Foods, Inc., Santa Barbara, Calif. Filed May 11, 1954.

Kettle Fresh!

For Canned Soups.
Use since Apr. 30, 1954.

SN 670,367. Mathieson Chemical Corporation, New York, N. Y., now by merger and change of name Olin Mathieson Chemical Corporation. Filed July 21, 1954.

The drawing is lined for red.
For Sweetening Agent—Namely, a Saccharin Preparation.
Use since June 17, 1954.

SN 672,079. Hubbard Milling Company, Mankato, Minn. Filed Aug. 23, 1954.

KING HUBBARD

Applicant claims ownership of Reg. Nos. 82,516 and 284,400.
For Wheat Flour.
Use since Jan. 1, 1936.

SN 674,908. Marion K. Summers, Brownstown, Ind. Filed Oct. 15, 1954.

For Food Flavoring Extracts, Food Colorings, Food Flavorings and Seasonings—Namely, Imitation Cocoanut, Strawberry, Egg-Nog Flavor, Bar B-Q Sauce Flavor, Angel Food Flavor, Almond, Orange, Orange Pineapple, Lemon, Liquid Onion, Liquid Garlic, Cardamom, Peppermint, Mint, Wintergreen, Lime, Raspberry, Rum Flavor, Rum and Butter Flavor, Rose, Liquid Mixed Spice, Anise, Sage, Celery, Kuemmel, Chili Powder, Maple, Black Walnut, Butter Flavor, Pumpkin, Banana, Pineapple, Cherry, Loganberry, Mixed Fruits, Poultry Seasoning, Mixed Spices, Cinnamon, Pickling Spice, Catsup Spice, and Vanilla.

Use since Oct. 1, 1928, on food flavoring extracts.

SN 674,909. Marion K. Summers, Brownstown, Ind. Filed Oct. 15, 1954.

Marion-Kay

For Food Flavoring Extracts, Food Colorings, Food Flavorings and Seasonings—Namely, Imitation Cocoanut, Strawberry, Egg-Nog Flavor, Bar B-Q Sauce Flavor, Angel Food Flavor, Almond, Orange, Orange Pineapple, Lemon, Liquid Onion, Liquid Garlic, Cardamom, Peppermint, Mint, Wintergreen, Lime, Raspberry, Rum Flavor, Rum and Butter Flavor, Rose, Liquid Mixed Spice, Anise, Sage, Celery, Kuemmel, Chili Powder, Maple, Black Walnut, Butter Flavor, Pumpkin, Banana, Pineapple, Cherry, Loganberry, Mixed Fruits, Poultry Seasoning, Mixed Spices, Cinnamon, Pickling Spice, Catsup Spice, and Vanilla.

Use since Oct. 1, 1928, on food flavoring extracts.

SN 676,509. Marion-Kay Products Co. Inc., Brownstown, Ind. Filed Nov. 12, 1954.

The word "Vanilla" is disclaimed apart from the mark as shown.

For Food Flavoring—Namely, Vanilla.

Use since Oct. 15, 1954.

SN 678,543. Lancaster Wholesale Grocery Company, Incorporated, Lancaster, Pa. Filed Dec. 17, 1954. Sec. 2(f).

Penn Dutch
BRAND

For Canned Fruits, Canned Fruit Cocktail, Canned Vegetables, Canned Beans With Pork, and Tomato Catsup.

Use since June 8, 1948, on canned vegetables and canned beans with pork.

SN 681,148. B. Fischer & Co., Inc., Bronx, N. Y. Filed Feb. 7, 1955.

BIG JOE

For Dog Food.

Use since Jan. 8, 1955.

CLASS 47

SN 669,445. Italian Swiss Colony, d. b. a. Shewan-Jones, San Francisco, Calif. Filed July 6, 1954. Sec. 2(f).

HARTLEY

Applicant claims ownership of Reg. No. 403,202.

For Wines.

Use since Oct. 27, 1941.

SN 673,472. Victor Place, Cadaujac pres Bordeaux, France. Filed Sept. 20, 1954.

Château Bouscaut

For Wines.

Use since 1926.

SN 677,734. D. W. Putnam Co., Inc., Hammondsport, N. Y. Filed Dec. 3, 1954.

Golden Age

Applicant claims ownership of the mark disclosed in Reg. Nos. 57,200 and 309,195.

For Champagne and Still Wines.

Use since January 1883.

CLASS 48

SN 665,843. Minneapolis Brewing Company, Minneapolis, Minn. Filed May 6, 1954.

Applicant claims ownership of Reg. Nos. 54,643, 415,764, and others.

For Beer.

Use since Mar. 29, 1954.

SN 666,978. Paulaner-Salvator-Thomasbrau, A. G., Munich, Germany. Filed May 24, 1954.

SALVATOR

Applicant claims ownership of Reg. Nos. 75,221 and 61,358 (expired).

For Beer.

Use since in the year 1760.

CLASS 49

SN 665,315. Soc. Com. Abel Pereira Da Fonseca (S. A. R. L.), Lisbon, Portugal. Filed Apr. 27, 1954.

For Brandy.
Use since June 1934.

SN 677,673. The Old Joe Distillery Company, d. b. a. Old Underoof Distilling Company, Lawrenceburg, Ky. Filed Dec. 2, 1954.

OLD
UNDEROOF

No claim is made to the exclusive right to the use of the word "Old" separate and apart from the mark as shown.
For Whiskey.
Use since January 1893.

SN 685,228. Standard Distillers Products, Inc., Baltimore, Md. Filed Apr. 27, 1955. Sec. 2(f).

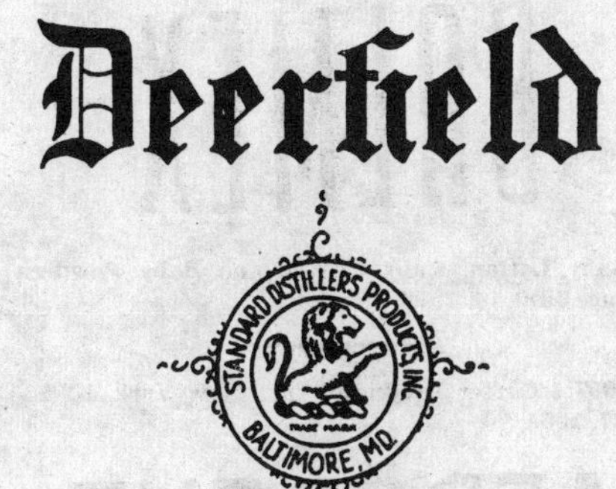

No claim is made to the words "Trade Mark" and "Baltimore, Md." apart from the mark as shown. Applicant claims ownership of Reg. No. 332,826.
For Rum.
Use since on or about Nov. 5, 1935.

CLASS 50

SN 636,595. Igelstroem-Oberlin, Inc., Massillon, Ohio. Filed Oct. 14, 1952.

For Non-Electrical Advertising Signs.
Use since July 30, 1952.

SN 644,354. Tropicraft, Inc., San Francisco, Calif. Filed Mar. 27, 1953. Sec. 2(f).

WOVEN-WOOD

For Material Composed of Parallel Wooden Strips Connected by Interwoven Threads for Use as Window Shades, Drapes, Wall Covering, Floor Screens, Floor Coverings, Canopies and the Like.
Use since in or about Apr. 1, 1947.

SN 662,844. Samuel Brown, d. b. a. Brown & Brown, Mobile, Ala. Filed Mar. 18, 1954. Sec. 2(f) as to "Weather-Tite."

Applicant claims ownership of Reg. Nos. 520,117 and 563,938.
For Canvas Tarpaulins.
Use since Feb. 1, 1954; and since Sept. 1, 1946, as to "Weather -Tite."

SN 662,868. The Jason Corporation, Hoboken, N. J. Filed Mar. 18, 1954.

JASON

For Plastic Sheeting, Quilted Plastic Cloth, Plastic Cloth Having the General Appearance of Artificial Leather.
Use since Nov. 1, 1947.

SN 662,988. Thomas R. Caton, d. b. a. Reproduction Research Laboratories, Lynbrook, N. Y. Filed Mar. 22, 1954.

COPPERITE

For Metal Plates Prepared for Processing Into Photolithographic Offset Printing Plates.
Use since Jan. 28, 1954.

SN 662,991. Thomas R. Caton, d. b. a. Reproduction Research Laboratories, Lynbrook, N. Y. Filed Mar. 22, 1954.

For Metal Plates Prepared for Processing Into Photolithographic Offset Printing Plates.
Use since Jan. 28, 1954.

SN 663,052. Picturepak, Inc., Los Angeles, Calif. Filed Mar. 22, 1954.

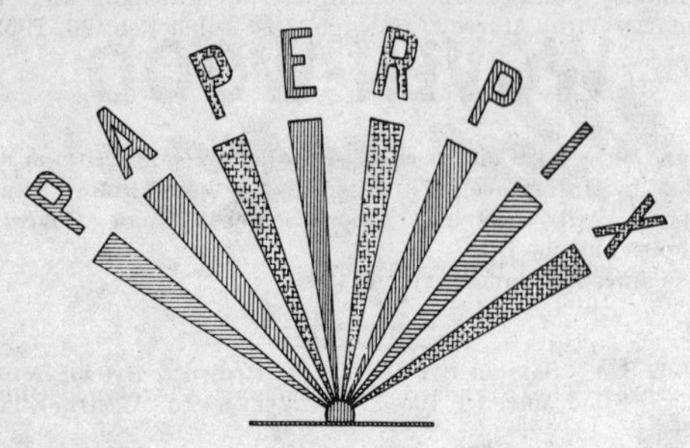

The drawing is lined for blue, red or pink, and yellow or gold.

For Picture Making Sets Using Colored Paper Consisting of Strips of Colored Paper, a Pair of Scissors, a Tweezer, a Pencil, an Eraser, and a Ruler.

Use since Jan. 10, 1954.

SN 664,664. Farmgard Products Company, Minneapolis, Minn. Filed Apr. 16, 1954.

ZEROGARD

For Collapsible Fish House.

Use since on or about Nov. 1, 1953.

SN 674,565. J. W. Clement Company, Buffalo, N. Y. Filed Oct. 11, 1954.

No claim is made to the words "Printing Plates" apart from the mark as shown.

For Printing Plates.

Use since Sept. 29, 1954.

SN 674,954. Will Ecker & Co., St. Louis, Mo. Filed Oct. 18, 1954.

MEMORY-TRAY

For Highly Polished Trays Upon Which Engagement Announcements, Wedding Invitations, and the Like Have Been Stamped.

Use since during June 1935.

CLASS 51

SN 655,583. The Kaynar Company, Los Angeles, Calif. Filed Oct. 30, 1953.

deltone

For Hair Cream.

Use since Aug. 31, 1953.

SN 657,438. Kabushiki Kaisha Fujino Shoten, Nishiyodogawa-ku, Osaka City, Japan, now by change of name Smoca Hamigaki Kabushiki Kaisha. Filed Dec. 4, 1953.

SMOCA

Priority under Sec. 44(d). Japanese application No. 25,775 filed Oct. 2, 1953, Reg. No. 464,404, dated Apr. 13, 1955.

For Dentifrices.

SN 664,898. Valmor Products Co., Chicago, Ill. Filed Apr. 20, 1954. Sec. 2(f).

SLICK-BLACK

Applicant claims ownership of Reg. No. 394,688.

For Hair Dressing Pomade Having Hair Coloring Properties.

Use since Dec. 12, 1940.

SN 667,295. Commercial Solvents Corporation, New York, N. Y. Filed May 28, 1954.

HEXAGON

Applicant claims ownership of Reg. Nos. 317,124 and 317,328.

For Grain Alcohol, Pure Commercial Grain Alcohol, and Anhydrous Grain Alcohol, for Pure Cologne Spirits.

Use since July 1, 1925.

SN 668,332. Macsil Inc., Camden, N. J. Filed June 16, 1954.

BALMEX

For Baby Lotion, Baby Cream, and Baby Powder.

Use since June 14, 1954.

SN 670,657. Carter Products, Inc., New York, N. Y. Filed July 27, 1954.

AERODENT

For Dentifrice.

Use since June 1, 1954.

SN 672,672. Laboratoires Valdor Parfums Chopin Societe a Responsabilite Limitee, d. b. a. Valdor, Puteaux (Seine), France. Filed Sept. 2, 1954.

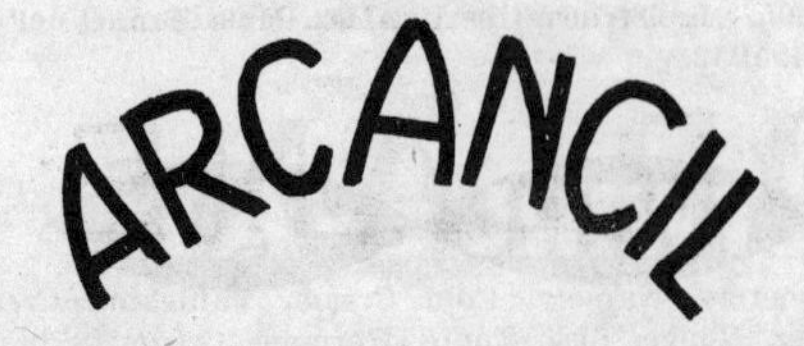

For Mascara.

Use since June 1935.

SN 672,673. Laboratoires Valdor Parfums Chopin Societe a Responsabilite Limitee, d. b. a. Valdor, Puteaux (Seine), France. Filed Sept. 2, 1954.

Guitare

The mark shown in the drawing is a French word, meaning "Guitar."
For Lipsticks and Rouge.
Use since June 1935.

SN 675,501. Parfums Weil Paris, S. A., Paris, France. Filed Oct. 26, 1954. Sec. 2(f).

No claim is made to the words "Parfums" and "Paris" apart from the mark as shown. Applicant claims ownership of the mark shown in expired Reg. No. 319,382.
For Perfumes, Face Powder, Talcum Powder, Cosmetic Skin Creams, Bath Salts, Face Lotions, Brilliantines, Lipstick, Rouge, Toilet Water.
Use since June 1933.

SN 676,099. Jacqueline Cochran, Inc., d. b. a. Jacqueline Cochran, Newark, N. J. Filed Nov. 5, 1954.

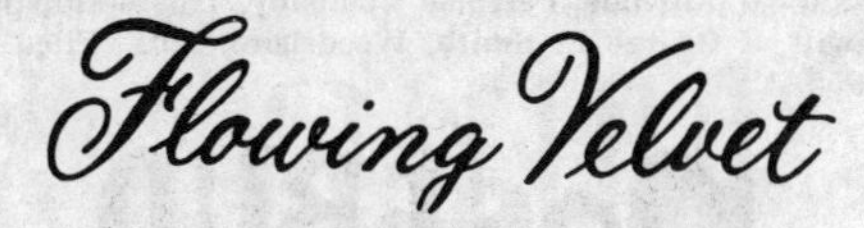

For Beautifying Skin Lotion.
Use since March 1949.

SN 678,340. Jean Patou Inc., New York, N. Y. Filed Dec. 14, 1954.

OLÉ
de
JEAN PATOU

Applicant claims ownership of Reg. Nos. 523,111 and 539,694.
For Perfumes and Toilet Waters.
Use since Dec. 1, 1954.

SN 679,370. Jean D'Albret, S. A. R. L., Paris, France. Filed Jan. 4, 1955.

ORLANE

For Cosmetics—Namely, Cold Cream, Vanishing Cream, Face Powder, Rouge, Skin Tonic, Hormone Cream, Skin Lotions, Camphor Lotion, Lipsticks, Eye Shadow.
Use since 1948.

CLASS 52

SN 642,859. James L. Younghusband, Chicago, Ill., to Thimothy Titus Morrow, Chicago, Ill. Filed Feb. 26, 1953.

Champoo

For Shampoo and Toilet Soap.
Use since June 20, 1952.

SN 664,097. The Standard Oil Company, Cleveland, Ohio. Filed Apr. 7, 1954.

SHOT-TREAT

Applicant claims ownership of Reg. Nos. 234,809 and 562,291.
For Graphite Pellets for Use in Cleaning Internal Combustion Engines.
Use since about Dec. 1, 1953.

SN 665,744. American Stores Company, Philadelphia, Pa. Filed May 5, 1954.

For Detergent in Powder Form for Use in Washing Clothing, Piece Goods, Dishes, and Other Washable Items.
Use since Feb. 17, 1954.

SN 665,757. Colgate-Palmolive Company, Jersey City, N. J. Filed May 5, 1954.

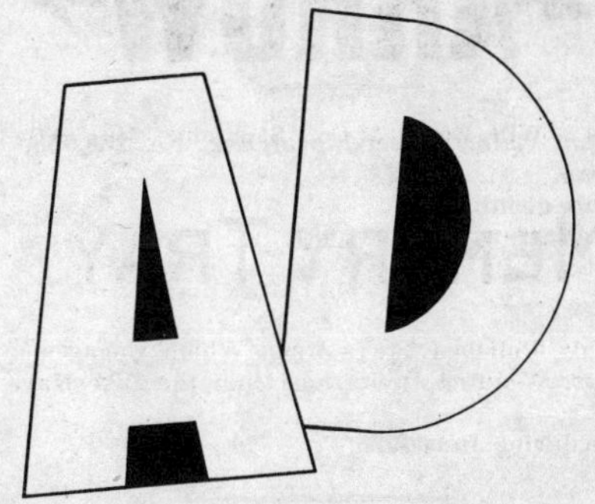

Applicant claims ownership of Reg. Nos. 402,937 and 407,780.
For Sudsing Cleaner, Cleanser, and Detergent.
Use since Apr. 1, 1954.

SN 668,344. Rehoboth Products Co. Inc., Rehoboth Beach, Del. Filed June 16, 1954.

BREAKWATER

For Bath and Toilet Soap.
Use since Sept. 4, 1953.

SN 669,624. Norda Essential Oil & Chemical Company, Inc., New York, N. Y. Filed July 8, 1954.

CHLORBISAN

For Germicidal Composition and Germicides for Use in Soap and Non-Soap Detergent Compositions.
Use since Apr. 1, 1954.

SN 671,217. Swift & Company, Chicago, Ill. Filed Aug. 5, 1954.

Rosamel

For Toilet Soap.
Use since in about 1914.

SN 675,637. Kosmos Electro-Finishing Research, Inc., Belleville, N. J. Filed Oct. 28, 1954.

KER-CHRO-MITE

For Acid Cleaning Solution.
Use since on or about Aug. 5, 1954.

SN 679,349. Swift & Company, Chicago, Ill. Filed Jan. 3, 1955. Sec. 2(f) as to "Swift's."

Swift's
ARROW

Applicant claims ownership of Reg. No. 288,958.
For Soap.
Use since about 1914.
Subj. to Intf. with SN 631,687.

SN 679,670. The Macco Products Company, Chicago, Ill. Filed Jan. 10, 1955.

For Metal Cleaning Compounds, Plate Cleaning Compounds, Burnishing Soaps—Namely, a Special Compounded Soap To Give a Better Finish, Compounds for Cleaning Porcelain Enamel, Paint Strippers, and Rust Removers.
Use since Apr. 1, 1931.

SN 680,395. Lehn & Fink Products Corporation, Bloomfield, N. J. Filed Jan. 24, 1955.

For Soap.
Use since about Aug. 18, 1954.

SN 683,139. Rosebud Perfume Company, trusteeship under the will of George F. Smith, Woodsboro, Md. Filed Mar. 9, 1955.

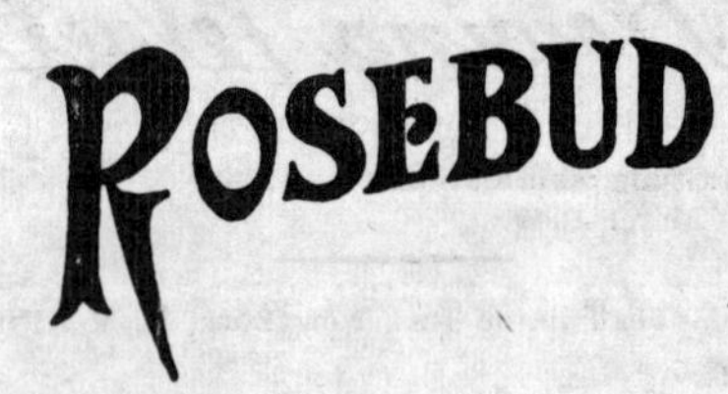

Applicant claims ownership of the mark shown in expired Reg. No. 71,878.
For Complexion and Toilet Soap.
Use since June 1, 1908.

SERVICE MARKS

CLASS 100

SN 651,544. Servo Corporation of America, New Hyde Park, N. Y. Filed Aug. 7, 1953. Sec. 2(f).

Applicant claims ownership of Reg. Nos. 542,274 and 559,985.

For Engineering and Consulting Services Rendered to Others on a Contract Basis—Namely, Conducting Surveys, Design Studies and Research Related to Electronics, Servomechanisms, and Computers.

Use since Apr. 1, 1948.

SN 655,308. Federation of Mutual Fire Insurance Companies, Chicago, Ill. Filed Oct. 26, 1953.

The drawing is lined for black.

For Disseminating Information and Advice Relating to Fire-Safety and Loss Prevention Practices in the Public Interest; Participating in National, State and Municipal Fire Prevention Committees and Their Fire-Safety Work; Assisting in the Preparation of National Fire Codes and Fire Standards of Safe Engineering Practices; and Furnishing or Assisting in Technical Fire Research and Education.

Use since Aug. 15, 1949.

SN 663,199. The Lummus Company, New York, N. Y. Filed Mar. 24, 1954. Sec. 2(f) as to "Lummus."

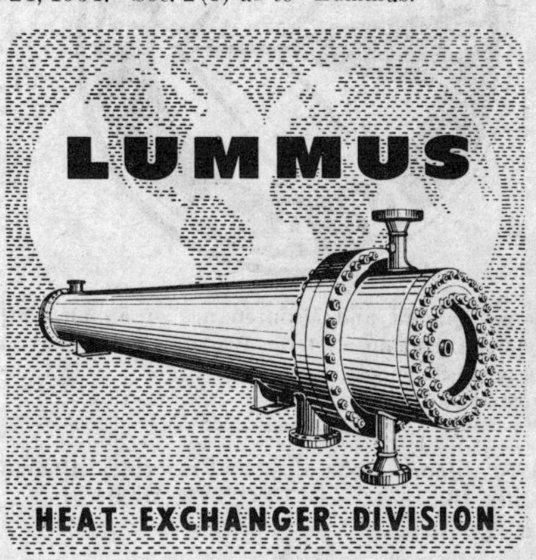

The drawing is lined for gray. The words "Heat Exchanger Division" are disclaimed apart from the mark as shown. Applicant claims ownership of Reg. Nos. 373,267, 589,237, and others.

For Engineering and Technical Services Pertaining to the Design and Development of Heat Exchange Equipment—Namely, Extraction Heaters, Boiler Blow-Down Heat Exchangers, Feed Water Heaters, Condensers, Drain Coolers, Evaporators, Heat Exchangers, Jacket Water Coolers, Lubricating Oil Coolers, Reboilers, Marine Condensers, Regenerators, Steam Generators, Steam Jet Refrigeration; and Power Plants, Chemical Plants, and Petroleum Refining Process Units Incorporating the Same, for Others.

Use since January 1951; and since 1913 as to "Lummus."

SN 667,327. National Grange of the Patrons of Husbandry, Washington, D. C. Filed May 28, 1954. Sec. 2(f).

The National Grange

For Developing Leadership, Improving Rural Life and Expanding Opportunities for Those Who Live by the Land.

Use since in 1876.

SN 667,427. North American Aviation, Inc., Los Angeles, Calif. Filed June 1, 1954.

MACH BUSTER

For Establishment and Maintenance of Membership in an Honorary Organization, and for Promoting the Advancement of Aeronautical Sciences.

Use since Mar. 13, 1954.

SN 667,933. Robert A. Cummings, Jr., d. b. a. Robert A. Cummings, Jr. and Associates, Pittsburgh, Pa. Filed June 9, 1954.

Sky-Mapping

For Aerial Mapping Surveys.

Use since May 1, 1954.

SN 670,119. North American Aviation, Inc., Los Angeles, Calif. Filed July 16, 1954.

For Establishment and Maintenance of Membership in an Honorary Organization, and for Promoting the Advancement of Aeronautical Sciences.
Use since June 1954.

SN 670,125. Sigma Alpha Epsilon Fraternity, Evanston, Ill. Filed July 16, 1954. COLLECTIVE MARK.

The Greek words appearing in the mark are "Sigma Alpha Epsilon." Applicant claims ownership of Reg. No. 538,680.
For Establishment and Maintenance of Membership in a National Collegiate Fraternity.
Use since March 1897.

SN 670,126. Sigma Alpha Epsilon Fraternity, Evanston, Ill. Filed July 16, 1954. COLLECTIVE MARK.

Applicant claims ownership of Reg. No. 538,680.
For Establishment and Maintenance of Membership in a National Collegiate Fraternity.
Use since March 1856.

SN 670,945. Arthur P. MacArthur, Milford, Pa. Filed Aug. 2, 1954.

For Purveying of Food, Beverages, and Other Refreshments to Travelers, Wayfarers and the General Public.
Use since July 9, 1954.

SN 674,310. Frank Robert Abbo, Washington, D. C. Filed Oct. 5, 1954.

For Restaurant Services.
Use since May 1, 1923.

CLASS 101

SN 668,125. Scripps-Howard Radio, Inc., Cleveland, Ohio. Filed June 11, 1954.

POOCH PARADE

For Advertising the Goods and Services of Others Through the Medium of a Television and Radio Animal Guest Star Program.
Use since May 26, 1951.

CLASS 102

SN 677,207. The Farm Bureau Mutual Automobile Insurance Company, Columbus, Ohio. Filed Nov. 24, 1954.

MINI-GROUP

For Underwriting Health and Accident Insurance.
Use since Oct. 13, 1954.

SN 681,473. Acacia Mutual Life Insurance Company, Washington, D. C. Filed Feb. 11, 1955.

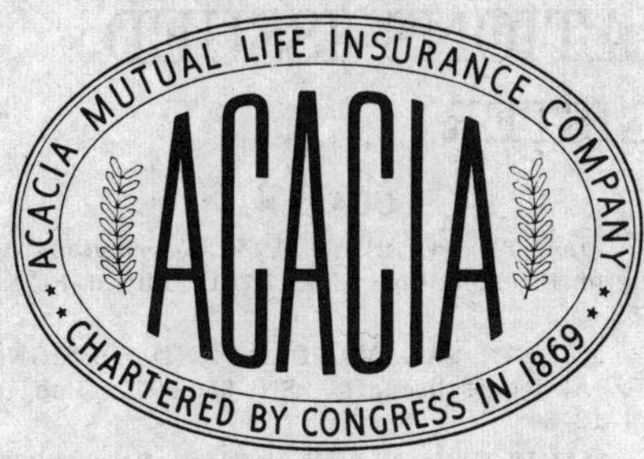

Applicant claims ownership of Reg. No. 580,097.
For Writing of Life Insurance.
Use since Feb. 10, 1955.

CLASS 103

SN 671,180. Monarch Dry Cleaners & Dyers, Inc., Chicago, Ill. Filed Aug. 5, 1954.

Applicant disclaims the words "Drapery Cleaning" apart from the mark as shown.
For Dry Cleaning Draperies.
Use since May 6, 1953.

CLASS 104

SN 664,999. Aircall, Inc., New York, N. Y. Filed Apr. 22, 1954.

A 30 MILE EXTENSION OF YOUR TELEPHONE BELL

For Radiopaging, a Service Consisting of Paging Subscribers by Radio.
Use since Mar. 15, 1954.

CLASS 106

SN 660,206. Graver Tank & Mfg. Co., Inc., East Chicago, Ind. Filed Jan. 27, 1954. Sec. 2(f).

Applicant claims ownership of Reg. Nos. 388,939 and 535,962.
For Fabrication of Water Treatment Equipment Generally on a Custom Basis.
Use since on or about Jan. 27, 1922.

SN 676,680. Tiarco Corporation, Newark, N. J. Filed Nov. 15, 1954.

TIARCORIZE

For Electrodeposition of Chromium Upon Metal Objects of Others.
Use since Oct. 5, 1954.

SN 676,681. Tiarco Corporation, Newark, N. J. Filed Nov. 15, 1954.

HARDALUME

For Electrodeposition of Chromium Upon Objects of Aluminum and Aluminum Alloys of Others.
Use since Oct. 5, 1954.

SN 676,682. Tiarco Corporation, Newark, N. J. Filed Nov. 15, 1954.

RIZZODIZE

For Electrodeposition of Chromium Upon Objects of Zinc, Zinc Alloys, Lead, and Lead Alloys of Others.
Use since Oct. 6, 1954.

SN 677,460. Tiarco Corporation, Newark, N. J. Filed Nov. 29, 1954.

SURE WEAR K

For Electrodeposition of Chromium Upon Dies and Cutting Tools of Others.
Use since Oct. 6, 1954.

SN 677,558. Fontaine Converting Works, Incorporated, Martinsville, Va. Filed Dec. 1, 1954.

HERTAINE

For Dyeing Synthetic Yarns and Fabrics, and Treating the Said Synthetic Yarns and Fabrics To Render Them Resistant to Pilling.
Use since June 25, 1954.

TRADEMARK REGISTRATIONS ISSUED

PRINCIPAL REGISTER

CLASS 1

611,601. WAHKEEN. Husman Hatchery. SN 658,456. Pub. 5–31–55. Filed 12–23–53.

611,602. FABRAY. The Stearns & Foster Company. SN 658,812. Pub. 5–31–55. Filed 12–30–53.

611,603. EXO-CHROME. Chromium Mining and Smelting Corporation. SN 673,646. Pub. 6–14–55. Filed 9–23–54.

CLASS 2

611,604. FEDERAL PRACTICAL HOUSEWARES AND DESIGN. Federal Tool Corp. SN 625,006. Pub. 6–7–55. Filed 2–15–52.

611,605. TICKETUBE. The Chapman Company. SN 626,855. Pub. 6–7–55. Filed 3–22–52.

611,606. SPORTSMAN. Gotham Industries, Inc. SN 631,158. Pub. 6–7–55. Filed 6–13–52.

611,607. GALARAMA. Garner & Co. SN 649,644. Pub. 5–31–55. Filed 7–1–53.

611,608. SUPER FRESHEEN. Dixie Wax Paper Company. SN 654,954. Pub. 5–31–55. Filed 10–19–53.

611,609. CORUKEG. Manhattan Container Corporation. SN 659,285. Pub. 6–7–55. Filed 1–11–54.

611,610. ROLL-A-CORD AND DESIGN. Oren D. Smith. SN 660,571. Pub. 5–31–55. Filed 2–3–54.

611,611. COFFEE CADDY AND DESIGN. Design Lab. Inc. SN 663,007. Pub. 5–31–55. Filed 3–22–54.

611,612. TRINKETRUNK. Emanuel Nathan, d. b. a. The Opticase Co., to The Opticase Company. SN 663,114. Pub. 6–7–55. Filed 3–23–54.

611,613. RAINBOW FOAM. Certified Creations. SN 663,438. Pub. 6–7–55. Filed 3–29–54.

611,614. DUBL-WAX. Bagcraft Corporation of America. SN 663,767. Pub. 6–7–55. Filed 4–2–54.

611,615. CURB-COP. The White-Giles Company. SN 665,459. Pub. 5–31–55. Filed 4–29–54.

611,616. PROTECTUBE. Rochester Paper Company. SN 665,643. Pub. 5–31–55. Filed 5–3–54.

611,617. ROLLA-BAG. Continental Can Company, Inc. SN 666,735. Pub. 5–31–55. Filed 5–20–54.

611,618. PERFO-ROLL. Continental Can Company, Inc. SN 666,737. Pub. 5–31–55. Filed 5–20–54.

611,619. PRE-FOLD. Atlas Plywood Corporation. SN 667,367. Pub. 6–7–55. Filed 6–1–54.

611,620. BLISTER-PAK. Merit Displays Co. SN 672,586. Pub. 5–31–55. Filed 9–1–54.

611,621. COLLAPS-A-TAINER. Republic Steel Corporation. SN 672,757. Pub. 5–31–55. Filed 9–3–54.

611,622. PLY-VENEER. Weyerhaeuser Timber Company. SN 673,321. Pub. 6–7–55. Filed 9–16–54.

611,623. PEN·E·NICKEL. Alla Products, Inc. SN 673,493. Pub. 5–31–55. Filed 9–21–54.

611,624. ECONOMATIC. American Linen Supply Co., sometimes d. b. a. Steiner Sales Company, Division of American Linen Supply Co. SN 673,496. Pub. 6–7–55. Filed 9–21–54.

611,625. CYPRESS. Gulf States Paper Corporation. SN 673,590. Pub. 5–31–55. Filed 9–22–54.

611,626. CHARRED OAK. Gulf States Paper Corporation. SN 673,591. Pub. 5–31–55. Filed 9–22–54.

611,627. KNOTTY PINE. Standard Packaging Corporation. SN 673,617. Pub. 6–7–55. Filed 9–22–54.

611,628. NOTTY PINE. Standard Packaging Corporation SN 673,618. Pub. 6–7–55. Filed 9–22–54.

611,629. "MY" TOOTH. Precision Dental Manufacturing Co. SN 674,091. Pub. 5–31–55. Filed 9–30–54.

611,630. RETRIPPER. Inland Container Corporation. SN 674,253. Pub. 6–7–55. Filed 10–4–54.

CLASS 4

611,631. DAINTY POLISH. C. E. Derrington, d. b. a. Derrington Products Co. SN 637,761. Pub. 6–7–55. Filed 11–7–52.

611,632. MARKE SALAMANDER AND DESIGN. Salamander Aktiengesellschaft. SN 650,007. Pub. 6–7–55. Filed 4–12–54.

611,633. MAGIK-OLA. Ed. W. Softley. SN 664,806. Pub. 6–7–55. Filed 4–19–54.

611,634. GRIFFIN MICROSHEEN. Griffin Manufacturing Co. Inc. SN 674,155. Pub. 5–31–55. Filed 10–1–54.

611,635. MICROSHEEN. Griffin Manufacturing Co. Inc. SN 674,156. Pub. 5–31–55. Filed 10–1–54.

611,636. PD AND DESIGN. The Cincinnati Milling Machine Co. SN 674,564. Pub. 6–7–55. Filed 10–11–54.

611,637. VINY LIFE LUSTRE. Vinylustre, Inc. SN 674,716. Pub. 6–7–55. Filed 10–12–54.

611,638. KITCHEN AID. The Hobart Manufacturing Company. SN 674,818. Pub. 6–7–55. Filed 10–14–54.

611,639. FILL-MARK. L. H. Tallmadge Co., Inc. SN 675,029. Pub. 6–7–55. Filed 10–18–54.

611,640. U-DO AND DESIGN (HUMAN HANDS). General Plywood Corporation. SN 675,143. Pub. 6–7–55. Filed 10–20–54.

CLASS 5

611,641. R RUTGERS AND DESIGN. Rutgerswerke-Aktiengesellschaft. SN 658,142. Pub. 6–7–55. Filed 12–17–53.

CLASS 6

611,642. CHLORO-CHEM. David Abelson, d. b. a. Abelson Bedding Co. SN 636,264. Pub. 6–7–55. Filed 10–7–52.

611,643. IRGACLAROL. Geigy Company, Inc., now by merger and change of name Geigy Chemical Corporation. SN 655,468. Pub. 6–7–55. Filed 10–28–53.

611,644. POULTRICIDE. King Research, Inc. SN 661,139. Pub. 6–7–55. Filed 2–16–54.

611,645. PQ. Philadelphia Quartz Company. SN 661,359. Pub. 6–7–55. Filed 2–19–54.

611,646. LORCO. Lord Chemical Corporation. SN 665,930. Pub. 6–7–55. Filed 5–7–54.

611,647. LORCO AND DESIGN. Lord Chemical Corporation. SN 665,931. Pub. 6–7–55. Filed 5–7–54.

611,648. SOLACET. Imperial Chemical Industries Limited. SN 666,112. Pub. 6–7–55. Filed 5–11–54.

611,649. POWERPLATE. Chemco Photoproducts Company, Inc. SN 666,571. Pub. 6–7–55. Filed 5–18–54.

611,650. BURNOK. T. F. Washburn Company. SN 672,471. Pub. 5–31–55. Filed 8–30–54.

611,651. GRAPHIDONE. Philip A. Hunt Company. SN 673,028. Pub. 5–31–55. Filed 9–10–54.

611,652. STA-FLUFF. A. E. Staley Manufacturing Company. SN 674,852. Pub. 6–7–55. Filed 10–14–54.

611,653. FIRON "A". Foote Mineral Company. SN 675,064. Pub. 6–7–55. Filed 10–19–54.

611,654. PARIFLUX. Foote Mineral Company. SN 675,065. Pub. 6–7–55. Filed 10–19–54.

611,655. FIRON "B". Foote Mineral Company. SN 675,067. Pub. 6–7–55. Filed 10–19–54.

CLASS 7

611,656. NATIONAL. Tom Stone Cordage Company. SN 643,247. Pub. 5–31–55. Filed 3–6–53.

CLASS 12

611,657. DEPENDON. Dependon Products Co. SN 648,007. Pub. 7–20–54. Filed 6–1–53.

611,658. "BONATE." Carl N. Beetle Plastics Corporation. SN 652,261. Pub. 6–7–55. Filed 8–25–53.

611,659. 400. Mycalex Corporation of America. SN 659,882. Pub. 6-7-55. Filed 1-21-54.

611,660. 410. Mycalex Corporation of America. SN 659,884. Pub. 6-7-55. Filed 1-21-54.

611,661. TERRASTYLE. The Mastic Tile Corporation of America. SN 664,164. Pub. 6-7-55. Filed 4-8-54.

611,662. KAOMUL. The Babcock & Wilcox Company. SN 666,903. Pub. 6-7-55. Filed 5-24-54.

611,663. METAL-LITE. American Heating & Appliance Company. SN 668,961. Pub. 6-7-55. Filed 6-28-54.

611,664. XTRAFLX. E. A. Chamberlain Limited. SN 671,415. Pub. 6-7-55. Filed 8-10-54.

611,665. KS-4. A. P. Green Fire Brick Company. SN 672,952. Pub. 6-7-55. Filed 9-9-54.

611,666. LUMISHADE. Mapes Industries, Inc. SN 675,297. Pub. 6-7-55. Filed 10-22-54.

611,667. BEAUTY ROLL. Orchard Brothers Incorporated. SN 675,416. Pub. 6-7-55. Filed 10-25-54.

611,668. SUNBEAM. Richard F. Horton, d. b. a. Aluminum Specialties. SN 678,114. Pub. 6-7-55. Filed 12-10-54.

611,669. CHAMCO. Chamberlin Company of America. SN 678,198. Pub. 6-7-55. Filed 12-13-54.

CLASS 13

611,670. SWISH. Swish Products Limited. SN 662,765. Pub. 6-7-55. Filed 3-16-54.

611,671. BELTREE. Garwood Metal Company, Inc. SN 665,351. Pub. 6-7-55. Filed 4-28-54.

611,672. CIRCLE SUPERIMPOSED ON LOCK WASHER. Illinois Tool Works. SN 673,234. Pub. 5-10-55. Filed 9-15-54.

611,673. ZIP ! IN EASY. Walter H. Jones, d. b. a. Lakeshore Products. SN 675,286. Pub. 5-3-55. Filed 10-22-54.

611,674. RED COLLAR WITHIN LOCK-NUT. Elastic Stop Nut Corporation of America. SN 675,976. Pub. 6-14-55. Filed 11-3-54.

CLASS 14

611,675. AIRCOLITE. Air Reduction Company, Incorporated. SN 665,165. Pub. 6-7-55. Filed 4-26-54.

CLASS 16

611,676. SEAL GRIP. The Verflex Co., Inc. SN 595,108. Pub. 6-7-55. Filed 4-1-50.

611,677. SEAL-PEEL. Seal-Peel, Inc. SN 624,054. Pub. 4-15-52. Filed 1-24-52.

611,678. PUBLIC SERVICE, ETC. AND DESIGN. Republic Paint and Varnish Co. SN 659,209. Pub. 6-7-55. Filed 1-8-54.

611,679. RAIN-SHIELD. Armstrong Cork Company. SN 664,726. Pub. 6-7-55. Filed 4-19-54.

611,680. COPACO. Cook Paint & Varnish Company. SN 679,215. Pub. 6-7-55. Filed 12-31-54.

CLASS 18

611,681. BIDROLIN. Armour and Company. SN 638,940. Pub. 6-7-55. Filed 12-4-52.

611,682. FASINETS. Pharmaceutical Products Co., Inc., to Chas. Pfizer & Co., Inc. SN 654,992. Pub. 6-7-55. Filed 10-19-53.

611,683. TRIVERMOL. Instituto Rosenbusch S. A. de Biologia Experimental Agropecuaria. SN 655,784. Pub. 6-7-55. Filed 10-19-54.

611,684. SALIZID. Nepera Chemical Co., Inc. SN 658,075. Pub. 6-7-55. Filed 12-16-53.

611,685. NAETENE. The Dietene Company. SN 658,705. Pub. 6-7-55. Filed 12-29-53.

611,686. AVINAR. Armour and Company. SN 659,023. Pub. 6-7-55. Filed 1-6-54.

611,687. OVINAR. Armour and Company. SN 659,027. Pub. 6-7-55. Filed 1-6-54.

611,688. GROLAC. Gooch Feed Mill Co. SN 661,432. Pub. 6-7-55. Filed 2-23-54.

611,689. COPPERIN "B." Jewell Pharmaceuticals, Inc. SN 661,899. Pub. 6-7-55. Filed 3-2-54.

611,690. NEOLITER. Don Baxter, Inc. SN 663,714. Pub. 6-7-55. Filed 4-1-54.

611,691. R12P. Rogers Grain & Feed Co. SN 665,783. Pub. 6-7-55. Filed 5-5-54.

611,692. THALOMYCIN. Schering Corporation. SN 666,601. Pub. 6-7-55. Filed 5-18-54.

611,693. BENETRYCIN. Julius Blackman Corporation, d. b. a. Supreme Pharmaceutical Company. SN 668,231. Pub. 6-7-55. Filed 6-15-54.

611,694. K5 KAFIVE. Nordmark-Werke Gesellschaft mit beschraenkter Haftung. SN 668,270. Pub. 6-7-55. Filed 6-15-54.

611,695. TAPE TAB. Hilltop Laboratories, Inc. SN 668,754. Pub. 6-7-55. Filed 6-23-54.

611,696. PINOLES. John O. Rose, d. b. a. Richards Brothers. SN 669,913. Pub. 6-7-55. Filed 7-13-54.

611,697. BSP. Otis E. Glidden & Company, Inc. SN 670,192. Pub. 6-7-55. Filed 7-19-54.

611,698. TRIOLANDREN. Ciba Limited. SN 672,797. Pub. 6-7-55. Filed 9-7-54.

611,699. ELECTROCORTEN. Ciba Limited. SN 672,798. Pub. 6-7-55. Filed 9-7-54.

611,700. TRIZATE. Warner-Hudnut, Inc., now by change of name Warner-Lambert Pharmaceutical Company. SN 673,212. Pub. 6-7-55. Filed 9-14-54.

611,701. EPILON. Wynlit Pharmaceuticals, Inc. SN 673,414. Pub. 6-7-55. Filed 9-17-54.

611,702. TOYTABS. Haver-Glover Laboratories, Inc. SN 674,969. Pub. 6-7-55. Filed 10-18-54.

611,703. AQUAMUNE. American Cyanamid Company. SN 675,196. Pub. 6-7-55. Filed 10-21-54.

611,704. NARZETS. Merck & Co., Inc. SN 675,301. Pub. 6-7-55. Filed 10-22-54.

611,705. SERPENESIN. E. S. Miller Laboratories, Inc. SN 676,327. Pub. 6-7-55. Filed 11-9-54.

CLASS 19

611,706. GRIFFIN. Griffin Wheel Company. SN 648,917. Pub. 6-7-55. Filed 6-17-53.

611,707. TURBOLECTRIC. Curtiss-Wright Corporation. SN 649,222. Pub. 6-7-55. Filed 6-23-53.

611,708. DRIV-EEZ. The Kampa Company, Inc. SN 659,929. Pub. 6-7-55. Filed 1-22-54.

611,709. CYCLE KING. Monark Silver King, Inc. SN 661,825. Pub. 5-31-55. Filed 3-1-54.

611,710. BMW. Bayerische Motoren Werke Aktiengesellschaft. SN 662,350. Pub. 6-7-55. Filed 3-10-54.

611,711. GOODWILL. General Motors Corporation. SN 666,842. Pub. 5-24-55. Filed 5-21-54.

611,712. DESIGN OF STORK. Storkline Furniture Corporation. SN 672,848. Pub. 5-31-55. Filed 9-7-54.

611,713. WEST-WOOD, ETC. AND DESIGN. West-Wood Products, Inc. SN 679,540. Pub. 6-7-55. Filed 1-6-55.

611,714. RETRACTO AND DESIGN. Design-Rite Company. SN 679,763. Pub. 6-7-55. Filed 1-12-55.

CLASS 21

611,715. MEALS-ON-WHEELS CRIMSCO INC. AND DESIGN. Crimsco, Inc. SN 638,195. Pub. 6-7-55. Filed 11-17-52.

611,716. REDI MIXER. Knapp-Monarch Company. SN 644,039. Pub. 6-7-55. Filed 3-23-53.

611,717. SPANMASTER. Spanmaster Crane Corporation of America. SN 645,614. Pub. 6-7-55. Filed 3-30-53.

611,718. WESTRON. Westron Corporation. SN 646,188. Pub. 6-7-55. Filed 4-29-53.

611,719. FLEETWOOD. Conrac, Inc. SN 646,738. Pub. 6-7-55. Filed 5-11-53.

611,720. OERLIKON AND DESIGN. Holding Intercito S. A., to Werkzeugmaschinenfabrik Oerlikon Bührle & Co., Administration Company. SN 652,859. Pub. 6-7-55. Filed 9-8-53.

611,721. WANDELUX. Wolf & Dessauer Company. SN 653,536. Pub. 6-7-55. Filed 9-21-53.

611,722. SFR AND DESIGN. Societe Francaise Radio-Electrique. SN 657,736. Pub. 6-7-55. Filed 12-9-53.

611,723. MAGNATRAP. Carter Carburetor Corporation. SN 657,878. Pub. 6-7-55. Filed 12-14-53.

611,724. STAGHOUND. Lee Rubber & Tire Corporation. SN 658,998. Pub. 6-7-55. Filed 1-5-54.

611,725. WRIST-ACTION. Burrell Corporation. SN 659,236. Pub. 6-7-55. Filed 1-11-54.

611,726. MUSICALE. Webster-Chicago Corporation. SN 659,583. Pub. 6-7-55. Filed 1-15-54.

611,727. NYCO AS OLD AS THE INDUSTRY. New York Coil Company, Inc. SN 660,426. Pub. 6-7-55. Filed 2-1-54.

611,728. DYMON-VANE. The Gabriel Company. SN 662,211. Pub. 6-7-55. Filed 3-8-54.

611,729. TEMPREX. Canadian Armature Works Inc. SN 662,514. Pub. 6-7-55. Filed 3-12-54.

611,730. INCREVOLT. Sorensen and Company, Inc. SN 662,682. Pub. 6-7-55. Filed 3-15-54.

611,731. ID IN CIRCLE DESIGN. International Distributors. SN 663,796. Pub. 6-7-55. Filed 4-2-54.

611,732. PARARITE. Essex Wire Corporation. SN 664,142. Pub. 6-7-55. Filed 4-8-54.

611,733. WASSCO AND DESIGN. Wassco Electric Products Corporation. SN 664,712. Pub. 6-7-55. Filed 4-16-54.

611,734. KW AND DESIGN. K. W. Battery Company. SN 665,770. Pub. 6-7-55. Filed 5-5-54.

611,735. PLANETARY. Paul D. Burgin, d. b. a. Burgin Battery and Auto Repair. SN 666,434. Pub. 6-7-55. Filed 5-17-54.

611,736. CERAFIL. Aerovox Corporation. SN 667,010. Pub. 6-7-55. Filed 5-25-54.

611,737. KITCHEN-POWER. Kitchen Power, Inc. SN 669,887. Pub. 6-7-55. Filed 7-13-54.

611,738. LYNN LIGHTNING. Vaco Products Company. SN 670,076. Pub. 6-7-55. Filed 7-15-54.

611,739. CHROMY (CARICATURE OF MAN). Edwin L. Wiegand Company, SN 670,143. Pub. 6-7-55. Filed 7-16-54.

611,740. UNI-BALLAST. McGraw Electric Company. SN 670,843. Pub. 6-7-55. Filed 7-30-54.

611,741. GROTESQUE DRAFTSMEN AT WORK. Donald M. May. SN 671,512. Pub. 6-7-55. Filed 8-11-54.

611,742. PACIFICA. Harry Jackson, d. b. a. Pacifica Interior Designs, to Pacifica Designs. SN 671,569. Pub. 6-7-55. Filed 8-12-54.

611,743. TOM THUMB. Automatic Radio Mfg. Co., Inc. SN 672,214. Pub. 6-7-55. Filed 8-25-54.

611,744. TRANSOLVER. Ketay Manufacturing Corporation, now by change of name Norden-Ketay Corporation. SN 673,031. Pub. 4-12-55. Filed 9-10-54.

611,745. COMPTOMETER. Felt and Tarrant Manufacturing Company. SN 674,237. Pub. 6-7-55. Filed 10-4-54.

611,746. START! AND GO! Hester Battery Manufacturing Company, Inc. SN 674,246. Pub. 6-7-55. Filed 10-4-54.

611,747. ELECTRIKBROOM. The Regina Corporation. SN 675,006. Pub. 5-31-55. Filed 10-18-54.

611,748. PULS-ALERT. National Alert, Inc. SN 675,904. Pub. 6-7-55. Filed 11-2-54.

611,749. G AND DESIGN. American Hard Rubber Company. SN 676,030. Pub. 6-7-55. Filed 11-4-54.

611,750. PERMERON. I-T-E Circuit Breaker Company. SN 676,115. Pub. 5-31-55. Filed 11-5-54.

611,751. WESTRIC AND DESIGN. Westric Battery Company. SN 676,694. Pub. 6-7-55. Filed 11-15-54.

611,752. WESTRIC CADMETIC AND DESIGN. Westric Battery Company. SN 676,695. Pub. 6-7-55. Filed 11-15-54.

611,753. FLASH IN A CIRCLE DESIGN. Westric Battery Company. SN 676,697. Pub. 6-7-55. Filed 11-15-54.

611,754. WESTRI-LITE. Westric Battery Company. SN 676,698. Pub. 6-7-55. Filed 11-15-54.

611,755. RECOSTAT. Regulator Equipment Corporation. SN 676,758. Pub. 6-7-55. Filed 11-16-54.

CLASS 22

611,756. GINASTA. Ginasta Co., to Ginasta Corporation of America. SN 665,494. Pub. 6-7-55. Filed 4-30-54.

611,757. DERWYN DE FOREST. Edw. K. Tryon Company. SN 670,997. Pub. 6-7-55. Filed 8-2-54.

611,758. JAY HARVEY. Edw. K. Tryon Company. SN 670,998. Pub. 6-7-55. Filed 8-2-54.

611,759. PITVEYOR. American Machine & Foundry Company. SN 675,250. Pub. 6-7-55. Filed 10-22-54.

611,760. RED-LOK. Aladdin Laboratories, Incorporated. SN 675,524. Pub. 6-7-55. Filed 10-27-54.

CLASS 23

611,761. SWORDSMAN AND REPRESENTATION OF SWORDSMAN. Edsbyns Industri Aktiebolag. SN 610,897. Pub. 6-7-55. Filed 3-6-51.

611,762. SPEED-O-TROL AND DESIGN. Link-Belt Speeder Corporation. SN 674,266. Pub. 6-7-55. Filed 10-4-54.

611,763. "QUICKSILVER." Kiekhaefer Corporation. SN 674,400. Pub. 6-7-55. Filed 10-6-54.

611,764. NATIONAL CRETE. National Foam System, Inc. SN 675,220. Pub. 6-7-55. Filed 10-21-54.

611,765. ROLL-A-SET. Sheldon M. Booth. SN 675,536. Pub. 6-7-55. Filed 10-27-54.

611,766. FASHION. The Singer Manufacturing Company. SN 675,922. Pub. 6-7-55. Filed 11-2-54.

611,767. CUT-RITE BITS. Dewey E. Joy, d. b. a. Cutter Bit Service Company. SN 676,504. Pub. 6-7-55. Filed 11-12-54.

611,768. C & B. Chisholm Industries, Inc. SN 676,785. Pub. 6-7-55. Filed 11-17-54.

CLASS 24

611,769. FLEX-I-DRI. Solar Corporation, d. b. a. Beam Manufacturing Company. SN 672,696. Pub. 6-7-55. Filed 9-2-54.

CLASS 26

611,770. METROCHORD. John Ridlon, d. b. a. Metrochord Company. SN 644,518. Pub. 5-24-55. Filed 3-31-53.

611,771. ELECTRO-MATIC. Eastern Industries, Incorporated. SN 645,885. Pub. 5-24-55. Filed 4-24-53.

611,772. BRONWILL. Bronwill Scientific, Inc. SN 648,284. Pub. 4-26-55. Filed 6-5-53.

611,773. NEOFLOW. Peerless Photo Products, Inc. SN 659,947. Pub. 6-7-55. Filed 1-22-54.

611,774. ISOLETTE. Agfa Camera-Werk Aktiengesellschaft. SN 659,968. Pub. 6-7-55. Filed 1-25-54.

611,775. KARAT. Agfa Camera-Werk Aktiengesellschaft. SN 659,969. Pub. 6-7-55. Filed 1-25-54.

611,776. MEGALUME. National Electrical Machine Shops, Inc. SN 662,330. Pub. 5-31-55. Filed 3-9-54.

611,777. HOUZE TONE AND DESIGN. L. J. Houze Convex Glass Company. SN 663,791. Pub. 6-7-55. Filed 4-2-54.

611,778. GRAVINER FIREWIRE AND DESIGN. Graviner Manufacturing Company Limited. SN 663,905. Pub. 6-7-55. Filed 4-5-54.

611,779. ELTRONIK AND DESIGN (CIRCLES). Blaupunkt Elektronik G. m. b. H. SN 664,505. Pub. 6-7-55. Filed 4-14-54.

611,780. FERROMETER. Eriez Manufacturing Company. SN 665,483. Pub. 6-7-55. Filed 4-30-54.

611,781. CALOTRON. Peerless Electric, Inc., now by change of name The Peerless Corporation. SN 665,634. Pub. 6-14-55. Filed 5-3-54.

611,782. MAGNI WHIRL. Blue M Electric Co. SN 667,776. Pub. 5-24-55. Filed 6-7-54.

611,783. STANLEY HANDYMAN AND DESIGN. The Stanley Works. SN 668,283. Pub. 5-24-55. Filed 6-15-54.

611,784. OPTI-MIKE. McGlynn Hays Industries, Inc. SN 668,769. Pub. 5-24-55. Filed 6-23-54.

611,785. MINOLTA-35. Chiyoda Kogaku Seiko Kabushiki Kaisha. SN 672,635. Pub. 5-24-55. Filed 9-2-54.

611,786. GOLD HERITAGE (DESIGN OF A CROWN). The Ednalite Optical Company, Inc. SN 672,647. Pub. 6-7-55. Filed 9-2-54.

611,787. MICRO-CLIP. Micro-Photo, Inc. SN 673,602. Pub. 5-24-55. Filed 9-22-54.

611,788. HERSEY. Hersey Manufacturing Company. SN 673,987. Pub. 5-31-55. Filed 9-29-54.

611,789. COPEASE. Copease Corporation. SN 674,138. Pub. 6-7-55. Filed 10-1-54.

611,790. KERN. Kern & Co. Ltd. SN 674,257. Pub. 6-7-55. Filed 10-4-54.

611,791. ANALYTE. Sal Iannelli, Inc., d. b. a. Crown Engineering and Sales Company. SN 674,334. Pub. 6-7-55. Filed 10-5-54.

611,792. ARITH-ME-KAT. Andrew G. Bertsik, d. b. a. Kounting Kat Co. SN 674,655. Pub. 6-7-55. Filed 10-12-54.

611,793. WHITEHALL. Quick-Set, Incorporated. SN 674,843. Pub. 6-7-55. Filed 10-14-54.

611,794. ANCYCLO AND DESIGN. Ancyclo Corporation of America. SN 674,927. Pub. 6-7-55. Filed 10-18-54.

611,795. POWERPILE. Minneapolis-Honeywell Regulator Company. SN 674,988. Pub. 6-7-55. Filed 10-18-54.

611,796. REG-U-FLOW AND DESIGN. Kerns Manufacturing Corporation. SN 675,148. Pub. 6-7-55. Filed 10-20-54.

611,797. MULTI-PULSER. Kay Electric Company. SN 675,398. Pub. 6-7-55. Filed 10-25-54.

611,798. M2-PLUS AND DESIGN. Herbert George Company. SN 675,678. Pub. 6-7-55. Filed 10-20-54.

CLASS 30

611,799. DIXIE DOGWOOD. Melvin H. Jacobs, to Marshall-Burns, Inc. SN 657,355. Pub. 12-14-54. Filed 12-3-53.

611,800. DIXIE DOGWOOD. Stetson China Co., Inc., to Marshall-Burns, Inc. SN 657,659. Pub. 12-14-54. Filed 12-8-53.

CLASS 31

611,801. HOME GUILD. Dap Incorporated. SN 654,528. Pub. 5-31-55. Pub. 10-12-53.

CLASS 32

611,802. LASTING BEAUTY. Luther B. Rush, d. b. a. Rush Mattress & Furniture Co. SN 654,844. Pub. 5-31-55. Filed 10-15-53.

611,803. STUMP "FOREMOST" AND DESIGN. Howard D. Stump, d. b. a. Stump Sales & Engineering Co. SN 656,275. Pub. 5-31-55. Filed 11-12-53.

611,804. LOUNGE-RITE. National Furniture Manufacturing Company, Inc. SN 658,304. Pub. 5-31-55. Filed 12-21-53.

CLASS 34

611,805. AIRLINER. Air Controls, Inc. SN 640,713. Pub. 5-31-55. Filed 1-14-53.

CLASS 35

611,806. PERMA-FLEX. Lawrence Process Company, Inc. SN 655,329. Pub. 9-21-54. Filed 10-26-53.

611,807. RAMAPO. Baldwin Belting, Inc. SN 667,665. Pub. 5-24-55. Filed 6-4-54.

611,808. CRISSCROSS BRAID. The Belmont Packing & Rubber Company. SN 679,079. Pub. 6-7-55. Filed 12-29-54.

CLASS 36

611,809. CUSHION RIM. Rudolf Muck. SN 664,241. Pub. 5-24-55. Filed 4-9-54.

CLASS 39

611,810. "PAT JOHNSON" (UNDERLINED). Johnson Garment Company. SN 632,525. Pub. 5-24-55. Filed 7-14-52.

611,811. PROMISE JR. Poirette Corsets, Inc. SN 651,288. Pub. 5-24-55. Filed 8-3-53.

CLASS 42

611,812. SHUR-TAPE. House Beautiful Curtains, Inc. SN 628,484. Pub. 10-21-52. Filed 4-22-52.

611,813. PERMA-THAL. Alamac Knitting Mills, Inc. SN 649,457. Pub. 5-31-55. Filed 6-29-53.

CLASS 43

611,814. ANGOR-GLO. Brant Yarns, Inc. SN 658,965. Pub. 5-31-55. Filed 1-5-54.

CLASS 44

611,815. PILLING-PHILA. The George P. Pilling & Son Company. SN 643,434. Pub. 6-7-55. Filed 3-10-53.

CLASS 45

611,816. WAVERLY. Superior Beverage Co., Inc. SN 626,480. Pub. 5-24-55. Filed 3-14-52.

611,817. REPRESENTATION OF BUFFALO, MOUNTAIN BACKGROUND WITHIN A HORSESHOE. Buffalo Rock Company, Inc. SN 657,106. Pub. 5-24-55. Filed 11-30-53.

611,818. FLOWING GOLD. Hart's Fruit Products Company, d. b. a. Brea Valley Processing Company, Citrus Products Company of California, Flowing Gold Citrus Products Company, and Brea Canning Company. SN 657,921. Pub. 5-31-55. Filed 12-14-53.

CLASS 46

611,819. TATC COSMOCRATIC AND DESIGN. Tenth Avenue Trading Corp. SN 603,089. Pub. 1-1-52. Filed 9-1-50.

611,820. G. S. L. GUARANTEED SHELF LIFE AND DESIGN. Peanut Corporation of America, d. b. a. Peanut Products Company. SN 614,464. Pub. 6-7-55. Filed 5-28-51.

611,821. Q-MAN AND DESIGN. Kuehmann Foods, Inc. SN 651,342. Pub. 6-7-55. Filed 8-4-53.

611,822. BLENDIPPED. Pangburn Company, Inc. SN 655,268. Pub. 6-7-55. Filed 10-23-53.

611,823. TEXO BOOSTER 54. Burrus Mills, Incorporated, d. b. a. Burrus Feed Mills. SN 655,637. Pub. 6-7-55. Filed 11-2-53.

611,824. OCEAN JOY. Alexander Aloff, d. b. a. Chicago Packing Company. SN 656,287. Pub. 6-7-55. Filed 11-13-53.

611,825. PROTERONI. A. Zerega's Sons, Inc. SN 656,686. Pub. 6-7-55. Filed 11-19-53.

611,826. THE FLOWER OF THE CROP. J. Aron & Company, Inc. SN 656,760. Pub. 6-7-55. Filed 11-23-53.

611,827. SWISS CHALET AND DESIGN. Joseph A. Reich, d. b. a. Swiss Chalet Products Co. SN 656,846. Pub. 6-7-55. Filed 11-23-53.

611,828. SEAL OF MINNESOTA. International Milling Company. SN 657,138. Pub. 6-7-55. Filed 11-30-53.

611,829. SUNRED. Sunred Corporation. SN 658,029. Pub. 6-7-55. Filed 12-15-53.

611,830. PICK OF THE PACK. Sam Reisfeld & Son. SN 658,862. Pub. 6-7-55. Filed 12-31-53.

611,831. PADRE BRAND ITALIAN STYLE AND DESIGN. Adolpho Minni, d. b. a. The Phoenix Macaroni Company. SN 659,695. Pub. 6-7-55. Filed 12-7-53.

611,832. LIGVAN. Flavor Corporation of America. SN 660,330. Pub. 6-7-55. Filed 1-29-54.

611,833. INSTANT REDI-LEMS IMITATION LEMON CONCENTRATE AND DESIGN. Seeman Brothers, Inc., to Airkem, Inc. SN 660,673. Pub. 6-7-55. Filed 2-5-54.

611,834. REDI-LEMS. Seeman Brothers, Inc., to Airkem, Inc. SN 660,674. Pub. 3-8-55. Filed 2-5-54.

611,835. HEINZ 57 IN KEYSTONE DESIGN. H. J. Heinz Company. SN 661,441. Pub. 6-7-55. Filed 2-23-54.

611,836. SUI GENERIS SINCE 1918 AND DESIGN. E. R. Jagenburg, Inc. SN 661,568. Pub. 6-7-55. Filed 2-24-54.

611,837. VITALMIN. Michael Shapiro, d. b. a. M. Shapiro. SN 662,824. Pub. 6-7-55. Filed 3-17-54.

611,838. GREENFIELD AND DESIGN (KEYSTONE). Hammond Standish & Co. SN 663,019. Pub. 6-7-55. Filed 3-22-54.

611,839. GRENNAN. American Bakeries Company. SN 663,334. Pub. 6-7-55. Filed 3-26-54.

611,840. GLORY OF NORWAY AND DESIGN. A/S Trondhjem Canning and Export Co. SN 663,339. Pub. 6-7-55. Filed 3-26-54.

611,841. RUDHARD'S. Rudhard Products, Inc. SN 663,520. Pub. 6-7-55. Filed 3-29-54.

611,842. ANTIQUE AUTOS. Haelan Laboratories, Inc., now by change of name Connelly Containers, Inc. SN 665,197. Pub. 6-7-55. Filed 4-26-54.

611,843. PRICE'S. Boyd Coffee Company. SN 665,556. Pub. 6-7-55. Filed 5-3-54.

611,844. PALACE. Palace Poultry Corp., d. b. a. Palace Products. SN 666,595. Pub. 6-7-55. Filed 5-18-54.

611,845. HENTEX. Henningsen, Inc. SN 666,755. Pub. 6-7-55. Filed 5-20-54.

611,846. SYNTOMATIC. Syntomatic Corporation. SN 669,823. Pub. 6-7-55. Filed 7-12-54.

611,847. RIVIERA. Caravetta Foods Co. SN 670,030. Pub. 6-7-55. Filed 7-15-54.

611,848. MIMBRES VALLEY AND DESIGN. Mimbres Valley Farmers Association, Inc. SN 670,770. Pub. 6-7-55. Filed 7-21-54.

611,849. 400 STABILIZER. American Maize-Products Company. SN 671,078. Pub. 6-7-55. Filed 8-4-54.

611,850. ALL-WIP. Milk Foods, Inc., U. S. A. SN 671,514. Pub. 6-7-55. Filed 8-11-54.

611,851. HY-KURE. Sterwin Chemicals Inc. SN 671,538. Pub. 6-7-55. Filed 8-11-54.

611,852. INHIBITOL. The General Emulsifier Corporation. SN 671,938. Pub. 6-7-55. Filed 8-19-54.

611,853. NOB HILL. Safeway Stores, Incorporated. SN 673,246. Pub. 6-7-55. Filed 9-15-54.

611,854. FLAVES. Smith Brothers, Inc. SN 673,483. Pub. 6-7-55. Filed 9-20-54.

611,855. THE JONES DAIRY FARM SAUSAGE MEAT AND DESIGN. Jones Dairy Farm. SN 673,596. Pub. 6-7-55. Filed 9-22-54.

611,856. BLUEBERRY ACRES AND DESIGN. Fred McDowell. SN 673,673. Pub. 6-7-55. Filed 9-23-54.

611,857. VALLEY VIEW. Valley View Packing Co., Inc., d. b. a. Valley View Packing Co. SN 674,210. Pub. 6-7-55. Filed 10-1-54.

611,858. PEEBLES' ANGEL-WHIP AND DESIGN. Western Condensing Company. SN 674,211. Pub. 6-7-55. Filed 10-1-54.

611,859. FAMOUS INN. Standard Brands Incorporated. SN 675,327. Pub. 6-7-55. Filed 10-22-54.

611,860. RUBY-KIST. Clement Pappas and Company, Inc. SN 675,500. Pub. 6-7-55. Filed 10-26-54.

611,861. McDANIEL. McDaniel & Sons, Inc. SN 676,212. Pub. 6-7-55. Filed 11-8-54.

611,862. CELEBRITY. The Borden Company. SN 676,463. Pub. 6-7-55. Filed 11-12-54.

611,863. VALCHRIS. Enoch S. Christoffersen, d. b. a. Christoffersen Poultry, Egg & Feed Market. SN 676,606. Pub. 6-7-55. Filed 11-15-54.

611,864. PHEZHEN AND DESIGN. Cambridge Grant Farm, Inc. SN 676,710. Pub. 6-7-55. Filed 11-16-54.

611,865. FOUR PEOPLE, COACH, ETC. Inn Maid Products, Inc. SN 677,570. Pub. 6-7-55. Filed 12-1-54.

611,866. CLASS LEADER. Eastern States Farmers' Exchange, Incorporated. SN 677,703. Pub. 6-7-55. Filed 12-3-54.

611,867. LOUDY'S AND DESIGN. Frank V. Loudy, d. b. a. Loudy Candy Co. SN 677,796. Pub. 6-7-55. Filed 12-6-54.

611,868. OMSTEAD'S. Omstead Fisheries Limited. SN 678,920. Pub. 6-7-55. Filed 12-24-54.

611,869. STEINFELD'S, ETC. AND DESIGN. Steinfeld's Products Company. SN 678,929. Pub. 6-7-55. Filed 12-24-54.

CLASS 48

611,870. FROM THE LAND OF SKY BLUE WATERS. Theo. Hamm Brewing Co. SN 649,423. Pub. 5-24-55. Filed 6-26-53.

CLASS 49

611,871. KORD AND DESIGN. Vychodoceske Lihovary, Narodni Podnik. SN 640,537. Pub. 5-24-55. Filed 1-8-53.

611,872. CLIFTON SPRINGS. Joseph S. Finch and Company. SN 656,499. Pub. 5-24-55. Filed 11-17-53.

CLASS 50

611,873. BOOT BUNNY AND DESIGN (REPRESENTATION OF A BOY). Arthur S. Cosler, Jr., d. b. a. Raravis Products Ltd. SN 656,054. Pub. 5-31-55. Filed 11-9-53.

611,874. YARNTURE. Art-N-Yarn Corporation. SN 662,409. Pub. 5-24-55. Filed 3-11-54.

CLASS 51

611,875. FRAC. Le Galion, Societe Anonyme. SN 657,719. Pub. 6-7-55. Filed 12-9-53.

611,876. ZEN. Trich-O-Matic Products, Inc. SN 659,328. Pub. 6-7-55. Filed 1-11-54.

611,877. LIDO VENICE PINK. Elizabeth Arden Sales Corporation. SN 659,519. Pub. 6-7-55. Filed 1-15-54.

611,878. JOLIE MADAME. Pierre Balmain. SN 660,182. Pub. 6-7-55. Filed 1-27-54.

611,879. VENT VERT. Pierre Balmain. SN 660,183. Pub. 6-7-55. Filed 1-27-54.

611,880. FLORAL DESIGN AND THE LETTER "O." C. H. Boehringer Sohn. SN 662,496. Pub. 6-7-55. Filed 2-25-54.

611,881. GUIDING STAR. Prince Matchabelli, Inc. SN 672,439. Pub. 6-7-55. Filed 8-30-54.

611,882. LILITH. Prince Matchabelli, Inc. SN 672,441. Pub. 6-7-55. Filed 8-30-54.

611,883. ODYSSEY. Prince Matchabelli, Inc. SN 672,442. Pub. 6-7-55. Filed 8-30-54.

611,884. REPRISE. Prince Matchabelli, Inc. SN 672,443. Pub. 6-7-55. Filed 8-30-54.

311,885. SOFT-LUSTRE. Schieffelin & Co. SN 674,848. Pub. 6-7-55. Filed 10-14-54.

CLASS 52

611,886. EC-54. The Pennsylvania Salt Manufacturing Company. SN 655,152. Pub. 6-7-55. Filed 10-21-53.

611,887. B·K—B·K. The Pennsylvania Salt Manufacturing Company. SN 656,448. Pub. 6-7-55. Filed 11-16-53.

611,888. MAGIKITCH'N. Magikitch'n Equipment Corporation. SN 662,804. Pub. 6-7-55. Filed 3-17-54.

611,889. TM-4. Winfield Brooks Company, Inc. SN 665,976. Pub. 6-7-55. Filed 5-7-54.

611,890. SOLVAY S AND DESIGN (REPRESENTATION OF AN ANCHOR). Allied Chemical & Dye Corporation. SN 668,293. Pub. 6-7-55. Filed 6-16-54.

611,891. RUSTRIPPER. Oakite Products, Inc. SN 670,607. Pub. 6-7-55. Filed 7-26-54.

611,892. WHITE KING. Los Angeles Soap Company. SN 671,840. Pub. 6-7-55. Filed 8-17-54.

611,893. DEED. Monsanto Chemical Company. SN 672,008. Pub. 5-31-55. Filed 8-20-54.

611,894. DISC-LUBE. Standard Record Company. SN 673,400. Pub. 6-7-55. Filed 9-17-54.

Service Marks

CLASS 100

611,895. BROOKMIRE. Brookmire Investors Service, Inc. SN 630,920. Pub. 6-7-55. Filed 6-9-52.

611,896. POSSIBILITIES UNLIMITED. Possibilities Unlimited, Inc. SN 640,804. COLLECTIVE MARK. Pub. 6-7-55. Filed 1-15-53.

611,897. FOUR CYLINDER CLUB OF AMERICA AND DESIGN. The Four Cylinder Club of America. SN 643,313. Pub. 6-7-55. Filed 3-9-53.

611,898. ADDING KNOW-WHY TO YOUR KNOW-HOW ETC. AND DESIGN. The Detroit Testing Laboratory, Inc. SN 653,123. Pub. 6-7-55. Filed 9-14-53.

611,899. AISC ETC. AND DESIGN. American Institute of Steel Construction, Inc. SN 656,288. Pub. 6-7-55. Filed 11-13-53.

611,900. EVERY SIRE PROVED GREAT. John Rockefeller Prentice, d. b. a. American Breeders Service. SN 657,959. Pub. 6-7-55. Filed 12-14-53.

611,901. GIRLS CLUBS OF AMERICA, INC. AND DESIGN. Girls Clubs of America, Inc. SN 674,884. Pub. 6-7-55. Filed 10-15-54.

CLASS 101

611,902. SOS. Stivers Office Service. SN 664,809. Pub. 6-7-55. Filed 4-19-54.

CLASS 102

611,903. "SCULLY ON YOUR INSURANCE, IS LIKE STERLING ON SILVER." George L. Scully, d. b. a. Geo. L. Scully Insurance Agency. SN 659,216. Pub. 6-7-55. Filed 1-8-54.

611,904. REPRESENTATION OF HORSE DRAWN FIRE ENGINE. Birmingham Fire Insurance Company. SN 670,159. Pub. 6-7-55. Filed 7-19-54.

611,905. TELE-TRIP AND DESIGN. Tele-Trip Policy Company, Inc. SN 670,763. Pub. 6-7-55. Filed 7-28-54.

CLASS 103

611,906. ORKIN ETC. AND DESIGN. Orkin Exterminating Company, Inc. SN 658,856. Pub. 6-7-55. Filed 12-31-53.

611,907. ORKIN-TOX. Orkin Exterminating Company, Inc. SN 658,857. Pub. 6-7-55. Filed 12-31-53.

611,908. VALETONE. One-Hour Valet of Atlanta, Georgia, Inc. SN 673,608. Pub. 6-7-55. Filed 9-22-54.

611,909. DEEP CLEANING AND DESIGN. Intrastate Spic and Span Dry Cleaners, Inc. SN 674,078. Pub. 6-7-55. Filed 9-30-54.

CLASS 104

611,910. GENERAL SYSTEM AND DESIGN. General Telephone Corporation. SN 671,893. Pub. 6-7-55. Filed 8-18-54.

611,911. KDDD DUMAS, ETC. AND REPRESENTATION OF A COWBOY. North Plains Broadcasting Corporation. SN 674,614. Pub. 6-7-55. Filed 10-11-54.

CLASS 105

611,912. CAMPBELL 66 EXPRESS ETC. AND DESIGN (CAMEL). Campbell Sixty-Six Express, Inc. SN 635,813. Pub. 6-7-55. Filed 9-26-52.

611,913. CAMPBELL 66 EXPRESS ETC. AND DESIGN (CAMEL). Campbell Sixty-Six Express, Inc. SN 635,814. Pub. 6-7-55. Filed 9-26-52.

611,914. MX. Michigan Express, Inc. SN 647,509. Pub. 6-7-55. Filed 5-22-53.

CLASS 106

611,915. INTERNATIONAL SERVICE AND DESIGN. International Steel Company. SN 641,929. Pub. 6-7-55. Filed 2-9-53.

611,916. KINETIX. Ketay Manufacturing Corporation, now by change of name Norden-Ketay Corporation. SN 649,595. Pub. 6-7-55. Filed 6-30-53.

611,917. CRINKLETEX. Cranston Print Works Company. SN 663,358. Pub. 6-7-55. Filed 3-26-54.

611,918. STEELFAB INC. AND SHIELD DESIGN. Steelfab, Inc. SN 664,807. Pub. 6-7-55. Filed 4-19-54.

CLASS 107

611,919. DANCING WATERS. Dancing Waters, Inc. SN 663,563. Pub. 6-7-55. Filed 3-30-54.

611,920. REPRESENTATION OF BELL. Telecourses, Incorporated. SN 664,101. Pub. 6-7-55. Filed 4-7-54.

611,921. LUNCHEON AT SARDI'S. Sardi's Enterprises, Inc. SN 673,867. Pub. 6-7-55. Filed 9-27-54.

SUPPLEMENTAL REGISTER

These registrations are not subject to opposition.

CLASS 1

611,922. Flex-O-Glass, Inc., d. b. a. Warp Bros., Chicago, Ill. SN 630,067. Filed P. R. 5-22-52. Am. S. R. 9-9-54.

For Transparent, Flexible, Plastic Sheet Material.
Use since Apr. 1, 1952.

611,923. The M. A. Hanna Company, Cleveland, Ohio. SN 670,480. Filed P. R. 7-23-54. Am. S. R. 6-28-55.

The Only Genuine
FRANKLIN COAL
of the
Susquehanna
Valley

For Coal.
Use since Apr. 23, 1954.

CLASS 12

611,924. Globe Roofing Products Co., Inc., Whiting, Ind. SN 668,659. Filed P. R. 6-22-54. Am. S. R. 7-5-55.

RAIN-TITE

For Roll Siding, Asphalt Roll Roofing, Insulated Roofing and Siding, Asphalt Shingles, Asphalt Siding, and Insulating Roofing and Siding for Building Structures.
Use since June 1, 1954.

611,925. The Beckman Supply Company, Hammond, Ind. SN 668,816. Filed P. R. 6–24–54. Am. S. R. 6–9–55.

For Rock Wool, Vermiculite and Fiberglass Building Insulation; Asphaltic and Bituminous Roofing and Building Siding; Lintels; Chimney Cleanouts, Ash Dumps, Coal Chutes, Metal Doorframes; Masonry Anchors, To Wit: Wall Ties, Flat and Rod Type Anchors, Flat Stone Anchors; Masonry Reinforcing Mesh, Bulk Aggregates for Concrete; Cement, Compounds Providing for an Expansion Joint Between Concrete Walls; Bricks, Masonry, Sand and Lime for Building Purposes; Mortar Cement; Lath, Plaster, Metal Corner Beads; Vitrified Clay Sewer Pipe and Fittings; Steel Window and Door Frames.
Use since Apr. 15, 1954.

CLASS 13

611,926. Norman V. Gibson, d. b. a. Gibson Manufacturing Company, Mission, Kans. SN 669,877. Filed P. R. 7–13–54. Am. S. R. 7–5–55.

Hang-a-board

For Hangers To Be Affixed to Doors or Walls for Holding Ironing Boards.
Use since June 4, 1954.

CLASS 18

611,927. John M. Yourek, d. b. a. Yourek Laboratories, Chicago, Ill. SN 662,907. Filed P. R. 3–18–54. Am. S. R. 1–22–55.

YOUR-ACHE

For Liniment.
Use since Nov. 22, 1953.

CLASS 23

611,928. Olson Industrial Products, Inc., Wakefield, Mass. SN 627,403. Filed P. R. 4–1–52. Am. S. R. 6–23–55.

For Gears and Gear Mechanisms, Including Change Speed Gear Heads and Speed Reducers Comprising Both Single Reduction and Double Reduction Helical Gear Speed Reducers.
Use since Sept. 1, 1947.

611,929. Dexter Folder Company, Pearl River, N. Y. SN 657,507. Filed P. R. 12–7–53. Am. S. R. 11–29–54.

MIEHLE-DEXTER

For Air Pumps.
Use since during July 1953.

611,930. Wright Power Saw and Tool Corp., Stratford, Conn. SN 667,278. Filed P. R. 5–27–54. Am. S. R. 2–25–55.

WRIGHT OILER

For Air Tool Lubricators for Introducing Oil in Measured Quantities Into an Air Line.
Use since June 29, 1953.

611,931. E. R. Wagner Manufacturing Company, Milwaukee, Wis. SN 669,836. Filed P. R. 7–12–54. Am. S. R. 6–29–55.

Rug Kleen

For Carpet Sweepers of the Mechanical Type.
Use since June 7, 1954.

CLASS 31

611,932. Kay-Gee Corp., Milwaukee, Wis. SN 669,772. Filed P. R. 7–12–54. Am. S. R. 6–30–55.

KOLD-MAT

For Mats for the Application of Cold to Parts of the Body and to Food To Be Preserved During Serving, the Mats Including Sheets of Sponge Rubber Partly Saturated With Water and Sealed Within a Water Impervious Flexible Envelope and Adapted To Be Frozen.
Use since May 8, 1953.

CLASS 34

611,933. O'Keefe and Merritt Company, Los Angeles, Calif. SN 626,108. Filed P. R. 3–7–52. Am. S. R. 11–5–54.

VANISHING SHELF

For Gas Ranges and Covers Therefor.
Use since October 1945.

CLASS 37

611,934. Orchard Paper Co., St. Louis, Mo. SN 653,497. Filed P. R. 9–21–53. Am. S. R. 8–20–54.

For Silver Anti-Tarnish Paper Storage Bags.
Use since July 15, 1953.

611,935. Fort Howard Paper Company, Green Bay, Wis. SN 654,396. Filed P. R. 10–8–53. Am. S. R. 12–13–54.

Page

For Toilet Tissue, Paper Napkins, and Paper Towels.
Use since Sept. 22, 1953.

CLASS 38

611,936. Crestwood Publishing Company, Inc., New York, N. Y. SN 646,531. Filed P. R. 5–6–53. Am. S. R. 7–6–55.

your Romance

For Periodical Publication.
Use since Apr. 29, 1953.

611,937. Kennedy Sinclaire, Inc., New York, N. Y. SN 669,816. Filed P. R. 7–12–54. Am. S. R. 5–18–55.

The ESTATE PLANNER

For Pamphlets Issued Periodically.
Use since April 1954.

611,938. Kennedy Sinclaire, Inc., New York, N. Y. SN 669,817. Filed P. R. 7–12–54. Am. S. R. 5–27–55.

The PENSION and PROFIT-SHARING Reporter

For Pamphlets Issued Periodically.
Use since September 1953.

611,939. The Miami Daily News, Inc., Miami, Fla. SN 671,849. Filed P. R. 8–17–54. Am. S. R. 6–29–55.

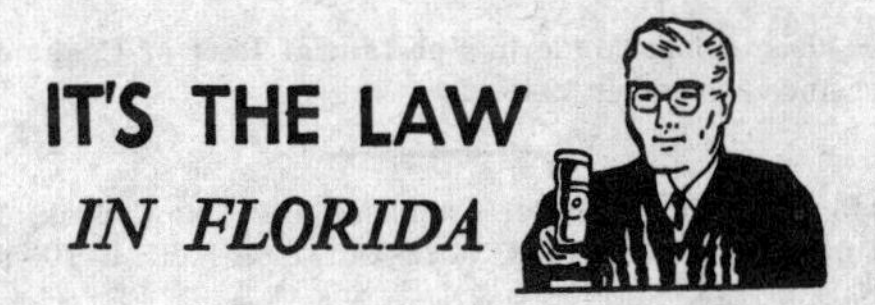

For Newspaper Column.
Use since May 10, 1954.

611,940. The Miami Daily News, Inc., Miami, Fla. SN 671,850. Filed P. R. 8–17–54. Am. S. R. 6–29–55.

For Newspaper Magazine Section.
Use since Jan. 10, 1954.

CLASS 39

611,941. Premier Knitting Co., Inc., New York, N. Y. SN 675,100. Filed P. R. 10–19–54. Am. S. R. 6–21–55.

Doublespun

For Ladies', Misses', Men's, and Children's Sweaters, and Ladies' and Misses' Knitted Suits and Knitted Ensembles Comprising a Pull Over Sweater and a Cardigan.
Use since May 17, 1954.

611,942. Standard Knitting Mills, Inc., Knoxville, Tenn. SN 677,999. Filed 12–8–54.

KRIS-KROS

For Men's and Boys' Knitted Underwear.
Use since Jan. 26, 1940.

611,943. Elgin Knit Sportswear, Inc., New York, N. Y. SN 676,796. Filed P. R. 11–17–54. Am. S. R. 6–21–55.

A Stanley BOUTIQUE

For Ladies' and Misses' Sweaters, Sweater Ensembles, Comprising Knitted Sweater and Skirt, Coats, Suits, Dresses, Blouses, Skirts, Slacks, Outer Shorts and Bathing Suits.
Use since June 7, 1954.

CLASS 42

611,944. Bernhard Altmann, New York, N. Y., to Massachusetts Textile Company (Bernhard Altmann) Inc., New York, N. Y. SN 628,984. Filed P. R. 5–1–52. Am. S. R. 9–24–54.

For Piece Goods Made in Substantial Part of Cashmere.
Use since November 1948.

611,945. House Beautiful Curtains, Inc., New York, N. Y. SN 668,663. Filed P. R. 6–22–54. Am. S. R. 6–10–55.

Frame-A-Bed

For Skirt Ruffles Used on Beds.
Use since June 7, 1954.

CLASS 43

611,946. The American Thread Company, New York, N. Y. SN 664,120. Filed P. R. 4–8–54. Am. S. R. 7–6–55.

DALTON

For Thread.
Use since May 1927.

CLASS 52

611,947. E. F. Drew & Co., Inc., New York, N. Y. SN 667,102. Filed P. R. 5–26–54. Am. S. R. 5–10–55

WET-N-CLEEN

For Dry Cleaning Compositions for Fabrics.
Use since July 6, 1953.

611,948. The Selig Company, Inc., Atlanta, Ga. SN 675,831. Filed P. R. 11–1–54. Am. S. R. 5–26–55.

VERI-KLEEN

For Cleaner, Reconditioner and Restorer for Terazzo and Other Floors.
Use since Aug. 10, 1953.

TRADEMARK REGISTRATIONS RENEWED

100,358. LAMSON. Cl. 23. 10–20–14.

100,359. LAMSON. Cl. 32. 10–20–14.

104,334. REPRESENTATION OF BEDSPRING WITH RECLINING FIGURES. Cl. 32. 5–18–15.

104,646. PILLSBURY'S XXXX ETC. AND DESIGN. Cl. 46. 6–8–15.

105,586. DESIGN OF KNIGHT. Cl. 46. 8–10–15.

105,835. EXCELLO. Cl. 6. 8–24–15.

106,249. SUPERIOR. Cl. 6. 10–19–15.

106,396. NOON HOUR. Cl. 46. 10–19–15.

106,471. KINGSFORD'S. Cl. 46. 10–19–15.

106,479. EXCELSIOR OTC ETC. AND DESIGN. Cl. 44. 10–19–15.

106,576. LEADER AND DESIGN. Cl. 16. 10–26–15.

106,668. DESIGN OF CIRCLE AND RECTANGLE. Cl. 16. 10–26–15.

107,189. OTCO CHAMPION AND DESIGN. Cl. 44. 11–16–15.

107,203. U S. Cl. 2. 11–16–15.

315,886. BORGES ETC. AND DESIGN. Cl. 47. 8–14–34.

316,488. PYRANIT. Cl. 21. 8–28–34.

320,590. GUIMARAENS' PORT. Cl. 47. 1–1–35.

320,591. GUIMARAENS & CO. Cl. 47. 1–1–35.

321,216. COUNTY FAIR. Cl. 49. 1–22–35.

321,277. STEEL STRONG. Cl. 37. 1–22–35.

321,567. FUNDADOR. Cl. 49. 2–5–35.

321,575. VERLEY BFA. Cl. 46. 2–5–35.

322,049. LA PROMESSE. Cl. 51. 2–26–35.

322,357. LE TRAITEMENT CORDAY. Cl. 52. 3–5–35.

323,934. AMERICAN CREAM. Cl. 49. 5–7–35.

323,937. SCHENLEY'S CREAM. Cl. 49. 5–7–35.

323,938. DOUBLE CREAM. Cl. 49. 5–7–35.

324,970. J. E. GROSJEAN ETC. AND DESIGN. Cl. 39. 6–4–35.

325,512. SYLPHCORD. Cl. 43. 6–25–35.

325,657. SYLPHCASE. Cl. 2. 7–2–35.

325,658. SYLVANIA SYLPHCASE. Cl. 2. 7–2–35.

325,716. AMERICAN PRIDE. Cl. 49. 7–2–35.

325,874. SYLVANIA SYLPHCASE AND DESIGN. Cl. 2. 7–9–35.

325,978. D. L. MOORE. Cl. 49. 7–9–35.

326,157. OLD GOLD. Cl. 1. 7–16–35.

327,086. EARLY TIMES. Cl. 49. 8–13–35.

327,260. HYDROIL. Cl. 15. 8–20–35.

327,894. CUBETTES. Cl. 23. 9–10–35.

328,076. ELBERT HOFFMAN PRODUCTS BELTEL AND DESIGN. Cl. 18. 9–17–35.

328,209. VACSUL. Cl. 15. 9–17–35.

328,268. MOBILGLOSS. Cl. 4. 9–24–35.

328,657. GLOBE AND DESIGN. Cl. 6. 10–1–35.

329,401. LLANO. Cl. 39. 10–29–35.

329,480. AVALUBE. Cl. 15. 10–29–35.

329,481. AVAGAS. Cl. 15. 10–29–35.

329,580. PACOLIZED. Cl. 35. 11–5–35.

329,989. SUN-MAID. Cl. 49. 11–19–35.

329,997. SUN-MAID. Cl. 47. 11–19–35.

TRADEMARK REGISTRATIONS CANCELED

Section 7

384,906. MEXICAN HATS. Cl. 46. 2–4–41.

Section 8

73,396. RISISTO. Cl. 16. 4–13–09.
85,661. REPRESENTATION OF AN ELEPHANT. Cl. 46. 3–5–12.
107,718. TAPLOW AND DESIGN. Cl. 22. 12–21–15.
112,215. BLUE TAG. Cl. 46. 8–22–16.
165,309. REGALEATHER. Cl. 50. 3–6–23.
202,362. MAIER AND DESIGN. Cl. 48. 8–18–25.
215,147. VICTORY. Cl. 40. 7–13–26.
216,942. NAVA-HUE. Cl. 39. 8–24–26.
225,132. FENOLE. Cl. 6. 3–15–27.
226,767. DIPLOMAT AND DRAWING LINED FOR BLUE AND GREEN. Cl. 46. 4–19–27.
228,930. GREAT PLAINS. Cl. 23. 6–14–27.
229,250. ENVOY BRAND AND CIRCULAR DESIGN LINED FOR BLUE AND RED. Cl. 46. 6–21–27.
230,466. SEA ROAMER. Cl. 46. 7–26–27.
231,159. TURKEY OF THE SEA. Cl. 46. 8–16–27.
239,843. MOSTLITE. Cl. 12. 3–13–28.
245,174. FRENCH SHRINER & URNER AND DESIGN. Cl. 39. 8–7–28.
245,412. BONNIE DOONE. Cl. 39. 8–14–28.
249,733. WIRTHMORE 16 PROTEIN AND DESIGN. Cl. 46. 11–20–28.
250,872. TWIN-GRIP. Cl. 39. 12–18–28.
252,728. GUILD OPTICAL PRODUCTS ETC. AND DESIGN. Cl. 26. 2–12–29.
271,885. DASHLEY. Cl. 39. 6–17–30.
279,852. LADY LUCK. Cl. 46. 2–3–31.
285,546. CHARLITE. Cl. 52. 7–28–31.
295,810. PEGGY PAREE ETC. AND DRAWING. Cl. 39. 7–12–32.
296,893. PERMAG ETC. AND DESIGN. Cl. 52. 8–23–32.
298,185. PERMAG ETC. AND DESIGN. Cl. 52. 10–18–32.
299,345. GOOD LUCK AND DESIGN LINED FOR RED, BLUE, AND YELLOW. Cl. 46. 11–29–32.
328,546. TAFFASWISH. Cl. 39. 10–1–35.
328,762. BUY-MORE AND DESIGN. Cl. 46. 10–1–35.
348,461. COACH & FOUR. Cl. 39. 7–27–37.
350,375. REGENT. Cl. 46. 9–28–37.
350,867. KIT'NSUEDE. Cl. 39. 10–12–37.
353,509. TREASURE LOOM AND DESIGN. Cl. 39. 1–11–38.
356,478. TUT TUT. Cl. 6. 4–26–38.
362,404. HEDGES & BUTLER 250 ETC. Cl. 49. 11–15–38.
364,062. SCRIPTOGRAM. Cl. 39. 1–17–39.
365,155. TUNABITS. Cl. 46. 2–21–39.
369,226. LOOMCRAFT SUPERIOR. Cl. 39. 7–18–39.
371,254. MEM-O-CLOCK. Cl. 37. 9–19–39.
371,752. KOLZOL. Cl. 15. 10–3–39.
372,645. LOOMCRAFT FROCKS. Cl. 39. 11–7–39.
372,723. BIG V AND DESIGN. Cl. 46. 11–14–39.
373,609. STARLETS. Cl. 39. 12–12–39.
383,379 MARDI GRAS. Cl. 33. 12–3–40.
393,942. 3 FACE ROCK AND DESIGN. Cl. 46. 3–10–42.
396,968. "DUST TO DUST". Cl. 26. 8–11–42.
397,065. SOUTHERN GEM. Cl. 46. 8–18–42.
398,757. GRAND PRIZE AND DESIGN. Cl. 48. 11–24–42.
401,356. SIGNATURE CRYSTAL. Cl. 33. 5–11–43.
408,911. PERSONALIZED. Cl. 33. 9–5–44.
409,551. TRAVALONG. Cl. 3. 10–10–44.
412,836. POPULARITY. Cl. 39. 3–27–45.
420,473. NOXTANE. Cl. 6. 4–16–46.
427,846. TROPIC-RAY. Cl. 26. 2–25–47.
433,752. S-O-R-O-A. Cl. 46. 10–28–47.
436,496. R. V. D. AND DESIGN. Cl. 46. 2–10–48.
436,696. GEOMETRICAL STRIPED DESIGN. Cl. 38. 2–17–48.
437,219. HEARTSTRINGS. Cl. 6. 3–9–48.
437,324. BANDLEY. Cl. 42. 3–16–48.
508,028. DANOLITE. Cl. 50. 3–29–49.
508,030. DAVOK. Cl. 39. 3–29–49.
508,031. "SNAP'R-ALL". Cl. 39. 3–29–49.
508,033. DILL-WILLIE AND DESIGN. Cl. 46. 3–29–49.
508,039. CLARK'S STEEL-TRUSS. Cl. 50. 3–29–49.
508,046. EVERY MEAL. Cl. 46. 3–29–49.
508,048. GREEN FLYER. Cl. 46. 3–29–49.
508,049. SALAMANDER. Cl. 46. 3–29–49.
508,050. 1847. Cl. 46. 3–29–49.
508,060. FLANDELL. Cl. 39. 3–29–49.
508,064. ROGER WILCO. Cl. 46. 3–29–49.
508,074. PERMASPUN AND DESIGN. Cl. 42. 3–29–49.
508,077. PERMA-LOT. Cl. 42. 3–29–49.
508,080. GRO-KWICK. Cl. 10. 3–29–49.
508,082. INSTANT HEAT. Cl. 34. 3–29–49.
508,084. MODERN. Cl. 26. 3–29–49.
508,087. TOWRY'S PLA-DOLL HOUSE AND DESIGN. Cl. 22. 3–29–49.
508,088. INVESTMENT GUIDES. Cl. 38. 3–29–49.
508,089. CARROLL BRAND. Cl. 38. 3–29–49.
508,090. MISS AMERICA. Cl. 26. 3–29–49.
508,094. CARDENAL GONZALEZ DE MENDOZA. Cl. 47. 3–29–49.
508,098. INSECT-O-POWDER. Cl. 6. 3–29–49.
508,099. FARNSWORTH. Cl. 47. 3–29–49.
508,100. MATRIS-DRI. Cl. 42. 3–29–49.
508,105. VITALIZED. Cl. 22. 3–29–49.
508,111. MAASDAM POW'R-PULL. Cl. 23. 3–29–49.
508,112. FIT-ALL. Cl. 50. 3–29–49.
508,114. EMAIL BARIL. Cl. 6. 3–29–49.
508,115. KIDDYKOOK. Cl. 22. 3–29–49.
508,116. 'RECORDA-CALL'. Cl. 23. 3–29–49.
508,120. LADY LYNN. Cl. 47. 3–29–49.
508,121. PUGLIESE. Cl. 47. 3–29–49.
508,122. THEY WEAR AND TAILOR BEST AND DESIGN. Cl. 42. 3–29–49.
508,124. HARRIMAN PRODUCTS CO. ETC. AND DESIGN. Cl. 4. 3–29–49.
508,125. MASKINERIETS POPULARE TIDD SKRIFT. Cl. 38. 3–29–49.
508,127. POPULÄRA MASKINERS. Cl. 38. 3–29–49.
508,129. READY SUDS. Cl. 4. 3–29–49.
508,130. TIDER X-SPONGE KLOTH. Cl. 1. 3–29–49.

Section 18

254,466. TUX AND DESIGN. Cl. 6. 3–26–29. Canc. 6154.
349,904. LIFE TIME CONSTRUCTION AND DESIGN. Cl. 21. 9–14–37. Canc. 6249.
353,886. WARMBAK. Cl. 39. 1–18–38. Canc. 6439.
363,155. ADORNO. Cl. 6. 12–13–38. Canc. 6428.
373,353. ACADEMY AWARD AND DESIGN. Cl. 39. 12–5–39. Canc. 5352.
378,327. LIN-O-SHEER. Cl. 42. 6–4–40. Canc. 6419.

393,963. B AND DESIGN. Cl. 48. 3–10–42. Canc. 6230.
429,779. TAM ETC. Cl. 39. 5–20–47. Canc. 6422.
528,997. REED REEL OVEN. Cl. 34. 8–15–50. Canc. 5732.
558,598. 1887 AND DESIGN. Cl. 48. 5–13–52. Canc. 6230.
595,729. NELVA ASSURES. Cl. 39. 9–21–54. Canc. 6442.
598,197. BEAU RESTA ETC. Cl. 32. 11–16–54. Canc. 6466.

Inadvertently Issued

608,924. FUL-LIFE. Cl. 4. 7–19–55.

TRADEMARK REGISTRATIONS AMENDED, DISCLAIMED, CORRECTED, ETC.

434,969. P D Q HI-ARC AND DESIGN. Cl. 21. 12–9–47. The Petrol Corporation. Standard Oil Company of California, San Francisco, Calif. Amended: In the statement, column 2, lines 5 and 6, 'No claim is made to the words "Hi-Arc" apart from the mark shown.' is deleted, and the drawing is amended to appear:

516,791. BAILEY. Cl. 26. 10–25–49. Bailey Meter Company, Cleveland, Ohio. Amended: In the statement, lines 19, 20 and 21, "air tracer contour control and duplicating attachments;" is deleted, and in line 24 "weighing scales," is deleted.

517,124. BAILEY METER. Cl. 26. 11–1–49. Bailey Meter Company, Cleveland, Ohio. Amended: In the statement, line 22, "weighing scales," is deleted.

517,125. BAILEY METER AND DESIGN. Cl. 26. 11–1–49. Bailey Meter Company, Cleveland, Ohio. Amended: In the statement, line 22, "weighing scales," is deleted.

525,281. WHINK. Cl. 52. 5–16–50. Whink Products Company, Eldora, Iowa. Amended: In the statement, line 6, "powdered soaps," is deleted; and in same line 6 and in line 7, after "liquid" *chemical* is inserted.

602,073. REPRESENTATION OF HUMAN MALE. Cl. 14. 2–15–55. Titan Metal Manufacturing Company, Bellefonte, Pa. Corrected: In the printed copy of the registration, column 2, lines 8 and 9, "June 1, 1915" should be *April 1, 1945*, and in line 10, "120,800" should be *288,569*.

602,281. REPRESENTATION OF HUMAN MALE. Cl. 13. 2–22–55. Titan Metal Manufacturing Company, Bellefonte, Pa. Corrected: In the printed copy of the registration, column 2, line 5, "June 1, 1915" should be *April 1, 1945*, and in line 6, "120,800" should be *288,569*.

608,161. POLY-PLY. Cl. 2. 7–5–55. The Greif Bros. Cooperage Corporation, Delaware, Ohio. Corrected: In the certificate, lines 2 and 3, and 13, and in the statement, column 1, line 1, name of registrant, for "The Grief Bros. Cooperage Corporation", each occurrence, read *The Greif Bros. Cooperage Corporation*.

608,398. BENSDORP. Cl. 46. 7–5–55. Bensdorp Inc., Bussum, Netherlands, assignor to Bensdorp, N. V., Bussum, Netherlands. Corrected: In the certificate, lines 2 and 3, address of registrant, for "Bussum, Netherlands, a corporation of the Netherlands" read *Boston, Massachusetts, a corporation of Massachusetts*; in the statement, column 1, line 1, for "(Dutch corporation)" read (*Massachusetts corporation*); line 2, strike out "Nieuwe Spiegelstraat 9,"; line 3, for "Bussum, Netherlands" read *Boston, Mass.*

REGISTRATIONS PUBLISHED UNDER SEC. 12(c)

The following marks registered under the act of 1905, or the act of 1881, are published under the provisions of section 12(c) of the Trademark Act of 1946. These registrations are not subject to opposition but are subject to cancellation under section 14 of the act of 1946.

CLASS 4

247,021. Sept. 18, 1928. Annabell H. Beggs, d. b. a. H-O Chemical Company, Denver, Colo. Pub. by William S. Beggs, Denver, Colo.

H-O

For Cleaning Fluid for Cleaning Gloves, Laces, Velvets, Chiffons, Silks, Carpets, Canvas, Suede, Buckskin, and Kid Gloves.

CLASS 5

411,444. Jan. 16, 1945. Oliver C. Eckel, d. b. a. Stic-Klip Mfg. Company, Cambridge, Mass. Pub. by Stic-Klip Manufacturing Co., Inc., Cambridge, Mass.

STIC - KLIP

For Adhesive Cement.

CLASS 13

388,239. June 17, 1941. Henry C. Abell, Asbury Park, N. J. Pub. by Schnitzer Alloy Products Company, Elizabeth, N. J.

An·cor·lox

For Lock Nuts.

389,671. Aug. 19, 1941. Henry C. Abell, Asbury Park, N. J. Pub. by Schnitzer Alloy Products Company, Elizabeth, N. J.

The word "Lox" is disclaimed apart from the mark as shown.
For Lock Nuts Used for Tightening Down Bolted Joints.

CLASS 19

443,717. Jan. 31, 1950. The Raleigh Cycle Company Ltd., Nottingham, England. Pub. by registrant.

The words "Nottingham England" are disclaimed apart from the mark as shown.
For Bicycles and Structural Parts Thereof.

CLASS 21

395,722. June 9, 1942. The Art Metal Company, Cleveland, Ohio. Pub. by registrant.

ART METAL

For Electric Lighting Fixtures.

CLASS 22

321,080. Jan. 15, 1935. Tricouni S. A., Geneva, Switzerland. Pub. by registrant.

TRICOUNI

For Skis, Ski Fastenings, Ski Sticks, Snow Shoes, Coasting Sleds, Roller Skates, Ice Skates, Tennis Rackets, Tennis Nets and Supports, Tennis Balls, Foot Balls, and Ping Pong Rackets and Balls.

CLASS 28

183,396. Apr. 29, 1924. Cohn & Rosenberger, Inc., New York, N. Y. Pub. by Coro, Inc., New York, N. Y.

For Jewelry for Personal Wear, Not Including Watches, Metal Pencils, Vanity Cases, Puff Boxes, Coin Purses, Mesh Bags and Metal Purses, All of Precious Metal or Plated With Precious Metal, and Necklace Sets Consisting of Beads of Precious Metal or Plated With Precious Metal or of Imitation Pearl, Composition, or Base Metal.

CLASS 37

104,710. June 15, 1915. Dennison Manufacturing Company, Framingham, Mass. Pub. by registrant.

THE TAG MAKERS

For Metal Fasteners for Tags, Blank Paper Cards for the Display of Jewelry, Cards for Correspondence, Cards for Presentations, Cards for Congratulations, Card Board, Adhesive Blank Labels, Blank Paper Seals, Envelopes, Pen Cases, Tissue Paper, Crepe Paper, and Wrapping Paper.

CLASS 38

211,570. Apr. 13, 1926. Concrete Publishing Company, Chicago, Ill. Pub. by Concrete Publishing Corporation, Chicago, Ill.

Concrete

For Magazines.

CLASS 42

309,981. Feb. 6, 1934. Atlas Mills, Inc., New York, N. Y. Pub. by Atlas Fabrics Corp., New York, N. Y.

Crepe Chiquita

For Textile Fabrics—Namely, Silk Piece Goods.

CLASS 45

318,612. Oct. 30, 1934. Good Humor Corporation of America, Brooklyn, N. Y. Pub. by Good Humor Corporation, Brooklyn, N. Y.

For Non-Alcoholic Maltless Beverages.

CLASS 47

323,811. Apr. 30, 1935. John Aquino Sons, Inc., New York, N. Y. Pub. by registrant.

Baronet

For Wines.

CLASS 48

312,834. May 8, 1934. The Croft Brewing Company, Boston, Mass. Pub. by Narragansett Brewing Company, Cranston, R. I.

For Ale.

CLASS 49

312,763. May 8, 1934. Kano Gomei Kaisha, Hyogo-ken, Japan. Pub. by Hakutsuru Sake Brewing Co., Ltd., Kobe. Japan.

For Japanese Sake.

LIST OF REGISTRANTS OF TRADEMARKS

A/S Trondhjem Canning and Export Co., Trondheim, Norway. 611,840, pub. 6-7-55. Cl. 46.
Abell, Henry C., Asbury Park, by Schnitzer Alloy Products Co., Elizabeth, N. J. 388,239, 12(c) pub. 9-6-55. Cl. 13.
Abell, Henry C., Asbury Park, by Schnitzer Alloy Products Co., Elizabeth, N. J. 389,671, 12(c) pub. 9-6-55. Cl. 13.
Abelson Bedding Co.: *See*—
Abelson, David.
Abelson, David, d. b. a. Abelson Bedding Co., Brooklyn, N. Y. 611,642, pub. 6-7-55. Cl. 6.
Adorno Mfg. Co., Seattle, Wash. 363,155, canc. Cl. 6.
Aerovox Corp., New Bedford, Mass. 611,736, pub. 6-7-55. Cl. 21.
Affiliated Distillers Brands Corp.: *See*—
Schenley Distributors, Inc.
Agfa Camera-Werk Aktiengesellschaft, Munich, Germany. 611,774-5, pub. 6-7-55. Cl. 26.
Air Controls, Inc., Cleveland, Ohio. 611,805, pub. 5-31-55. Cl. 34.
Airkem, Inc.: *See*—
Seeman Brothers, Inc.
Air Reduction Co., Inc., New York, N. Y. 611,675, pub. 6-7-55. Cl. 14.
Aladdin Laboratories, Inc., Minneapolis, Minn. 611,760, pub. 6-7-55. Cl. 22.
Alamac Knitting Mills, Inc., New York, N. Y. 611,813, pub. 5-31-55. Cl. 42.
Alla Products, Inc., Valparaiso, Ind. 611,623, pub. 5-31-55. Cl. 2.
Allied Chemical & Dye Corp., New York, N. Y. 611,890, pub. 6-7-55. Cl. 52.
Aloff, Alexander, d. b. a. Chicago Packing Co., Chicago, Ill. 611,824, pub. 6-7-55. Cl. 46.
Altmann, Bernhard, to Massachusetts Textile Co. (Bernhard Altmann) Inc., New York, N. Y. 611,944. Cl. 42.
Aluminum Specialties: *See*—
Horton, Richard F.
Aluminum Specialty Co., Manitowoc, Wis. 508,115, canc. Cl. 22.
American Bakeries Co., Chicago, Ill. 611,839, pub. 6-7-55. Cl. 46.
American Breeders Service: *See*—
Prentice, John R.
American Cyanamid Co., New York, N. Y. 611,703, pub. 6-7-55. Cl. 18.
American Hard Rubber Co., New York, N. Y. 611,749, pub. 6-7-55. Cl. 21.
American Heating & Appliance Co., Philadelphia, Pa. 611,663, pub. 6-7-55. Cl. 12.
American Institute of Steel Construction, Inc., New York, N. Y. 611,899, pub. 6-7-55. Cl. 100.
American Linen Supply Co., sometimes d. b. a. Steiner Sales Co., Division of American Linen Supply Co., Chicago, Ill. 611,624, pub. 6-7-55. Cl. 2.
American Machine & Foundry Co., New York, N. Y. 611,759, pub. 6-7-55. Cl. 22.
American Maize-Products Co., New York, N. Y. 611,849, pub. 6-7-55. Cl. 46.
American Spirits Inc., to Schenley Import Corp., New York, N. Y. 325,716, ren. 7-2-55. Cl. 49.
American Thread Co., The, New York, N. Y. 611,946. Cl. 43.
American Viscose Corp.: *See*—
Sylvania Industrial Corp.
Ancyclo Corp. of America, New York, N. Y. 611,794, pub. 6-7-55. Cl. 26.
Apex Specialties Co.: *See*—
Carson, George M.
Aquino, John, Sons, Inc., New York, N. Y. 323,811, 12(c) pub. 9-6-55. Cl. 47.
Arden, Elizabeth, Sales Corp., New York, N. Y. 611,877, pub. 6-7-55. Cl. 51.
Armour and Co., Chicago, Ill. 611,681, pub. 6-7-55. Cl. 18.
Armour and Co., Chicago, Ill. 611,686-7, pub. 6-7-55. Cl. 18.
Armstrong Cork Co., Lancaster, Pa. 611,679, pub. 6-7-55. Cl. 16.
Aron, J., & Co., Inc., New York, N. Y. 611,826, pub. 6-7-55. Cl. 46.
Art Metal Co., The, Cleveland, Ohio. 395,722, 12(c) pub. 9-6-55. Cl. 21.
Atlas Fabrics Corp.: *See*—
Atlas Mills, Inc.
Art-N-Yarn Corp., Rochester, N. Y. 611,874, pub. 5-24-55. Cl. 50.
Atlas Mills, Inc., by Atlas Fabrics Corp., New York, N. Y. 309,981, 12(c) pub. 9-6-55. Cl. 42.
Atlas Plywood Corp., Boston, Mass. 611,619, pub. 6-7-55. Cl. 2.
Attorney General of the United States: *See*—
Bosch, Robert, Aktiengesellschaft.
Automatic Radio Mfg. Co., Inc., Boston, Mass. 611,743, pub. 6-7-55. Cl. 21.
Axelman, Sam, d. b. a. The Travalong Co., Chicago, Ill. 409,551, canc. Cl. 3.
Babcock & Wilcox Co., The, New York, N. Y. 611,662, pub. 6-7-55. Cl. 12.
Bagcraft Corp. of America, Chicago, Ill. 611,614, pub. 6-7-55. Cl. 2.
Bailey Meter Co., Cleveland, Ohio. 516,791. Am. 7(d). Cl. 26.
Bailey Meter Co., Cleveland, Ohio. 517,124-5. Am. 7(d). Cl. 26.
Bakers Engineering & Equipment Co., Kansas City, Kans. 528,997, canc. Cl. 34.
Baldwin Belting, Inc., New York, N. Y. 611,807, pub. 5-24-55. Cl. 35.
Balmain, Pierre, Paris, France. 611,878-9, pub. 6-7-55. Cl. 51.
Baril, Marcel, Paris, France. 508,114, canc. Cl. 6.
Barnes, Kenneth B., Tulsa, Okla. 327,260, ren. 8-20-55. Cl. 15.
Barnes, Kenneth B., Tulsa, Okla. 329,480-1, ren. 10-29-55. Cl. 15.
Baxter, Don, Inc., Glendale, Calif. 611,690, pub. 6-7-55. Cl. 18.
Bayerische Motoren Werke Aktiengesellschaft, Munich, Germany. 611,710, pub. 6-7-55. Cl. 19.
Beam Mfg. Co.: *See*—
Solar Corp.
Beckman Supply Co., The, Hammond, Ind. 611,925. Cl. 12.
Beetle, Carl N., Plastics Corp., Fall River, Mass. 611,658, pub. 6-7-55. Cl. 12.
Beggs, Annabell H., d. b. a. H-O Chemical Co., by W. S. Beggs, Denver, Colo. 247,021, 12(c) pub. 9-6-55. Cl. 4.
Beggs, William S.: *See*—
Beggs, Annabell H.
Belmont Packing & Rubber Co., The, Philadelphia, Pa. 611,808, pub. 6-7-55. Cl. 35.
Bensdorp Inc., Bussum, Netherlands, to Bensdorf, N. V., Bussum, Netherlands. 608,398, cor. Cl. 46.
Bensdorp, N. V.: *See*—
Bensdorp Inc.
Benson, Christian L., to Benson Fish Co., Inc., Chicago, Ill. 106,396, ren. 10-19-55. Cl. 46.
Benson Fish Co., Inc.: *See*—
Benson, Christian L.
Berghoff Brewing Corp., Fort Wayne, Ind. 393,963, canc. Cl. 48.
Berghoff Brewing Corp., Fort Wayne, Ind. 558,598, canc. Cl. 48.
Bertsik, Andrew G., d. b. a. Kounting Kat Co., La Grange, Ill. 611,792, pub. 6-7-55. Cl. 26.
Birmingham Fire Insurance Co., Birmingham, Ala. 611,904, pub. 6-7-55. Cl. 102.
Bisceglia Brothers Wine Co., Fresno, Calif. 508,121, canc. Cl. 47.
Blackman, Julius, Corp., d. b. a. Supreme Pharmaceutical Co., Jersey City, N. J. 611,693, pub. 6-7-55. Cl. 18.
Blaupunkt Elektronik G. m. b. H., Berlin-Wilmersdorf, Germany. 611,779, pub. 6-7-55. Cl. 26.
Bliss Steel Products Corp., East Syracuse, N. Y. 239,843, canc. Cl. 12.
Blue M Electric Co., Chicago, Ill. 611,782, pub. 5-24-55. Cl. 26.
Boehringer, C. H., Sohn, Ingelheim am Rhine, Germany. 611,880, pub. 6-7-55. Cl. 51.
Booth, Sheldon M., South Haven, Mich. 611,765, pub. 6-7-55. Cl. 23.
Borden Co., The, New York, N. Y. 611,862, pub. 6-7-55. Cl. 46.
Bosch, Robert, Aktiengesellschaft, Stuttgart, Germany, to Attorney General of the United States, Washington, D. C. 316,488, ren. 8-28-54. Cl. 21.
Boyd Coffee Co., Portland, Oreg. 611,843, pub. 6-7-55. Cl. 46.
Brea Canning Co.: *See*—
Hart's Fruit Products Co.
Brea Valley Processing Co.: *See*—
Hart's Fruit Products Co.
Brent Yarns, Inc., New York, N. Y. 611,814, pub. 5-31-55. Cl. 43.
Brice-Waldrop Pickle Co., Denison, Tex. 508,033, canc. Cl. 46.
Bronwill Scientific, Inc., Rochester, N. Y. 611,772, pub. 4-26-55. Cl. 26.
Brookmire Investors Service, Inc., New York, N. Y. 611,895, pub. 6-7-55. Cl. 100.
Brown-Forman Distillers Corp.: *See*—
Brown-Forman Distillery Co., Inc.
Brown-Forman Distillery Co.: *See*—
Brown-Forman Distillery Co., Inc.
Brown-Forman Distillery Co., Inc., to Brown-Forman Distillery Co., to Brown-Forman Distillers Corp., Louisville, Ky. 327,086, ren. 8-13-55. Cl. 49.
Buffalo Rock Co., Inc., Birmingham, Ala. 611,817, pub. 5-24-55. Cl. 45.
Burgin Battery and Auto Repair: *See*—
Burgin, Paul D.
Burgin, Paul D., d. b. a. Burgin Battery and Auto Repair, Kansas City, Mo. 611,735, pub. 6-7-55. Cl. 21.
Burrell Corp., Pittsburgh, Pa. 611,725, pub. 6-7-55. Cl. 21.
Burris Mfg. Co., Inc., Lincolnton, N. C. 598,197, canc. Cl. 32.
Burrus Feed Mills: *See*—
Burrus Mills, Inc.
Burrus Mills, Inc., d. b. a. Burrus Feed Mills, Fort Worth, Tex. 611,823, pub. 6-7-55. Cl. 46.
Cambridge Grant Farm, Inc., Ashburnham, Mass. 611,864, pub. 6-7-55. Cl. 46.
Cameo Curtains Inc., New York, N. Y. 508,077, canc. Cl. 42.

Campbell Sixty-Six Express, Inc., Springfield, Mo. 611,912–13, pub. 6–7–55. Cl. 105.
Canadian Armature Works Inc., Montreal, Quebec, Canada. 611,729, pub. 6–7–55. Cl. 21.
Canadian Chamois & Leather Corp., New York, N. Y. 353,886, canc. Cl. 39.
Caravetta Foods Co., Chicago, Ill. 611,847, pub. 6–7–55. Cl. 46.
Carson, George M., d. b. a. Apex Specialties Co., New York, N. Y. 508,090, canc. Cl. 26.
Carter Carburetor Corp., St. Louis, Mo. 611,723, pub. 6–7–55. Cl. 21.
Certified Creations, New York, N. Y. 611,613, pub. 6–7–55. Cl. 2.
Chamberlain, E. A., Ltd., Nailsworth, England. 611,664, pub. 6–7–55. Cl. 12.
Chamberlin Co. of America, Detroit, Mich. 611,669, pub. 6–7–55. Cl. 12.
Chapman Co., The, Hammond, Ind. 611,605, pub. 6–7–55. Cl. 2.
Charlotte Chemical Laboratories, Inc., Charlotte, N. C. 285,546, canc. Cl. 52.
Charters-Davis Co., Corona, Calif. 226,767, canc. Cl. 46.
Charters-Davis Co., Corona, Calif. 229,250, canc. Cl. 46.
Chemco Photoproducts Co., Inc., Glen Cove, N. Y. 611,649, pub. 6–7–55. Cl. 6.
Chicago Packing Co.: *See*—
Aloff, Alexander.
Chisholm Industries, Inc., Lynn, Mass. 611,768, pub. 6–7–55. Cl. 23.
Chiyoda Kogaku Seiko Kabushiki Kaisha, Higashi-Ku, Osaka, Japan. 611,785, pub. 5–24–55. Cl. 26.
Christoffersen, Enoch S., d. b. a. Christoffersen Poultry, Egg & Feed Market, Turlock, Calif, 611,863, pub. 6–7–55. Cl. 46.
Christoffersen Poultry, Egg & Feed Market: *See*—
Christoffersen, Enoch S.
Chromium Mining and Smelting Corp., Chicago, Ill. 611,603, pub. 6–14–55. Cl. 1.
Ciba Ltd., Basel, Switzerland. 611,698–9, pub. 6–7–55. Cl. 18.
Cincinnati Milling Machine Co., The, Cincinnati, Ohio. 611,636, pub. 6–7–55. Cl. 4.
Citrus Products Co. of California: *See*—
Hart's Fruit Products Co.
Clark, J. R., Co., The, Minneapolis, Minn. 508,039, canc. Cl. 50.
Cohn & Rosenberger, Inc., by Coro, Inc., New York, N. Y. 183,396, 12(c) pub. 9–6–55. Cl. 28.
Concrete Publishing Co., by Concrete Publishing Corp., Chicago, Ill. 211,570, 12(c) pub. 9–6–55. Cl. 38.
Concrete Publishing Corp.: *See*—
Concrete Publishing Co.
Connelly Containers, Inc.: *See*—
Haelan Laboratories, Inc.
Conrac, Inc., Los Angeles County, Calif. 611,719, pub. 6–7–55. Cl. 21.
Continental Can Co., Inc., New York, N. Y. 611,617–18, pub. 5–31–55. Cl. 2.
Continental Distilling Corp., Philadelphia, Pa. 321,216, ren. 1–22–55. Cl. 49.
Cook Paint & Varnish Co., Kansas City, Mo. 611,680, pub. 6–7–55. Cl. 16.
Copease Corp., New York, N. Y. 611,789, pub. 6–7–55. Cl. 26.
Corn Products Refining Co.: *See*—
National Starch Co.
Corn Products Refining Co., New York, N. Y. 105,835, ren. 8–24–55. Cl. 6.
Corn Products Refining Co., New York, N. Y. 328,657, ren. 10–1–55. Cl. 6.
Coro, Inc.: *See*—
Cohn & Rosenberger, Inc.
Cosler, Arthur S., Jr., d. b. a. Raravis Products Ltd., Columbus, Ohio. 611,873, pub. 5–31–55. Cl. 50.
Cranston Print Works Co., Cranston, R. I. 611,917, pub. 6–7–55. Cl. 106.
Crestwood Publishing Co., Inc., New York, N. Y. 611,936. Cl. 38.
Crimsco, Inc., Kansas City, Mo. 611,715, pub. 6–7–55. Cl. 21.
Croft Brewing Co., The, Boston, Mass., by Narragansett Brewing Co., Cranston, R. I. 312,834, 12(c) pub. 9–6–55. Cl. 48.
Crown Engineering and Sales Co.: *See*—
Iannelli, Sal, Inc.
Cube Steak Machine Co., Inc., Boston, to Needham Mfg. Co. Inc., Needham Heights, Mass. 327,894, ren. 9–10–55. Cl. 23.
Curtiss-Wright Corp., Woodridge, N. J. 611,707, pub. 6–7–55. Cl. 19.
Cutter Bit Service Co.: *See*—
Joy, Dewey E.
Danbury Rubber Co., Inc., Danbury, Conn. 508,028, canc. Cl. 50.
Dancing Waters, Inc., New York, N. Y. 611,919, pub. 6–7–55. Cl. 107.
Dap Inc., Chicago, Ill. 611,801, pub. 5–31–55. Cl. 31.
Davies & Meacham, to The Patterson-Sargent Co., to The Patterson-Sargent Co., Cleveland, Ohio. 106,576, ren. 10–26–55. Cl. 16.
De Lisle, Nina: *See*—
Norris, Elizabeth.
Dennison Mfg. Co., Framingham, Mass. 104,710, 12(c) pub. 9–6–55. Cl. 37.
Dependon Products Co., Chicago, Ill. 611,657, pub. 7–20–54. Cl. 12.
De Pinna, A., Co., Inc., New York, N. Y. 373,609, canc. Cl. 39.
Derrington, C. E., d. b. a. Derrington Products Co., Jackson, Miss. 611,631, pub. 6–7–55. Cl. 4.
Derrington Products Co.: *See*—
Derrington, C. E.
Design Lab. Inc., Denver, Colo. 611,611, pub. 5–31–55. Cl. 2.
Design-Rite Co., Brainerd, Minn. 611,714, pub. 6–7–55. Cl. 19.
Detmer, Bruner & Mason, Inc., New York, N. Y. 508,122, canc. Cl. 42.
Detroit Testing Laboratory, Inc., The, Detroit, Mich. 611,898, pub. 6–7–55. Cl. 100.
Dexter Folder Co., Pearl River, N. Y. 611,929. Cl. 23.
Dietene Co., The, Minneapolis, Minn. 611,685, pub. 6–7–55. Cl. 18.
Dixie Wax Paper Co., Dallas, Tex. 611,608, pub. 5–31–55. Cl. 2.
Domecq, Pedro, S. A.: *See*—
Domecq, Pedro, y Cia.
Domecq, Pedro, y Cia, to Pedro Domecq, S. A., Jerez de la Frontera, Spain. 321,567, ren. 2–5–55. Cl. 49.
Downey, C. L., Co., The, Hannibal, Mo. 321,277, ren. 1–22–55. Cl. 37.
Drew, E. F., & Co., Inc., New York, N. Y. 611,947. Cl. 52.
Eastern Industries, Inc., East Norwalk, Conn. 611,771, pub. 5–24–55. Cl. 26.
Eastern States Farmers' Exchange, Inc., West Springfield, Mass. 611,866, pub. 6–7–55. Cl. 46.
Eckel, Oliver C., d. b. a. Stic-Klip Mfg. Co., by Stic-Klip Mfg. Co., Inc., Cambridge, Mass. 411,444, 12(c) pub. 9–6–55. Cl. 5.
Econo-Pak Register Co.: *See*—
Smith, Roy R.
Ednalite Optical Co., Inc., The, Peekskill, N. Y. 611,786, pub. 6–7–55. Cl. 26.
Edsbyns Industri Aktiebolag, Edsbyn, Sweden. 611,761, pub. 6–7–55. Cl. 23.
Elastic Stop Nut Corp. of America, Union, N. J. 611,674, pub. 6–14–55. Cl. 13.
Elgin Knit Sportswear, Inc., New York, N. Y. 611,943. Cl. 39.
Emerson-Brantingham Corp., Rockford, Ill. 228,930, canc. Cl. 23.
Eriez Mfg. Co., Erie, Pa. 611,780, pub. 6–7–55. Cl. 26.
Essex Wire Corp., Detroit, Mich. 611,732, pub. 6–7–55. Cl. 21.
Euclid Candy Co. of California, Inc., San Francisco, Calif. 279,852, canc. Cl. 46.
Federal Tool Corp., Chicago, Ill. 611,604, pub. 6–7–55. Cl. 2.
Felt and Tarrant Mfg. Co., Chicago, Ill. 611,745, pub. 6–7–55. Cl. 21.
Fenole Chemical Co.: *See*—
Powell, Wylie R.
Finch, Joseph S., & Co., Schenley, Pa., to Schenley Industries, Inc., New York, N. Y. 323,937, ren. 5–7–55. Cl. 49.
Finch, Joseph S., and Co., Schenley, Pa., and New York, N. Y. 323,938, ren. 10–26–55. Cl. 49.
Finch, Joseph S., and Co., New York, N. Y. 611,872, pub. 5–24–55. Cl. 49.
Fine Art Apron Co. Inc., New York, N. Y. 508,100, canc. Cl. 42.
Fit-All Mannequin Base: *See*—
Springer, Donald E.
Flavor Corp. of America, Chicago, Ill. 611,832, pub. 6–7–55. Cl. 46.
Flex-O-Glass, Inc., d. b. a. Warp Bros., Chicago, Ill. 611,922. Cl. 1.
Flowing Gold Citrus Products Co.: *See*—
Hart's Fruit Products Co.
Foote Mineral Co., Berwyn, Pa. 611,653–5, pub. 6–7–55. Cl. 6.
Fort Howard Paper Co., Green Bay, Wis. 611,935. Cl. 37.
Four Cylinder Club of America, The, Glendale, Calif. 611,897, pub. 6–7–55. Cl. 100.
French, Shriner & Urner, Boston, Mass. 245,174, canc. Cl. 39.
French, Shriner & Urner, Boston, Mass. 250,872, canc. Cl. 39.
Fuld Brothers, Inc., Baltimore, Md. 608,924, canc. Cl. 4.
Gabriel Co., The, Cleveland, Ohio. 611,728, pub. 6–7–55. Cl. 21.
Gallagher-Thomas Chemical Co., San Francisco, Calif. 254,466, canc. Cl. 6.
Gargiulo & Amendola, Inc., New York, N. Y. 433,752, canc. Cl. 46.
Garner & Co., New York, N. Y. 611,607, pub. 5–31–55. Cl. 2.
Garwood Metal Co., Inc., Garwood, N. J. 611,671, pub. 6–7–55. Cl. 13.
Geigy Chemical Corp.: *See*—
Geigy Co., Inc.
Geigy Co., Inc., New York, N. Y., now by merger and change of name Geigy Chemical Corp. 611,643, pub. 6–7–55. Cl. 6.
Gemal Handkerchief Co., New York, N. Y. 364,062, canc. Cl. 39.
General Emulsifier Corp., The, New York, N. Y. 611,852, pub. 6–7–55. Cl. 46.
General Motors Corp., Detroit, Mich. 611,711, pub. 5–24–55. Cl. 19.
General Plywood Corp., Louisville, Ky. 611,640, pub. 6–7–55. Cl. 4.
General Telephone Corp., New York, N. Y. 611,910, pub. 6–7–55. Cl. 104.
Gerber & Co., Thoune, Switzerland, to Gerber & Co., Inc., New York, N. Y. 105,586, ren. 8–10–55. Cl. 46.
Gerber & Co., Inc.: *See*—
Gerber & Co.
Gibson Mfg. Co.: *See*—
Gibson, Norman V.
Gibson, Norman V., d. b. a. Gibson Mfg. Co., Mission, Kans. 611,926. Cl. 13.
Ginasta Co., to Ginasta Corp. of America, New York, N. Y. 611,756, pub. 6–7–55. Cl. 22.
Ginasta Corp. of America: *See*—
Ginasta Co.
Girls Clubs of America, Inc., Springfield, Mass. 611,901, pub. 6–7–55. Cl. 100.

Glenmore Novelty Co., Inc. : *See*—
Tedoff, Samuel R.
Glidden, Otis E., & Co., Inc., Waukesha, Wis. 611,697, pub. 6-7-55. Cl. 18.
Globe Roofing Products Co., Inc., Whiting, Ind. 611,924. Cl. 12.
Godchaux, Leon, Clothing Co., Ltd., The, New Orleans, La. 271,885, canc. Cl. 39.
Gooch Feed Mill Co., Lincoln, Nebr. 611,688, pub. 6-7-55. Cl. 18.
Good Humor Corp. : *See*—
Good Humor Corp. of America.
Good Humor Corp. of America, by Good Humor Corp., Brooklyn, N. Y. 318, 612, 12(c) pub. 9-6-55. Cl. 45.
Gotham Industries, Inc., New York, N. Y. 611,606, pub. 6-7-55. Cl. 2.
Gould Storage Battery Corp., to National Battery Co., St. Paul, Minn. 349,904, canc. Cl. 21.
Graviner Mfg. Co. Ltd., London, England. 611,778, pub. 6-7-55. Cl. 26.
Green, A. P., Fire Brick Co., Mexico, Mo. 611,665, pub. 6-7-55. Cl. 12.
Greif Bros Cooperage Corp., The, Delaware, Ohio. 608,161, cor. Cl. 2.
Griffin Mfg. Co. Inc., Brooklyn, N. Y. 611,634-5, pub. 5-31-55. Cl. 4.
Griffin Wheel Co., Chicago, Ill. 611,706, pub. 6-7-55. Cl. 19.
Gro-Cord Rubber Co. : *See*—
Lima Cord Sole and Heel Co., The.
Guimaraens & Co., Villa Nova de Gaia, Portugal. 320,590-1, ren. 1-1-55. Cl. 47.
Gulf Brewing Co., Houston, Tex. 398,757, canc. Cl. 48.
Gulf States Paper Corp., Tuscaloosa, Ala. 611,625-6, pub. 5-31-55. Cl. 2.
H-O Chemical Co. : *See*—
Beggs, Annabell H.
Haelan Laboratories, Inc., Philadelphia, Pa., now by change of name Connelly Containers, Inc. 611,842, pub. 6-7-55. Cl. 46.
Hakutsuru Sake Brewing Co., Ltd. : *See*—
Kano Gomei Kaisha.
Hamm, Theo., Brewing Co., St. Paul, Minn. 611,870, pub. 5-24-55. Cl. 48.
Hammond Standish & Co., Detroit, Mich. 611,838, pub. 6-7-55. Cl. 46.
Hanna, M. A., Co., The, Cleveland, Ohio. 611,923. Cl. 1.
Hansen Glove Corp., Milwaukee, Wis. 348,461, canc. Cl. 39.
Hansen, O. C., Mfg. Co., Milwaukee, Wis. 216,942, canc. Cl. 39.
Hansen, O. C., Mfg. Co., Milwaukee, Wis. 245,412, canc. Cl. 39.
Harriman, Alfred Earl, d. b. a. Harriman Products Co., Los Angeles, Calif. 508,124, canc. Cl. 4.
Harriman Products Co. : *See*—
Harriman, Alfred Earl.
Harris Soap Co., Buffalo, N. Y. 508,129, canc. Cl. 4.
Hart's Fruit Products Co., d. b. a. Brea Valley Processing Co., Citrus Products Co. of California, Flowing Gold Citrus Products Co., and Brea Canning Co., Brea, Calif. 611,818, pub. 5-31-55. Cl. 45.
Haver-Glover Laboratories, Inc., Kansas City, Mo. 611,702, pub. 6-7-55. Cl. 18.
Hawley & Hoops, New York, N. Y. 384,906, canc. Cl. 46.
Hedges & Butler, Ltd., London, England. 362,404, canc. Cl. 49.
Heinz, H. J., Co., Pittsburgh, Pa. 611,835, pub. 6-7-55. Cl. 46.
Henningsen, Inc., New York, N. Y. 611,845, pub. 6-7-55. Cl. 46.
Herbert George Co., Chicago, Ill. 611,798, pub. 6-7-55. Cl. 26.
Hersey Mfg. Co., South Boston, Mass. 611,788, pub. 5-31-55. Cl. 26.
Hester Battery Mfg. Co., Inc., Nashville, Tenn. 611,746, pub. 6-7-55. Cl. 21.
Hider Equipment Corp., New York, N. Y. 508,130, canc. Cl. 1.
Hilltop Laboratories, Inc., Minneapolis, Minn. 611,695, pub. 6-7-55. Cl. 18.
Hobart Mfg. Co., The, Troy, Ohio. 611,638, pub. 6-7-55. Cl. 4.
Hoffman, C. Elbert, Philadelphia, Pa. 328,076, ren. 9-17-55. Cl. 18.
Holding Intercito S. A., to Werkzeugmaschinenfabrik Oerlikon Bührle & Co., Administration Co., Zurich-Oerlikon, Switzerland. 611,720, pub. 6-7-55. Cl. 21.
Hollywood Piece Dye Works, Inc. : *See*—
Passaic Piece Dye Works.
Horton, Richard F., d. b. a. Aluminum Specialties, Seattle, Wash. 611,668, pub. 6-7-55. Cl. 12.
House Beautiful Curtains, Inc., New York, N. Y. 611,812, pub. 10-21-52. Cl. 42.
House Beautiful Curtains, Inc., New York, N. Y. 611,945. Cl. 42.
Houze, L. J., Convex Glass Co., Point Marion, Pa. 611,777, pub. 6-7-55. Cl. 26.
Hunt, Philip A., Co., Palisades Park, N. J. 611,651, pub. 5-31-55. Cl. 6.
Husman Hatchery, Newman, Calif. 611,601, pub. 5-31-55. Cl. 1.
I-T-E Circuit Breaker Co., Philadelphia, Pa. 611,750, pub. 5-31-55. Cl. 21.
Iannelli, Sal, Inc., d. b. a. Crown Engineering and Sales Co., Harrison, N. J. 611,791, pub. 6-7-55. Cl. 26.
Illinois Tool Works, Chicago, Ill. 611,672, pub. 5-10-55. Cl. 13.
Imperial Chemical Industries Ltd., Millbank, London, England. 611,648, pub. 6-7-55. Cl. 6.
Imperial Glass Corp., Bellaire, Ohio. 383,379, canc. Cl. 33.
Imperial Glass Corp., Bellaire, Ohio. 401,356, canc. Cl. 33.
Imperial Glass Corp., Bellaire, Ohio. 408,911, canc. Cl. 33.
Inland Container Corp., Indianapolis, Ind. 611,630, pub. 6-7-55. Cl. 2.
Inn Maid Products, Inc., Millersburg, Ohio. 611,865, pub. 6-7-55. Cl. 46.
Instant Heat Oil Burner Corp. Brooklyn, N. Y. 508,082, canc. Cl. 34.
Instituto Rosenbusch S. A. de Biologia Experimental Agropecuaria, Buenos Aires, Argentina. 611,683, pub. 6-7-55. Cl. 18.
International Distributors, Chicago, Ill. 611,731, pub. 6-7-55. Cl. 21.
International Milling Co., Minneapolis, Minn. 611,828, pub. 6-7-55. Cl. 46.
International Steel Co., Evansville, Ind. 611,915, pub. 6-7-55. Cl. 106.
Intrastate Spic and Span Dry Cleaners, Inc., Milwaukee, Wis. 611,909, pub. 6-7-55. Cl. 103.
Jackson, Harry, d. b. a. Pacifica Interior Designs, to Pacifica Designs, Oakland, Calif. 611,742, pub. 6-7-55. Cl. 21.
Jacobs, Melvin H., to Marshall-Burns, Inc., Chicago, Ill. 611,799, pub. 12-14-54. Cl. 30.
Jagenburg, E. R., Inc., Brooklyn, N. Y. 611,836, pub. 6-7-55. Cl. 46.
Jelke, John F., Co., Chicago, Ill. 299,345, canc. Cl. 46.
Jewell Pharmaceuticals, Inc., Mount Vernon, N. Y. 611,689, pub. 6-7-55. Cl. 18.
Jewett & Sherman Co., Milwaukee, Wis. 85,661, canc. Cl. 46.
Johnson Garment Co., Chicago, Ill. 611,810, pub. 5-24-55. Cl. 39.
Johnson, Walter H., Candy Co., Chicago, Ill. 508,064, canc. Cl. 46.
Jones Dairy Farm, Fort Atkinson, Wis. 611,855, pub. 6-7-55. Cl. 46.
Jones, Walter H., d. b. a. Lakeshore Products, Woodinville, Wash. 611,673, pub. 5-3-55. Cl. 13.
Joseph & Feiss Co., The, Cleveland, Ohio. 329,401, ren. 10-29-55. Cl. 39.
Joy, Dewey E., d. b. a. Cutter Bit Service Co., Christopher, Ill. 611,767, pub. 6-7-55. Cl. 23.
K. W. Battery Co., Skokie, Ill. 611,734, pub. 6-7-55. Cl. 21.
Kalb, Voorhis & Co., New York, N. Y. 508,088, canc. Cl. 38.
Kampa Co., Inc., The, Minneapolis, Minn. 611,708, pub. 6-7-55. Cl. 19.
Kano Gomei Kaisha, Hyogo-Ken, Japan, by Hakutsuru Sake Brewing Co., Ltd., Kobe, Japan. 312,763, 12(c) pub. 9-6-55. Cl. 49.
Kay Electric Co., Pine Brook, N. J. 611,797, pub. 6-7-55. Cl. 26.
Kay-Gee Corp., Milwaukee, Wis. 611,932. Cl. 31.
Kennedy, Sinclaire, Inc., New York, N. Y. 611,937-8. Cl. 38.
Kern & Co. Ltd., Aarau, Switzerland. 611,790, pub. 6-7-55. Cl. 26.
Kerns Mfg. Corp., Long Island City, N. Y. 611,796, pub. 6-7-55. Cl. 26.
Ketay Mfg. Corp., New York, N. Y., now by change of name Norden-Ketay Corp. 611,744, pub. 4-12-55. Cl. 21.
Ketay Mfg. Corp., New York, N. Y., now by change of name Norden-Ketay Corp. 611,916, pub. 6-7-55. Cl. 106.
Kiekhaefer Corp., Cedarburg, Wis. 611,763, pub. 6-7-55. Cl. 23.
King Research, Inc., Brooklyn, N. Y. 611,644, pub. 6-7-55. Cl. 6.
Kirsch, Benjamin : *See*—
Minneapolis Bedding Co.
Kitchen Power, Inc., Cincinnati, Ohio. 611,737, pub. 6-7-55. Cl. 21.
Knapp-Monarch Co., St. Louis, Mo. 611,716, pub. 6-7-55. Cl. 21.
Kounting Kat Co. : *See*—
Bertsik, Andrew G.
Kruger, David O., d. b. a. Kruger Surgical Co., Los Angeles, Calif. 508,030, canc. Cl. 39.
Kruger Surgical Co. : *See*—
Kruger, David O.
Kuehmann Foods, Inc., Toledo, Ohio. 611,821, pub. 6-7-55. Cl. 46.
Lakeshore Products : *See*—
Jones, Walter H.
Lamson Co., The, Boston, Mass., to Lamson Corp., Syracuse, N. Y. 100,358, ren. 10-20-54. Cl. 23.
Lamson Co., The, Boston, Mass., to Lamson Corp., Syracuse, N. Y. 100,359, ren. 10-20-54. Cl. 32.
Lamson Corp. : *See*—
Lamson Co., The.
Lawrence Process Co., Inc., Lawrence, Mass. 611,806, pub. 9-21-54. Cl. 35.
Lederer, Henry, and Brother, Inc., Providence, R. I. 371,254, canc. Cl. 37.
Lee Rubber & Tire Corp., Conshohocken, Pa. 611,724, pub. 6-7-55. Cl. 21.
Le Galion, Societe Anonyme, Neuilly-sur-Seine, France. 611,875, pub. 6-7-55. Cl. 51.
Liberoff, William, d. b. a. Tropical Importing Co., Miami, Fla. 427,846, canc. Cl. 26.
Lima Cord Sole and Heel Co., The, to Gro-Cord Rubber Co., Lima, Ohio. 324,970, ren. 6-4-55. Cl. 39.
Link-Belt Speeder Corp., Cedar Rapids, Iowa. 611,762, pub. 6-7-55. Cl. 23.
Lionel Trading Co. Inc., to Parfumerie Internationale Corday, to Parfums Corday, Inc., New York, N. Y. 322,357, ren. 3-5-55. Cl. 52.
Lord Chemical Corp., York, Pa. 611,646-7, pub. 6-7-55. Cl. 6.
Los Angeles Soap Co., Los Angeles, Calif. 611,892, pub. 6-7-55. Cl. 52.
Loudy Candy Co. : *See*—
Loudy, Frank V.

Loudy, Frank V., d. b. a. Loudy Candy Co., Bristol, Va. 611,867, pub. 6–7–55. Cl. 46.
Maasdam, Felber, d. b. a. Maasdam Pow'R-Pull, Oak Grove, Oreg. 508,111, canc. Cl. 23.
Maasdam Pow'R-Pull: *See—*
Maasdam, Felber.
Magikitch'n Equipment Corp., Quakertown, Pa. 611,888, pub. 6–7–55. Cl. 52.
Magnuson Products Corp., Brooklyn, N. Y. 296,893, canc. Cl. 52.
Magnuson Products Corp., Brooklyn, N. Y. 298,185, canc. Cl. 52.
Maier Brewing Co., Inc., Los Angeles, Calif. 202,362, canc. Cl. 48.
Manhattan Container Corp., Hazel Park, Mich. 611,609, pub. 6–7–55. Cl. 2.
Mapes Industries, Inc., Lincoln, Nebr. 611,666, pub. 6–7–55. Cl. 12.
Marshall-Burns, Inc.: *See—*
Jacobs, Melvin H.
Stetson China Co., Inc.
Marvlo Mills, Inc., New York, N. Y. 378,327, canc. Cl. 42.
Massachusetts Textile Co. (Bernhard Altmann) Inc.: *See—*
Altmann, Bernhard.
Mastic Tile Corp. of America, The, Newburgh, N. Y. 611,661, pub. 6–7–55. Cl. 12.
May, Donald M., d. b. a. May Engineering Co., North Hollywood, Calif. 611,741, pub. 6–7–55. Cl. 21.
May Engineering Co.: *See—*
May, Donald M.
McDaniel & Sons, Inc., Somerton, Ariz. 611,861, pub. 6–7–55. Cl. 46.
McDowell, Fred, Neptune, N. J. 611,856, pub. 6–7–55. Cl. 46.
McGlynn Hays Industries, Inc., Belleville, N. J. 611,784, pub. 5–24–55. Cl. 26.
McGraw Electric Co., Milwaukee, Wis. 611,740, pub. 6–7–55. Cl. 21.
Merck & Co., Inc., Rahway, N. J. 611,704, pub. 6–7–55. Cl. 18.
Merit Displays Co., Paterson, N. J. 611,620, pub. 5–31–55. Cl. 2.
Metrochord Co.: *See—*
Ridlon, John.
Miami Daily News, Inc., The, Miami, Fla. 611,939–40. Cl. 38.
Michigan Express, Inc., Grand Rapids, Mich. 611,914, pub. 6–7–55. Cl. 105.
Micro-Photo, Inc., Cleveland, Ohio. 611,787, pub. 5–24–55. Cl. 26.
Milk Foods, Inc., U. S. A., Modesto, Calif. 611,850, pub. 6–7–55. Cl. 46.
Miller, E. S., Laboratories, Inc., Los Angeles, Calif. 611,705, pub. 6–7–55. Cl. 18.
Mimbres Valley Farmers Assn., Inc., Deming, N. Mex. 611,848, pub. 6–7–55. Cl. 46.
Minneapolis Bedding Co., Minneapolis, Minn., to B. Kirsch, Brooklyn, N. Y. 104,334, ren. 5–18–55. Cl. 32.
Minneapolis-Honeywell Regulator Co., Minneapolis, Minn. 611,795, pub. 6–7–55. Cl. 26.
Minni, Adolpho, d. b. a. The Phoenix Macaroni Co., Phoenix, Ariz. 611,831, pub. 6–7–55. Cl. 46.
Modern Optics, Inc., Houston, Tex. 508,084, canc. Cl. 26.
Monark Silver King, Inc., Chicago, Ill. 611,709, pub. 5–31–55. Cl. 19.
Monsanto Chemical Co., St. Louis, Mo. 611,893, pub. 5–31–55. Cl. 52.
Muck, Rudolf, New York, N. Y. 611,809, pub. 5–24–55. Cl. 36.
Mycalex Corp. of America, Clifton, N. J. 611,659–60, pub. 6–7–55. Cl. 12.
Narragansett Brewing Co.: *See—*
Croft Brewing Co., The.
Nathan, Emanuel, d. b. a. The Opticase Co., to The Opticase Company, Newark, N. J. 611,612, pub. 6–7–55. Cl. 2.
National Alert, Inc., Colorado Springs, Colo. 611,748, pub. 6–7–55. Cl. 21.
National Battery Co.: *See—*
Gould Storage Battery Corp.
National Electrical Machine Shops, Inc., Silver Spring, Md. 611,776, pub. 5–31–55. Cl. 26.
National Foam System, Inc., West Chester, Pa. 611,764, pub. 6–7–55. Cl. 23.
National Furniture Mfg. Co., Inc., Evansville, Ind. 611,804, pub. 5–31–55. Cl. 32.
National Starch Co., to Corn Products Refining Co., New York, N. Y. 106,249, ren. 10–19–55. Cl. 6.
National Starch Co., to Corn Products Refining Co., New York, N. Y. 106,471, ren. 10–19–55. Cl. 46.
Needham Mfg. Co. Inc.: *See—*
Cube Steak Machine Co., Inc.
Nelva, Rose, New York, N. Y. 595,729, canc. Cl. 39.
Nepera Chemical Co., Inc., Nepera Park, Yonkers, N. Y. 611,684, pub. 6–7–55. Cl. 18.
New York Coil Co., Inc., Phoenixville, Pa. 611,727, pub. 6–7–55. Cl. 21.
Norden-Ketay Corp.: *See—*
Ketay Mfg. Corp.
Nordmark-Werke Gesellschaft mit beschraenkter Haftung, Hamburg, Germany. 611,694, pub. 6–7–55. Cl. 18.
Norris, Elizabeth, d. b. a. Nina de Lisle, New York, N. Y. 437,219, canc. Cl. 6.
North Plains Broadcasting Corp., Dumas, Tex. 611,911, pub. 6–7–55. Cl. 104.
Oakite Products, Inc., New York, N. Y. 611,891, pub. 6–7–55. Cl. 52.
Ohio Truss Co., The, to Surgical Appliance Industries, Inc., Cincinnati, Ohio. 106,479, ren. 10–19–55. Cl. 44.
Ohio Truss Co., The, to Surgical Appliance Industries, Inc., Cincinnati, Ohio. 107,189, ren. 11–16–55. Cl. 44.
O'Keefe and Merritt Co., Los Angeles, Calif. 611,933. Cl. 34.
Olson Industrial Products, Inc., Wakefield, Mass. 611,928. Cl. 23.
Omstead Fisheries Ltd., Wheatley, Ontario, Canada. 611,868, pub. 6–7–55. Cl. 46.
One-Hour Valet of Atlanta, Georgia, Inc., Atlanta, Ga. 611,908, pub. 6–7–55. Cl. 103.
Opticase Co., The: *See—*
Nathan, Emanuel.
Opticase Company, The: *See—*
Nathan, Emanuel.
Orchard Brothers Inc., Rutherford, N. J. 611,667, pub. 6–7–55. Cl. 12.
Orchard Paper Co., St. Louis, Mo. 611,934. Cl. 37.
Orkin Exterminating Co., Inc., Atlanta, Ga. 611,906–7, pub. 6–7–55. Cl. 103.
Oxnard Canners, Inc., Port Hueneme, Calif. 393,942, canc. Cl. 46.
Oxnard Canners, Inc., Port Hueneme, Calif. 397,065, canc. Cl. 46.
Pacific Marine Supply Co., Seattle, Wash. 329,580, ren. 11–5–55. Cl. 35.
Pacifica Designs: *See—*
Jackson, Harry.
Pacifica Interior Designs: *See—*
Jackson, Harry.
Palace Poultry Corp., d. b. a. Palace Products, New York, N. Y. 611,844, pub. 6–7–55. Cl. 46.
Palace Products: *See—*
Palace Poultry Corp.
Pangburn Co., Inc., Fort Worth, Tex. 611,822, pub. 6–7–55. Cl. 46.
Pappas, Clement, and Co., Inc., Cedarville, Township of Lawrence, Cumberland County, N. J. 611,860, pub. 6–7–55. Cl. 46.
Paraffine Paint Co., The, San Francisco, Calif. 73,396, canc. Cl. 16.
Parfumerie Internationale Corday: *See—*
Lionel Trading Co. Inc.
Parfums Corday, Inc.
Parfums Corday, Inc.: *See—*
Lionel Trading Co. Inc.
Parfums Corday, Inc., to Parfumerie Internationale Corday, to Parfums Corday, Inc., New York, N. Y. 322,049, ren. 2–26–55. Cl. 51.
Passaic Piece Dye Works, Passaic N. J., now by change of name Hollywood Piece Dye Works, Inc. 508,074, canc. Cl. 42.
Patterson-Sargent Co., The: *See—*
Davies & Meacham.
Patterson-Sargent Co., The, Cleveland, Ohio. 106,668, ren. 10–26–55. Cl. 16.
Peanut Corp. of America, d. b. a. Peanut Products Co., Des Moines, Iowa. 611,820, pub. 6–7–55. Cl. 46.
Peanut Products Co.: *See—*
Peanut Corp. of America.
Peerless Corp., The: *See—*
Peerless Electric, Inc.
Peerless Electric, Inc., New York, N. Y., now by change of name The Peerless Corp. 611,781, pub. 6–14–55. Cl. 26.
Peerless Photo Products, Inc., Shoreham, N. Y. 611,773, pub. 6–7–55. Cl. 26.
Penley, Eugene W., Auburn, Maine. 112,215, canc. Cl. 46.
Pennsylvania Salt Mfg. Co., The, Philadelphia, Pa. 611,886–7, pub. 6–7–55. Cl. 52.
People's Outfitting Co., Detroit, Mich. 295,810, canc. Cl. 39.
Pepper, Jas. E., & Co., Lexington, Ky., and New York, N. Y., to Schenley Distillers, Inc., New York, N. Y. 325,978, ren. 7–9–55. Cl. 49.
Petrol Corp., The, Standard Oil Co. of California, San Francisco, Calif. 434,969. Am. 7(d). Cl. 21.
Pfizer, Chas., & Co., Inc.: *See—*
Pharmaceutical Products Co., Inc.
Pharmaceutical Products Co., Inc., Hato Rey, P. R., to Chas. Pfizer & Co., Inc., New York, N. Y. 611,682, pub. 6–7–55. Cl. 18.
Philadelphia Quartz Co., Philadelphia, Pa. 611,645, pub. 6–7–55. Cl. 6.
Phoenix Macaroni Co., The: *See—*
Minni, Adolpho.
Philling, George P., & Son Co., The, Philadelphia, Pa. 611,815, pub. 6–7–55. Cl. 44.
Pillsbury Flour Mills Co., to Pillsbury Mills, Inc., Minneapolis, Minn. 104,646, ren. 6–8–55. Cl. 46.
Pillsbury Mills, Inc.: *See—*
Pillsbury Flour Mills Co.
Plymouth Wholesale Dry Goods Corp., New York, N. Y. 508,089, canc. Cl. 38.
Poirette Corsets, Inc., New York, N. Y. 611,811, pub. 5–24–55. Cl. 39.
Popular Mechanics Co., Chicago, Ill. 508,125, canc. Cl. 38.
Popular Mechanics Co., Chicago, Ill. 508,127, canc. Cl. 38.
Possibilities Unlimited, Inc., Cleveland, Ohio. 611,896, pub. 6–7–55. Cl. 100.
Powell, Wylie R., d. b. a. Fenole Chemical Co., Jacksonville, Fla. 225,132, canc. Cl. 6.
Precision Dental Mfg. Co., Chicago, Ill. 611,629, pub. 5–31–55. Cl. 2.
Premier Knitting Co., Inc., New York, N. Y. 611,941. Cl. 39.
Prentice, John R., d. b. a. American Breeders Service, Chicago, Ill. 611,900, pub. 6–7–55. Cl. 100.
Prince Matchabelli, Inc., New York, N. Y. 611,881–4, pub. 6–7–55. Cl. 51.
Quick-Set, Inc., Skokie, Ill. 611,793, pub. 6–7–55. Cl. 26.
Raleigh Cycle Co. Ltd., The, Nottingham, England. 443,717, 12(c) pub. 9–6–55. Cl. 19.
Raravis Products Ltd.: *See—*
Cosler, Arthur S., Jr.

Regina Corp., The, Rahway, N. J. 611,747, pub. 5–31–55. Cl. 21.
Regulator Equipment Corp., Paterson, N. J. 611,755, pub. 6–7–55. Cl. 21.
Reich, Joseph A., d. b. a. Swiss Chalet Products Co., Colorado Springs, Colo. 611,827, pub. 6–7–55. Cl. 46.
Reisfeld, Sam, & Son, New Orleans, La. 611,830, pub. 6–7–55. Cl. 46.
Republic Paint and Varnish Co., Chicago, Ill. 611,678, pub. 6–7–55. Cl. 16.
Republic Steel Corp., Cleveland, Ohio. 611,621, pub. 5–31–55. Cl. 2.
Richards Brothers : *See*—
Rose, John O.
Ridlon, John, d. b. a. Metrochord Co., Chicago, Ill. 611,770, pub. 5–24–55. Cl. 26.
Riverbank Wine Co., Boston, Mass. 508,099, canc. Cl. 47.
Robinson & Golluber, New York, N. Y. 412,836, canc. Cl. 39.
Rochester Paper Co., Rochester, Mich. 611,616, pub. 5–31–55. Cl. 2.
Rogers Grain & Feed Co., Ainsworth, Nebr. 611,691, pub. 6–7–55. Cl. 18.
Rose, John O., d. b. a. Richards Brothers, Marietta, Ohio. 611,696, pub. 6–7–55. Cl. 18.
Rubel, Jack, New York, N. Y. 429,779, canc. Cl. 39.
Rudhard Products, Inc., Buffalo, N. Y. 611,841, pub. 6–7–55. Cl. 46.
Rush, Luther B., d. b. a. Rush Mattress & Furniture Co., Alexandria, La. 611,802, pub. 5–31–55. Cl. 32.
Rush Mattress & Furniture Co. : *See*—
Rush, Luther B.
Ruskin Vegetable Distributors, Ruskin, Fla. 436,496, canc. Cl. 46.
Rutgerswerke-Aktiengesellschaft, Frankfurt am Main, Germany. 611,641, pub. 6–7–55. Cl. 5.
Safeway Stores, Inc., Oakland, Calif. 611,853, pub. 6–7–55. Cl. 46.
St. Albans Grain Co., St. Albans, Vt. 249,733, canc. Cl. 46.
Salamander Aktiengesellschaft, Kornwestheim, Wurttemberg, Germany. 611,632, pub. 6–7–55. Cl. 4.
Salant & Salant, Inc., New York, N. Y. 107,203, ren. 11–16–55. Cl. 2.
Salinas Marketing Cooperative, Salinas, Calif. 508,048, canc. Cl. 46.
Sanchez Romate Hermanos S. A., Jerez de la Frontera, Spain. 508,094, canc. Cl. 47.
Sardi's Enterprises, Inc., New York, N. Y. 611,921, pub. 6–7–55. Cl. 107.
Scadron, Louis H., New York, N. Y. 252,728, canc. Cl. 26.
Schenley Distillers, Inc. : *See*—
Pepper, Jas. E., & Co.
Schenley Distributors, Inc., to Affiliated Distillers Brands Corp., New York, N. Y. 323,934, ren. 5–7–55. Cl. 49.
Schenley Import Corp. : *See*—
American Spirits Inc.
Schenley Industries, Inc. : *See*—
Finch, Joseph S., & Co.
Schering Corp., Bloomfield, N. J. 611,692, pub. 6–7–55. Cl. 18.
Schieffelin & Co., New York, N. Y. 611,885, pub. 6–7–55. Cl. 51.
Schneierson, I., & Sons, Inc., New York, N. Y. 328,546, canc. Cl. 39.
Schneierson, I., & Sons, Inc., New York, N. Y. 350,867, canc. Cl. 39.
Schneierson, I., & Sons, Inc., New York, N. Y. 353,509, canc. Cl. 39.
Schneierson, I., & Sons, Inc., New York, N. Y. 369,226, canc. Cl. 39.
Schneierson, I., & Sons, Inc., New York, N. Y. 372,645, canc. Cl. 39.
Schnitzer Alloy Products Co. : *See*—
Abell, Henry C.
Schnur & Cohan, Inc., New York, N. Y. 373,353, canc. Cl. 39.
Scully, Geo. L., Insurance Agency : *See*—
Scully, George L.
Scully, George L., d. b. a. Geo. L. Scully Insurance Agency, Port Arthur, Tex. 611,903, pub. 6–7–55. Cl. 102.
Sea Island Mfg. Co., Inc., New York, N. Y., to Tempest Shirt Mfg. Co., Inc., Jesup, Ga. 508,060, canc. Cl. 39.
Seal-Peel, Inc., New York, N. Y. 611,677, pub. 4–15–52. Cl. 16.
Seeman Brothers, Inc., to Airkem, Inc., New York, N. Y. 611,833, pub. 6–7–55. Cl. 46.
Seeman Brothers, Inc., to Airkem, Inc., New York, N. Y. 611,834, pub. 3–8–55. Cl. 46.
Selig Co., Inc., The, Atlanta, Ga. 611,948. Cl. 52.
Shapiro, M. : *See*—
Shapiro, Michael.
Shapiro, Michael, d. b. a. M. Shapiro, Brooklyn, N. Y. 611,837, pub. 6–7–55. Cl. 46.
Sheffield Co., The, New London, Conn. 356,478, canc. Cl. 6.
Shepard Point Packing Co., Seattle, Wash. 230,466, canc. Cl. 46.
Singer Mfg. Co., The, Elizabeth, N. J. 611,766, pub. 6–7–55. Cl. 23.
Smith Brothers, Inc., Poughkeepsie, N. Y. 611,854, pub. 6–7–55. Cl. 46.
Smith, Oren D., Englewood, N. J. 611,610, pub. 5–31–55. Cl. 2.
Smith, Roy R., d. b. a. Econo-Pak Register Co., Kansas City, Mo. 508,116, canc. Cl. 23.
Smith & Sons Corp., Buffalo, N. Y. 215,147, canc. Cl. 40.
Smucker, J. M., Co., The, Orrville, Ohio. 508,046, canc. Cl. 46.
Sociedade dos Vinhos Borges & Irmao, Lda., to Sociedade dos Vinhos Borges & Irmão, S. A. R. L., Villa Nova de Gaia, Portugal. 315,886, ren. 8–14–54. Cl. 47.
Sociedade dos Vinhos Borges & Irmao, S. A. R. L. : *See*—
Sociedade dos Vinhos Borges & Irmão, Lda.
Societa Vinicola Torino, Inc., New York, N. Y. 508,120, canc. Cl. 47.
Societe Francaise Radio-Electrique, Paris, France. 611,722, pub. 6–7–55. Cl. 21.
Socony Mobil Oil Co., Inc. : *See*—
Socony-Vacuum Oil Co., Inc.
Socony-Vacuum Oil Co., Inc., to Socony Mobil Oil Co., Inc., New York, N. Y. 328,209, ren. 9–17–55. Cl. 15.
Socony-Vacuum Oil Co., Inc., to Socony Mobil Oil Co., Inc., New York, N. Y. 328,268, ren. 9–24–55. Cl. 4.
Softley, Ed. W., Butte, Mont. 611,633, pub. 6–7–55. Cl. 4.
Solar Corp., d. b. a. Beam Mfg. Co., Milwaukee, Wis. 611,769, pub. 6–7–55. Cl. 24.
Sorensen and Co., Inc., Stamford, Conn. 611,730, pub. 6–7–55. Cl. 21.
Southwest Mfg. Co., Little Rock, Ark. 508,105, canc. Cl. 22.
Spanmaster Crane Corp. of America, South Gate, Calif. 611,717, pub. 6–7–55. Cl. 21.
Springer, Donald E., d. b. a. Fit-All Mannequin Base, Jacksonville, Fla. 508,112, canc. Cl. 50.
Staley, A. E., Mfg., Co., Decatur, Ill. 611,652, pub. 6–7–55. Cl. 6.
Standard Brands Inc., New York, N. Y. 611,859, pub. 6–7–55. Cl. 46.
Standard Knitting Mills, Inc., Knoxville, Tenn. 611,942. Cl. 39.
Standard Oil Co. of California : *See*—
Petrol Corp., The.
Standard Packaging Corp., Chicago, Ill. 611,627–8, pub. 6–7–55. Cl. 2.
Standard Record Co., Long Island City, N. Y. 611,894, pub. 6–7–55. Cl. 52.
Stanley Works, The, New Britain, Conn. 611,783, pub. 5–24–55. Cl. 26.
Stearns & Foster Co., The, Lockland, Cincinnati, Ohio. 611,692, pub. 5–31–55. Cl. 1.
Steelfab, Inc., Castroville, Calif. 611,918, pub. 6–7–55. Cl. 106.
Steiner Sales Co., Division of American Linen Supply Co. : *See*—
American Linen Supply Co.
Steinfeld's Products Co., Portland, Oreg. 611,869, pub. 6–7–55. Cl. 46.
Sterwin Chemicals Inc., New York, N. Y. 611,851, pub. 6–7–55. Cl. 46.
Stetson China Co., Inc., Lincoln, to Marshall-Burns, Inc., Chicago, Ill. 611,800, pub. 12–14–54. Cl. 30.
Stic-Klip Mfg. Co. : *See*—
Eckel, Oliver C.
Stivers Office Service, Chicago, Ill. 611,902, pub. 6–7–55. Cl. 101.
Stone, Tom, Cordage Co., Chariton, Iowa. 611,656, pub. 5–31–55. Cl. 7.
Storkline Furniture Corp., Chicago, Ill. 611,712, pub. 5–31–55. Cl. 19.
Stump, Howard D., d. b. a. Stump Sales & Engineering Co., South Bend, Ind. 611,803, pub. 5–31–55. Cl. 32.
Stump Sales & Engineering Co. : *See*—
Stump, Howard D.
Sun-Maid Raisin Growers of California, Fresno, Calif. 329,989, ren. 11–19–55. Cl. 49.
Sun-Maid Raisin Growers of California, Fresno, Calif. 329,997, ren. 11–19–55. Cl. 47.
Sunred Corp., Winona, Minn. 611,829, pub. 6–7–55. Cl. 46.
Superior Beverage Co., Inc., Gary, Ind. 611,816, pub. 5–24–55. Cl. 45.
Supreme Pharmaceutical Co. : *See*—
Blackman, Julius.
Surgical Appliance Industries, Inc. : *See*—
Ohio Truss Co., The.
Swish Products Ltd., Small Heath, Birmingham, England. 611,670, pub. 6–7–55. Cl. 13.
Swiss Chalet Products Co. : *See*—
Reich, Joseph A.
Sylvania Industrial Corp., New York, N. Y., and Fredericksburg, Va., to American Viscose Corp., Philadelphia, Pa. 325,512, ren. 6–25–55. Cl. 43.
Sylvania Industrial Corp., New York, N. Y., and Fredericksburg, Va., to American Viscose Corp., Philadelphia, Pa. 325,657–8, ren. 7–2–55. Cl. 2.
Sylvania Industrial Corp., Fredericksburg, Va., and New York, N. Y., to American Viscose Corp., Philadelphia, Pa. 325,874, ren. 7–9–55. Cl. 2.
Sylvania Industrial Corp., Fredericksburg, Va., and New York, N. Y., to American Viscose Corp., Philadelphia, Pa. 326,157, ren. 7–16–55. Cl. 1.
Syntomatic Corp., New York, N. Y. 611,846, pub. 6–7–55. Cl. 46.
Tallmadge, L. H., Co., Inc., Cranston, R. I. 611,639, pub. 6–7–55. Cl. 4.
Tedoff, Samuel R., d. b. a. Glenmore Novelty Co., Inc., Brooklyn, N. Y. 508,031, canc. Cl. 39.
Telecourses, Inc., Richmond, Va. 611,920, pub. 6–7–55. Cl. 107.
Tele-Trip Policy Co., Inc., New York, N. Y. 611,905, pub. 6–7–55. Cl. 102.
Tempest Shirt Mfg. Co., Inc. : *See*—
Sea Island Mfg. Co., Inc.
Tenth Avenue Trading Corp., New York, N. Y. 611,819, pub. 1–1–52. Cl. 46.
Textileather Co., New York, N. Y. 165,309, canc. Cl. 50.
Thomas, M. & W., Co., New York, N. Y. 437,324, canc. Cl. 42.
Titan Metal Mfg. Co., Bellefonte, Pa. 602,073, cor. Cl. 14.
Titan Metal Mfg. Co., Bellefonte, Pa. 602,281, cor. Cl. 13.
Towry, M. Forrest, d. b. a. Towry Products Co., Tulsa, Okla. 508,087, canc. Cl. 22.

Towry Products Co.: *See*—
Towry, M. Forrest.
Travalong Co., The: *See*—
Axelman, Sam.
Trich-O-Matic Products, Inc., Fond du Lac, Wis. 611,876, pub. 6–7–55. Cl. 51.
Tricouni S. A., Geneva, Switzerland. 321,080, 12(c) pub. 9–6–55. Cl. 22.
Tropical Importing Co.: *See*—
Liberoff, William.
Tryon, Edw. K., Co., Philadelphia, Pa. 611,757–8, pub. 6–7–55. Cl. 22.
Underwood, Howard G., Borger, Tex. 396,968, canc. Cl. 26.
United States News Publishing Corp., Washington, D. C. 436,696, canc. Cl. 38.
Vaco Products Co., Chicago, Ill. 611,738, pub. 6–7–55. Cl. 21.
Valley View Packing Co.: *See*—
Valley View Packing Co., Inc.
Valley View Packing Co., Inc., d. b. a. Valley View Packing Co., San Jose, Calif. 611,857, pub. 6–7–55. Cl. 46.
Van Camp Sea Food Co. Inc., Terminal Island, Calif. 231,159, canc. Cl. 46.
Van Camp Sea Food Co., Inc., Terminal Island, Calif. 350,375, canc. Cl. 46.
Van Camp Sea Food Co., Inc., Terminal Island, Calif. 365,155, canc. Cl. 46.
Vanderford, Thomas S., Aberdeen, Idaho. 372,723, canc. Cl. 46.
Vegetable Oil Products Co., Inc., Wilmington, Calif. 371,752, canc. Cl. 15.
Verflex Co., Inc., The, Carlstadt, N. J. 611,676, pub. 6–7–55. Cl. 16.
Verley, Albert, & Co.: *See*—
Verley, Albert, Inc.
Verley, Albert, Inc., Chicago, Ill., to Albert Verley & Co., New York, N. Y. 321,575, ren. 2–5–55. Cl. 46.
Vinylustre, Inc., Newark, N. J. 611,637 pub. 6–7–55. Cl. 4.
Vychodoceske Lihovary, Narodni Podnik, Pardubice, Czechoslovakia. 611,871, pub. 5–24–55. Cl. 49.
Wagner, E. R., Mfg. Co., Milwaukee, Wis. 611,931. Cl. 23.
Wagner, John, & Sons, Philadelphia, Pa. 508,049–50, canc. Cl. 46.
Wanamaker, John, New York, New York, N. Y. 107,718, canc. Cl. 22.
Warner-Hudnut, Inc., New York, N. Y., now by change of name Warner-Lambert Pharmaceutical Co. 611,700, pub. 6–7–55. Cl. 18.
Warner-Lambert Pharmaceutical Co.: *See*—
Warner-Hudnut, Inc.
Warp Bros.: *See*—
Flex-O-Glass, Inc.
Washburn, T. F., Co., Chicago, Ill. 611,650, pub. 5–31–55. Cl. 6.
Wassco Electric Products Corp., Joliet, Ill. 611,733, pub. 6–7–55. Cl. 21.
Webster-Chicago Corp., Chicago, Ill. 611,726, pub. 6–7–55. Cl. 21.
Werkzeugmaschinenfabrik Oerlikon Bührle & Co., Administration Co.: *See*—
Holding Intercito S. A.
Western Condensing Co., Petaluma, Calif. 611,858, pub. 6–7–55. Cl. 46.
Western Vegetable Distributors, Denver, Colo. 328,762, canc. Cl. 46.
Westhauser Nurseries: *See*—
Westhauser, Rexford E.
Westhauser, Rexford E., administrator of the estate of William Westhauser, deceased, d. b. a. Westhauser Nurseries, Sawyer, Mich. 508,080, canc. Cl. 10.
Westhauser, William: *See*—
Westhauser, Rexford E.
Westric Battery Co., Denver, Colo. 611,751–4, pub. 6–7–55. Cl. 21.
Westron Corp., New York, N. Y. 611,718, pub. 6–7–55. Cl. 21.
Westwood Products, Inc., Cassopolis, Mich. 611,713, pub. 6–7–55. Cl. 19.
Weyerhaeuser Timber Co., Tacoma, Wash. 611,622, pub. 6–7–55. Cl. 2.
Whink Products Co., Eldora, Iowa. 525,281. Am. 7(d). Cl. 52.
White-Giles Co., The, Decatur, Ala. 611,615, pub. 5–31–55. Cl. 2.
Wiegand, Edwin L., Co., Pittsburgh, Pa. 611,739, pub. 6–7–55. Cl. 21.
Wilson, Andrew, Inc., Springfield, N. J. 508,098, canc. Cl. 6.
Winfield Brooks Co., Inc., Woburn, Mass. 611,889, pub. 6–7–55. Cl. 52.
Wolf & Dessauer Co., Fort Wayne, Ind. 611,721, pub. 6–7–55. Cl. 21.
Wright Power Saw and Tool Corp., Stratford, Conn. 611,930. Cl. 23.
Wyandotte Chemicals Corp., Wyandotte, Mich. 420,473, canc. Cl. 6.
Wynlit Pharmaceuticals, Inc., New York, N. Y. 611,701, pub. 6–7–55. Cl. 18.
Yourek, John M., d. b. a. Yourek Laboratories, Chicago, Ill. 611,927. Cl. 18.
Yourek Laboratories: *See*—
Yourek, John M.
Zerega's, A., Sons, Inc., Fair Lawn, N. J. 611,825, pub. 6–7–55. Cl. 46.

U. S. GOVERNMENT PRINTING OFFICE: O——1955